COMPLETE BOOK
OF
WELDING

COMPLETE BOOK
OF
WELDING

Gower A. Kennedy
Lethbridge Community College, Alberta

HOWARD W. SAMS & CO., INC.
INDIANAPOLIS · KANSAS CITY · NEW YORK

Complete Book of Welding explores virtually all known welding methods to date. It presents a comprehensive overview of the many welding processes available and provides practical experiences and instructions for commonly-used welding procedures. The book is designed to be beneficial to both new and experienced welders alike.

Procedures are discussed for welding a majority of both ferrous and nonferrous metals, as well as plastics. Information concerning both arc and oxy-acetylene welding is presented in addition to thorough discussions of tungsten inert gas, metal inert gas, and other forms of contemporary welding practice. Covered in individual chapters are hardsurfacing, flux-core arc welding, and inspecting and testing of completed welds. Concluding individual chapters include definitions of common welding terms and a comprehensive chapter on the conversion to metric measure for the welding trade.

The book is divided into six basic sections. Any one of the sections may be used individually. However, the sections and chapters, when studied in order, are arranged to provide comprehensive coverage of current welding technology.

ACKNOWLEDGEMENTS

The information presented in *Complete Book of Welding* represents an effort to compile a composite text of material on the current development of the welding art. The firms and individuals listed below provided valuable additions to the author's personal experiences, gained while welding commercially and training and supervising welders.

Airco Welding Products Division
Aircom Metal Products, Inc.
Province of Alberta, Apprenticeship Board and Tradesmen's Qualifications
Alcan Canada Products, a Division of Aluminum Company of Canada, Ltd.
Aluminum Company of America
American Cast Iron Pipe Company
American Optical Corporation
American Society for Testing and Materials
American Welding Society, Inc.
Ampco Metal, Inc.
Arcair of Canada Ltd.
Arcos Corporation
Arvin Automation, Inc.
Armco Steel Corporation

Bausch & Lomb, Scientific Instrument Division
Bernard Company
Bethlehem Steel Corporation
Bethlehem Steel Export Company of Canada, Limited
Bridgeport Brass Company, Division of National Distillers and Chemical Corporation
Canadian Liquid Air Ltd.
Cast Iron Pipe Research Association
Chemtron Corporation
Department of Health, Public Health Division, Ontario, Canada
The Dow Chemical Company
Eddy Products Corp.
Eutectic Institute for the Advancement of Maintenance and Repair Welding Techniques
The Falstrom Company
Federal Tool Engineering Company, Subsidiary of Arvin Industries Inc.
Fibre-Metal (Canada) Limited
Fyr-Fyter Company of Canada, Ltd.
General Electric Company

Handy & Harman of Canada, Ltd.
Hardfacing Alloys Limited
Hercules Welding Products Co.
Hobart Brothers Company
The International Nickel Company, Inc.
Kierk's Metallizing and Machine, Ltd.
Laramy Products Co., Inc.
Lenco, Inc.
The Lincoln Electric Company
Liquid Carbonic Corporation
Magnaflux Corporation
Mapp® Products Division, Air Reduction
  Company, Inc.
Miller Electric Manufacturing Co.
Nickel of Canada
The Norton Company of Canada, Limited
Patton, Mr. W. J.

Proline Pipe Equipment Ltd.
The Reid-Avery Co.
Safety Supply Company
Sciaky Bros., Inc.
Sellstrom Manufacturing Ltd.
Smith-Roles Ltd.
Smith Welding Equipment, Division of
  Tescom Corporation
Stoody Company
Turner Corporation
Union Carbide Canada Ltd.
Union Carbide Corporation,
  Linde Division
United States Steel Corporation
Victor Equipment Company
Welding Engineer Publications, Inc.
Westinghouse Electric Corporation

To: Olive and Irene

TABLE OF CONTENTS

Section I—Introduction to Heat in Metalworking

Section II—Oxy-Acetylene Welding

Section III—Arc Welding

Section IV—Tungsten Inert Gas (TIG) Welding

Section V—Other Welding Processes

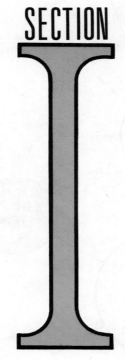

SECTION I

introduction to heat in metal working

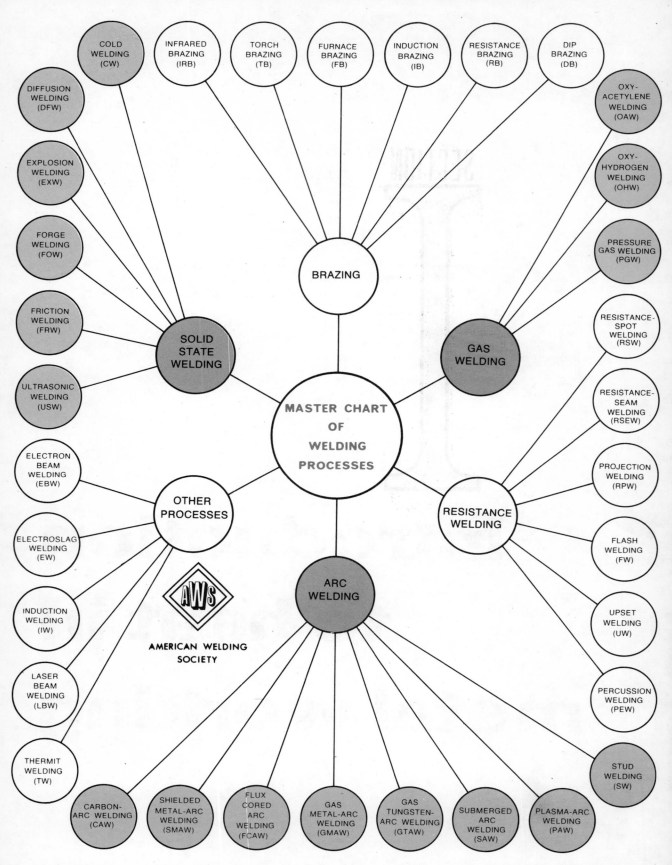

A complete welding process chart. (Courtesy of The American Welding Society.)

Introduction to Welding

From the time that man first discovered metals, he has strived to adapt these metals to be used for a better society. The advances in *metallurgy,* the art of extracting metals from ores, have made our modern way of living possible.

Today's increased knowledge of electronics, along with the new universe of atomic research, give technologists and scientists a field of great opportunity, coupled with great responsibility. The art of producing new metals, as well as *joining* metals, is as important for survival in our space-age society as it was to the people of ancient times.

Although steel is the basic metal of our industrial economy, today's industry demands special types of other metals. In order to keep pace with this demand, scientific advances must be made not only in the search for new metals, but also in the processes needed to join these metals.

The earliest method of joining two pieces of metal into a unit was heating both pieces to the melting temperature and allowing the fluid to melt together and solidify. This was a process known as *forge welding.* This union by melting, used in the welding process and known as *fusion,* has changed little from ancient times. Welding processes other than heat have been developed over the years to improve the fusion of metals for fabrication

purposes, especially in the welding industry. Welding methods have been developed using sound, and, within recent years, light from the laser, which generates electromagnetic waves by means of the natural oscillations of atoms.

Originally, joining metals by welding was done mainly to repair damaged or worn metal parts. Improvements in metal-joining processes have made it economically more feasible to manufacture, fabricate, and reinforce with modern high-strength steel, as well as other metals and their alloys, than was possible less than a decade ago. Most assemblies today whose component parts are joined by welding, known as a *weldment,* would have been bolted or riveted a few years ago.

Today, welding is well on its way to becoming a science, through the steady progress and improvement in the development of scientific methods, equipment, and welding accessories. These have evolved from the knowledge gained by using various welding procedures through the years.

Since many of these recent methods are seldom used for commercial purposes today, they will be dealt with briefly, and only the fundamental and basic principles of application will be presented. The more modern methods of shielded arc welding and oxyacetylene welding will also be covered.

Welding Methods

Forge Welding

The *forge* was a type of furnace fired by coal or other fuel whereby metals could be heated, and, by hammering or other pressure, could be fused together as a unit. Fig. 1-1 shows a typical early forge. Methods by

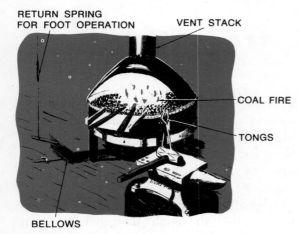

Fig. 1-1. Equipment for an early forge.

which pressure is applied to the portions of the metal pieces (which are heated to a soft, pliable condition), are hammering, rolling, or die welding. In manual hammering, the pieces to be joined are placed on an anvil and struck at the point of desired fusion by manual

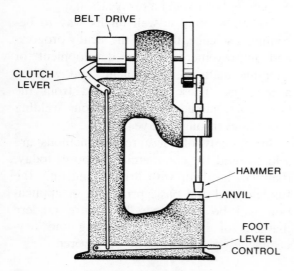

Fig. 1-2. Trip hammer.

hammer blows. Mechanical hammering devices are also used. These are commonly known as trip hammers, which can be steam, pneumatic or hydraulic. Fig. 1-2 shows a mechanical hammering device.

Roll Welding

The heated metal pieces are forced between rolls which create the necessary joining pressure in *roll welding*. In die welding, pressure is exerted on the heated portion by hydraulic pressure with various shaped mandrels. Fig. 1-3 shows pressure exerted by means of dies.

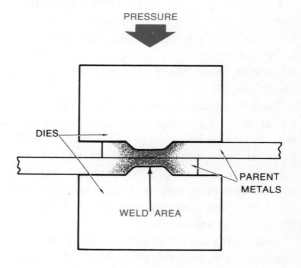

Fig. 1-3. The die welding process.

When heating the ends of two pieces to be welded, as in Fig. 1-4, uniform heat

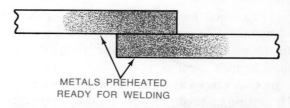

Fig. 1-4. Preheated metals, ready for welding.

throughout the heated ends is essential. Overheated portions can become burned and brittle, while under-heating the areas to be joined will retard a complete union of the pieces when they are hammered or placed under pressure.

Oxidation, the reaction when the air's oxygen attacks hot, exposed, carbon steels, creates a poor bond between the portions to be fused. To aid in overcoming oxidation, a substance is sprinkled on the ends to be joined. This additive lowers the melting temperature of the oxides so that they become fluid enough to be squeezed out under welding pressure. The additive is known as a *flux* and is quite common in many welding processes. Borax is used as a flux and is used essentially when forge welding higher carbon steels, while ordinary fine silica sand can be used as a flux on low-carbon steels.

Cold Welding

This is a cold pressure process suited for ductile (pliant) metals categorized as nonferrous. Nonferrous metals (metals other than iron) can be joined with comparative ease and speed using cold welding especially if the points of contact are free of foreign subtances and oxide film.

Nonferrous metals such as aluminum and its *alloys* (mixtures of at least two metals), cadmium, silver, nickel, and zinc can be easily cold welded. Moreover, satisfactory results have been obtained in joining dissimilar metals such as copper and aluminum. With enough pressure from tool dies, the materials flow into a unit. See Fig. 1-5.

Since that portion of the parts to be placed under pressure must be free of foreign matter, cleaning prior to cold welding is necessary. A rotary brush attached to a motor or grinding machine is a most effective method. The formation of the weld is impaired by other methods of cleaning, such as grinding and filing, where foreign matter could impregnate the surfaces. Chemical cleaning procedures can also be unsatisfactory, especially where after-washing is required, because the washing forms undesirable oxides on the surface.

The length of the pressure time, whether exerted manually or mechanically, has little effect on the strength of the weld; but the design of the dies must provide the amount of compression necessary to reduce the original thickness of the materials by just enough to assure a sound weld.

Since nonferrous materials placed under pressure have little effect on the wear of the dies, chromium manganese tool steel for die making is most suitable. Cold welding tool dies are also designed to give a special finished shape to the weld area, as in Fig. 1-6.

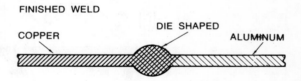

Fig. 1-6. Finished cold weld of dissimilar metals.

Thermit Welding

One of the earlier forms of metal joining was *thermit welding.* Although classified as *cast,* (a material formed by pouring heated metals into molds) thermit welding has excellent physical properties.

In the thermit welding process, fusion is produced (with or without the application of pressure) by the heat developed by a superheated metal in liquid form coming in contact with the parts to be welded. The filler material of superheated metal is brought to

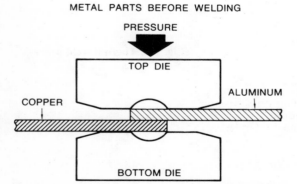

Fig. 1-5. Cold welding pressure exerted by dies.

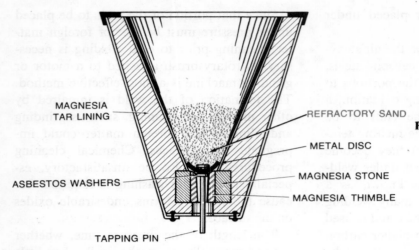

MAGNESIA TAR LINING

REFRACTORY SAND

METAL DISC

MAGNESIA STONE

ASBESTOS WASHERS

MAGNESIA THIMBLE

TAPPING PIN

Fig. 1-7. Thermit crucible cutaway.

liquid form by means of a chemical reaction between a metal oxide and aluminum; this is known as the *thermit reaction.*

The thermit mixture, consisting of metal oxide and fine aluminum, can also include other alloying elements for increasing the strength of the weld joint. The reaction of this mixture takes place in a thermit *crucible,* as shown in Fig. 1-7.

The ignition of the mixture in the crucible is brought about by a highly flammable barium peroxide powder, known as the thermit *starter.* A single reaction has produced as much as two tons of material during a very short period of time.

This superheated material is poured from the crucible into a sand form surrounding the broken parts. Similar to flow welding, the broken parts become fluid from contacting the superheated thermit mixture. When the mixture and parts reach fusion temperature, the crucible flow is shut off. Solidification then takes place, completing the welding process as shown in Fig. 1-8.

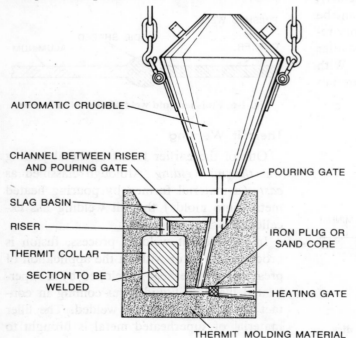

AUTOMATIC CRUCIBLE

CHANNEL BETWEEN RISER AND POURING GATE

POURING GATE

SLAG BASIN

RISER

IRON PLUG OR SAND CORE

THERMIT COLLAR

SECTION TO BE WELDED

HEATING GATE

THERMIT MOLDING MATERIAL

Fig. 1-8. Thermit weld process.

Cadwelding

Like thermit welding, another fusion welding process with limited applications is *cadwelding*. Here, molten aluminum oxide slag and copper are produced through the reduction of the copper oxide by aluminum.

One use of cadwelding is in cable connections, which are welded satisfactorily by placing the ends of the cables to be spliced in a machined graphite mold mounted in a hand clamped frame assembly, as shown in Fig. 1-9.

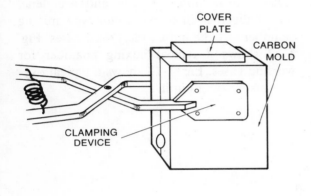

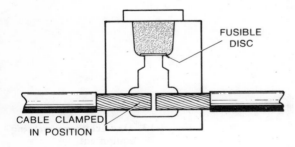

Fig. 1-9. Preparation for cadwelding—a hand-clamped frame assembly.

A cartridge, containing the powdered copper oxide and aluminum, is poured over a small steel disc placed in the mold, the ends of the cables having been clamped in the mold. When the starting powder is ignited, a reaction turns the copper into a molten liquid, which in turn melts the steel disc. The disc then liquifies and flows over the cable ends, fusing them together. When the cable is removed from the clamped crucible mold,

it takes on the appearance shown in Fig. 1-10. Cadwelding is often used in various industrial applications, either copper to copper or steel to copper.

FINISHED WELD

Fig. 1-10. Finished cadweld.

Flow Welding

A molten filler metal, the same composition as the metal to be repaired, is poured over the surface of the broken base metal in *flow welding*. The temperature of the base metal is raised by the continuous flow of the molten filler metal over the parts to be welded, until the base metal accepts the filler metal as a bond.

Flow welding is used in the repair of heavy, thick, sectional, nonferrous parts. The filler metal, which continually runs off the horizontal base metal, must be available in such quantities and at such temperatures, so that the parent metal will eventually soften to the point where base metal fusion, or acceptance, takes place. Once this point is reached, the filler metal run-off spout is closed by a

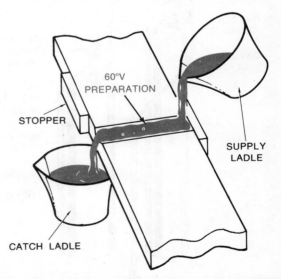

Fig. 1-11. The flow welding process, using a 60°V preparation.

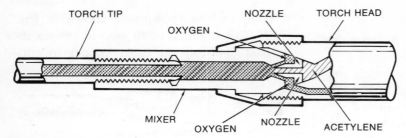

TORCH TIP   NOZZLE   TORCH HEAD

OXYGEN

MIXER   OXYGEN   NOZZLE   ACETYLENE

Fig. 1-12. Mixing chamber for gas welding.

stopper. This allows the filler metal to rise higher than the base metal. By having the reinforcement higher than the base metal, a sound weld is assured. Fig. 1-11 illustrates a common flow welding procedure.

Preparing the parts to be joined is essential in this welding procedure. Since this welding technique is used in heavy sectional breakdowns, a single-Vee butt joint with a 60° included angle is most satisfactory, as shown in Fig. 1-11.

Gas Welding

Flame welding, as it is sometimes called, makes welding of materials possible by using ignited gases to form a flame, commonly called *gas welding.* Oxygen and acetylene, mixed together in a carburetor-type mixing chamber, are the most widely used gases. Fig. 1-12 shows a typical mixing chamber for welding gases. Fig. 1-13 shows a typical oxygen-gas welding outfit.

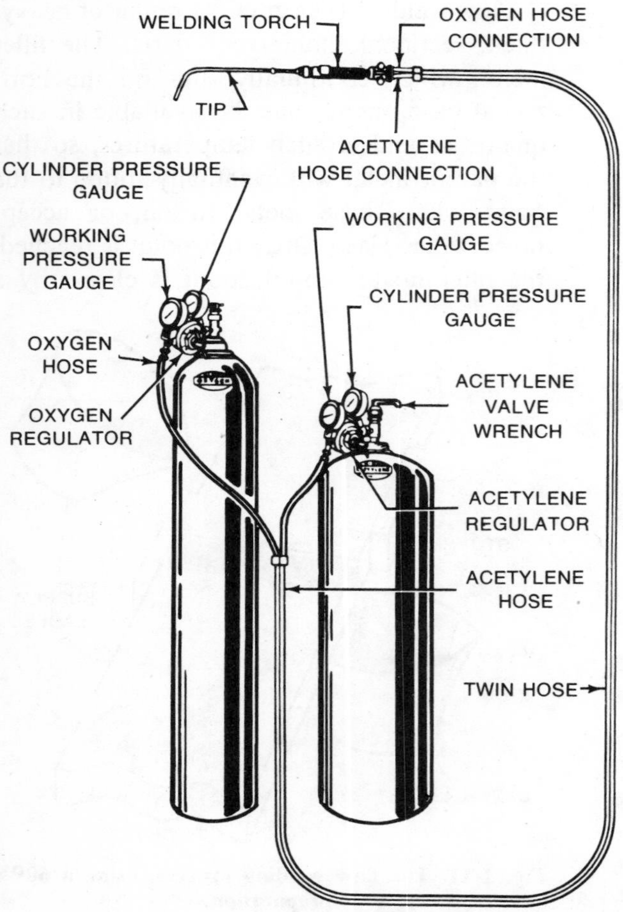

WELDING TORCH   OXYGEN HOSE CONNECTION

TIP

ACETYLENE HOSE CONNECTION

CYLINDER PRESSURE GAUGE

WORKING PRESSURE GAUGE

WORKING PRESSURE GAUGE

CYLINDER PRESSURE GAUGE

OXYGEN HOSE

ACETYLENE VALVE WRENCH

OXYGEN REGULATOR

ACETYLENE REGULATOR

ACETYLENE HOSE

TWIN HOSE →

Fig. 1-13. Oxy-acetylene welding equipment.

Under various other conditions where higher temperatures are not required, hydrogen, propane, Mapp®, and other gases can replace acetylene. These gases have been found satisfactory where torch cutting has been required.

General use of the oxy-acetylene process for welding and cutting did not become a regular welding procedure until the early part of the twentieth century. In the gas welding process, the oxygen and acetylene meet, and, after mixing, they undergo combustion, producing one of the hottest known flames, estimated to be approximately 6300°F.

Using an oxy-actylene flame, welding is possible with or without a filler metal. The filler metal, if needed, is known as the welding rod. Without the use of filler metal, parts can be fused together by bringing the parts' contact portions to a melting temperature where the molten metal from both portions of the parts flows together. When the mixture is left to solidify, the parts become a unit as shown in Fig. 1-14.

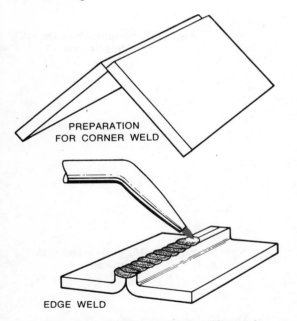

Fig. 1-14. Welding without a filler rod.

With gas welding, it is also possible to heat ferrous metal to a red heat, after which

a stream of pure oxygen is blown onto the heated metal, causing the metal to separate in a cutting manner. This cutting by oxidization will not affect cast iron or ferrous metals the same way as it does steel. Oxidization is the chemical uniting of a substance with oxygen. Hot iron is attacked more rapidly by oxygen than cold material or nonferrous metals, so the hot iron disintegrates rapidly where the oxygen is blown against it.

Using the oxy-acetylene flame, pressure gas welding can also be introduced. The *closed joint* pressure method unites two pieces as in Fig. 1-15.

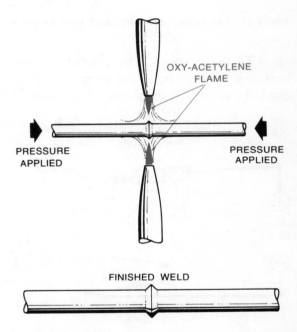

Fig. 1-15. The process of making a cold pressure oxy-acetylene weld.

The faces of the parts to be joined are machined and butted together while held securely in a press. Heat from the oxy-acetylene flame is directed on the joint at a high temperature just below the melting temperature. When the material near the joint becomes "plastic," the press exerts pressure to the weld faces, resulting in upsetting. This upsetting yields an increase in the volume of the material around the point of fusion be-

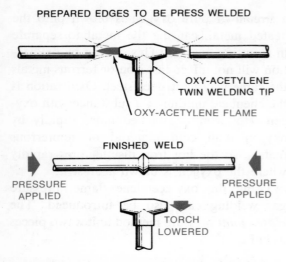

Fig. 1-16. The open joint method of cold pressure welding.

cause of the pressure. The combination of heat *and pressure* bond the pieces together. Since the melting temperature of the parts to

be joined has not been reached, the weld made by the oxy-acetylene closed-joint pressure method cannot be classified as a fusion-type weld.

The *open joint* pressure method, as shown in Fig. 1-16, is similar to the above method, except that the machined ends of the parts to be joined are kept apart, and the torch heat is directed on the exposed, machine-faced ends.

Unlike the closed-end pressure method, melting the face ends is required before pressure is applied to upset the joint and create a fusion type weld. Carefully preparing the ends of the parts to be joined by these methods is important for a strong, homogeneous weld joint.

Carbon Arc Welding

This was one of the earliest forms of welding. Here, an arc is formed between the

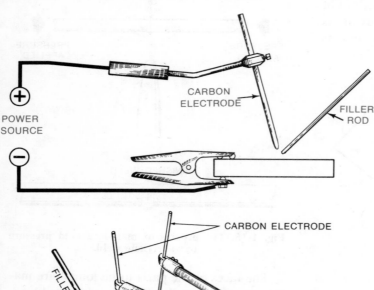

Fig. 1-17. Single-carbon-arc stick welding process.

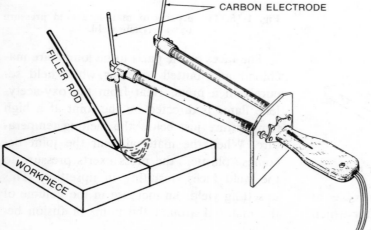

Fig. 1-18. Two-carbon stick process.

work to be heated and a carbon stick (or *electrode*), each of which is connected to a different pole of an electric current, as in Fig. 1-17. This is known as *carbon-arc welding*.

The carbon electrode is a nonmetallic element found in the rough as graphite, or as a member of the petroleum, coal and other carbonate family. During carbon-arc welding, it is surrounded by a coil carrying a welding current so that an arc is squeezed or compressed by an electromagnetic field, giving arc stability. Another carbon-arc process was produced by using two carbon sticks, each attached to the current terminals to create an arc between the carbon electrodes when the current was introduced, as shown in Fig. 1-18. Fig. 1-19 shows a typical carbon-arc dual stick holder.

Today the carbon arc welding process is used for cutting and gouging in metal preparation before welding. This process, which is commercially known as *Arc-air*, will be described in another chapter.

Initially, the carbon-arc process was used for welding both ferrous and nonferrous materials. Even today, some brazing, (a form of soldering with a relatively infusible alloy,) is used where gas welding or other heating methods are nonexistent.

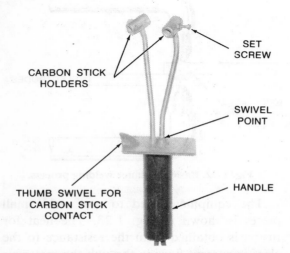

SET SCREW

CARBON STICK HOLDERS

SWIVEL POINT

THUMB SWIVEL FOR CARBON STICK CONTACT

HANDLE

Fig. 1-19. Holder used for two-carbon stick process.

Metallic Arc Welding

The carbon arc welding machine, which made use of the carbon electrodes, was followed, near the end of the 19th century, by a welding arc process using a consumable metallic rod. This rod was used up as it fused with the parts being welded and became a filler metal. *Metal arc welding* for joining took many years of developmental study.

The arc is a spark formed between the work and the electrode. The work is attached to a ground cable clamp from one pole of an electric current, whereas the other pole of the electric current is connected to the electrode (Fig. 1-20). When the electrode comes into contact with the work, which is attached to the ground clamp, an arc forms at the tip of the electrode, where the temperature reaches just under 7000°F. See Fig. 1-21.

Manual metallic arc welding was closely followed by semi-automatic and automatic machines which required little or no manual manipulation of the electrode holder for welding. These machines are today's maintenance tools which are very necessary for the future of any industrial nation.

Not only the machines, but accessories such as electrodes required considerable scientific and mechanical research. The first electrodes were bare, consumable wires which caused striking and arc stabilization problems.

Metallic arc-weld deposits were greatly improved with the discovery of the electrode coatings, commonly called *flux*. On heating, flux forms a gaseous shield around the weld. This improves mechanical properties of the weld because the outside air is blocked out. Two of the elements blocked out were oxygen and nitrogen, which could adversely affect the weld if permitted to enter the high temperature areas where they would form oxides and nitrides. Oxides are low in tensile strength and ductility, while nitrides make the steel brittle; both can change the metal's normal properties.

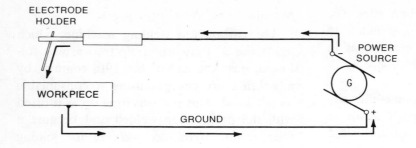

Fig. 1-20. A simple arc welding circuit.

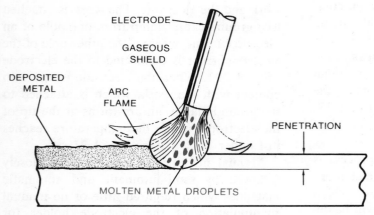

Fig. 1-21. The basic arc welding process.

Today, there is a variety of coated electrodes for use on direct or alternating current machines, which are described fully in the chapter on electrodes. These rods, with their variety of flux coating, make it possible to alter the deposit metal composition and property. Alloys can be added to flux covering for the purpose of depositing an alloyed weld; however, nonferrous metals or very highly alloyed metals must have a major portion of the required additional materials in the core wire.

Resistance Welding

This form of welding is used mainly in the industrial field for mass production, and encompasses the common processes of spot welding and seam welding. In resistance welding, the parts to be welded are fused by the combined use of heat and pressure. The heat is generated by means of contacting electrodes which carry the current through the pieces to be welded.

Spot welding—This is a resistance welding process where fusion of the metal is produced when the parts of the material to be joined are pressed together between *electrodes* (positive and negative poles in an electric circuit). See Fig. 1-22.

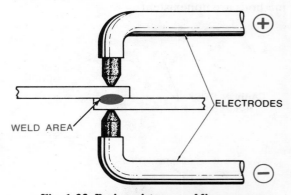

Fig. 1-22. Basic resistance welding process.

The equipment used to spot weld small pieces is shown in Fig. 1-23. The heat for fusion is obtained from the resistance to the electric current flowing through the materials to be joined. On some resistance welding machines, the electrodes slide along the surface of the work intermittently for spot

Fig. 1-23. A typical floor model spot welder.

welding and seam welding purposes. The electrodes may, for job versatility, be cylinders, clamps, rotating bars, or wheels (Fig. 1-24).

Seam Welding—This is similar to spot welding and produces a series of welds along a joint, as in Fig. 1-25. These welds can be controlled for intermittent or continuous

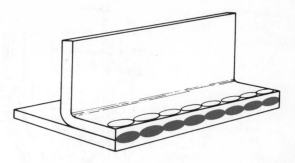

Fig. 1-25. Seam welding along a joint.

fusion, depending on the job requirements. Pressure, applied to the pieces to be welded by the clamping action of the electrodes, causes the current to flow through the electrode contact area. The pressure applied by the electrodes can be controlled by manual, motor driven, or hydraulic means, with the length of current hold time and heat settings fully automatic for proper fusion of various metal thicknesses.

Submerged Arc Welding

This welding method was designed for use in fabricating metal components. It is either an automatic or semiautomatic process, allowing higher heats and faster welding speeds. Unlike shielded arc welding, the rod or coil wire and the flux are introduced into

Fig. 1-24. Wheel type resistance welding machine. (Courtesy of Arvin Automation, Inc.)

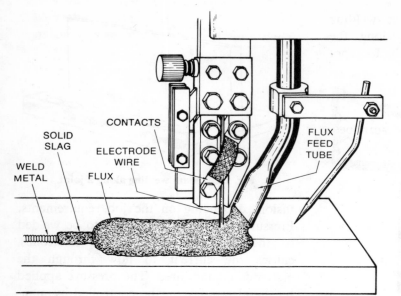

CONTACTS

SOLID
SLAG

ELECTRODE
WIRE

WELD
METAL

FLUX

FLUX
FEED
TUBE

Fig. 1-26. The welding head used in submerged arc welding.

the weld area separately. The flux, made up of special, conductive, high-resistance granules, is added in such quantity along the line of travel that the arc, molten metal, and base metal are completely covered or submerged. Besides the complete protection from the atmosphere and the cooling factor supplied, varying the chemical makeup of the flux can adapt submerged arc welding to a variety of metals and alloys and to various types of joint design. Flux that does not melt during the welding process can be reclaimed and reused. The coil wire or rod is fed mechanically through an electrical contacting "head," and, because of the precise automatic control of arc and welding speed, currents as high as 2900 amps can be controlled.

The welding head, as shown in Fig. 1-26, can be mounted on different carriage devices so that welds on other than straight line applications can be performed. The submerged arc is commonly used in fabricating; the head is mounted to move automatically along an overhead track in a straight direction. The same straight line welds can be achieved if the head is mounted in a permanent position and the structure being welded is moved along automatically beneath it. Box framing (Fig. 1-27) can be manufactured

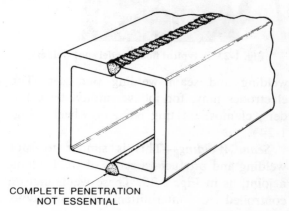

COMPLETE PENETRATION
NOT ESSENTIAL

Fig. 1-27. Box framing welded with submerged arc.

easily and economically this way. As is the case in box framing, the use of higher currents and faster welding speeds will reduce weld shrinkage and limit the distortion factor of any welded structure. Without the skills of the manual arc welding operator, one-pass welds on metal plates from $\frac{1}{16}$ to $\frac{1}{4}$ inch thick can be done with the submerged arc method by correctly adjusting the automatic electrical controls and then supervising the operation.

Induction Welding

Fusion with or without pressure is produced by the heat generated through the resistance set up by the parts to be welded, when current passes through them during *induction welding*. In this process, a heating coil acts as a transformer's primary coil, while the work surface acts as the short-circuited secondary coil of the same transformer.

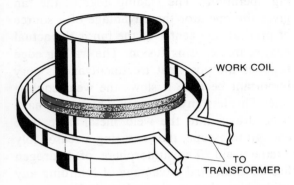

Fig. 1-28. Edge seam welded by the induction process.

A transformer, unlike an induction coil, lacks a metallic contact between the coils. It is an apparatus for generating currents by electromagnetic induction and usually consists of two coils of insulated wire. One of the coils is called the primary, and the other is the secondary. A current is induced in the secondary coil by rapid automatic making and breaking of the primary coil's circuit, which induces voltage in the secondary coil. When a resistor (a device causing resistance in an electrical circuit) is placed across the ends of the secondary coil, current will flow, and the resistor will become heated.

When the metal to be welded is placed in the heating coil (which acts as the primary of a transformer), the work becomes a type of short-circuited secondary transformer terminal. It becomes hot. With increased frequency in the current, increased heating results, and makes welding possible, as in Fig. 1-28.

Welding steels up to 1/8″ thick and slightly over 2″ in diameter has been done satisfactorily using induction welding. Nonferrous metals of high conductivity are not conducive to this type of welding process. This method, however, is found useful as a heating process for brazing and soldering on intricate light materials, especially where large quantities would make the initial cost practical.

Atomic Hydrogen Welding

Advances in the field of electric welding have brought about the development of the *atomic hydrogen welding* process, in both manual and automatic models. In this weld-

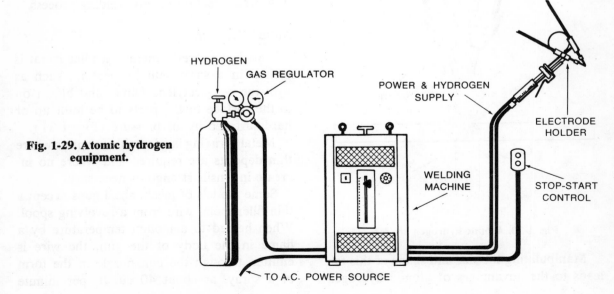

Fig. 1-29. Atomic hydrogen equipment.

ing process, a transformer-powered arc is maintained and manipulated between two tungsten electrodes and is operated within an atmosphere of supplied hydrogen. Fig. 1-29 shows the equipment necessary for the manual operation of the atomic welding process. This method of welding is unique, in that the heat generated by the arc does not perform the actual fusion of the metals being welded, as is usually the case in other arc procedures. Instead, the heat of the arc causes a *separation* of the hydrogen atoms, or a transformation from molecular to atomic hydrogen. This reaction brings about the absorption of heat by these free atoms. Upon making contact with the relatively cold metal surface being welded, the atoms of hydrogen recombine, which causes a *release* of this same absorbed heat energy. It is the heat transfer brought about by the separation and recombination of hydrogen atoms that supplies the intense heat required for the melting of the weld metals.

The force of the arc and hydrogen gas while operating causes a more or less horseshoe-shaped fan to be projected on the work, as illustrated in Fig. 1-30.

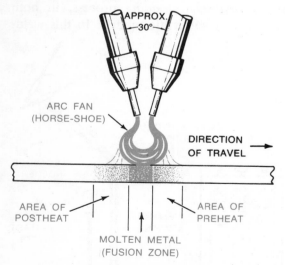

Fig. 1-30. Atomic hydrogen arc fan.

Manipulating the size and shape of this fan leads to the advantages of atomic hydrogen welding. Even at its finest adjustment, or *silent arc,* welding speeds about twice as fast as oxy-acetylene welding can be successfully achieved.

The fan, as can be seen in the illustration, performs more than one of the standard welding operations. The leading edge of the fan gives the operator the advantage of a source of preheat; the center of the fan is the actual heat source causing fusion. The trailing edge supplies the post-heat treatment and is very important because it slows the weld cooling rate. This lets any hydrogen which may have been drawn into the weld metal have time enough to escape, thus ensuring a reasonably ductile weld. The atmosphere of hydrogen has an additional advantage of retarding any oxidization that would ordinarily be present.

The atomic hydrogen process is suitable for welding metals of varying gauges, merely by "tuning," or manipulating, the arc fan. Practically all ferrous and nonferrous metals (or alloys) can be welded easily. Because the operator can control carbon "burn out," sound welding of hard-to-weld metals can be achieved. Using a filler rod with about 30% more carbon than the metals to be welded will produce a weld with properties identical to those of the parent metals. This result is a desirable feature of any welding process.

Metal Spraying

This is a process whereby a filler metal is melted in a spray gun by heating, such as with the oxy-acetylene flame, and blown on to the surface of the parts to be built up or hard surfaced by air pressure (Fig. 1-31).

Metal spraying is adaptable for parts where thin deposits are required and where no increase in tensile strength is necessary.

Some models of mechanized guns accept a thin filler metal wire from a revolving spool. When heated to a molten temperature by a flame in the body of the gun, the wire is emitted through the gun nozzle in the form of a spray, at about 40 cu. ft. per minute

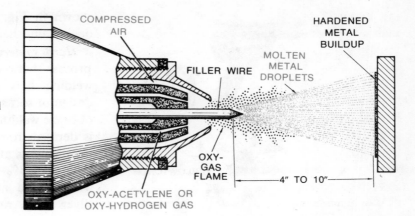

COMPRESSED AIR

FILLER WIRE

MOLTEN METAL DROPLETS

HARDENED METAL BUILDUP

OXY-GAS FLAME

4" TO 10"

OXY-ACETYLENE OR OXY-HYDROGEN GAS

Fig. 1-31. Metal spraying gun and sample procedure.

(cfm) air pressure. Powders and other granulated materials can also be emitted by using various other types of guns.

The material to be sprayed should be prepared by grinding the surface so that it will accept the molten particles. On hitting the cold prepared surface, the molten particles flatten and lock together mechanically in various interlocking patterns. The finer the spray, the better is the opportunity of having a nonporous bond which will help close tolerance machining.

Metal spraying can be used to restore broken or worn parts, which makes this application, today, essentially a machinery repair and hardsurfacing procedure.

Metal Surfacing

The welding processes used for surfacing are, in most cases, performed by the oxyacetylene or arc welding procedures. *Metal surfacing* requires adding a different material to a metallic body to extend the usefulness of the body.

Since wear on a metallic body is usually caused by a group of factors, rather than a single factor, the welder must analyze the causes contributing to wear before selecting suitable surfacing materials. Causes of deterioration which require metal surfacing may fall into one or more of the following categories: shock and impact, abrasion, corrosion, erosion and friction.

Soldering, Brazing and Braze Welding

Whether the operator is *soft soldering, hard soldering, brazing* or *braze welding,* the principle behind the procedures is the same. While the purposes behind soldering and brazing differ in some respects, the processes will be mentioned together at this time.

In any of these operations, the base or parent metals are never heated to their melting point. The joint is obtained through the proper preheating of the parent metal to a temperature which will allow the molten filler metal to *tin* (flow smoothly on) the surfaces of the weld area. Using flux is necessary to adequately clean the joint. The strength of the joint is gained in the initial tinning operation; additional filler metal is added only if required. Filler metals in all processes are of a nonferrous composition. Typical tools used in soldering are shown in Fig. 1-32.

Soft soldering is more important to the sheet metal industry than to welding, so it is mentioned here merely because of its similarity to the other processes. The filler metal is usually a combination of tin and lead in varying proportions; for example, 60/40 solder is made up of 60% tin and 40% lead. Other elements, such as bismuth or antimony, can be alloyed into the mixture, if changes in melting temperatures are desired. The composition of the solder to be used is determined by the type and use of the metal

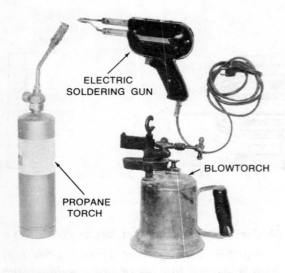

Fig. 1-32. Heating equipment used for various soldering procedures.

ELECTRIC SOLDERING GUN

PROPANE TORCH

BLOWTORCH

and the position of the joint to be made. Soft soldering is of little value where strength is required and is usually used as a sealing application, thus making heavy build-up unnecessary. Because of the low melting range of the "soft solders" (between 350° to 800°F), the use of soldering coppers, as

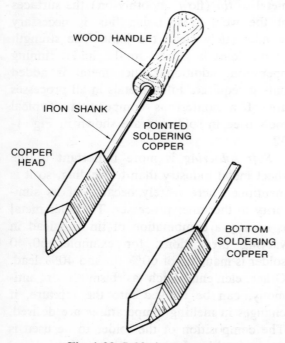

WOOD HANDLE

IRON SHANK

COPPER HEAD

POINTED SOLDERING COPPER

BOTTOM SOLDERING COPPER

Fig. 1-33. Soldering coppers.

shown in Fig. 1-33, is usually preferred to open flame heat sources.

Hard soldering, or silver soldering, is the process between soft soldering and braze welding. It is a process which uses the joint design of soft soldering and the flame process of braze welding. The name "silver soldering" is derived from the fact that silver is one of the elements alloyed into the filler metal. As little as 10% silver content is enough to warrant the mixture being classified as a silver alloy, rather than being named after some other metal which is present in larger amounts. Brazing other filler alloys, such as copper- or nickel-based alloys, is quite similar to silver soldering. Like soft soldering, "silver brazing," as it is sometimes called, can be used as a sealer, but is superior because of its high weld strength and resistance to chemical corrosion. Because of the strength obtainable and its high electrical conductivity, silver brazing is found to be a very satisfactory method of making electrical connections when fabricating or repairing electrical equipment. The oxy-acetylene flame is a good heat source for this process, since heats of around 1100°F are required. As illustrated in Fig. 1-34, the proper joint de-

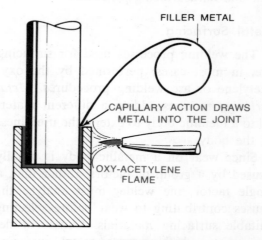

FILLER METAL

CAPILLARY ACTION DRAWS METAL INTO THE JOINT

OXY-ACETYLENE FLAME

Fig. 1-34. Silver brazing.

sign, plus the exact heat, will produce welds with strengths up to 60,000 psi.

Brazing and *braze welding* are almost one and the same. Both require that the parent metals be preheated to the exact tinning temperature. However, brazing depends on the capillary action of the molten filler metal, which in this case is bronze, with the parent metals. In this regard, brazing and the soldering processes are alike.

The difference between the two processes is in the type of joint being made. If the joint required can be obtained through a flowing of molten bronze between the prepared parts, then the process would be one of brazing. If additional build up in the weld area is required, then one would use braze welding. Preparing the parent metals for braze welding is like preparing for fusion welding. Fillet, groove, and plug or slot preparations are commonly used. Braze welding, while of some value in the fabricating industry, has a more practical use in repair and maintenance work.

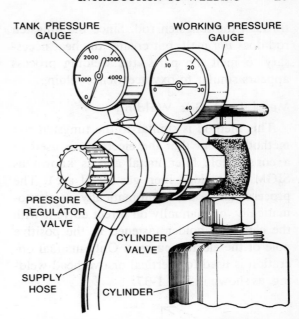

Fig. 1-36. Inert gas supply for TIG welding.

to be welded and a tungsten wire electrode, as shown in Fig. 1-35.

The electrode holder is built in such a way as to allow inert gas to flow out and around the rod to form a gaseous shield near the arc field. Water is also circulated through the holder for cooling. The inert gas, supplied in high-pressure cylinders, leaves the cylinder through an attached regulator (Fig. 1-36) similar to the ones used on oxygen cylinders.

Refractory metals (metals difficult to fuse) have been used in the inert welding process as electrode material. They are less suitable than *tungsten*, a steel gray element of the chromium group, because they erode easily. The rate that tungsten erodes is so slow that the rod is classified as nonconsumable. Tungsten electrodes, when attached to the negative pole of a current, can withstand much higher currents than if operated on the positive pole. However, since the cleaning action of the arc is better on direct current positive, this polarity is used, except where excessive amperage is required. Alternating current is used if excessive amperage is required.

The metal filler rod, along with the parts to be welded, is melted by the arc established

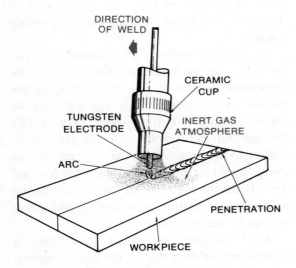

Fig. 1-35. TIG welding arc.

Tungsten Inert Gas Welding

The pieces being welded are protected from the atmosphere by a continuous stream of inert gas consisting of *argon* or *helium* during *inert gas arc welding*. The electric arc creates the necessary heat between the work through the tungsten rod. Since the tungsten rod does *not* melt, but creates the heat necessary to melt the parts, the welding process appears similar to oxy-acetylene welding.

Metal Inert Gas Welding

This method is similar to the tungsten rod method, except that the electrode is used as a consumable filler metal and is known as SIGMA (shielded inert-gas metal arc). The process can be used automatically, semiautomatically, or manually on direct current with the electrode holder fastened to the positive pole of the electric current. Only manual operation is used for vertical or overhead welding, as shown in Fig. 1-37.

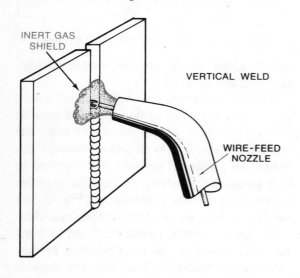

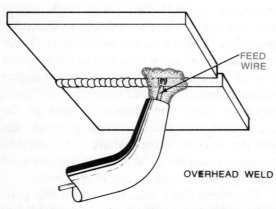

Fig. 1-37. Metal Inert Gas (MIG) welding samples.

Inert gas welding can be used to repair most commercial metals, but the procedures outlined are especially preferred for hard-to-weld materials such as stainless steel and aluminum. Normally, these are adversely affected by atmospheric elements during more common welding processes.

Plasma-Arc Welding

This welding process consists of a plasma jet produced by electrically heating plasma, forming gases such as nitrogen or hydrogen. The resulting ionized jet contains an equal number of positive ions and electrons. A special copper-alloyed torch is used to withstand the high temperature created by the ejection of the plasma jet when the arc is struck. The arc is struck only within the torch between the water-cooled copper anode (positive pole) and the tungsten cathode (negative pole). This *nontransferred arc* does not come into contact with the work piece; it is expelled from the gun nozzle in the form of a flame, making metal spraying possible.

In another method, *transferred arc,* the cathode is in the torch and the anode is the workpiece, similar to the arc striking in the metallic arc process. This method can be converted into a cutting torch (Fig. 1-38) for use on nonferrous and alloyed steels, similar to electron-beam. During cutting, the energy density from the plasma arc is focused directly on the metal to be cut, and, with the proper manipulation, hole cutting and gouging can be accomplished.

Although there are many advantages in plasma-arc welding due to the high temperature established through the arc, further research will be necessary to make it practical for general industrial use. Fig. 1-39 illustrates how a Linde PT-9 standard torch works.

Laser and Electron Beam Welding

Today, industry is developing the newest space-age welding technique; drilling and

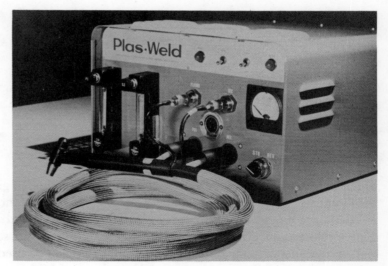

Fig. 1-38. Plasma arc cutting equipment. (Courtesy of Canadian Liquid Air.)

cutting by using the most modern tools of technology—laser beams and electron beams.

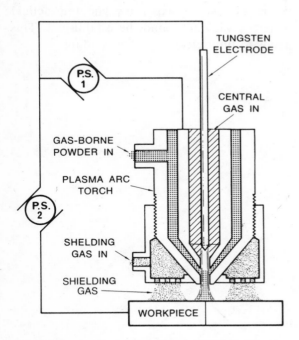

Fig. 1-39. A type of standard plasma arc torch. (Courtesy of Union Carbide Canada Ltd.)

The Laser—Drilling the hardest of materials, including diamonds, is easily accomplished today by means of a directed beam called a *laser*. Welding hard-to-fuse materials, of thread size or smaller thickness, is now

possible. In the not-too-distant future, the welder will become familiar with both laser and electron beam welding, which today are all but unknown.

Since the first laser was developed, different approaches have been taken to convert energy into laser light, but basically all methods rely on an active material for energy. In the earlier lasers, where a cylindrical ruby crystal was used for energy, two mirrors were placed on opposite ends of the ruby. Each mirror reflected the light back and forth through the crystal until the energy reached a point where it passed as a laser beam through a selected mirror, which had been very lightly silvered for less reflective power.

When this laser light, which is produced in a type of heartbeat pulsation, is focused through a type of lens, atoms send out a single narrow beam of high amplitude photons (a quantum of radiant energy) which hit only a tiny area with such frequency as to penetrate with little expended energy (Fig. 1-40).

Since the energy from a laser beam can be controlled and located more closely and efficiently than an arc or flame, welding high precision instruments is more suitable by laser. Laser welds are completed quickly, and heat loss by conduction through the metals being welded is minimized by the high rate at

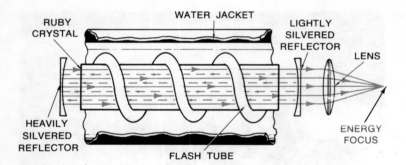

RUBY CRYSTAL — WATER JACKET — LIGHTLY SILVERED REFLECTOR — LENS — HEAVILY SILVERED REFLECTOR — FLASH TUBE — ENERGY FOCUS

Fig. 1-40. Laser light as used to weld.

which the beam energy is delivered to the weld joint.

Welding gold, silver, and copper is easily done by laser, since they can withstand the high instantaneous temperatures of the laser pulsating beams without the expulsion of the metal at the point of contact. Also, since they heat quickly due to their high thermal conductivity, laser welding on these metals is more feasible and practical today.

The *electron beam* is an intense, high-velocity beam. With this method, a hot metal cathode gives off electrons (particles of charged negative electricity) which are focused and speeded up to a considerable degree by a series of electrodes. Electron beams can be used only in a vacuum, since air molecules have the power to scatter electrons out of the beam. Under vacuum-controlled conditions, the electron beam will pinpoint an area even more accurately than the laser, but such a process can be expensive and impractical where a vacuum cannot be established. This tends to limit electron beam welding.

2

Metallurgy and Metallography

The subject of *metallurgy* and *metallography* is presented to help the welder become familiar with the various expressions and terms used by his trade. To present a complete study of metallurgy in a single chapter is impossible, but fairly complete coverage of topics related to welding has been attempted.

Metallurgy can be described as *the science and art necessary to extract metals from their ores* for the purpose of adapting these metals to the use of industry. Ore is the mineral deposits from which we mine the metals, and it sometimes forms part of a rock.

Metallography, on the other hand, is known as the *branch* of metallurgy dealing with *the study of a metal* and the *constitution and structure* of its alloys. By means of metallography, the grain structures, which influence the metal's properties, are studied to determine how the different properties are affected by the metal's grain structure.

The Atom

Before going into the details of this subject, the welder should have some knowledge of atoms' effect on metallic substances. The atom is so very small that it cannot be seen under a microscope, and yet it has all the chemical properties of an element.

An element is made up of many atoms, and all the atoms of the same element (for example, iron) are identical in size and structure. The atoms in iron will differ in size and structure from those in copper. Therefore, iron differs from copper due to the atomic structure.

The arrangement and motion of atoms determine how the matter behaves. The atoms in a *solid,* such as iron, are slower-moving and packed closer together than are the atoms in a *gas* or a *liquid.* The atoms, when packed together with other atoms, are responsible for the characteristic properties of a metal. The atom is known as the fundamental unit from which the *grain structure,* shown in Fig. 2-1,

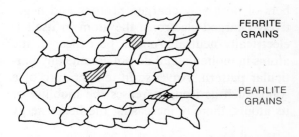

Fig. 2-1. An example of metallic grain structure.

of a metal is formed. The grain structure depends on the makeup of the individual atom, along with the manner in which all atoms are grouped together. Fig. 2-2 shows an example of the grain structure in actual steel.

Each atom is made up of small electrical particles with a positively-charged *nucleus* (a

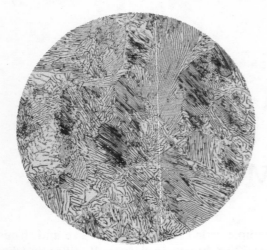

Fig. 2-2. Grain growth in high-carbon steel. (Courtesy of Bethlehem Steel.)

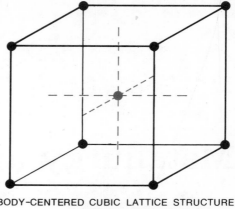

BODY-CENTERED CUBIC LATTICE STRUCTURE
(STEEL AT NORMAL TEMPERATURE)

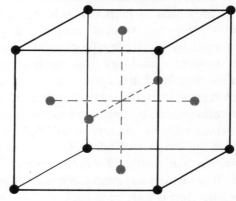

FACE-CENTERED CUBIC LATTICE STRUCTURE
(STEEL AT TEMPERATURE RAISED ABOVE 1600°)

Fig. 2-4. Space lattices.

central core). The nucleus contains two fundamentals known as the *proton* and the *neutron*. The proton has a positive charge, and the neutron has no charge.

Outside the nucleus, yet within the shell of the atom, are negatively-charged particles called *electrons*. If the nucleus is visualized as the sun, and the electrons thought of as planets orbiting about the sun, the atom can be more readily understood (Fig. 2-3).

Because every atom contains equal numbers of positively charged protons and negatively charged electrons, the atom in itself is electrically neutral. As can be expected, the atoms in molten metal do not assume any particular pattern. Any rise in temperature of a solid, such as iron, increases the mobility of its atoms, thereby increasing the heat energy

which, in turn, helps to keep the atoms separated. With a temperature decrease, the heat energy created by the mobility of the atoms is decreased. As the mobility of the atoms is retarded by lowering the temperature, the

Fig. 2-3. Structure of an atom.

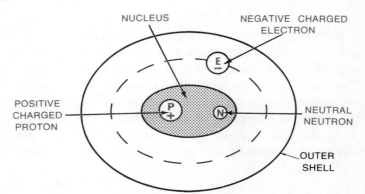

NUCLEUS

NEGATIVE CHARGED ELECTRON

POSITIVE CHARGED PROTON

NEUTRAL NEUTRON

OUTER SHELL

force which tended to keep the atoms apart is also retarded to a point where each atom binds to the next in a regular pattern. A molten metal, on cooling, becomes solid because the mobility of its atoms slows down.

Space Lattices

The pattern formed by the slowed atoms is a symmetrical arrangement known as a *space lattice*, as in Fig. 2-4. This holds the atoms about the same distance from one another and allows only a limited freedom of movement.

All metals and their alloys have a form known as a *crystal structure*. This structure is brought about by the attraction of tiny crystalline grains, each a completed homogeneous solid unique to the element being solidified. The tiny crystals bind together in turn to form a larger particle, known to the welder as the *grain*. The grain, unlike the crystalline particle, is usually not symmetrical (regular in structure) because the uneven surfaces are formed by each crystal binding in an irregular manner during crystallization, as in Fig. 2-5. Crystals can become strained (Fig. 2-6) as

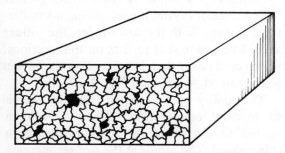

Fig. 2-5. Grain Structure—Metal heated and quenched slowly.

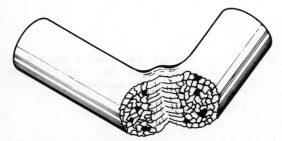

Fig. 2-6. Crystal grain structure strained.

stress is placed on the metal being used, but a recrystallization can take place with old crystals being made into new ones by hot working the metal.

Properties of Metals

The effect of both high and low temperatures on characteristics of metallic elements, due to the atomic structure, is probably not fully understood by the welder at this point. However, the welder should recognize that the effects of temperature can create problems (which could be easily solved) and can offer advantages (which would be beneficial), if he is well-informed on the subject. Therefore, it will be necessary for the welder to be familiar with the *mechanical properties* of metals in order to determine the ability of the repaired parts to withstand breaking force.

Knowing about *chemical properties* of metals is also necessary if a welder is to understand how corrosion and oxidation can be dealt with. Knowing the *physical properties* of metals, including melting points and magnetic properties, must also be considered for good welding.

Physical, chemical, and mechanical properties will be discussed to help make the welder aware of the changes taking place during the treatment of various metals and their alloys, as they are presented in the chapter.

Physical Properties

Many characteristics of metal, such as its color and magnetic properties, can be easily identified by sight or easy tests. Most of these are *physical properties*.

Magnetic properties are limited to the ferromagnetic metals. Cobalt, nickel, and certain alloys of these metals can be extremely magnetic, like iron. They are able to be attracted by a *lodestone*, which is a variety of magnetite that shows *polarity* (a quality of having opposite poles) and acts like a magnet. It is thought that these materials react magnetically due to the motion of the electrons

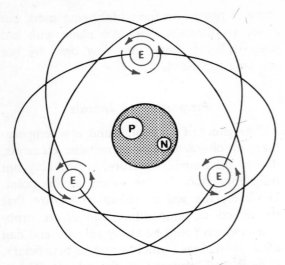

Fig. 2.7. *Electrons* spin while orbiting the nucleus.

within the shell of the atom, as in Fig. 2-7. Since the number of electrons with positive spin is not counteracted by an equal number of negative spins, the result is a magnetic *moment* (moving force) for each atom.

Color, as a *physical property,* describes the surface appearance of a metal. The light of a given wavelength gives a sensation to the human eye in the form of a color. The appearance of a metal is the ability of the metal to reflect or absorb light of various wavelengths (Fig. 2-8).

The *melting point* is the temperature at which a solid substance changes into a liquid. Atoms, when grouped in an orderly geometric arrangement known as the *pattern,* or *lattice,* have the property of *rigidity,* as mentioned previously. This condition is known as the *crystalline,* or *solid,* state.

The revolution of the electrons around the nucleus of the atom, and the spinning of the electrons on their own axes, creates energy. This energy is affected by the closeness of the atoms within any system, as well as by temperature and pressure changes. If the pattern is distorted by any change, the atoms are free to vibrate in new positions, increasing the internal energy.

The increase in internal energy caused by a higher temperature change can also increase the volume of a solid. If the temperature is increased to the point where the action of the atoms allows them to escape from the attraction of other atoms, a liquid is formed.

The quantity of energy given up by any system when it solidifies can equal the energy taken in by the system on melting and is known as the *heat of fusion.* The system is at its melting point when the rate of melting equals the rate of freezing. Fig. 2-9 shows the melting point of metals and alloys.

As the liquid cools, the energy of the atoms decreases to a point where atom attracts atom for grouping, until a geometric arrangement is made for the formation of a crystal. The more rapidly a liquid is cooled, the greater the number of crystal nuclei, giving a smaller-sized crystal. With this knowledge, the welder should be aware that repairs on thin sections, which cool faster than heavier sections, have a finer-grained structure (Fig. 2-10).

The ability of a metal to conduct heat is its *thermal conductivity.* When a solid is heated, as in welding, the action of the atoms is increased. The atoms at the surface increase

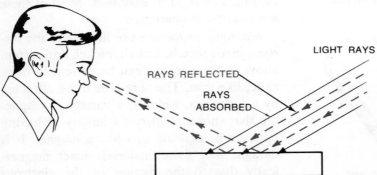

LIGHT RAYS

RAYS REFLECTED

RAYS ABSORBED

Fig. 2-8. The color of metal is determined by light reflection and absorption.

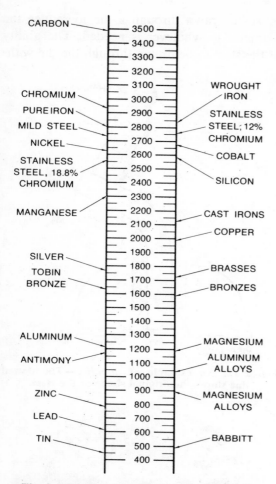

Fig. 2-9. Melting points of metals and alloys.

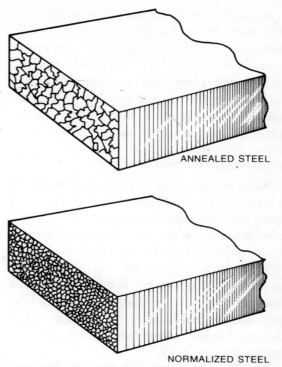

ANNEALED STEEL

NORMALIZED STEEL

Fig. 2-10. Grain structure depends upon the rate of cooling.

in vibration and thereby transfer heat and action to the layer of atoms immediately below them. The surface atoms must give up heat continuously until the inner core of atoms reaches a temperature equal to the total environment. The surface metal is, therefore, never up to temperature until the entire mass of the atoms is in equilibrium with the environment. Therefore, the speed with which the heat is transferred or conducted from one part of the mass to another is the *thermal conductivity*.

Chemical Properties

Corrosion, oxidation, and *reduction* are listed as chemical properties. Corrosion is the chemical attack on a metal, usually by atmos-

pheric conditions, which causes the metal to waste away. The combination of an oxide, such as iron rust, is known as oxidation; also, a combination of an oxide with oxygen forms a higher oxide. Partial or complete removal of oxygen from an oxide is known as reduction.

Oxygen is a highly reactive element that combines readily with other elements in steel, especially at high temperatures, forming undesirable oxides and gases. When oxygen combines with iron, compounds are formed that lower the mechanical properties of the metal. The new oxygen-formed compounds produce what are known as *inclusions* in the weld. These inclusions are nonmetallic matter. The ease with which oxygen combines with metals to form oxides complicates the welding process. Since iron oxide is of a lower density and melts at a lower temperature than steel, any iron oxide formed during welding iron will separate and form a film on top of the molten metal.

Regardless of its manufactured grade, steel will reabsorb oxygen during welding, unless methods are adapted to shield the heated areas from the air. There are various methods of *reducing* (nonoxidizing) the atmosphere surrounding the molten puddle; these methods will be described in the chapters covering fluxes and welding methods.

Mechanical Properties

Since the mechanical properties determine a metal part's ability to withstand tearing or breaking when subjected to force, they are considered to be the most important properties from the welder's point of view.

Brittleness—Any metal that will fracture suddenly, without slight bending and under low stress, can be classified as brittle. *Brittleness,* such as that found in cast iron, is one of the mechanical properties a welder must know. See Fig. 2-11.

Ductility—A metal with good *ductility* can be deformed by force without breakage. This allows the metal to be cold worked, hammered, or pressed out into various shapes, as in Fig. 2-12. Fig. 2-13 shows a magnification of ductile steel. The picture was taken through a microscope.

One application where ductility is very important is manufacturing wire. Here, the steel

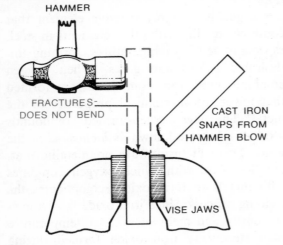

Fig. 2-11. An example of *brittleness*—The dotted line shows the original shape of the cast iron.

must be drawn through a die to reduce the diameter to whatever is needed. The ability of the steel to be drawn through the die with-

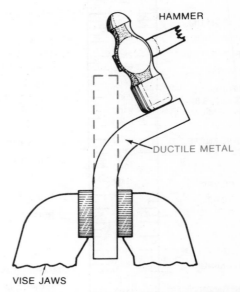

Fig. 2-12. An example of *ductility*—The dotted line shows the original shape of the steel.

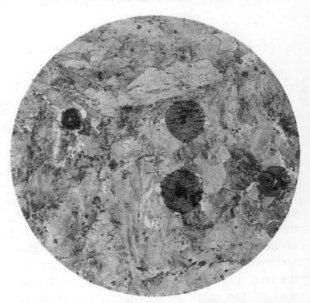

Fig. 2-13. Typical structure of ductile iron, magnified 250 times. (Courtesy of Bethlehem Steel.)

out breaking depends on its ductility. Fig. 2-14 shows a cutaway of a die that is being used to form wire as the steel bar is drawn through.

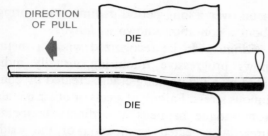

Fig. 2-14. Smaller diameter wire being formed by drawing through a die.

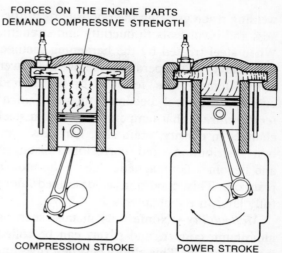

Fig. 2-15. The internal combustion engine parts must have a great deal of *compressive strength*.

Compressive Strength—Another mechanical property of metal is *compressive strength*. Metals with much compressive strength are able to withstand a force over a large surface area without deforming. The force must be applied slowly and evenly, unlike the sharp force used to test for brittleness.

A good example of compressive strength is an internal combustion engine, where the engine block and/or cylinder head must be able to withstand the engine's *compression*. Although able to withstand over 200 lb./sq. in. pressure inside a cylinder, a cast iron cylinder head may crack if dropped one or two feet to a concrete floor. Fig. 2-15 shows the compression stroke of a common gasoline engine. The engine parts must have a great deal of compressive strength.

Tensile Strength—A property known also as maximum strength is more commonly referred to as *tensile strength*. This is the property of a metal that describes the maximum normal load *per unit area* a material can withstand without failure. Tensile strength is measured in thousands of pounds per square inch; when carbon is added, the tensile strength of steel will go up steadily as the carbon is increased to 0.83%. Any carbon addition after 0.83% will cause the tensile strength to decrease, as shown in Fig. 2-16.

Toughness—Metals must also withstand sudden shock, a property known as *toughness*. Toughness is sometimes described in the

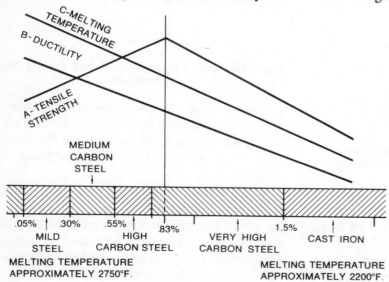

Fig. 2-16. How steel is affected by the addition of carbon. (A) Tensile strength increases as carbon content goes up to 0.83%. If carbon is added above 0.83%, tensile strength decreases. (B) As carbon is added, ductility is decreased. (C) As carbon is added, melting temperature decreases.

welding trade as being the opposite of brittleness and involves both ductility and strength. When steel treated by the hardening method is reheated to temperatures below the lower critical temperature, followed by the desired cooling, the steel becomes tough. *Critical temperature* is that temperature where a steel changes in density, grain size, hardness, carbon distribution, and corrosion resistance; and becomes nonmagnetic due to the atom's reaction. This condition is described more fully later in the chapter.

Malleability—Nonferrous metals, such as aluminum, copper, and silver, can be rolled into thin sheets. This ability best describes another mechanical property, known as *malleability*. This is a metal's ability to be deformed by forging, pressing, or rolling, without a noticeable increase in its resistance to deform. See Fig. 2-17.

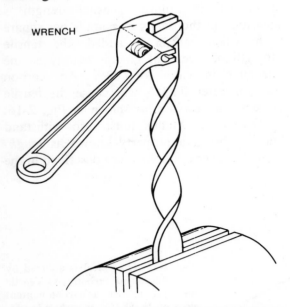

Fig. 2-17. *Malleability*—**The ability of a metal to be worked cold without noticeable resistance.**

Impact Strength—The ability of a metal to withstand shock without fracturing is known as *impact strength*. Impact is a type of wear caused by blows such as a hammering effect. Resistance to impact requires toughness in a metal. Continuous hammering, or impact, on

metal over a long period of time could bring about a condition known as *fatigue*.

Fatigue can be recognized when a metal shows progressive fracturing, usually indicated by a lengthening crack, created by continuous stress. Failure of steels or other metals from fatigue happens at ordinary temperatures. A repeated application of the load, which creates stress on a minor flaw in the metal, can cause the flaw to grow to the point where the material weakens and fails. Corrosion *does* accelerate fatigue, but is not responsible for starting it. See Fig. 2-18.

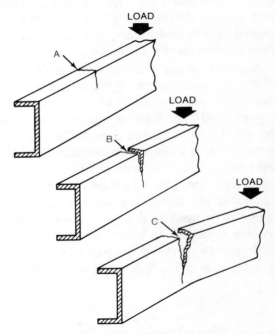

Fig. 2-18. *Fatigue Failure*—**A, B, and C show the progressing flaw.**

Stress—When a metal is placed under *stress*, the metal may bend like an elastic. This bending is known as *elasticity*, if the metal returns to its original shape when the stress is relieved.

When ferrous alloys are stressed, they first suffer elastic deformation, and, if the stress is increased to a high enough degree, the metals wil deform like a plastic. Increased temperatures would lower the amount of stress necessary to deform these metals.

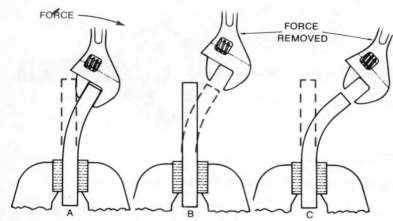

Fig. 2-19. The *Elastic Limit* demonstrated on a steel bar. (A) Force applied to the metal bar. (B) Force removed—Bar returns to its original shape because the elastic limit had not been exceeded. (C) Force removed—Bar does not return to its original shape because the force exceeded the elastic limit of the material.

Elastic limit is the maximum stress point that will not cause permanent deformation and will allow the metal to fly back to its original shape, as in Fig. 2-19.

Creep—Once the elastic limit has been reached, any small, steady force exerted over a long period of time will create a slow, permanent change in the material's form, known as *creep*.

When metals are heated to very high temperatures they lose their shape, due to the pull created by the temperature change. Creep has been described as a slow deformation in a metallic substance caused by a small, steady force, just below the elastic limit, acting for a long period of time. See Fig. 2-20.

Fig. 2-20. *Creep*—The red line shows the original shape of the spring. Continuous weight has caused deformation of the spring.

Tensile strength and hardness were the only two mechanical properties that early engineers used in designing for safety. As mechanical power increased, separation of materials under load increased after prolonged periods of use. This became known as *fatigue*, which today accounts for 90% of all mechanical breakdowns. Therefore, the *fatigue strength* of metal becomes one of the most important mechanical properties facing the modern welder.

Since all metals and alloys will creep if subjected to stress at certain minimum temperatures, a knowledge of this property is also necessary. Today, high stresses, combined with high temperatures, are increasing in many industries. A scientific study to find new materials that can withstand the required stress at the required temperature, thereby increasing efficiency, becomes a necessity. The age of space travel makes the property of *creep* as important to tomorrow's welder as *fatigue* was to engineers before the steam engine.

All these properties will be referred to periodically, to describe the suitability of metals and their alloys during welding.

Metal Testing Procedures

Selecting and processing metals for a certain job depends on their mechanical properties. If the welder understands the significance of such tests, he becomes better equipped to handle various welding jobs. Here are some testing measurements.

A welder may use several new tools to test the physical structures, chemical composition, and/or flaws of materials, even though he knows little about the physical principles of testing instruments.

Controlling and measuring temperatures can become as important to the welder as it

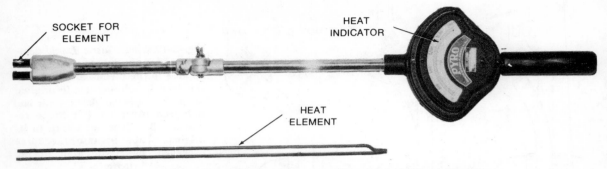

SOCKET FOR ELEMENT

HEAT INDICATOR

HEAT ELEMENT

Fig. 2-21. A dismantled Pyrometer, used for measuring heat closely.

is to the furnace operator producing metals or alloys. While heat treating alloys, it is sometimes important to control temperature very closely. A *pyrometer,* shown in Fig. 2-21, is used for this purpose and can be adapted to operate automatically. A pyrometer is an instrument to measure high degrees of heat. The *resistance thermometer* is a type of pyrometer for measuring relatively low temperatures.

Tension Test

The capacity of a material to withstand a static load is determined by a *tension,* or *compression, test.* Forces which are applied gradually to the piece of material and remain essentially constant after being applied are known as static, or dead, loads. A welded

TEST RAM

WELDED BAR

COMPRESSION

WELD JOINT

EXPANSION

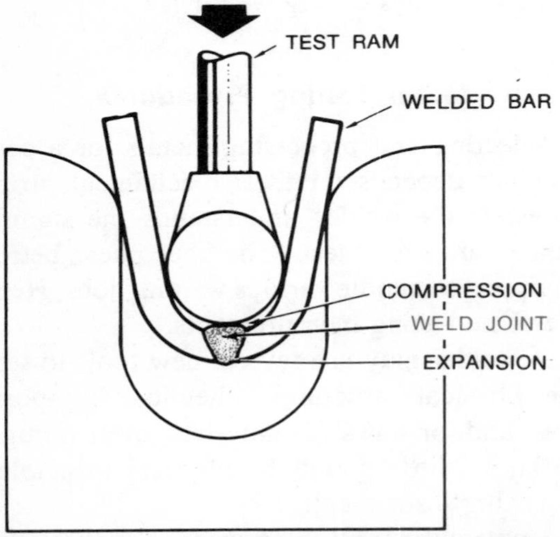

Fig. 2-22. Metals testing: This welded bar has been tested for *expansion* and *compression.*

bar in Fig. 2-22 has been tested for both expansion and compression.

Brittle materials are often weak in tension because of small cracks, some of which cannot be seen under a microscope. The low tensile strength of such materials varies from sample to sample, but they can be fairly strong when under compression. Therefore, they are used in compression applications where their strengths are much higher. Cast iron, a brittle material, can withstand stress under compression better than under tension, as shown in Fig. 2-23.

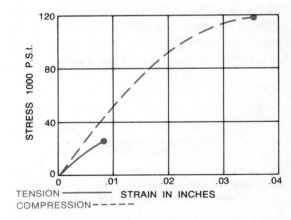

Fig. 2-23. Cast iron's ability to withstand compression and tension forces.

If a stressed body does not regain its original shape when the stress is released, the body has undergone *plastic deformation.* Plastic deformation never occurs during a compression test, because the test increases the cross-sectional area of the piece being tested.

Very ductile materials are seldom tested in compression because the material being tested is constrained where it contacts the

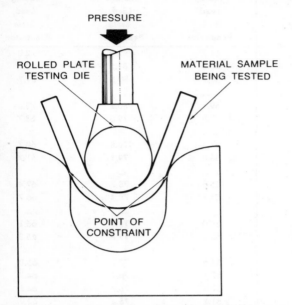

Fig. 2-24. Attempting to test a ductile material in compression.

rolled plate of the die. This constraint will complicate stress distribution and make accurate testing difficult. See Fig. 2-24.

Hardness Test

Information about how a metal can resist permanent deformation is gained by a *hard-*

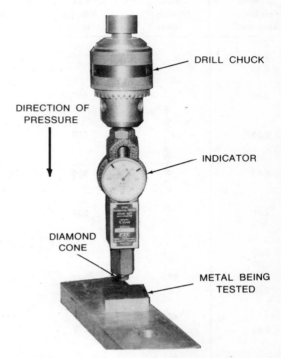

Fig. 2-25. Using a Rockwell Hardness Tester in a drill press.

Fig. 2-26. A pressure-applying machine used in hardness testing.

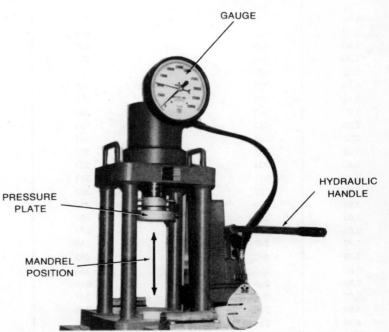

Table 2-1. Hardness Conversion For Steel.

	Brinell Hardness Number			Rockwell Hardness Number		
Brinell Indentation Diameter, mm.	10-mm. Standard Ball, 3000-kg. Load	10-mm. Hultgren Ball, 3000-kg. Load	10-mm. Carbide Ball, 3000-kg. Load	C Scale 150-kg. Load, Brale Penetrator	A Scale 60-kg. Load, Brale Penetrator	D Scale 100-kg. Load, Brale Penetrator
2.35	No Hardness Values	......	682	61.7	82.2	72.0
2.40		......	653	60.0	81.2	70.7
2.45		......	627	58.7	80.5	69.7
		......	......	......	......	......
2.50		601	......	59.1	80.7	70.0
		......	601	57.3	79.8	68.7
		......	......	......	......	......
2.55		578	......	57.3	79.8	68.7
		......	578	56.0	79.1	67.7
		......	......	......	......	......
2.60		555	......	55.6	78.7	67.4
		......	555	54.7	78.4	66.7
		......	......	......	......	......
2.65		534	......	54.0	78.0	66.1
		......	534	53.5	77.8	65.8
		......	......	......	......	......
2.70		514	......	52.5	77.1	65.0
		......	514	52.1	76.9	64.7
	495	......	......	51.6	76.7	64.3
2.75	......	495	......	51.1	76.4	63.9
	......	......	495	51.0	76.3	63.8
	477	......	......	50.3	75.9	63.2
2.80	......	477	......	49.6	75.6	62.7
	......	......	477	49.6	75.6	62.7
	461	......	......	48.8	75.1	61.9
2.85	......	461	......	48.5	74.9	61.7
	......	......	461	48.5	74.9	61.7
	444	......	......	47.2	74.3	61.0
2.90	......	444	......	47.1	74.2	60.8
	......	......	444	47.1	74.2	60.8
2.95	429	429	429	45.7	73.4	59.7
3.00	415	415	415	44.5	72.8	58.8
3.05	401	401	401	43.1	72.0	57.8
3.10	388	388	388	41.8	71.4	56.8
3.15	375	375	375	40.4	70.6	55.7
3.20	363	363	363	39.1	70.0	54.6
3.25	352	352	352	37.9	69.3	53.8
3.30	341	341	341	36.6	68.7	52.8
3.35	331	331	331	35.5	68.1	51.9
3.40	321	321	321	34.3	67.5	51.0
3.45	311	311	311	33.1	66.9	50.0
3.50	302	302	302	32.1	66.3	49.3
3.55	293	293	293	30.9	65.7	48.3
3.60	285	285	285	29.9	65.3	47.6
3.65	277	277	277	28.8	64.6	46.7
3.70	269	269	269	27.6	64.1	45.9
3.75	262	262	262	26.6	63.6	45.0
3.80	255	255	255	25.4	63.0	44.2
3.85	248	248	248	24.2	62.5	43.2
3.90	241	241	241	22.8	61.8	42.0
3.95	235	235	235	21.7	61.4	41.4
4.00	229	229	229	20.5	60.8	40.5

ness test. Hardness is measured by how much resistance the material has to an *indenter* (Fig. 2-25). The indenter is made of a material much harder than the materials to be tested, such as a diamond, hardened steel, or tungsten carbide particles.

For testing purposes, the indenter is pushed slowly into the surface of the metal being tested. All indenter tests must be taken at some distance from the edge of the material being tested. The test results are analyzed by using a table designed for a known applied load, compared with the depth, or cross-sectional area, of the impression. An example of such a table is Table 2-1.

The hardness of a metal can be tested by the Brinell hardness, Rockwell hardness, Scleroscope hardness, or Micro hardness processes. Most testers exert a *controlled* pressure on a sample of the metal. An example of such a tester is pictured in Fig. 2-26.

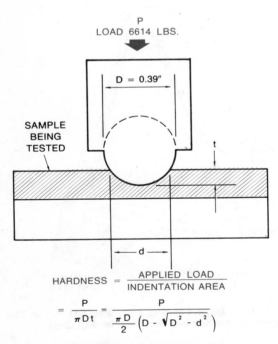

$$\text{HARDNESS} = \frac{\text{APPLIED LOAD}}{\text{INDENTATION AREA}}$$

$$= \frac{P}{\pi D t} = \frac{P}{\frac{\pi D}{2}\left(D - \sqrt{D^2 - d^2}\right)}$$

Fig. 2-27. Brinell hardness testing.

Brinell Hardness Test—In the *Brinell* test, a steel ball is pressed into the metal being tested, to create an indentation. The diameter of the indentation can be measured. Usually, a load of 3000 kg. is applied for a period of 10 to 30 seconds. The ball, known as the *penetrator,* is put under a 500-kg. load for metals softer than steel. See Fig. 2-27.

A microscope with a calibrated *reticle* (cross hairs for accuracy) measures the diameter of the impression made in the metal. This measurement corresponds to a certain Brinell hardness number found in Table 2-1.

Rockwell Hardness Test—The *Rockwell* test is similar to the Brinell test, except that the depth of penetration is read directly from a dial on the testing machine.

A *Rockwell C* hardness machine, used to test steels, uses a diamond cone. This cone is pushed slowly by a minor load of 10 kg., to seat the point of the cone. The major load, consisting of a 150-kg. weight, is then applied (Fig. 2-28.) The difference in the two pene-

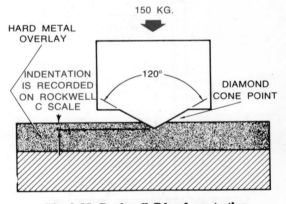

Fig. 2-28. Rockwell C hardness testing.

trations is then read from the dial. In the *Rockwell B* hardness procedure, a hardened steel ball about $\frac{1}{16}$ inch in diameter is used with a 100-kg. major load, as in Fig. 2-29.

Scleroscope Hardness Tests—The *Scleroscope* method measures the hardness of a material by the height that a diamond-faced hammer rebounds off the test piece from a given height. Most Scleroscope testing apparatuses have a quick, readable hardness dial attached. This machine, although more difficult to operate than the previous two testers, is easily transported to test large, immovable

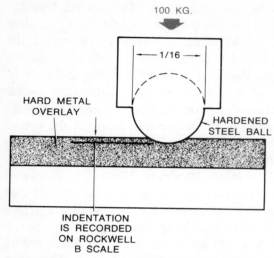

Fig. 2-29. Rockwell B hardness testing.

parts. Fig. 2-30 shows the Scleroscope procedure.

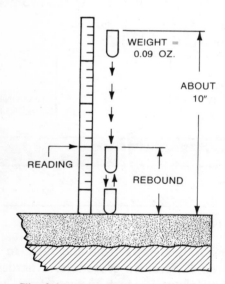

Fig. 2-30. Shore Scleroscope testing.

Micro Hardness Test—This is used to test the hardness of single steel crystals. A four-sided diamond pyramid, under a controlled load, marks the specimen at a highly polished point. The penetration depth is measured in *microns* (one micron equals .000039″), and readings can be calibrated to .002″ wide. Heat treatment effects can be evaluated by the *micro hardness test*.

Impact Test

Some indication of a materials' toughness under shock is given by an *impact test*. The part to be tested is notched and then placed across parallel bars (Fig. 2-31). A heavy

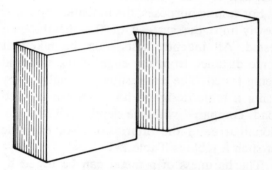

Fig. 2-31. Metal bar notched for impact testing.

pendulum, known as the *impulse load,* is allowed to drop on the specimen from a known height. See Fig. 2-32. The energy absorbed

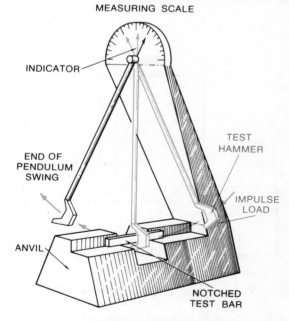

Fig. 2-32. Impact testing unit. The test bar in this figure is the notched test bar in Fig. 2-31.

in fracture can be calculated by the difference between the initial and final heights of the pendulum, along with its weight. Internal stress raisers, such as internal cracks and

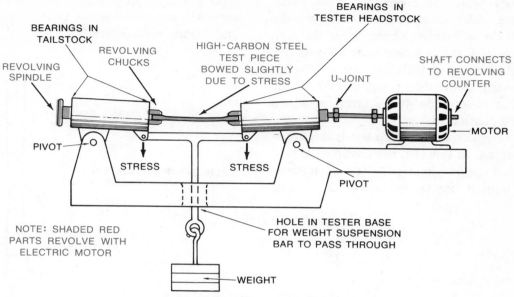

Fig. 2-33. Fatigue testing equipment.

grain boundary inclusions, have various effects on the impact test results.

The Fatigue Test

How many stresses a standard dimension metal can safely endure for a given number of cycles is determined by the *fatigue test*. Although a material will withstand high stress over long periods under *static* loading, it may not withstand the same stress *recurring* in *cycles*. Materials under static load are measured only by the yield point and ultimate tensile strength. A common fatigue test is one where a sample is loaded in pure bending, as in Fig. 2-33, and is rotated. With each rotation, all the positions on the circumference of the revolving piece go through one state of compression and one of tension.

During each revolution, there is a complete cycle of stress reversal. This procedure is continuous, completed many thousands of times per minute, and recorded on an indicator. Different loads are used on different specimens. In each case, the number of cycles before failure is noted. Data is then plotted to present an endurance limit, wherein the stress is on a vertical axis, measured in lb/sq. in,

versus the logarithm of the number of cycles to failure, as in Fig. 2-34.

The fatigue strength is increased by surface compressive stresses. These are produced by mechanical or chemical hardening on the surface of a specimen. A welder should know that corrosion *lowers* the fatigue strength, as does a steady tensile stress, while a steady compressive stress *raises* the fatigue strength.

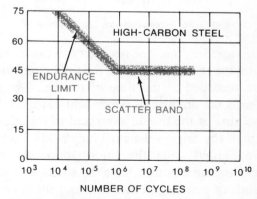

Fig. 2-34. Fatigue failure curve for high carbon steel.

Stress Rupture and Creep Test

Materials subjected to high temperatures for long periods of time can be evaluated by a *stress rupture and creep test*. When materials are under constant stress, they are deformed for an indefinite time. When the temperature of a metal is less than 40% of the absolute melting temperature, creep does not create any noticeable problem. When the temperature exceeds 40%, creep problems must be considered.

An extension of the creep test is the stress-rupture test. In this test, the specimen is held under an applied load at a definite temperature until it breaks, as in Fig. 2-35. The

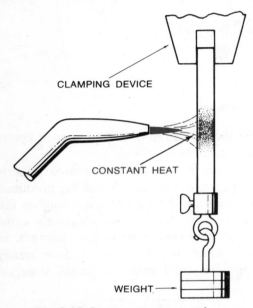

Fig. 2-35. Stress-rupture test under way.

load applied, the time necessary to cause fracture, the temperature, and the final elongation are recorded at the time of testing. *Elongation* is the amount of permanent extension in the area of the break, expressed in a percentage of the original length (Fig. 2-36) Curves are plotted from the information into the graphs. Some graphs have the stress, in thousands of pounds, on one axis and the time of fracture on the other axis.

Metal Identification

Welding methods vary with the type of material being repaired. Therefore, the welder

must practice recognizing metals and alloys, especially between various types of cast iron and carbon-content steels.

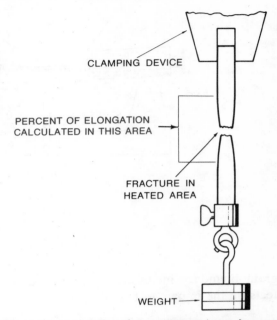

Fig. 2-36. Calculating the permanent extension percentage in the area of the break.

The welder should become more familiar, for practical purposes, with the following *identification tests* than with the previously mentioned *metallurgy* and *metallography* tests. Better welding will be possible when the operator is able to detect differences in metals. Since immediate identification is not always possible, a process of elimination, through the following tests, will help make identification possible.

Appearance

The new welder with little experience will probably have some knowledge of various common metals; this makes recognition possible by a *visual test*, using an identification chart such as Table 2-2. The appearance includes not only the *form* in which the material is produced but also the *weight, shape,* and *color*. The description of metals in this chapter will include color, to help the welder identify metals through elimination.

Table 2-2. Identification of Metals by Appearance.

	Alloy Steel	Copper	Brass and Bronze	Aluminum and Alloys	Monel Metal	Nickel	Lead
Fracture	medium gray	red color	red to yellow	white	light gray	almost white	white; crystalline
Unfinished Surface	Dark gray; relatively rough; rolling or forging lines may be noticeable	Various degrees of reddish brown to green due to oxides; smooth	Various shades of green, brown, or yellow due to oxides; smooth	Evidences of mold or rolls; very light gray	Smooth; dark gray	Smooth; dark gray	Smooth; velvety; white to gray
Newly Machined	Very smooth; bright gray	Bright copper red color dulls with time.	Red through to whitish yellow; very smooth.	Smooth; very white.	Very smooth; light gray	Very smooth; white	Very smooth; white

	White Cast Iron	Gray Cast Iron	Malleable Iron	Wrought Iron	Low-Carbon Steel and Cast Steel	High-Carbon Steel
Fracture	Very fine silvery white silky crystalline formation	dark gray	dark gray	bright gray	bright gray	very light gray
Unfinished Surface	Evidence of sand mold; dull gray	Evidence of sand mold; very dull gray	Evidence of sand mold; dull gray	Light gray; smooth	Dark gray; forging marks may be noticeable; cast—evidences of mold	Dark gray; rolling or forging lines may be noticeable
Newly Machined	Rarely machined	Fairly smooth; light gray	Smooth surface; light gray	Very smooth surface; light gray	Very smooth; bright gray	Very smooth; bright gray

Courtesy of Linde Division, Union Carbide Corp.

Even a new welder recognizes the advantages of a visual test to distinguish between copper and aluminum. Cast iron and malleable cast iron, covered in this chapter, can be distinguished by their appearance if both are fractured. The surfaces of the parts of the malleable cast look like the surfaces of the broken cast, except that a silvery-colored frame outlines the malleable surface parts, as in Fig. 2-37.

Chip Test

Metal *chips,* removed from metal by a cold chisel, can help identify the metal, as shown in Table 2-3. Wrought iron, which contains little, if any, of the hardening element *carbon,* can be easily cut with a chisel, giving a smooth, continuous chip. Cast iron, on the other hand, contains at least 1.7% carbon

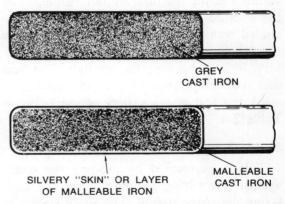

GREY CAST IRON

SILVERY "SKIN" OR LAYER OF MALLEABLE IRON

MALLEABLE CAST IRON

Fig. 2-37. Malleable cast iron has a "skin" not found on grey cast iron.

and is not as easy to chip, since the chips break off in small pieces. Malleable cast iron will not break into particles like cast iron. However, malleable cast iron cannot be chipped as easily as can wrought iron.

Table 2-3. Identification of Metals by Chips.

	Copper	Brass and Bronze	Aluminum and Alloys	Monel Metal	Nickel	Lead
Appearance Of Chip	Smooth chips; saw edges where cut	Smooth chips; saw edges where cut	Smooth chips; saw edges where cut	Smooth edges	Smooth edges	Any shaped chip can be secured because of soft-ness
Size Of Chip	Can be continuous if desired	Can be continuous if desired	Can be continuous if desired	Can be continuous if desired	Can be continuous if desired	Can be continuous if desired
Facility of Chipping	Very easily cut	Easily cut; more brittle than copper	Very easily cut	Chips easily	Chips easily	Chips so easily it can be cut with penknife

	White Cast Iron	Gray Cast Iron	Malleable Iron	Wrought Iron	Low-Carbon Steel And Cast Steel	High-Carbon Steel
Appearance Of Chip	Small broken fragments	Small, partially broken chips but possible to chip a fairly smooth groove	Chips do not break short as in cast iron	Smooth edges where cut	Smooth edges where cut	Fine grain fracture; edges lighter in color than low-carbon steel
Size of Chip		⅛ in.	¼-⅜ in.	Can be continuous if desired	Can be continuous if desired	Can be continuous if desired
Facility Of Chipping	Brittleness prevents chipping a path with smooth sides	Not easy to chip because chips break off from base metal	Very tough, therefore harder to chip than cast iron	Soft and easily cut or chipped	Easily cut or chipped	Metal is usually very hard, but can be chipped

Courtesy of Linde Division, Union Carbide Corp.

Metals can be distinguished from one another by using the chip test, even though the surfaces of the fracture are not visible, as in the fracture test. However, if cast steel is included with the above three materials, a new test, known as the *spark test*, may be necessary. This is because the wrought iron and cast steel chips look very much alike. The spark test will give a variety of sparks, as in Table 2-4.

Spark Test

This is performed by pushing the metal against a revolving grinding wheel and analyzing the *spark stream*. The spark stream is created by particles of metal torn away by the grinding wheel while in contact with the metal. The particles become so hot, due to grinding, that they glow as they fly from between the metal and the wheel (Fig. 2-38). The type of grinder usually used for a spark test is shown in Fig. 2-39.

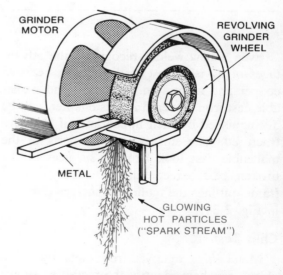

GRINDER MOTOR

REVOLVING GRINDER WHEEL

METAL

GLOWING HOT PARTICLES ("SPARK STREAM")

Fig. 2-38. The spark test on a bar of metal.

Each metal, and its various alloying elements, give off different patterns and colors when in contact with a grinding wheel. With experience, a welder using the spark test will

Table 2-4. Spark Characteristics Chart

Metal	Volume of Stream	Relative Length of Stream, Inches†	Color of Stream Close to Wheel	Color of Streaks Near End of Stream	Quantity of Spurts	Nature of Spurts
1. Wrought iron	Large	65	Straw	White	Very few	Forked
2. Machine steel (AISI 1020)	Large	70	White	White	Few	Forked
3. Carbon tool steel	Moderately large	55	White	White	Very many	Fine, repeating
4. Gray cast iron	Small	25	Red	Straw	Many	Fine, repeating
5. White cast iron	Very small	20	Red	Straw	Few	Fine, repeating
6. Annealed mall. iron	Moderate	30	Red	Straw	Many	Fine, repeating
7. High speed steel (18-4-1)	Small	60	Red	Straw	Extremely few	Forked
8. Austenitic manganese steel	Moderately large	45	White	White	Many	Fine, repeating
9. Stainless steel (Type 410)	Moderate	50	Straw	White	Moderate	Forked
10. Tungsten-chromium die steel	Small	35	Red	Straw*	Many	Fine, repeating*
11. Nitrided Nitralloy	Large (curved)	55	White	White	Moderate	Forked
12. Stellite	Very small	10	Orange	Orange	None	
13. Cemented tungsten carbide	Extremely small	2	Light Orange	Light Orange	None	
14. Nickel	Very small**	10	Orange	Orange	None	
15. Copper, brass, aluminum	None					

†Figures obtained with 12" wheel on bench stand and are relative only. Actual length in each instance will vary with grinding wheel, pressure, etc. *Blue-white spurts. **Some wavy streaks.

**Table 2-4. Characteristics of the sparks generated by grinding common metal samples.
(Courtesy of The Norton Company.)**

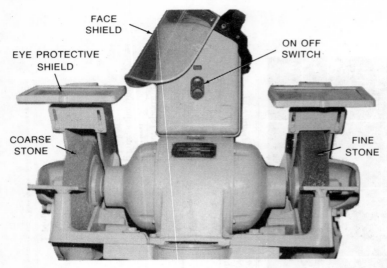

Fig. 2-39. A type of grinder used to spark test metals.

be able to identify not only the metal, but its alloying elements as well.

Nonferrous metals (described later in the chapter) cannot be identified by this method, due to the lack of any appreciable spark stream. The spark test, however, would help to separate nonferrous and ferrous groupings.

Another advantage of spark testing is its ability to distinguish between monel metal and stainless steel. Monel and other metals high in copper or nickel *will not* spark; however, stainless steel *will* spark when held against a revolving grinding wheel.

The spark stream consists of (1) Tail Section, (2) Center Stream Section, and (3) Wheel Sparks (Fig. 2-40). The flying action

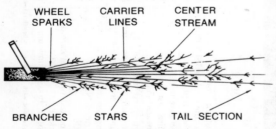

Fig. 2-40. The parts of a spark stream.

of the stream, along with the color and appearance of each of its three main parts, gives clues to the tested metal's composition.

The stream can be further broken down for describing and analyzing. Various elements affect the lines carrying the sparks. These continuous lines, from the contact point of the grinding wheel to the tail section, are called *carrier lines*. The carrier lines can break out at any point, depending on the alloying elements, in forms shaped as sprigs, branches, spearheads, blocks, and stars. When a common metal-working file is on contact with the grinding wheel, a good example of star bursting is presented. More carbon content in metals will increase the volume and intensity at the spark stream tail, as in Fig. 2-41.

Welders are advised to obtain samples of various metals and their alloys to observe the various changes in the spark stream, especially the size, number, intensity, and shape of the stars at the tail of the stream.

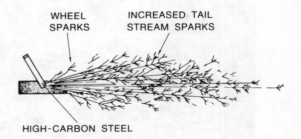

Fig. 2-41. Analyzing the spark stream of a high-carbon steel.

Melting Temperature

The *melting temperature* of a metal can also help identify the metal, if proper procedures and equipment are used to correctly recognize the melting point. Some alloys, although a mixture of two or more elements, have a narrow melting range and can be grouped with pure metals that melt at a single temperature. Various types of marking pencils are available to be used on metal about to be heated. These will show a visible change in the mark on the metal at the designated temperature for that pencil. Should the identity of a metal be assumed, a quick check can be made by looking at a melting temperature table, as in Fig. 2-9, and then marking the metal with a heat indicator pencil designated for a temperature just below the melting point of the test metal. When the marked metal is heated, if the metal melts shortly after the reaction is seen on the pencilled mark, the metal's melting temperature has been found.

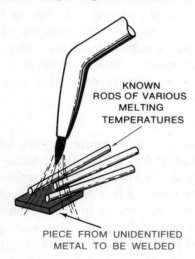

Fig. 2-42. Choosing a rod to weld on unidentified metal—whichever rod melts at the same time as the piece to be welded will be the most suitable to make the weld.

In some cases, it is possible to take a flame test of a material, such as an aluminum alloy, at the same time and under the same conditions as samples of various filler metals that would be used to weld the material. Without any knowledge of the material's identity, a weld can sometimes be completed by using the filler metal which melted under the flame

at the same time as the material being repaired. See Fig. 2-42.

Magnet Test

By using a *magnet,* the experienced welder can classify a metal into one of three groups. This would help identify its general make-up for a required weld. Magnets may be found in the natural state, already magnetized, like *lodestone.* Or, they may be artificially magnetized by placing them in the field of another magnet or in a magnetizing area, called a *magnetized field,* caused by electric current (Fig. 2-43). This type of magnet is called an *electromagnet.*

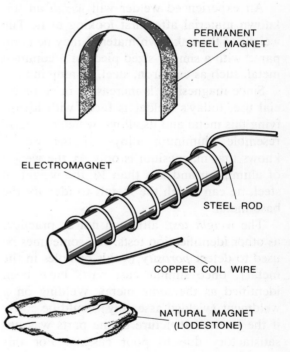

Fig. 2-43. Types of magnets.

After the welder has made the magnet test, he puts the metal into one of the three groups: (1) Nonmagnetic, (2) Magnetic, and (3) Slightly Magnetic.

The first group, the *nonmagnetic* group, will be described later in the chapter. In the group are metals such as monel and stainless steel, containing at least 18% chromium and

8% nickel. Zinc, aluminum, copper (and their alloys) are also part of this group.

The second group, *magnetic,* are the low-carbon and low-alloy steels, along with wrought iron and the cast irons. Pure nickel and stainless steels with less than 17% chromium are also magnetic.

The third group, known as the *slightly magnetic* group, reacts to a magnet in proportion to their alloying elements. For this reason, monel, stainless steels, and highly nickeled alloy materials could show a slight reaction to the magnetic test.

Weight Test

An experienced welder will *weigh* an unknown material after first looking at it. The weight of the unknown material may be compared with a similar-sized piece of a common metal, such as cast iron, steel, or aluminum.

Since magnesium is increasing in commercial use, today's welder is faced with identifying this metal and its alloys, which so nearly resemble aluminum alloys. If the welder knows that magnesium is only ⅔ the weight of aluminum and less than ¼ the weight of steel, he can weigh the metals to identify the base metal.

The *weight test,* although not as practical as other identification tests, can sometimes be used to detect *porosity* (voids or holes in the metal), once similar cast parts have been identified as the *same* metal. Welding on a weldment (welded assembly) could be costly if the internal structure of the parts was unsatisfactory due to poor casting. For this reason, the weight test, using a scale, becomes a practical instrument for testing castings after the metal's weight is determined.

How Metals Are Made

The processed, finished metals that we use daily cannot be found in the ground. Miners, for example, are not able to dig a ¾″ round, cold-rolled steel rod out of the earth. Many processes go between mining the *ore* and pro-

ducing bare metal pieces. Here is a basic outline of different metals and how they are made.

The Industrial Revolution was based on *iron,* the principal metal of the era, because *steel,* made by expensive methods, was very rare. Iron is second to aluminum as the earth's most common metal. It has been estimated that 5% of the earth's crust consists of iron. Iron is extracted from mineral formations high in iron and low in other elements, called *impurities.* Iron is found only in combination with other elements because it is highly active chemically, even being affected by moist air.

The oxides are used mainly to produce iron for commercial purposes. Since iron ore is a natural mineral deposit, it can give up this type of useful metal through different processing methods. The greatest amount of iron ore refined this way is known as *hematite,* or ferric oxide, which consists of over 70% iron. Another iron ore, known as *limonite,* contains a little less than 60% iron; another is *magnetite,* containing just over 72% iron.

Pig Iron

Pig iron is a hard, brittle, impure form of iron, obtained by chemically reducing iron ore compounds. This reduction (separation) is done by *smelting* in a *blast furnace.* Smelting is a refining process in which iron ore is melted to separate its parts. The resultant pig iron, used for making iron and steel, must be kept low in sulphur and phosphorus, since these elements create brittleness in a finished product.

Blast Furnace. This is the basic tool used to turn iron ore into the first form of iron, pig iron. The first step in making pig iron is *charging* the blast furnace. This is done by pouring *iron ore, coke, flux,* and *air* into the furnace. See Fig. 2-44.

Coke is a fuel obtained by distilling the easily burned parts from coal. *Flux* is used to lower the melting temperature of aluminum and silicon oxides, which are found mixed

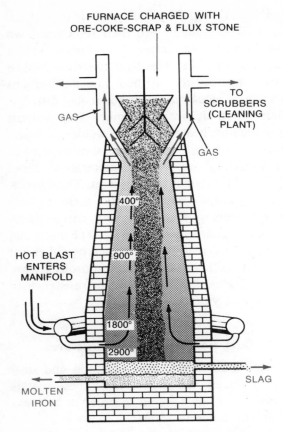

FURNACE CHARGED WITH
ORE-COKE-SCRAP & FLUX STONE

GAS

TO
SCRUBBERS
(CLEANING
PLANT)

GAS

400°

900°

HOT BLAST
ENTERS
MANIFOLD

1800°

2900°

SLAG

MOLTEN
IRON

Fig. 2-44. Cutaway of the blast furnace.

together as an impurity in iron ore known as *gangue*. The flux consists of *limestone* (mainly calcium carbonate) and *dolomite* (calcium carbonate and magnesium carbonate), to remove impurities. The flux also makes the *slag* (liquid residue separated in the reduction of ore) more fluid. *Fluospar* (calcium fluoride) is a flux used in the manufacture of steel, since it increases the slag's *fluidity* (ability to flow) without changing the chemical properties. This allows the impurities to be easily removed from the molten iron.

The charge is loaded at the top of the blast furnace when the ignited coke has raised the temperature to about 400° F. The coke, while burning, gives off fast-rising carbon monoxide. The carbon monoxide reacts with the falling charge, reducing the iron in the ore to ferrous oxide and iron. As the carbon monoxide comes in contact with the charge,

it is cooled to the point where carbon, in the form of soot, and carbon dioxide are formed. The carbon penetrates the ore's surface, and by the time the charge falls halfway to the bottom of the stack, the carbon in the ore lumps reduces any remaining ferrous oxide into metallic iron. The remaining carbon is dissolved by the iron, lowering the melting temperature. This allows the new iron to become a spongy (rather than a solid) mass.

At this point, the temperature changes the limestone into carbon dioxide and lime. The lime combines with the aluminum and silicon oxides to form a slag, while the carbon dioxide reacts with the coke to form carbon monoxide.

As the reduced ore falls down the furnace, the increased temperature and carbon monoxide accelerate the reaction to a point where ferr*ic* oxide becomes ferr*ous* oxide and metallic iron. Most furnaces are built to slow down the travel of the charge between the loading area and the *tuyeres* (the pipe through which the air is forced) to about eight hours.

Before the charge reaches the tuyeres, the pasty slag consists of ferrous oxide. The oxygen, in the form of a hot blast, reacts with the coke to form carbon dioxide at the tuyeres. This carbon dioxide, in turn, reacts once more with additional coke and is reduced to carbon monoxide.

Between the tuyeres and the bottom of the furnace, the last of the ferrous oxide is reduced to iron. Here the iron and slag form a liquid, and the reduced elements of phosphorus, silicon, and manganese enter the iron. The liquid slag formation, consisting of the lime with coke ash and gangue, trickles onto the hearth with the molten iron. Here, the insoluble slag floats to the top, since it is the lighter of the two. The molten metal, now called *pig iron,* is allowed to remain for several hours before tapping.

In most furnaces, about 250 tons of pig iron are drawn off every six hours. To keep the slag from rising as high as the tuyeres and plugging them, the slag is tapped off more often. The pig iron, when tapped, is allowed to run into molds where it is stored until needed. Pig iron consists of from 3½% to 4¼% carbon.

Wrought Iron

The type of iron used for industrial purposes until steel was produced more economically by modern processes was *wrought iron*. Wrought iron is a ferrous material, very low in carbon and exceptionally pure. The glass-like, noncorroding, iron silicate slag is more closely associated with the iron itself, when combined, than would be an alloy such as steel.

When pig iron is melted in a ferrous oxide-lined furnace with additional iron oxides, the carbon content of the pig iron is practically removed through oxidation. During this process, the other impurities are also oxidized. This leaves the silicon to react in such a manner as to produce iron silicate slag.

The Puddling Furnace—The furnace used to produce wrought iron, as in Fig. 2-45, is

the liquid slag. The balls, which can weigh up to 700 pounds, are then removed from the furnace and squeezed, to remove all of the slag possible. The balls are then broken down in order to roll them into flat sections known as *muck bars*. For better slag distribution within the iron, the muck bars are cut into lengths, heated to fusion welding temperature, and rolled. The slag, now distributed as fibers throughout the iron, amounts to about 3% of the wrought iron's weight. This fibrous makeup keeps corrosion on the surface of the metal, making the metal very corrosion-resistant by guarding the interior. The metal can be machined readily and formed by either hot or cold methods. Due to the fibrous nature of the material, it has excellent resistance to shock and vibration.

Cast Iron and Steel Structures

Before explaining how cast iron and steel are produced, the structures of these metals as solids will be covered in order to help explain the different terms.

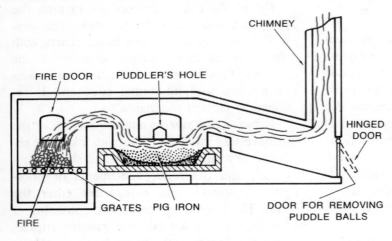

Fig. 2-45. Puddling furnace. Used in the production of wrought iron.

known as a *puddling furnace*. This is because the mixture is puddled, or stirred, to speed up oxidation with a flame above the metal; this furnace is similar to a baker's oven. The heat reaction during the process eventually slows to the point where the refined iron gathers into a sticky, spongy ball saturated with

Allotropy is a term used to describe the property of an element that can exist in more than one form of space lattice structure. Iron is one of the few metals that may exist in any one of three allotropic forms. The diamond is an allotropic substance whose form is an allotrope of carbon. The temperature change on the carbon is responsible for turning the carbon into a diamond. Pure iron, as such, is shown in Fig. 2-46.

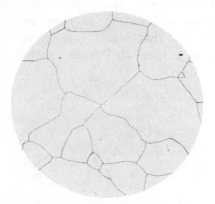

Fig. 2-46. Pure iron, magnified 250 times. (Courtesy of Bethlehem Steel.)

The critical point, as mentioned previously, is the temperature where changes take place in a metal as it is heated from 70° F to its melting point.

Alpha iron is magnetic, ductile, and soft. At 70° F, it is composed of body-centered cubic space lattice structures, as shown in Fig. 2-47. At room temperature, the body-

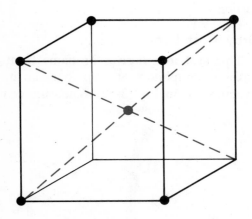

Fig. 2-47. Body-centered cubic unit cell.

centered cubic space lattice could include chromium, tungsten, molybdenum, and vanadium, as well as iron. When alpha iron is heated over 1420° F, it is no longer magnetic but retains its body-centered cubic struc-

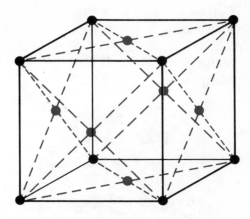

Fig. 2-48. Face-centered cubic unit cell.

ture. Since it is no longer magnetic, it can be considered as having a different allotropic form, now known as *beta* iron or *nonmagnetic alpha* iron.

The structure remains in its original space lattice form until it is heated to 1670° F. At this point, the crystalline structure changes to the face-centered form, as in Fig. 2-48, and its new allotropic form is known as *gamma* iron. Gamma iron is nonmagnetic; the density of the grain structure becomes greater than that of the body-centered cubic structure as the volume of the iron contracts.

The next critical, or transformation, point is around 2550° F. Here, the body-centered form takes over from the faced-centered cubic lattice and is now known as a magnetically *delta* iron. The next critical point is where the iron absorbs enough heat to become a liquid. At this point, the temperature is about 2795° F, and welding is possible.

Cast Iron

Iron, alloyed with at least 1.7% carbon and varying quantities of sulfur, phosphorus, manganese, and silicon, is known as *cast iron*. One form of cast iron, magnified 200 times, is shown in Fig. 2-49.

Cast iron welding *coupons* are shown in Fig. 2-50. Cast iron coupons are test pieces for welding, and contain these elements in the following amounts:

Carbon	2.75%
Phosphorus	0.15%
Silicon	0.80%
Sulfur	0.10%
Manganese	0.60%

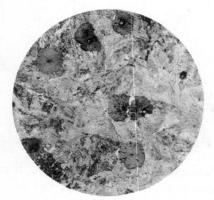

Fig. 2-49. Cast iron, magnified 200 times. (Courtesy of Bethlehem Steel.)

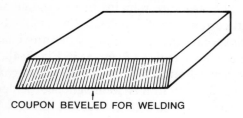

COUPON BEVELED FOR WELDING

WELDING PREPARATION

Fig. 2-50. Cast iron welding *coupons*, beveled to prepare for a test weld.

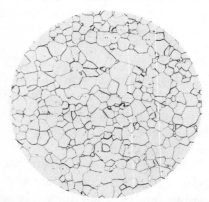

Fig. 2-51. Ferrite part of the cast iron matrix, magnified 100 times. (Courtesy of Bethlehem Steel.)

The *matrix* (the principal materials binding the particles of cast iron together) consists of *ferrite* and *pearlite*. Ferrite (Fig. 2-51) contains less than 0.05% carbon and is sometimes referred to as *pure iron,* while pearlite has thin plates (as shown in Fig. 2-52) in a laminated form created by the

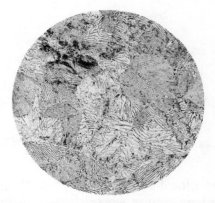

Fig. 2-52. Pearlite, magnified 500 times. (Courtesy of Bethlehem Steel.)

mixture of ferrite and *cementite*. The cementite, called iron carbide, contains iron and about 6.8% carbon. The hard, brittle, cementite sometimes occurs in needle form within the grains of steel, if the steel exceeds 0.83% carbon.

Gray Cast Iron

This cast iron derives its name from the flakes of graphite that help to make up its structure, as in Fig. 2-53. When the cemen-

Fig. 2-53. Graphite in gray cast iron, magnified 500 times (Courtesy of Bethlehem Steel.)

tite decomposes, a form of free carbon known as *graphite* is released along with the iron. The amount of graphite in cast iron does depend on the carbon content of the iron, but depends primarily on the cooling rate. The graphite content also depends on the amount of silicon in the iron. Silicon is a non-metallic element occuring in the earth's crust, second only to oxygen in abundance.

Ferrite's capacity to dissolve the carbon is reduced if silicon is prevalent, since silicon is soluble in the ferrite. Graphitization, producing a softer iron, is created because the ferrite does not reduce the carbon. Heavy sections of cast iron, high in carbon or silicon, have an open-grained interior, creating a weakness within the metal. *Sulfur,* although usually classed as an impurity, will impede the formation of graphite through a chemical reaction. *Manganese,* when used as an alloy, imparts high strength and density to cast iron. *Phosphorus,* another classical impurity, weakens the cast iron and makes it more fluid while

molten, although it has little effect on the graphite structure.

Due to its low cost, gray cast iron is used for various machine parts where impact and stress conditions are at a minimum.

The Cupola Furnace—Gray cast iron can be produced by melting scrap iron and steel with pig iron in what is known as a *cupola furnace.* See Fig. 2-54. In order to obtain a definite composition in the finished material, calculating the ingredients' carbon content is necessary. During the cupola furnace remelting process, various changes in the metal's elements make carbon content important.

The cupola is started by lighting a coke fire in the bottom part of the furnace. When the fire is well under way, coke, pig iron, and limestone are loaded through the charging door in the top of the furnace, in alternate layers. The cast iron's quality depends on the mixture of the scrap cast iron and steel, which is added with the pig iron. Scrap iron, when added to the pig iron, will usually lower the carbon content of the finished product.

When the temperature is raised, due to increased air circulation, the metals melt and drop to the bottom of the furnace. The metal can be tapped (drawn off) from the bottom of the cupola in its new form, cast iron. *Chilled cast iron,* which has a structure somewhat like pearlite and cementite, is produced by a rapid cooling of gray cast iron (Fig. 2-55).

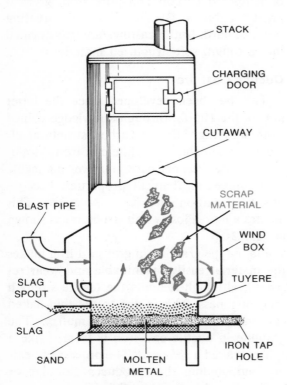

Fig. 2-54. Cupola furnace.

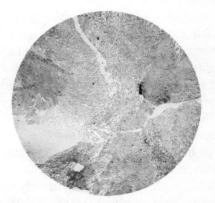

Fig. 2-55. Pearlite, magnified 500 times. (Courtesy, Nickel of Canada.)

A type of *slag*, formed by melting limestone and the coke's ash residue, floats on top of the molten metal. The slag can be tapped by itself through the slag spout.

White Cast Iron

The structure of *white cast iron* is pearlite and cementite, since carbon is not found as graphite due to its low carbon and silicon content. White cast iron is hard and brittle and, in most instances, produced to make malleable cast iron. White cast iron can be made in cupolas or air furnaces where the charge of iron and steel scrap, mixed with the pig iron, is heated indirectly by the fuel. Since the charge is not in direct contact with the fuel, carbon from the fuel will not be able to penetrate the molten metal, making it easy to calculate the carbon content of the finished metal.

Malleable Cast Iron

If white cast iron is annealed, the resultant material is *malleable cast iron*. Annealing (which will be covered more fully later) is a heating operation, followed by slow cooling. The temperature and the rate of cooling depend on the material being treated, as well as the desired results. Annealing decomposes the iron carbide, and the characteristic structure of the white cast iron (pearlite and cementite) is converted into a mixture of graphite and ferrite. A small mass of rounded, or irregular-shaped, carbon known as *globular* or *temper* carbon, forms the graphite of malleable cast iron, as shown in Fig. 2-56.

White cast iron *castings* (the metal's shape after pouring molten cast iron into molds) are placed in cast iron pots and then packed with sand or some other substance to conserve the heat. Fire clay is also placed over the tops of the pots to seal the metal from the oxidizing effect of heated gases which accumulate after the furnace is fired. When the furnace ovens are loaded with castings, they are sealed and allowed to heat to about 1600° F. After remaining at this temperature for about 50

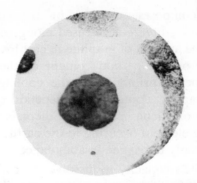

Fig. 2-56. Malleable cast iron, magnified 500 times. (Courtesy of Bethlehem Steel.)

hours, they are slowly cooled to about 1300° F. The temperature is kept at this critical point for 16 to 20 hours, after which the oven heat is reduced to just under 1200° F. At this temperature, the oven doors are opened so that the cast pots containing the metal can be cooled naturally. The resulting cast iron, magnified 500 times, is shown in Fig. 2-56.

Although the finished product is not really workable, its mechanical structure, consisting of tempered carbon and pure iron, gives it great toughness. By the above heat-treating process, hard, brittle castings are transformed into a tough, soft-structured material.

Ductile Cast Iron

This has been developed since the latter part of the 1950's, through knowledge gained from studying different carbon structural effects on malleable and white cast iron. Modular graphite (temper carbon) gives malleable irons higher mechanical properties, because of smaller carbon particles in the matrix. This creates a more closely knit structure, as shown in Fig. 2-56.

Flake-type graphite in gray cast iron makes the material easily machinable and wear resistant, although it limits the tensile strength and toughness. This is due to the thin-flaked carbon scattered through the matrix. The graphite in ductile cast iron is formed like a sphere, called a *spheroid*. The process of heating and cooling that produces globular or rounded carbon-iron particles is known as *spheroidizing*. The ductile-cast spheroid, being a polycrystalline particle (an aggregate of crystal grains), is stronger than the single-grain or fine-flake aggregate found in malleable cast iron.

Ductile cast iron has fluidity and, like other cast irons, is easily machined. Like steel, it can be poured into molds, gives good ductility, and has high yield strength and high elasticity. All of these qualities make it suitable for various machinery parts not suited for standard cast irons.

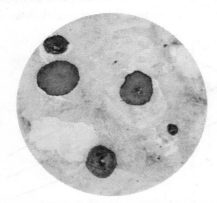

Fig. 2-57. Ductile iron—Pearlite. Magnified 250 times. Chemical composition consists of 3½% carbon and 1% nickel.

To make ductile cast iron (Fig. 2-57), a nickel-magnesium with high nickel content is added to the molten cast iron metal, giving the finished material its spheroidal graphite structure.

Steel

An alloy of iron with no more than 1.7% carbon content is called a *steel*. By varying the carbon content, heat treating, or adding other elements, steel can be produced for several applications. Unhardened, low-carbon steel can be so soft and ductile that bending it can be done by hand, without tools. High-carbon steel, when heated at high temperature and quenched in a liquid such as cold water or oil, will become hard enough to cut other steels. See Fig. 2-58. In the same way, steel can be made brittle enough to break under impact, like gray cast iron.

Fig. 2-58. High-carbon steel, quenched to harden. (Courtesy of International Nickel.)

The tensile strength of steel is measured in thousands of pounds per square inch. Tensile strength means how much a steel can withstand, under stress, before breaking. It can vary from 40,000 pounds to just over 500,000 pounds, depending on the treatment and composition.

Number Indications for Steel Carbon Content—A plain carbon steel has carbon as the only alloying element. Since a number index system is used to designate steel's composition, a plain carbon steel could be expressed as a number 1090. The *first* digit of the number indicates the type to which the steel belongs. In the case of an uncomplicated alloying process, the *second* digit indicates the percentage of the main alloy, other than carbon. The *final two digits* indicate the carbon content of the material. The carbon content is expressed in hundredths of 1%, such as 0.90%, or 0.73%.

Therefore, in the number 1090 designation, the number 1 indicates plain carbon steel; the second digit, zero, describes a predominant alloy *other than carbon* as nonexistent (0%); the final two digits indicate a carbon content of 0.90%.

Plain carbon steel contains from 0.05% to 1.7% carbon (Fig. 2-59) with small amounts of other elements, consisting of silicon, manganese, phosphorus, and sulfur. The strength, ductility, and hardness of the steel depend on the amount of carbon present. A low-carbon steel can be deformed by forging or rolling, due to its ductility. The higher the carbon content, the lower the ductility, due to the carbon increasing the metal's hardness. This hardness will make a steel more resistant to abrasion, penetration, or plastic deformation. Tensile strength increases as carbon is added until the carbon content reaches 0.83%, at which point the tensile strength starts to fall off slightly. Ductility, on the other hand, continues to decrease as the carbon content of steel is increased. See Fig. 2-59.

when heated, a temperature change just below the melting point would allow any free carbon to penetrate the surface of the steel, creating a surface hardness.

Forging steels, those with carbon content from 0.30% to 0.60%, can be heat treated more satisfactorily than cold-rolled steel. The high carbon content of the forging steels makes increased hardness and strength possible under the proper heat-treating process.

Tool steels contain at least 0.60% carbon combined with 0.50% silicon and manganese. Tools made from this material lose their

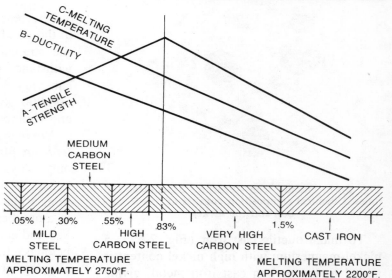

Fig. 2-59. How steel qualities change as carbon is added.

A carbon content increase over 0.90% ensures the steel's resistance to wear, which, in turn, sacrifices such properties as strength and toughness.

Carbon Steel Classifications — Classifications known as *machining steels, forging steels,* and *tool steels,* each having a different range of carbon content, divide up the broad group of *carbon steels.*

Cold-rolled steel is produced by removing the scale from hot-rolled steel, and then subjecting the steel to a cold-rolling process. Cold-rolled steel can be classified as a *machining steel* due to its low carbon content, which has a range of from 0.10% to 0.20%. If this material is placed in contact with solids, liquids, or gases that give off carbon

hardness at high temperatures, usually caused during high speed machine work. The surface of these steels can be made very hard while the core remains soft (Fig. 2-60) through a process known as *case-hardening.* This process will be explained under the treatment of steels.

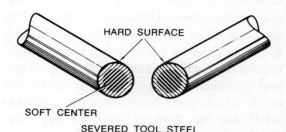

Fig. 2-60. Cross-section of tool steel—0.60% carbon and 0.50% silicon and manganese.

Production of Steel

A welder needs a basic understanding of how steel is made. It is important that all impurities be eliminated when making steel, so steel production will be viewed from this point.

The *Bessemer Process* was the first method used to control steel's composition. Where the raw materials are low in sulfur and phosphorus, a process known as the acid Bessemer is used. This acid process does not reduce sulfur or phosphorus and is not used for refining pig iron, which is high in phosphorus. The *Basic Bessemer,* a basic oxygen steel-making process, or the *Basic Open-Hearth Furnace* are used for raw materials containing much sulfur or phosphorus.

Although the *Open-Hearth* method is more expensive to operate than the Bessemer process, the steel produced is more completely deoxidized. This gives the steel less porosity, since the elements in the air that are detrimental to steel (oxygen and nitrogen) are not trapped in the core of the steel.

Another method of steel production uses the electric arc and is known as an *Electric Furnace.* A variation of the Bessemer process is the *L. D. Process,* using pure oxygen instead of air. This eliminates the nitrogen side effects that occur in the Bessemer process, where air containing both oxygen *and* nitrogen is blown in. Each of these methods will be discussed in greater detail.

The welder should know what effects various elements have on plain carbon steels before he tries to understand how steel is produced.

During the steel-making process, *sulfur* comes mostly from the coke and should be initially eliminated, as far as possible, in the blast furnace. The welder will notice a bad influence of high sulfur content in steel, especially if fabrications show a surface imperfection. Steel which contains more than 0.12% sulfur contains iron sulfide. This melts during forging and hot-rolling and creates a cracking or severing effect known as *hot shortness.* Hot shortness can be eliminated if manganese is added. The manganese combines with the sulfur to produce manganese sulfide, which (on heating) forms fibers in plastic form, holding the metal intact during rolling.

The range of *manganese* in plain carbon steel is from 6% to 9%. This acts as a deoxidizer during steel refining and gives the steel a higher tensile strength.

If a steel has over 0.05% *phosphorus,* the cold steel could be brittle, due to the coarse grain structure that phosphorus produces. This condition is known as *cold shortness.* Like sulfur, phosphorus can improve the machining qualities of steel; also, if phosphorus is alloyed with low-carbon steel, corrosion-resistance is improved.

Silicon acts as a deoxidizer during the smelting process and, as an element of steel, increases the tensile strength.

The Bessemer Process—This process turns the crude product of the blast furnace into a cheap, low-grade steel that is in great demand for various structural shapes. This type of material, known as Bessemer steel, is made by forcing air through molten iron in a converter.

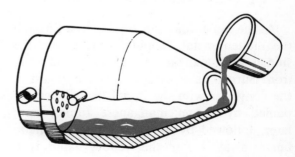

Fig. 2-61. Charging the Bessemer Converter.

The converter, as shown in Fig. 2-61, is a tilting, pear-shaped, steel-plated vessel. The interior is lined with heat-resisting bricks and clay to withstand the heat generated by the molten pig iron. The converter is charged (molten pig iron poured in) through the converter's open top, after the converter has been placed in a horizontal position. The converter is placed upright when it is charged, and the molten metal rests on the perforated top of what is known as the *wind box*. Air is blown into the wind box at a pressure ranging from 22 to 28 psi and at a rate of about 35,000 cfm. The air pressure keeps the molten pig iron from entering the wind box through the perforated holes, as shown in Fig. 2-62.

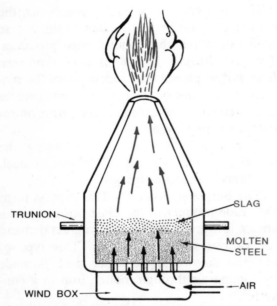

Fig. 2-62. Operating a Bessemer Converter.

TRUNION

SLAG

MOLTEN STEEL

WIND BOX

AIR

At the same time, the air is forced through the molten metal, helping to eliminate the silicon and manganese present in the pig iron. Burned gases escape from the mouth of the converter with the initial flames, accompanied by brown manganese fumes, and, later, followed by a yellow flame as silicon burns away. After the yellow flame is a brilliant white flame, as carbon combustion takes place.

In about 20 minutes, the flames diminish to the point where steel can be poured into a ladle by tilting the converter. The resulting material, consisting of the original phosphorus and sulfur present in the pig iron charge, is now practically free of all the carbon, manganese, and silicon. The oxygen absorbed by the metal from the air blast is eliminated just before entering the ladle by adding an alloy of iron and manganese. The manganese combines with the oxygen to form slag.

Since a large amount of carbon is removed, recarburizing is necessary, but only in the percentage for the grade of steel required. Recarburization is done by adding carbon in the form of coke dust. The Bessemer process is fast and cheap, especially when small amounts of different variety steels are required.

The Open-Hearth Process—This process got its name because its hearth (floor) is open to the flames that melt the metal. It can process steel scrap as well as pig iron. About 90% of the steel produced in the United States is made by this method.

Steels can be produced in the open-hearth furnace either by the acid or basic process. The charge for the *acid* process uses pig iron or scrap that is low in sulfur and phosphorus. Additional pig iron is sometimes added to control the carbon content of the finished product. The charging process is complete, although small amounts of limestone may be added later to thin the acid slag.

The *basic* process uses limestone or burned lime before and during the blow. Phosphorus is eliminated and the slag remains basic because of these additives. Also, lining the sides and bottom of the furnace with materials like dolomite or magnesite would allow the raw charging materials to contain large amounts of phosphorus and sulfur and still retain a basic reaction. The dolomite used in the lining withstands high temperatures and is a mixture of magnesium and calcium. The magnesite, a natural magnesium carbonate, also resists attack by the basic slag.

All this takes place in an open-hearth furnace, as in Fig. 2-63. The furnace is a completely enclosed brick structure, built to withstand temperatures as high as 3000° F. The charge, consisting of iron ore, pig iron, scrap material, and limestone, is placed in the charging doors. On heating, the limestone combines with the impurities in the charge

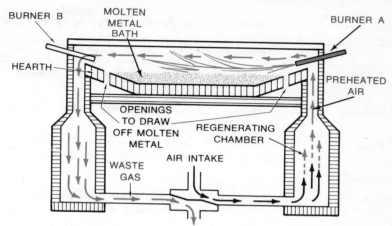

Fig. 2-63. Open-Hearth Furnace.

ALTERNATING FIRING BY MEANS OF BURNERS A AND B

and forms a slag. At each end of the hearth is a burner, which can be fired by various fuels. Only one burner is fired at a time; the flames cross the metal on the hearth, taking the combustible fumes with it through a port on the opposite end. The port is connected to fire-bricked regenerating chambers which become heated and, in turn, heat the incoming air directed to the other burner. When this chamber no longer gives up enough heat, the firing procedure is reversed to the other end of the hearth. Reversing the firing takes place every 10 or 15 minutes over a period of 12 hours, after which the back of the furnace is tapped and the molten steel is allowed to flow into large vats or ladles. The molten slag, flowing on top of the metal, is allowed to spill over into *slag pots*. An alloying substance can now be added to give the steel the required composition.

The molten steel can now be poured into molds known as *ingots*. A mold holding from 5 to 25 tons of molten steel is used in the open-hearth method. When the liquid steel solidifies in the molds, the molds are removed, and the first solid steel is in the form of ingots.

Electric Furnace—Fig. 2-64 shows an electric furnace used to produce steel in which various alloying elements are to be added. A good selection of basic alloying elements is shown in Table 2-5.

The heat to melt the charge in an electric furnace is generated by an electric arc, providing a temperature up to 3200° F. Since

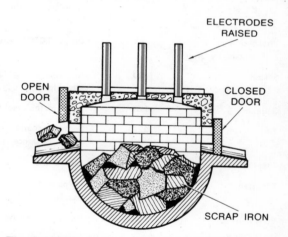

Fig. 2-64. Electric Arc Furnace being charged with scrap iron—electrodes raised.

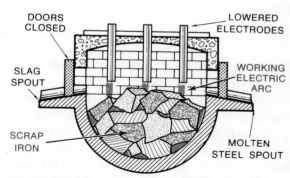

Fig. 2-65. Electric Arc Furnace. Electrodes lowered just before turning on the current.

Table 2-5. Principal Alloying Elements of Steel.

Element	Melting Point (°C)	Reason for Use	Application
Aluminum	658	Deoxidizes and refines grain. Removes impurities.	Little aluminum remains in steel.
Chromium	1615	Improves hardness of the steel in small amounts.	Stainless steels, tools, and machine parts.
Cobalt	1467	Adds to cutting property of steel, especially at high temperatures.	High speed cutting tools.
Copper	1082	Retards rust.	Sheet and plate materials.
Lead	327	Lead and added tin form a rust resistant coating on steels.	Machinery parts.
Manganese	1245	Prevents hot shortness by combining with sulfur. Deoxidizes. Increases toughness and abrasion-resistance.	Bucket teeth. Rails and switches.
Molybdenum	2535	Increases ductility, strength and shock resistance.	Machinery parts and tools.
Nickel	1452	In large amounts—resists heat, adds strength, toughness, and stiffness to steel.	Stainless steels. Acid-resistant tools and machinery parts.
Phosphorus (Provided by ore)	43	Up to 0.05% increases yield strength.	Some low-alloy steels.
Silicon	1420	Removes the gases from steel. Adds strength.	Precision castings.
Sulphur	120	Adds to the steel's machinability.	Some machined pieces.
Tin	232	Forms a coating on the steel for corrosion-resistance.	Cans and pans.
Titanium	1800	Cleans and forms carbide.	Used in low-alloy steels.
Tungsten	3400	Helps steel retain hardness and toughness at high temperatures.	For magnets and high-speed cutting tools.
Vanadium	1780	Helps to increase strength and ductility.	Springs, tools, and machine parts.
Zinc	420	Forms a corrosion-resistant coating on steel.	Wire, pails and roofing.
Zirconium	1850	Deoxidizes, removing oxygen and nitrogen. Creates a fine grain.	Machine parts and tools.

fuel and air are not required, oxidation of alloying elements is reduced.

The furnace consists of a swivel-type steel shell, lined with heat-resisting brick and capable of holding up to 100 tons. The current reaches the steel through three heavy electrodes, which can be raised or lowered through the top of the furnace. The slag can be poured from a spout on the bottom of one side of the shell by the swivel action of the furnace. By the same swivel, molten steel can be poured from the opposite side into what is known as a *teeming ladle*. See Fig. 2-65.

When the furnace is charged, the electrodes are lowered toward the scrap and the current is switched on. The intensity of the heat caused by the arc between the electrodes and the charge depends on the distance between the electrodes and the charge. The shorter the arc, the faster the cold scrap steel becomes molten.

After the mass is molten, iron ore is added to oxidize the phosphorus, carbon, manganese, and silicon. The slag that forms the molten metal is removed by wooden rakes. A new slag formation is created by adding calcium oxide and an iron-silicon alloy to remove excessive oxygen and reduce the remaining sulfur. Depending on the type of steel required, various alloys are added to the melt. Aluminum pellets in the ladle during tapping will counteract oxidation by air.

The L. D. Process—Named after its inventor, Donowitz of Linz, Austria, the *L. D. Process* uses a lined furnace built like a Bessemer converter. The tilting vessel makes it possible to load the furnace in a horizontal position (Fig. 2-66). When the furnace vessel

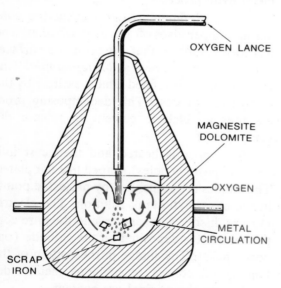

Fig. 2-66. L. D. Oxygen Furnace.

is placed in an upright position, a jet of pure *oxygen* is directed from the top through a pipe. When the oxygen strikes the molten metal, it burns out the impurities. The resultant product is tapped by tilting the vessel. This steel has no nitrogen, which does impregnate steel during the *air* jet of the Bessemer process.

Cast Steel

When molten steel of various carbon contents is poured into a suitably shaped mold, the material produced has a finished, desired shape. This material is called *cast steel*. Unlike cast iron, this steel has good shock resistance, toughness, ductility, and high strength. When this steel is cut by a chisel, the resultant chips will be ductile. The interior of the metal is silvery in color, rather than gray like cast iron. Welding cast steel, explained in a later chapter, is different from welding cast iron.

Copper

This metal has long been used in making some tools. Recent discoveries by archaeologists indicate that crude tools were fashioned from *copper* as early as 4000 B.C.

Copper is especially resistant to corrosion and oxidation. It has a reddish-brown lustre color which can give its alloys a yellow or reddish color. When copper is cooled from its molten state, it expands as it becomes a solid; when heated near its melting point, it becomes brittle like glass. It is a tough, ductile metal which can be rolled, pressed, and forged without cracking into various shapes and thicknesses.

Cast copper has a tensile strength of about 20,000 psi, which can be increased to about 65,000 psi by cold working. Cold working, like hammering, will permanently deform the copper below its recrystallization temperature. Heat treating cannot harden copper, but work-hardened copper can be reheated and placed in cold water, which will anneal (soften) the metal.

Copper is found in low grade ores in the form of *sulfide*. The ores are classified as *chalcocite* (copper sulfide); *chalcopyrite*, (iron with a double sulfide of copper); and *bornite*, (a brittle sulfide and iron combination).

Copper Production—To mine copper, ores are ground to a fine, sand-like mixture and placed in a froth flotation tank, similar in action to a washing machine, as shown in Fig. 2-67.

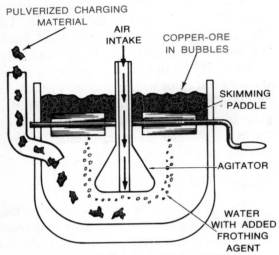

PULVERIZED CHARGING MATERIAL

AIR INTAKE

COPPER-ORE IN BUBBLES

SKIMMING PADDLE

AGITATOR

WATER WITH ADDED FROTHING AGENT

Fig. 2-67. Froth flotation tank.

Mineral particles placed in the froth tank rise to the top and hold onto gas bubbles created by the air blown into the water. Various desired metallic sulfides can then be collected and removed as they are lifted to the surface. Adding lime to the mixture keeps iron sulfide from rising with the copper. Pine oil or cresylic acid (a product of creosote) added to the water will keep the bubbles intact, especially at the surface, where the particles are suspended by the bubbles in the form of a froth. The froth is gathered and filtered, after which it is dried for roasting in a hearth-type furnace. The roasting process burns off volatile (explosive) oxides formed by sulfur, arsenic, and antimony.

Further copper concentration is done by smelting the remaining materials in a furnace about 2700° F. Unwanted material turns to slag during this process, with iron forming ferrous sulfide and copper forming cuprous sulfide. The sulfides in the molten state will eventually join to form *matte,* a solution of sulfide ores containing up to 45% copper. Matte is heavier than slag and is tapped from the bottom of the furnace. The slag, which

consists of some iron oxide, alumina, silica, and lime, is then removed.

The matte is placed in a basic lined converter similar to the Bessemer, as shown in Fig. 2-62, where iron sulfide is decomposed and any remaining iron is oxidized. This is done by forced air within the converter, which also takes out sulfur dioxide.

Violet flames from the top of the converter indicate iron oxidation (now part of the slag), with only pure cuprous sulfide in the furnace. At this time, the air blast is shut off and the slag is poured off by tipping the converter.

The cuprous sulfide is now heated in the converter to be decomposed, so that the sulfur burns out as sulfur dioxide, leaving only metallic copper. The copper is now poured into ladles, to be refined later by the furnace or electrolytic process.

Copper Refining—After leaving the converter, copper gives off gas as it solidifies. The gas forms bubbles that break at the surface, causing what is known as *blister copper.* This type of copper is refined almost entirely by the electrolytic process. This decomposing process uses an electric current to release the gas.

The copper is heated and then cast into *anodes* (positive poles) shaped like flat plates. The anodes are suspended by projected points in the electrolytic tank solution. The electric current dissolves the anode, and the pure copper released is deposited on the cathode (opposite pole). The sludge in the bottom of the tank could consist of precious metals, depending on the original ore content.

The copper cathode can now be formed according to commercial needs. This is now the purest commercial grade and is known as 99.9% pure electrolytic copper. This type of copper can be further deoxidized by using phosphorus; this produces an oxygen-free copper, which, unlike electrolytic copper, will shrink during casting. In this method, copper may be refined for specific casting needs.

Aluminum

This metal is found chemically combined with oxygen and other elements. The principal source of aluminum is *bauxite,* a white-to-reddish, clay-like, aluminum hydroxide of varying composition. Aluminum is a light, corrosion-resistant metal with high thermal conductivity. It cannot be smelted directly by carbon. Strength and ductility are increased when the temperature is lowered, while hotter temperatures, well below the melting point, will decrease the metal's strength. The surface of aluminum is oxidized readily by a thin film which hinders welding unless special preparations (explained under welding) are used. Fortunately, though, this thin oxidized surface film will prevent further oxidation.

known as *calcination,* in which the aluminum oxide (alumina) is powdered, makes it possible to electrolytically decompose the alumina into aluminum.

The alumina is dissolved in molten *cryolite,* a fluoride of sodium and aluminum, at 1000° C. The cell containing the bath is a double-lined steel box with the inner lining consisting of carbon, which acts as the negative electrode (cathode) in the electrolytic reaction. Carbon anodes (positive) are placed above the cryolite and alumina. As current passes between the poles, the alumina becomes oxygen and molten aluminum. The aluminum sinks through the lighter cryolite to the bottom of the cell, as in Fig. 2-68, while the oxygen is attracted to the carbon anodes,

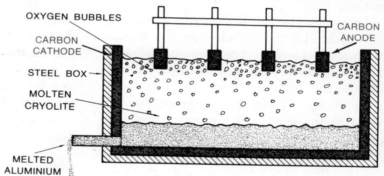

Fig. 2-68. Electrolytic reduction cell.

OXYGEN BUBBLES
CARBON CATHODE
STEEL BOX →
MOLTEN CRYOLITE
CARBON ANODE
MELTED ALUMINIUM

Fabricating aluminum is possible by every method used in industry. Since it is only about half as heavy as iron, it is widely used for structural members, such as beams, when alloyed for strength. Pure aluminum has a tensile strength of around 13,000 psi, but after alloying and a suitable heat treatment, the tensile strength can be raised to about 80,000 psi.

Aluminum Production—To produce aluminum, bauxite is ground into a fine powder and placed under pressure in a hot solution of of sodium hydroxide, forming sodium aluminate. The sodium aluminate cools in a precipitating tank where aluminum hydroxide is formed. The aluminum hydroxide is placed in rotary kilns at high temperature, to form water and aluminum oxide. This procedure,

which are gradually consumed. The cryolite remains practically unchanged, and only alumina needs to be added to the molten bath at intervals to keep the cell producing. Molten aluminum can be tapped from a spout in the bottom of the cell.

Magnesium

This is the lightest metal used today in large quantities. *Magnesium's* industrial usage was not recognized until after the beginning of the 20th century. Its weight is about ⅕ the weight of copper, and less than ⅔ the weight of aluminum. Magnesium can be fabricated by all industrial methods. Deep, high-speed machining is done more easily on magnesium than with other metals. The metal can be hot worked, but rapid oxidation makes certain welding and foundry methods necessary. Although cold working of magnesium is limited, alloying the material will give tensile strengths up to 50,000 psi.

Sources of Magnesium—Dolomite and magnesite are main sources of magnesium. Sea water and natural brines also contain magnesium. Where sea water is used, calcium hydroxide is added to precipitate magnesium hydroxide, which in turn settles. Then, the magnesium hydroxide is filtered and dissolved in hydrochloric acid to make magnesium chloride. Sulfuric acid is then used to precipitate any calcium left in the magnesium chloride, as a sulfate of calcium. This sulfate and some sodium chloride are again filtered, leaving magnesium chloride to be dried into a granular solid.

An electrolytic cell is used with a bath of sodium chloride and calcium chloride. Molten magnesium chloride is continuously added to the bath and, through an electrolytic action, floats to the cell's top where the magnesium is dipped out by ladles. This procedure for removing the magnesium from sea water is known as the *Dow Electrolytic Cell*. See Fig. 2-69.

The other two sources of magnesium—dolomite and magnesite—are produced by other processes. Dolomite, a carbonate of calcium and magnesium, uses a ferrosilicon process to produce aluminum. The dolomite is converted by *calcining* (heating at high temperatures) oxides of the two metals, and it is pulverized and pressed into pellets with a mixture of ferrosilicon. The pellets are heated under a vacuum of mercury (a heavy, silver-colored, metallic element as a liquid) at ordinary temperatures, after being placed tightly in nickel-chromium tubes. Crystals are formed in the end of these 12″ diameter tubes; after being removed, the crystals are melted down. Using a flux during this process eliminates the impurities.

Magnesite (magnesium carbonate) uses an electrolytic process known as the *Elektren Process*. Magnesium oxide is produced by the calcination of the magnesite. Pulverized coal is mixed with the oxide and moistened with a magnesium chloride solution. The mixture is then made into briquettes (fine material pressed into rectangular shapes) in a rotary kiln (oven). From the kiln they go directly to a furnace and are heated in a chlorine gas to make magnesium chloride. Electrolysis then extracts metallic magnesium from magnesium chloride in an enclosed cell.

Zinc

This metal resists oxidation by a protective surface, formed by initial oxidization similar to the oxidation on aluminum. Zinc and lead

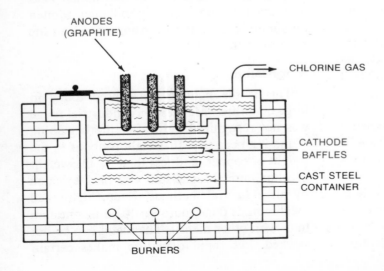

Fig. 2-69. Dow Electrolytic Cell.

will be covered only briefly, since their use to the welder is limited. Zinc is used primarily for die castings, alloying copper to make brass, and coating iron sheets to resist corrosion.

Zinc ore is crushed and roasted by a procedure similar to that for copper ores. Metallic zinc extracted by an electrolytic process is of the highest purity and is especially useful for die casting. These castings, however, are not suitable where temperatures could exceed 200° F.

When iron sheets are coated with surface zinc, the electrolytic process of the zinc protects even a surface break from corrosion. When the iron is pickled in a diluted acid to remove surface impurities and then dipped in molten zinc, the process is known as *galvanizing*. This process can build up the surface zinc to a greater thickness, especially on wrought iron. Molten zinc can be sprayed on various materials by a process called *metallizing*.

Lead

This metal is extracted from its ores by smelting in an open-hearth furnace, blast furnace, or reverberatory furnace. The open-hearth furnace is used to smelt high-grade concentrates. Ordinary atmosphere and moisture does not corrode lead.

Other important metals are recovered from their ores by metallurgical processes. Since the welder's major repair problems will be with steel, copper, aluminum, magnesium, cast iron, and all their alloys, other metals will be covered under a chapter dealing with welding various unusual metals.

Heat Treatment of Metals

The welder must be familiar with how to *anneal* (soften), *harden,* and *temper* metals. Each of these terms will be explained more clearly.

The structure of a *solid* metal is more dense than when it is a *liquid*. When the metal is solid, the atoms in the crystals making up the solid are packed in an orderly fashion into the space lattice, as shown in Fig. 2-70. This or-

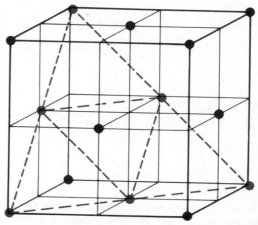

Fig. 2-70. Space lattice.

derly form of packing makes the metal's solid structure very dense. When the metal is liquid, the random movement of the atoms due to heat gives the form a less dense structure.

Heat Treatment

A combination of hot and cold operations to obtain desirable properties in a metal or an alloy is known as *heat treatment.*

In plain carbon steels with specified carbon content, the physical properties depend on what form of the carbon is present; therefore, the heat-treating effect depends on how the treatment changes the carbon distribution.

Regardless of the carbon content, when carbon steel is heat-treated to harden or soften it, the steel must initially change into a solid solution. Steels below 0.83% carbon are heated to the point where this solid solution consists of homogeneous *austenite,* as in Fig. 2-71. In steels above 0.83% carbon content (Fig. 2-71), heating follows the same pattern, except that the steel is not heated to where homogeneous austenite is formed. Uniform heating above the critical temperature, if held constant until the carbon is dissolved and diffused throughout the metal, will produce this homogeneous austenite. If this uniform heat is maintained for too long a period, especially on ordinary steels, a coarse-grained material is produced.

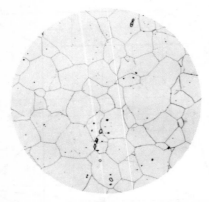

Fig. 2-71. Photograph of austenite, taken through a microscope. Magnified 100 times. (Courtesy of Bethlehem Steel.)

The structure and hardness of a finished piece of steel depends on the method and rate of cooling. Full annealing is the result of cooling at a very slow rate, a process which can turn austenite (iron carbide) into pearlite. A full annealing process on steels below 0.83% (hypoeutectoid, Fig. 2-72) presents a

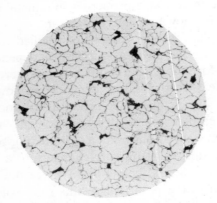

Fig. 2-72. Photograph of *Hypoeutectoid* steel, taken through a microscope. Polished and etched with nitol. Magnified 200 times. (Courtesy of Bethlehem Steel.)

body structure of pearlite and *ferrite,* while those steels over 0.83% (hypereutectoid) will become pearlite and *cementite* when fully annealed. Full annealing, then, will be affected by the carbon content of the steel.

Hardness of Steel

Steel's *hardness* depends on the transformed structure of the pearlite which would be present if full annealing was performed. During heat treatment, any transformation of the austenite into pearlite, as shown in Fig. 2-73, takes a certain time. This makes it possible to produce various structures by adjusting the time and procedures of cooling. When the transformation process is completed, the structure will be stable and will undergo no further change when cooled to room temperature.

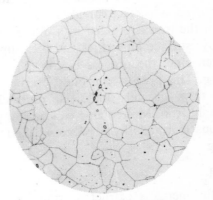

(A) Austenite, magnified 100 times.

(B) Pearlite, magnified 1000 times.

Fig. 2-73. Photographs of steel during heat treating. (Courtesy of Bethlehem Steel.)

Annealing

The temperature and rate of cooling depend on the material being heated and the treatment's purpose during *annealing*. Annealing can refine a grain structure and make steel more ductile. Steels which are stressed due to casting or welding can be relieved of these stresses by proper annealing. Steels that are difficult to machine due to hardness can be softened by proper annealing.

When heated, carbon steels eventually reach a point where the steel becomes non-magnetic. In welding, this point is considered the *critical point*. For full annealing, a steel should be placed in a sealed container and heated to at least 100° F. *above* the critical range. The temperature should remain constant for one hour for each inch of maximum section of the steel being softened. Warpage can be eased by heating if the heat is applied gradually, and slow heating will insure a more even temperature between the surface and the interior of the metal. After required heating, slow cooling transforms the austenite into pearlite. When the transformation is complete, a material that is free of internal stresses and has high ductile qualities will remain.

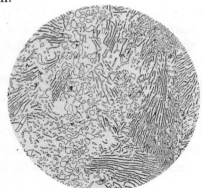

Fig. 2-74. Photograph of partly spheroidized pearlite, magnified 500 times. (Courtesy, Nickel of Canada.)

The welder should also be familiar with *spheroidizing*. This process of heating and cooling produces the globular, or rounded, form of carbide, making it possible to soften high carbon steel. In this process, the temperature used for prolonged heating is just *below* the critical point; slow cooling follows, and a steel such as that in Fig. 2-74 is produced.

Normalizing

Stresses set up in steel due to the heat of welding or through cold working can be *normalized*. Steel which has been normalized is harder than annealed steel and has a greater tensile strength, while it retains a relatively soft and ductile condition. High carbon steels and some alloy steels, however, become so hard through normalizing that they cannot be machined.

Steels are heated to slightly higher temperatures for normalizing than for annealing. The relatively short time necessary to transform austenite is all that is necessary for normalizing. Cooling the steel in still air at room temperature, immediately after removing the steel from the furnace, allows some pearlite to form. Cool, circulating air could harden the steel, a condition easily recognized and understood by any welder after studying steel hardening. Normalizing is a satisfactory method used on work-hardened or fatigue-welded metals to increase their life expectancy. An example of normalized metal is the ferrite and pearlite shown in Fig. 2-75.

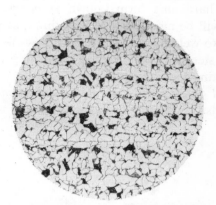

Fig. 2-75. Photograph of normalized ferrite and pearlite, magnified 200 times. (Courtesy, Nickel of Canada.)

Hardening

The *hardness* of any steel composition depends on how fast the steel is cooled after being heated. A metal's hardness will vary from its surface to its center, as shown in Fig. 2-76, depending on the length of quench-

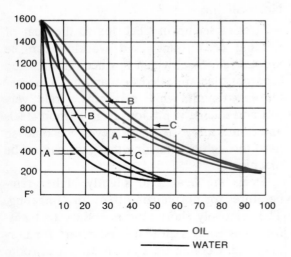

Fig. 2-76. Cooling rates during quenching. Figures are for 1″ round iron. (A) Surface of metal. (B) Midway to center. (C) Center of metal.

ing and the method used. This cooling, known as *quenching*, is done by putting the heated metal in air, oil, water, or brine. *Brine* is a salt and water solution used in place of water to make a more uniform hardness over the material's entire surface.

Air Quenching—Quenching in *air* is a mild hardening form known as *normalizing* and will be mentioned several times, as it applies to welding and postheating. Some high-alloy steels can be hardened by air to produce a cutting tool steel.

Oil Quenching—Metals cooled in *oil* will not distort as much as metals cooled in water,

due to the slower quench rate of oil. The viscosity (ability to flow) of oil should allow free circulation around the material being cooled. Oil quenching methods can be used satisfactorily on hypereutectoid steel and in various low-alloy steels for full hardening. Fig 2-77 shows a quenched and tempered martensite steel.

Water Quenching—At temperatures below 100° F, *water* is used to quench a wide range of carbon and medium-carbon, low-alloy, steels. Because of water's rapid quenching effect, low-carbon steels can be satisfactorily hardened. Water's quenching effect is increased by spraying it on the heated metal or agitating it as the metal is immersed. Cooling rates by oil and water are shown in Fig. 2-76.

Brine—When *brine* comes in contact with hot steel, it creates agitation that will break up any bubbles which could create soft spots by insulating the steel. Brine is unlike water because brine has about 10% salt in it. A water quench will transform the *outside* of the metal into martensite more rapidly than it does the *inside*, as shown in Fig. 2-78. Transforming austenite into martensite increases the metal's *volume*, which could cause cracking when the martensite reached the part's interior.

Interrupted Quenches—Most welders will have little practical use for information on *interrupted quenches,* but the information will help to better understand steel structure and formation.

Interrupted quenches suppress austenite's being transformed into pearlite, while avoiding martensite formation as well. The solution for quenching purposes is kept at required temperatures, and the metal remains immersed until the required formation is attained. Agitation during quenching is limited, and the use of molten salt makes it possible to harden finished parts.

Austenite can be transformed into *bainite,* a structure between pearlite and martensite, as in Fig. 2-79. When steel is cooled suddenly

Fig. 2-77. Photograph of martensite, quenched and tempered. Magnified 500 times. (Courtesy, Nickel of Canada.)

Fig. 2-79. Photograph showing the structure of fine pearlite and upper bainite in martensitic matrix. Magnified 1000 times.

WHITE
MARTENSITE

400° F
TEMPERED
MARTENSITE

1200° F
TEMPERED
MARTENSITE

Fig. 2-78. Photograph of martensite at different levels of tempering. Magnified 1000 times. (Courtesy of Bethlehem Steel.)

to about 800°F and held at about 500°F for a period of time, bainite is formed. This tempering procedure, known as *austempering*, gives greater ductility and toughness to steel with little distortion.

Martempering means quenching in molten salt at a temperature *above* the point where martensite forms. When the temperature equalizes throughout the metal due to a longer quenching time, a martensitic structure is formed, as shown in Fig. 2-78. Martempering is faster than austempering and is more satisfactory for hardening heavy parts.

Tempering

After a steel is hardened by quenching, it could fail under a work ioad due to residual (internal) stress. If the metal has been fully hardened, it could be too hard and brittle to do a satisfactory job.

Austenite has a tendency to transform into martensite. This, in turn, will eventually contract and fail. Since austenite is denser than martensite, as shown in Fig. 2-80, any transformation to a less dense structure would eventually cause failure. To stabilize the austenite, *tempering* will make a tougher, less brittle formation with lower internal stresses.

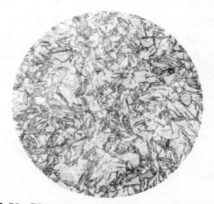

Fig. 2-80. Photograph of high-carbon steel, (martensite and retained austenite). Magnified 500 times. (Courtesy of International Nickel.)

Fig. 2-81. Tempered martensite microstructure, quenched and tempered at 950°F. Magnified 500 times. (Courtesy of International Nickel.)

Tempering requires reheating the hardened material before it has cooled to room temperature. Reheating forms carbide crystals from the carbon released from the martensite; the material which remains is a *tempered martensite microstructure* (Fig. 2-81).

Tempering requires both skill and experience, since the steel's composition depends on the time and temperature required to produce a satisfactory structure. The reheating temperature must be between 300°F and the critical temperature; however, the reheating temperature depends on the nature of the steel and the allowed hardness reduction. Steels with the same carbon content, but different percentages of alloying elements, require different tempering procedures. Toughness is usually higher in steels where higher tempering heats have been possible. The welder's interest in tempering, therefore, stems from the fact that tempering makes steel *tough,* a property necessary to give steel the ability to withstand a load without breaking.

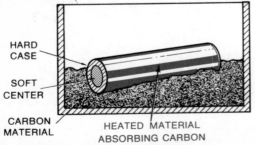

Fig. 2-82. Case-hardening process.

HARD CASE

SOFT CENTER

CARBON MATERIAL

HEATED MATERIAL ABSORBING CARBON

Surface Hardening

The inside of low-carbon steel retains both ductility and toughness while the surface is hard enough to resist abrasion and wear, if it has gone through *surface hardening.*

The outside layer of steel, known as the *case,* can be hardened if permitted to absorb carbon while the interior, or *core,* remains soft, as in Fig. 2-82. Plain carbon steels with a carbon content between 0.10% and 0.20% are able to absorb carbon while resisting any brittle, coarse-grained composition. After carburization, heat treatment and quenching can be used; tempering may follow, if required.

The material to be hardened can be cleaned and packed in a metal box with what is called a *carbonaceous* material. This material gives off carbon when burned. These carbonaceous materials may be charcoal, hides of animals, wood, or other types of commercial carburizing compounds that give off carbon monoxide when heated. Then, a chemical reaction creates carbon, which penetrates the case of the steel. The penetration depth depends on the temperature and the soak time. To better control the depth and hardness of the case, a process known as *gas carburizing,* using carburizing gases such as methane, propane, or butane, can be used. This process requires a more skilled operator than other methods.

Nitriding

Another process which uses nitride-forming alloying elements such as chromium, molybdenum, and aluminum is called *nitriding.* Nitriding not only produces hardness and-wear-resistant qualities, but it also gives the metal resistance to fatigue and corrosion. Because further heat-treatment is unnecessary, little distortion is experienced. In this process, an increase in temperature decreases the hardness.

Cyaniding

A form of case hardening, *cyaniding* produces a thin case on steel when the steel is immersed in a molten cyanide (poisonous white compound) bath. The 1500°F bath produces carbon dioxide for carburizing the immersed steel to a depth of about 0.015″. The resultant material consists of a fine-grained, low-carbon, core and a hard martensite case.

If an *oxy-acetylene* flame is passed over the parts to be hardened and an air or water stream is then blown on the parts at the proper temperature, a hard case on the parts will result, as shown in Figs. 2-83 and 2-84.

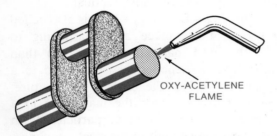

Fig. 2-83. Part being heated before quenching.

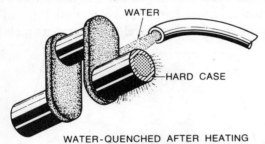

Fig. 2-84. The same part being water-quenched after heat was removed.

This procedure does not require skilled labor and can be used on both steel and gray cast iron. Since carbon and nitrogen are not added in this process, steels with higher carbon contents are necessary. Unlike plain carbon steels with low carbon content hardened by carburizing, the steels for flame-hardening must have a carbon content of 0.4% to 0.5%.

Metal Treating Review

The processes just covered can be reviewed and summarized to organize these treatments as a welder needs them.

Preheating—Applying heat before welding to compensate for internal residual stress or distortion is known as *preheating*. At the same time, preheating improves the quality of the weld, when possible.

Preheating can be done by the oxy-acetylene torch or other fuel-burning preheating torch. This method is satisfactory for small areas or isolated localities, but a forge fire or permanent preheating furnace should be used when the complete unit must be preheated.

Preheating lowers the cooling rate of the finished weld, helping eliminate breakage in the weld zone and helping eliminate major crack formation. Recommended preheat temperatures are shown in Table 2-6.

Table 2-6. Recommended Preheat Temperatures

Steel	% Carbon Content	°F. Temperature
Mild Steel	0.05 - 0.30	250
Medium-Carbon	0.30 - 0.55	600
High-Carbon	0.55 - 0.75	700
Very High-Carbon	0.75 - 1.5	800+
Manganese Steel	12.0	Not required
Stainless (18% Chromium and 8% Nickel)		Not required

Postheat—Heating the metal *after* welding or cutting is called *postheating*. It is done to temper the heat-affected zone adjacent to the weld and to relieve some residual stress.

Annealing—Heating and cooling the material relatively slowly to remove the stress, as in postheating, is called *annealing*. It will make hard, ferrous metals soft and therefore alter such physical properties as ductility and toughness. See Fig. 2-85.

The type of material and the reason for annealing will determine the temperature and rate of cooling. The slower the cooling, the softer the material will be when cold. A good guide for the heating temperature is to a point just above the critical point. If the heated material is placed in hot ashes, asbestos, or lime, the cooling is retarded. *Nonferrous* ma-

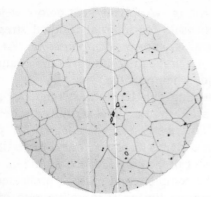

Fig. 2-85. Photograph of austenite, annealed at 1900°F. (Courtesy of International Nickel.)

terials, such as copper, must not be confused with *ferrous annealing,* since they are softened be heating and quenching in cold water.

Stress Relieving—Residual stress (internal stress) created through welding must be eliminated by postheating. The temperature for *stress relieving* is always *below* the critical range, while the temperature for *annealing* and *normalizing* is always *above* the critical range. Stress relieving should not be done in the critical range, since this creates distortion and changes the grain structure and dimensions. This could cause a defective weld. Plain carbon and low-alloy steel must soak up (absorb) carbon at varying temperatures, usually around 1100° F to 1250° F, depending on the material's thickness. A slow rate of heating and cooling are very important in stress relieving.

Materials showing distortion after welding are usually free of residual stress. When *re*heating to correct dimensional defects, however, *peening* should follow stress relieving.

Hardening—Steels below 0.83% carbon content can be heated above the critical range and quenched in air, water, or oil to form martensite, producing high strength, wear resistance, and hardness (Fig. 2-86). To develop full hardness, the cooling rate varies, depending on the steel's composition. Steels below 0.30% carbon can be somewhat hardened, but it is usually not practical.

Tempering—Drawing, which is also called *tempering,* takes place after hardening to make the hard material tough, rather than brittle. The hardened steel is reheated to a temperature below the critical temperature. Reheating breaks down the martensite to a ferrite matrix of iron carbide particles. With a lower carbon content in the steel, toughness will be predominant, as there is less hardness when a higher temperature is used for tempering. Postheating a weld also tempers metal

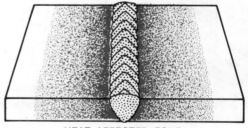

HEAT-AFFECTED ZONE
OXY-ACETYLENE WELD

(A) Oxy-acetylene weld area.

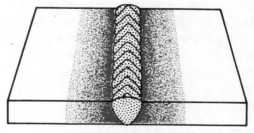

HEAT-AFFECTED ZONE
ELECTRIC (ARC) WELD

(B) Arc weld area.

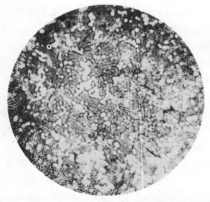

Fig. 2-86. Tempered martensite, magnified 1100 times. (Courtesy of Bausch and Lomb, Inc.)

Fig. 2-87. How welding methods vary the tempering effect.

adjacent to the weld and relieves stress. Oxy-acetylene welding feathers out the tempering area over a greater area than arc welding, as shown in Fig. 2-87. Usually, postheating after oxy-acetylene welding is not necessary.

Normalizing—The welded material, especially near the weld area, is heated above the critical range and allowed to cool in the air, well out of drafts, in *normalizing*. This improves the previously-described mechanical properties by refining the grain size (Fig. 2-88).

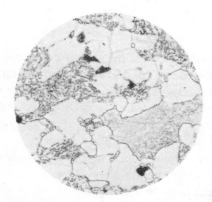

Fig. 2-88. Spheroidized pearlite, normalized at 1300°F for 4 hours. Magnified 1000 times. (Courtesy of International Nickel.)

Flame Hardening

To *flame harden,* the oxy-acetylene torch first heats a small surface from 1/16″ to 1/2″ deep, to a temperature *above* the critical range. Then, the metal is quenched in air, water, brine, or oil. The heated surface is immediately hardened, and the underlying portion is in the original condition. Only the areas necessary for wear resistance can be treated, leaving all other areas machinable and ductile. The physical properties are changed and the chemical composition remains constant during flame hardening.

Carbon steels of 0.40% to 0.50% carbon content are easily flame hardened, but steels above 0.50% could fail from cracking while quenching. Cast steel can be flame hardened, although it is not as suitable for flame hardening as is forged steel. Flame hardening cast iron and alloy steel depends entirely on the metal's makeup.

Case Hardening—This is a form of carburizing and is similar to flame hardening. During case hardening, however, only the outside of the material is hardened. Carbon is added to the surface of the metal, leaving the interior unchanged. Only wrought iron and low-carbon steels can absorb carbon through case hardening, whereby the surface is converted into a tool or high-carbon steel.

The material to be treated is placed in an enclosed container with high-carbon material. The container is heated to about 1800° F, and the steel soaks up the freed carbon. The thickness of the case, or the depth of carbon penetration, depends on the length of heating time.

After case hardening, the material can be hardened by another hardening procedure, but care must be taken not to burn out the added carbon. Potassium cyanide, burnt leather, bones, and charcoal are good materials for supplying the carbon.

Other heat treating procedures will be covered in chapters on welding certain metals and their alloys, as each form of heat treating pertains to the welding trade.

3

Alloying Elements

An *alloy* is a metallic material formed by mixing two or more chemical elements while in the molten state. Usually, metallic elements will dissolve into a common mixture at their molten temperatures. However, these same elements, on cooling, may become partially or wholly insoluble (unable to be dissolved).

Alloying elements are added to base metals for a variety of reasons. Examples would include: to increase the strength, to increase the corrosion-resistance, and/or to lower the melting point of the base metal.

Alloys are usually produced by mixing the required elements in the liquid state and then allowing them to solidify. Iron can, however, be made into steel by a process known as *infusion*.

Infusion is a method of soaking iron in carbon without raising the temperature of the iron to the boiling point, to produce the resultant alloy. Case-hardening is the name given to this process, discussed under *heat treating*.

Since the alloy usually has different properties than the components, it is possible to use metals for industrial use which, by themselves, would be too soft for most commercial purposes. The importance of these alloys in industry depends primarily on the internal structure of the metal that results when the mixture changes from liquid to solid. This, in turn, determines the alloy's usefulness.

In certain cases, important components of an alloy may be *Metalloid* (a nonmetallic element resembling the metal in some of its properties), *Arsenic* (a steel-gray, volatile, poisonous element), and/or *Antimony* (a silver-white, hard, crystalline element).

Alloys are classified according to the number of components they contain. A binary alloy contains two components, while ternary and quaternary alloys would contain three and four elements, respectively. Table 3-1 contains examples of binary, nonferrous alloys.

Iron and steel, like other metallic elements, are made up of *atoms*—the smallest units of an element—arranged in a precise geometric order. When in the gas or liquid state, the atoms join together to form a random arrangement of molecules. A *molecule* is the simplest structural unit that displays the characteristic physical and chemical properties of any compound.

The atoms of a pure metal, through their orderly arrangement, form what is known as *crystal lattices*. These lattices can take several forms; the three most common forms are: (1) The *Face-Centered Cubic* (FCC), (2) The *Body-Centered Cubic* (BCC), and, (3) The *Close-Packed Hexagonal* (CPH), as shown in Fig. 3-1.

Iron is known to have a cubic-type crystal with an atom at each corner and one in the

Table 3-1. Typical Industrial Nonferrous Alloys.

Percentage of Alloying Metals		Finished Product
Copper-Zinc		
85%	15%	
70%	30%	Brass
65%	35%	
60%	40%	
Copper-Tin		
95%	5%	Tin Bronze
90%	10%	
Copper-Aluminum		
95%	5%	Aluminum Bronze
90%	10%	
Copper-Nickel		
98%	2%	
90%	10%	
80%	20%	Copper-Nickel
70%	30%	
60%	40%	
Copper-Nickel		
30%	70%	Monel
Nickel-Chromium		
76%	15%	Inconel
Copper-Zinc-Nickel		Nickel Silver
Lead-Tin		
67%	33%	
50%	50%	Solder
33%	67%	
5%	95%	
Aluminum-Copper		
Aluminum-Magnesium		Aluminum Alloys
Magnesium-Manganese		
Magnesium-Aluminum		Magnesium Alloys

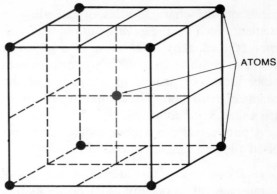

(A) Body-centered cubic. Crystal structure of molybdenum, iron (900° C) chromium, sodium, potassium, barium, and lithium. The basic cube, with an atom at each corner and one in the geometric center.

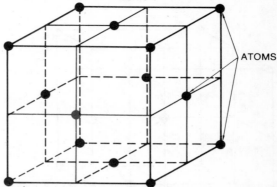

(B) Face-centered cubic. Crystal structure of metals such as nickel, gold, silver, copper, lead, aluminum, and calcium.

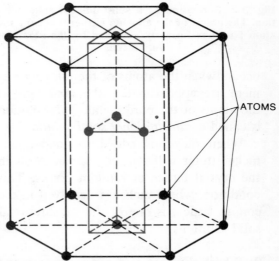

(C) Close-packed hexagonal. Crystal structure of magnesium, beryllium, zinc, cadmium, and cobalt.

Fig. 3-1. Atom arrangement in the crystal lattice.

exact center of the cube, as shown in Fig. 3-1A.

The *crystal* or *grain,* as it is sometimes called, consists of the formation of these atom-centered cubes. Iron, at room temperature and up to 900° C, is the *body-centered cubic,* or *Alpha,* as it is sometimes called. From 900° C to 1400° C, it changes to the *face-centered cubic.* The iron reverts back to the *body-centered cubic* once the temperature exceeds 1400° C.

The crystal lattice of a pure metal can always dissolve some amount of an alloying element or elements. The specific amount depends on the characteristics of the alloying element, such as the *relative atom size, valence* (the capacity of an atom or a group of atoms to combine in specific proportions with other atoms or groups of atoms), and the *chemical attraction*. These mixtures of atoms are called *solid solutions*.

A pure metal can accommodate the atoms of an alloying element in one of two ways:

1. By allowing the atoms of the alloying element to occupy atom positions formerly occupied by atoms of the pure metal, as shown in Fig. 3-2. This type of solution

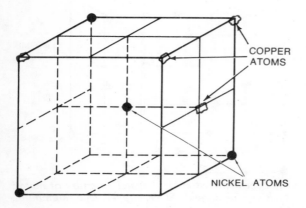

Fig. 3-2. Formation of a substitutional solid solution. The atoms of the alloying elements occupy the atom positions formerly occupied by the atoms of pure metal.

occurs when the atoms of the alloying element are approximately the *same size* as the atoms of the pure metal. This solution is called a *substitutional solid solution*.

When nickel is added to copper, the nickel atoms will replace copper atoms in the crystal lattice at random points. This condition, where nickel is dissolved in the copper lattice, is one of the substitutional solid solutions.

2. Where the atoms of a dissolving metal are *smaller* than those of the parent lattice, a new location of in-between points

(interstitial) is taken over within the lattice, rather than replacing the atoms of the solvent metal. This new lattice structure is known as an *interstitial solid solution*. (See Fig. 3-3.)

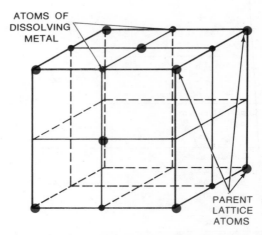

Fig. 3-3. Formation of an interstitial solid solution.

If more atoms of alloying elements are added than can be accommodated in the lattice of the pure metal, the "extra" atoms combine with the atoms of the pure metal to form a different crystal lattice, and the alloy con-

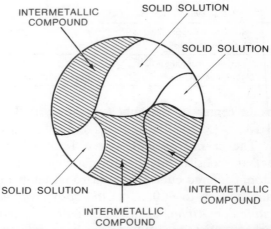

Fig. 3-4. A cross-section of composite materials. Atoms of the alloying element are forced to combine with the atoms of the pure metal to form a different crystal lattice.

tains a mixture of particles having the two crystal lattices, as shown in Fig. 3-4. These types of alloys are called *composite materials*.

Varieties of Alloys

Before explaining the effects of alloying elements on ferrous and nonferrous metals, it is important for the welder to have some knowledge of the important varieties of alloys.

The different substances which make up an alloy can dissolve to make a homogeneous mixture. When the alloy reaches a liquid, it is classed as a solution. If the metals that compose the alloy are soluble in both solid and liquid states, they will remain in a homogeneous mixture even after solidification. This condition is known as the *solid solution*.

To review, the atoms of one metal can replace some of the atoms of another metal in the space lattice. This takes place in solid solution. The microstructure of a solid solution resembles pure metal, since it contains only one type of crystal and grain structure. Nickel and copper, and nickel and chromium, are two alloys of metal mutually soluble as both liquids and solids.

Some alloys of metal are mutually soluble in the liquid state and partially soluble in the solid state. Tin and lead are metals which form two types of solid solutions, depending on the percentage ratio of the metals. The maximum solubility of tin in lead, in the solid state, is reached when the content of tin does not exceed 19.5%. This is known as an *alpha solution*.

Beta solutions of lead in tin have a composition of 2.5% lead and the balance tin. This ratio represents the maximum solubility of lead in tin in the solid state.

Most of the low-melting-point solders are alloys of tin and lead. Although 95% of solder could consist of lead, the lead *by itself* would not be satisfactory for soldering because it cannot *tin* (or *wet*, as it is sometimes known) the surface of the metals being joined.

Although tin and lead melt at definite temperatures in their pure form, 450° F and 621° F respectively, their alloys can be produced to melt over a range of temperatures, as can most alloys.

Solidus and Liquidus Temperatures

To understand Table 3-2, the student must be familiar with the terms *solidus* and *liquidus* temperatures.

Table 3-2. Tin-Lead Composition.

ASTM No.	Tin %	Lead %	Solidus Temp. °F	Liquidus Temp. °F
5A	5	95	572	596
10A	10	90	514	573
15A	15	85	437	553
20A	20	80	361	535
25A	25	75	"	511
30A	30	70	"	491
40A	40	60	"	455
50A	50	50	"	421
60A	60	40	"	374
70A	70	30	"	328

The tin-lead alloys begin to solidify at temperatures known as the *liquidus temperature,* and they will continue to solidify until freezing is completed at the lower *solidus temperature.* Where an alloy is mixed so as to have the same solidus and liquidus temperatures, that is, a definite melting point, it is known as an *eutectic composition.* The eutectic composition for an alloy of tin and lead is 61.9% tin and 38.1% lead, and its melting point is 361° F. Other tin-lead compositions with melting points are shown in Table 3-2.

Cadmium and bismuth are alloys mutually soluble in the liquid state but essentially insoluble in the solid state. *Eutectic alloy,* previously mentioned and found useful for soldering and welding, is produced by the reaction of cadmium and bismuth and acts similarly to pure metal during temperature changes. The melting points of pure cadmium and pure bismuth are 610° F and 520° F, respectively. Adding bismuth to cadmium lowers the melting point below 610° F, but if cadmium is added to a mass of bismuth, the melting point would be progressively lower than the pure bismuth. A common minimum melting point of 241° F for an eutectic alloy is obtained by alloying 40% cadmium and 60% bismuth, which, in the solid form, consists of a mixture of fine crystals of nearly pure bismuth and cadmium.

In any alloy, the element with the greatest concentration crystallizes first, when the alloy is cooled rapidly from its liquid form. This would create nonuniform crystals, since the composition of the alloy differs in concentration. A heterogeneous structure that is caused by rapidly cooling the alloy is known as a *cored crystal*.

Since most welding situations deal with repairing an alloy of carbon and iron, it is a good idea to look at the appearance and grain structure of this material.

It would require many volumes to cover the metallurgical information on steel and its alloys. However, the following information should meet the immediate needs of a welder.

Principal Alloying Elements of Steel

The principal alloying elements of steel and their melting points were listed in Table 2-4. The properties of steel are changed so much by alloying elements that is is impossible to generalize about the resultant material. Earlier, it was seen that carbon, added to plain carbon steels, will change the properties of the steel. Phosphorous, manganese, silicon, and sulfur are also present in plain carbon steel. If other suitable alloying elements are added to various types of steel, increased toughness and hardness can be produced, along with properties resisting wear and corrosion.

Through a microscope, grains or crystals in metals look different, especially during heat treating. For example, plain carbon steels through the microscope have white-colored grains along with dark gray or black-colored grains.

Darker-colored grains are composed of alternate layers of iron carbide and ferrite (iron) which, when combined, are known as *pearlite* grains (Fig. 3-5). These pearlite

Fig. 3-5. Pearlite in annealed 0.80% carbon steel, magnified 1000 times, showing lamination of iron and darker iron carbide. (Courtesy of United States Steel.)

grains have a 0.80% carbon content, while the iron carbide (a hard and brittle chemical compound) has a carbon content of 6.5%.

A steel with a carbon content of approximately 0.80% carbon is known as a *pearlitic steel*. This metal is a combination of both a hard and brittle compound and a weak and soft metal. On annealing (softening), the steel becomes much higher in tensile strength than the iron carbide or ferrite composing it.

Steels with a carbon content of less than 0.80% are known as *hypoeutectoid* steels, while steels with a carbon content over 0.80% are known as *hypereutectoid*.

In the hypoeutectoid steel there is an excess of ferrite (iron), and the microstructure is made up of some white grains together with the dark pearlite grains.

The color of the grain structure varies according to the percentage of the elements in the steel. If there is no carbon, the metal will consist of all white-colored ferrite grains, as shown in Fig. 3-6. If the metal contains more than 0.80% carbon, then the carbide of iron will exceed the amount of ferrite.

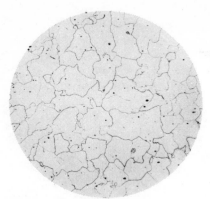

Fig. 3-6. Photograph of ferrite, magnified 100 times.
(Courtesy of United States Steel.)

Aluminum

This is a light, bluish-white, malleable, ductile, metallic element. It strongly resists oxidation and is found chiefly in bauxite.

Adding aluminum to steel gives the new metal surface stability against oxidation and corrosion, while raising the thermal critical point. Austenite is more refined when aluminum is used as an alloying element, since grain growth is slowed by the aluminum oxide forming fine particles.

A welder may find an infusible scum, incapable of being used or melted, while welding steel. This creates *porosity* (tiny holes in the steel) and is caused by a steel high in aluminum content. Steels alloyed with aluminum up to 0.2% can, however, be welded successfully with cellulose-covered electrodes, which are discussed in another chapter.

Chromium

This is a grayish-white, very hard, metallic element used in making special steels.

The critical range of temperatures is increased by adding chromium to steel. Steel can be hardened to a point three times greater than the same carbon content could produce in carbon steel, by adding chromium. At the same time, the steel would be less brittle. Toughness of steel is increased by the grain-refining characteristic of chromium. Chromium, a moderate carbide-forming element, increases both the hardness and toughness of steel at the same time. Steel containing chromium can be identified during a spark test by the small, very brilliant, stars with uniform shapes at the end of the carrier lines, as shown in Fig. 3-7.

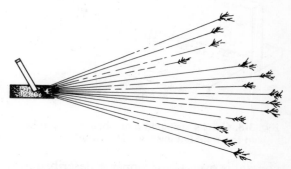

Fig. 3-7. Spark test for high-chromium steel.

A welder will experience little fusion trouble in ordinary chromium-alloyed steels. Although chromium tends to oxidize, successful welding can be done on steels containing up to 30% chromium. Welding chromium and chromium-nickel steels will be covered in a later chapter.

Cobalt

This is a tough metallic element, seldom found in the free state. As an alloy, lustrous gray *cobalt* strengthens ferrite, and its addition to steel forms a weak carbide formation. The hardening ability of steel is reduced when alloyed with cobalt; but when tungsten is added, the ferrite is hardened as in Fig. 3-6. Tools made of this dual-hardened material will withstand red heat.

Copper

Reddish-colored *copper* is a ductile, metallic element; it is an excellent conductor of heat and electricity. As an alloying element of steel, copper helps protect the metal's surface from oxidation and corrosion, and lowers its critical points.

Pure copper is easily cut, nonmagnetic, and does not spark when held against a grinding wheel. Copper alloyed up to 1% with steel will not create serious welding problems.

Cracks found in a finished weld, as shown in Fig. 3-8, can be attributed to copper if the

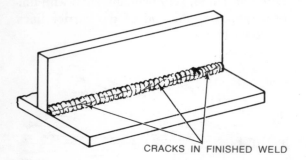

CRACKS IN FINISHED WELD

Fig. 3-8. Cracks visible in a finished weld can be due to the weld having over 1% copper content.

content is over 1%. Copper is described more fully under nonferrous alloys.

Lead

This is a soft, inelastic, malleable, ductile, bluish-gray, metallic element. When alloyed with steel and additional tin, it forms a rust-resistant coating on steel.

Manganese

A metallic element that is hard and brittle, *manganese* will rust like iron, but it is not magnetic. Manganese is grayish-white, tinged with red. When alloyed with steel, up to 14%, it can produce hard, ductile, manganese steel. Manganese acts as a deoxidizer and will combine with sulfur to prevent *hot shortness*.

Sulfur in solid solution with steel causes hot shortness, meaning that the steel is brittle at temperatures above 800° F. Adding manganese, however, causes the intermetallic compound *manganese sulfide* to form. The manganese sulfide removes the sulfur from the solid solution, preventing brittleness that could cause cracking.

Steels over 0.20% carbon can be increased in hardness by adding manganese to form martensite, as in Fig. 3-9. A brilliant spark stream of yellow is visible in the carrier lines when steels alloyed with 1% or more manganese are placed against a grinding wheel. See Fig. 3-10.

Fig. 3-9. Photograph of martensite, taken through a microscope. (Courtesy of Bausch & Lomb, Inc.)

Fig. 3-10. Spark test for manganese materials, giving off brilliant yellow streams.

Molybdenum

Steel alloyed with as little as 15% *molybdenum* gives a detached spearhead at the ends of the carrier lines when given the spark test, as shown in Fig. 3-11. Molybdenum is a

Fig. 3-11. Spark test for molybdenum-alloyed steel. This produces detached spearheads at the end of the spark stream.

silver-white, hard, metallic element. Steels alloyed with molybdenum can be joined easily and satisfactorily by either arc or oxyacetylene welding.

A complex carbide of iron and molybdenum is formed by alloying iron and molybdenum, which adds toughness to steel and at the same time raises the critical temperature. The hardening of steel alloyed with molybdenum is more effective when the proper heat is followed by oil or air quenching. Next to carbon, the alloy molybdenum has the greatest hardening effect on steel. Nickel and chromium are usually added with molybdenum to give the steel further desirable properties.

Nickel

Added to steel, *nickel* increases the steel's critical range of temperatures; also, due to its high solubility, nickel increases strength without decreasing ductility, regardless of the carbon content. It is a hard, ductile, magnetic, silver-white, metallic element. The strength and toughness of steel can be increased with approximately 5% nickel, without increasing the steel's hardness.

When 12% or more nickel is added to steel, quench-hardening is impossible. Oil and water quenching to harden under 12% nickel-alloyed steels has been done satisfactorily. Chromium is also added with nickel to give steel not only toughness and ductility, but also hardness and resistance to wear. Steels alloyed with nickel can be joined satisfactorily by arc or oxy-acetylene welding. When such steels, alloyed with at least 0.50% nickel, are spark-tested, orange sparks shaped like blocks or wedges are visible near the end of the carrier lines. These sparks are shown below in Fig. 3-12.

Fig. 3-12. Nickel-alloyed steel being spark tested. Note the orange block- or wedge-shaped sparks at the end of the carrier lines.

Phosphorus

A soft, yellowish, nonmetallic element found in steel, *phosphorus* is usually provided by the ore. Usually, the content is kept below 0.04%, since it segregates and decreases the ductility of the steel as the percentage is increased. Welding of such steels is satisfactory if the phosphorus content does not exceed 0.15%. Welding steels with more than 0.15% phosphorus cannot usually be done successfully.

Silicon

Steels containing *silicon* will produce short carrier lines of white flashing stars when spark tested, as in Fig. 3-13. Silicon is a non-

Fig. 3-13. Silicon added to steel produces a spark test of short carrier lines and white flashing stars.

metallic element, added during the steel-making process for deoxidizing. As an alloying element of steel, as much as 5% may be added, raising the critical temperature range. Adding silicon does not form carbides but strengthens the ferrite. Steels alloyed with a higher percentage of silicon are hard and brittle yet free from internal stress. Steels alloyed with reasonable amounts of silicon are satisfactorily joined by both arc and oxy-acetylene welding.

Sulfur

Steels containing *sulfur,* a pale-yellow, nonmetallic element, can be welded satisfactorily if manganese is distributed evenly throughout the steel to form manganese sulfide. Sulfur can be classed as an impurity because it makes the steel brittle. The content of sulfur in most steels is kept lower than 0.05% for this reason.

The presence of *some* sulfur in steel, however, will improve machining and metal-cutting properties. This type of steel, known as *free machining steel,* contains 0.12% or more sulfur, which can create welding problems. Welding steels with high sulfur content will be covered under welding steel and its alloys.

Tin

A white, malleable, metallic element of low tensile strength, *tin* is used as a coating on iron or steel to prevent rusting. Tin plate is a sheet iron or sheet steel coated with tin. The sheet steel is etched and treated in *pickling*

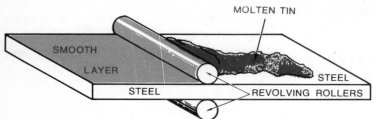

Fig. 3-14. The process of tin plating.

baths (water and sulfuric acid) to provide a surface to be wetted with molten tin. *Etching* is attacking the surface of a metal with a chemical agent that has a different form of attack on the different metal crystals. *Pickling* removes the scale from the sheet iron or steel, after which the acid is removed by first washing the sheets in pure water and then in a lime water dip. A simple lime water dip is made by mixing lime and water.

The sheet iron or steel is then passed through molten tin, spread evenly by rollers to form a thin layer of iron-tin alloy, followed by a thick layer of pure tin (Fig. 3-14). Today, the use of tin for coating sheet iron has, in many cases, been replaced by using cheaper alloys, such as zinc. Also, by taking advantage of oil industry by-products such as teflon, tin's costs can be saved.

Titanium

A dark gray metallic element resembling tin and silicon, *titanium* is usually used as a cleaning agent, or to form carbide. The amount of titanium usually present in carbon steels and low-alloy steels has little effect on welding these metals.

Titanium is used in various types of stainless steel to prevent chromium depletion due to chromium carbide formation. Titanium, being more attracted to carbon than is chromium, forms titanium carbides before chromium carbides are formed, preventing chromium depletion.

Titanium tends to slow the grain growth in steels and it combines with nitrogen in the steel, removing the nitrogen as *titanium nitride*.

Tungsten

This is a steel-gray, brittle, and heavy metallic element. Like molybdenum, *tungsten* forms hard carbides, preserving tool steel hardness at high temperatures. This metallic element will somewhat dissolve in ferrite, creating toughness and hardness and making the alloy suitable for various tool steel components. Tungsten is usually combined with other alloying elements, to produce a steel capable of staying hard at red heat. Using tungsten for welding electrodes will be covered under *Tungsten Inert Gas* (TIG) welding. Tungsten shows no sparks and short carrier lines when given the spark test shown in Fig. 3-15.

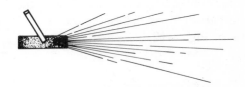

Fig. 3-15. Tungsten steel spark tested. Notice the lack of sparks and short carrier lines.

Vanadium

When alloyed with steel, silver-white, metallic *vanadium* increases steel's hardening ability while restraining grain growth. A steel with less than 0.05% vanadium will weld fairly easily.

Vanadium acts as a powerful deoxidizing agent and forms finely divided particles of vanadium oxide to act as nuclei for crystallizing while the metal solidifies. This creates a fine-grained structure. Vanadium alloy steel will resist softening by tempering. Vanadium over 0.15% in steel acts almost like molybdenum under the spark test. The spearheads are similar, but, unlike molybdenum, they are well-defined and attached to the carrier line ends. High-speed steel, also known as high-speed tool steel and self-hardening steel, contains vanadium as the carbide-former.

Zinc

Used as a coating for iron and steel, *zinc* is a brittle, hard, bluish-white metal. Zinc coating is similar to tin coating. The sheet metal is dipped in molten zinc and, like tinned sheets, acts to protect against rust. As the zinc cools, it adheres to the iron sheets and forms crystals to give a spotted bluish-white design. This process is known as *galvanizing;* welding on galvanized metal will be described under oxy-acetylene welding.

Zirconium

This is a metallic element that is prepared as a black, uncrystallized powder; as steel-gray shiny scales resembling graphite; or as a crystalline scale-like form resembling antimony. When added to molten steel, *zirconium* acts as a strong deoxidizer, removing nonmetallic inclusions as well as oxygen and nitrogen. Steel containing 0.10% zirconium will have an easily welded, fine-grained structure.

Principal Industrial Nonferrous Alloys

Typical industrial, nonferrous alloys, as listed in Table 3-1, can be studied like the preceding steel-alloying elements. The change in ductility, hardness, and strength of most nonferrous alloys cannot be brought about by heat treating. Cold-working through rolling or hammering, to cause permanent deformation of metals below their recrystallization temperature, will harden most nonferrous materials and their alloys. Alloying some nonferrous metals can also increase the hardness and tensile strength of the new material. Like steel, aluminum bronze, containing about 90% copper and 10% aluminum, can be heat-treated satisfactorily. Other nonferrous alloys, such as beryllium-copper, magnesium-

zinc, and alloys of titanium, can also be hardened by heat treating.

Copper

Tough pitch *copper,* which contains about 0.03% oxygen, forms the purest commercial grades of copper, known as *electrolytic.* This type of copper produces the best castings because it is 99.9% pure.

Using phosphorous during copper production eliminates copper oxides by deoxidizing and produces what is known as *oxygen-free copper.* Arsenical copper, containing 0.50% arsenic, can be cold worked better than pure copper, providing greater hardness and tensile strength. Copper, when alloyed with such elements as tin, silicon, lead, zinc, and nickel, produces useful industrial materials. Beryllium and copper can be alloyed to produce materials with high strength, as well as parts with surface hardness. Tellarium copper, 99.49% pure, is shown in Fig. 3-16.

TELLARIUM COPPER

CU	99.49%
TE	.50%
P	.01%

Fig. 3-16. Tellarium copper photographed through a microscope. (Courtesy of Bridgeport Brass.)

Brass

An alloy made essentially of copper and zinc, *brass* is readily machinable and has a wide range of physical properties, depending on the ratio of the added alloying metals. Corrosive agents can attack stress cracks in

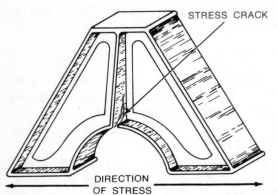

Fig. 3-17. Corrosion attacking a stress crack in a brass supporting bracket.

castings, as shown in Fig. 3-17. This condition is known as *season cracking,* and usually occurs when the brass is over 15% zinc.

Added to brass, lead can improve machinability but creates a loss in ductility and strength. Fig. 3-18 shows a leaded brass. Al-

PHOTOMICROGRAPH OF LEADED BRASS

CU	61.25%
ZN	35.35%
PB	3.40%

Fig. 3-18. Leaded brass, taken through a microscope. The black, rounded particles are lead. (Courtesy of Bridgeport Brass.)

though brass resists most corrosive agents, not over 3% aluminum added to the brass will improve corrosion resistance. The increased aluminum content will reduce ductility and make a stronger, but more porous, material.

Alpha Brasses—Those brasses containing up to 36% zinc, which are best hardened by cold working, are known as *alpha brasses.*

Hardened alpha brass can also be annealed by heating. Annealed alpha brass appears in Fig. 3-19.

Fig. 3-19. Typical annealed alpha brass, magnified 250 times.

An alloy of 85% copper and 15% zinc is known as *red brass*. This alloy is stronger than pure copper and resists season cracking. Brass pipe is made from the alpha brasses containing the highest zinc content, usually 35%.

Alpha-Beta Brasses—Brasses having a higher zinc content than common brasses are known as *alpha-beta brasses*. The zinc content could be as much as 46%. These brasses have greater strength but less ductility than the alpha brasses. When the ratio of copper to zinc reaches 60/40, a strong, low-ductile metal known as *muntz* is formed. When 0.5% manganese is added to the alpha-beta brasses and the copper content ranges from 50 to 60%, a high-tensile-strength alloy known as manganese bronze is formed, suitable for casting. Manganese bronze, with just over 1% iron, is shown in Fig. 3-20.

Bronze

A reddish-brown alloy, *bronze* is made of copper and tin. It is more costly, stronger, and more ductile than brass. Bronze usually has relatively high percentages of tin and lead,

PHOTOMICROGRAPH OF MANGANESE BRONZE

CU	59.00%
ZN	38.85%
FE	1.10%
SN	.75%
MN	.30%

Fig. 3-20. Manganese bronze photographed through a microscope. Magnified 200 times. (Courtesy of Bridgeport Brass.)

PHOTOMICROGRAPH OF PHOSPHOR BRONZE

CU	95.40%
SN	4.45%
P	.15%

Fig. 3-21. Phosphor bronze, magnified 200 times. This sample is about 95% copper and 4% tin. (Courtesy of Bridgeport Brass.)

with or without small amounts of zinc and other metals such as nickel.

Copper-tin bronzes containing less than 12% tin are very ductile, and if the tin content is kept below 6%, the bronze can be worked hot. Bronzes containing more than 6% tin are very resistant to corrosion and are high in tensile strength, but they can only be worked cold.

Alloys consisting of copper and tin are known as *tin bronze;* when phosphorous is added to deoxidize the alloy, it becomes known commercially as *phosphor bronze,* as seen in Fig. 3-21. This type of bronze can be made with high tensile strength by adding more tin. Phosphorus, added in quantities up to 0.35%, increases the hardness and wear-resistance of the bronze. Phosphor bronze can be produced to resist season cracking, fatigue, and corrosion. Adding lead increases the metal's lubricating qualities and plasticity. These qualities are important for its commercial use in manufacturing bearings (Fig. 3-22). These bronze bearings are sometimes called *bushings.* Lead and zinc will lower the strength of the phosphor bronze, but they make it easier to machine after casting.

Aluminum Bronze—High copper alloys of 4 to 10% aluminum, with lower percentage additions of nickel, managanese, iron and silicon, make up what is known as *aluminum bronze.*

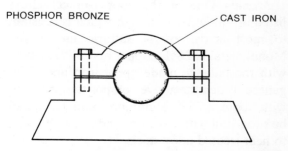

PHOSPHOR BRONZE   CAST IRON

Fig. 3-22. Phosphor bronze bearing, sometimes called a *bushing.*

This has higher wear and corrosion resistance than copper-tin bronzes. Adding nickel increases the tensile strength, making *aluminum-bronze* a commercial casting material.

Silicon Bronze—Copper-silicon alloys are easily welded and consist of iron, manganese, tin, and zinc with 1.6% to 3% silicon. These copper-silicon alloys, known as *silicon bronze,* can be machined more easily by adding lead. Everdur® is the trade name for silicon bronze composed of 1% manganese, 3% silicon, and 96% copper.

Beryllium Bronze—Copper with as much as 2.75% beryllium has only moderate tensile strength when not hardened, and is known as *beryllium bronze*. By heat treating, beryllium bronze can be hardened to such a high degree that tools capable of cutting (comparable to steel cutting tools) can be manufactured satisfactorily from it. Resilience and resistance to corrosion and fatigue make this material a valuable commercial product.

Other Copper-Nickel Alloys

Cupro-nickel is an alloy of 70% copper and 30% nickel. This *copper-nickel* alloy has high ductility and strength, but it is unsuitable for casting.

Another well-known copper-nickel alloy is *nickel silver,* or German silver, which has a dense, fine-grained structure and is highly corrosion resistant. A typical composition of nickel silver consists of 66% copper, with the balance equally divided between zinc and nickel.

Monel—One of the copper-nickel alloys listed in Table 3-1 is *monel*, which is resistant to most acids, except sulfuric and nitric. Monel metal is 28% copper and 67% nickel, with the balance made up of iron and manganese. When the above composition is mixed with about 2.75% aluminum, a material can be produced with hardness and strength equal to heat-treated alloy steels.

Inconel—This is an alloy of chromium, nickel, iron, and other elements. Annealed *inconel* has a higher tensile strength than ordinary steel. By cold working, inconel can be produced with 185,000 psi tensile strength.

Solder

A fusible alloy used for joining metallic surfaces, *solder* has been used for some time. It is applied in a molten state, as shown in Fig. 3-23. The most effective solder is made by alloying lead and tin in the percentages listed in Table 3-1. Because of its plasticity

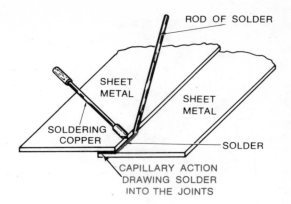

Fig. 3-23. Soldering sheet metal lap joints.

and working qualities, this solder produces the best joints in metal.

Fig. 3-24. Microphoto of a cast aluminum alloy magnified 500 times.

Aluminum Alloys

Alloying elements are added to *aluminum* to increase the metal's strength. Aluminum alloys, Fig. 3-24, can be classified as *wrought* or *casting* alloys, and consist mainly of one or more of these elements: magnesium, copper, manganese, nickel, silicon. The corrosion-resistance of pure aluminum is somewhat lessened by adding alloying elements.

Wrought Alloys—These are available in various forms and shapes such as bars, rods, sheets, and pipe. Some high-strength *wrought alloy* sheets are coated with pure aluminum or other materials that resist corrosion. These coated sheets are known as clad sheets. Further information will be given to the *clad metal process* at the end of the chapter. Wrought alloys, which cannot be hardened by heat treating, will respond somewhat to cold working.

Duralumin—A wrought alloy of aluminum, which can be hardened by heating and normal quenching, is known as *duralumin*. Duralumin consists of magnesium and manganese at 0.5% each, 4% copper, and the balance of aluminum. It can be heat-treated to a tensile strength of 55,000 psi.

Casting alloys are produced by alloying silicon and other elements with copper to insure not only good casting qualities, but good machinability. Magnesium-aluminum alloys have better physical properties than either aluminum-copper or aluminum-silicon alloys. Alloys with high strength, hardness, and machinability can be produced when about 10% magnesium is added.

Higher mechanical properties are obtained by adding chromium, but the new material creates welding problems. These, and their solutions, will be explained in another chapter.

Some of the aluminum alloys (with their commercial designation) are listed in Table 3-3. These alloys will be referred to in a later chapter on the weldability of metals.

Table 3-3. Aluminum Alloys and Their Commercial Designations.

Aluminum Alloy	Commercial Designation	
Aluminum-manganese	3 S	Wrought alloy
Aluminum-magnesium-chromium	52 S	Wrought alloy
Aluminum-silicon-magnesium	53 S	Wrought alloy
	61 S	
	63 S	
Aluminum-copper-magnesium-manganese	14 S	Wrought alloy
	17 S	
	24 S	
	A17 S	

Magnesium Alloys

The alloying elements commonly used with *magnesium* are aluminum, zinc, and manganese. Alloys of magnesium that are susceptible to heat treatment contain over 6% aluminum. Aluminum content between 3.5% and 10% increases both the strength and the hardness of the alloy. Adding zinc and manganese increases the alloy's corrosion-resistance. Wrought and cast magnesium alloys are available and can be heat treated by methods similar to those used with aluminum.

Various magnesium alloys have a common designation system. This form of describing alloys is based on chemical compositions, and has been used to eliminate any confusion with other metals. Usually, the first letter of the main alloying elements (the element having the highest percentage) is listed first. Only the two main alloys are designated by letters, followed by each alloy's percentage, rounded off to a whole number. A commercial magnesium alloy consisting of 3% aluminum, 1% zinc, and 0.3% manganese would be designated AZ31. Welders will find this information helpful when welding magnesium and magnesium alloys, to be studied later.

Clad Metals

Methods other than alloying are being used to protect metal against corrosion. One of these is applying a protective coating of either metallic or nonmetallic material.

In the cladding process, the base metal is placed between the coating material and then rolled in sheets, producing what is known as a *clad metal,* shown in Fig. 3-25. Duralumin, an aluminum alloy coated with pure aluminum, is known as *alclad.*

Metallic coatings are produced on base metal by electro-deposition, as shown in Fig. 3-26.

The *hot-dipping technique* (discussed with galvanizing) uses a bath of molten coating metal. In *spraying,* a coating metal is sprayed in fine particles on the prepared surface of the work by compressed air. Spraying is, therefore, both flexible and portable. Zinc and aluminum have been used satisfactorily

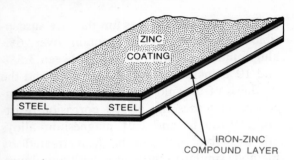

Fig. 3-25. Galvanized steel with two iron-zinc compound layers and an outer coating of zinc.

as coating materials for spraying bridges and oil refinery equipment. See Fig. 3-27.

When a metal of iron or steel is heated to a point below its melting temperature in a powder of zinc, the powdered zinc permeates the surface of the iron or steel to form a coating. This procedure is known as *sherardising,* as shown in Fig. 3-28. When chromium or aluminum powder is used during

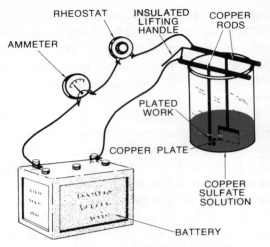

Fig. 3-26. Electrodeposition.

this process, the process is referred to as *chromising* or *calorising,* respectively.

Nonmetallic coatings (such as porcelain enamel, a type of opaque glass fused on steel

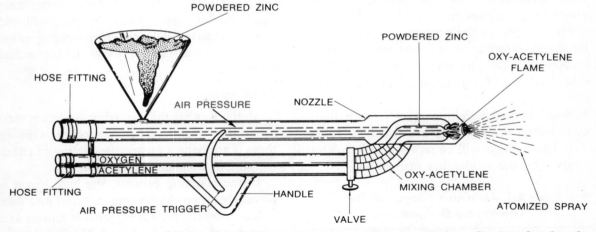

Fig. 3-27. Portable spray gun for spraying galvanized coatings. The oxy-acetylene flame melts the zinc being blown by, blowing molten zinc from the nozzle.

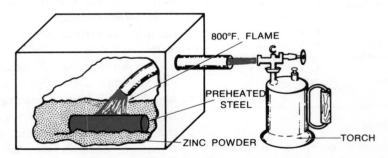

Fig. 3-28. Sherardising—zinc coating of steel.

or iron at red heat,) are another form of corrosion-resistant material. Other nonmetallic materials for coating are paints and enamels.

Welding various types of clad metals will be covered in a later chapter, along with welding dissimilar metals.

Regardless of the method of mixing metals, alloys can be distinguished by their general characteristics. Alloys are generally harder and usually melt at a lower temperature than their principal constituents. They are, however, less malleable.

A welder will better understand alloys if he groups them according to their major constituents. For example, some important groups of iron alloys are cast iron, cast steel, carbon steel, low alloy-high strength steel, tool steel, die steel, and alloy steels.

Alloy steels, such as stainless steel, will be fully discussed later. The two most common groups are *ferrous alloys* and *nonferrous* alloys. Iron is the basic constituent of any *ferrous* alloy; any metal *except* iron is the basic constituent of any *nonferrous* alloy.

4

Welding Heat Effects

Most welders will have problems such as distortion, expansion, and contraction, created by heat during welding. Practice, knowledge, and experience with such problems will help eliminate some of the unsatisfactory conditions caused by heat.

Heat Transfer

Heat is defined as the *energy* of *atom* or *molecule* motion in *solids*. Heat is capable of being transmitted through solids and fluids by *conduction,* through fluids by *convection,* and through empty space by *radiation.* These are the three types of *heat transfer.*

Radiation

Transmitting energy in the form of rays or particles is known as *radiation.* Light and heat rays, in the form of visible, infra-red and ultra-violet rays, radiate light and heat from the welding arc. Fig. 4-1 illustrates heat transfer by radiation.

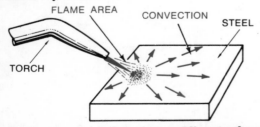

Fig. 4-1. Heat rays from the welding torch and from the metal show heat transfer through *radiation.*

Conduction

Heat *conductivity* is transmitting heat from one part of an object to another part (Fig. 4-2). The rate of conductivity varies with different materials and is known as the *thermal conductivity rate.*

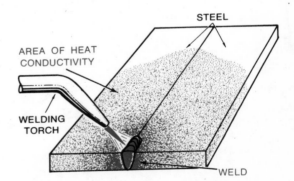

Fig. 4-2. Areas of heat *conductivity.*

A welder must have some knowledge of heat conductivity so he can adapt methods to allow for expansion and contraction while welding. For example, the higher a metal's thermal conductivity, the larger the volume of heat required to allow for the heat loss taking place in the surrounding areas, including the atmosphere.

Although aluminum melts at a lower temperature than mild steel, it is necessary to preheat the aluminum (especially in large size plates), because its thermal conductivity is higher than that of steel.

Thermal Conductivity is, therefore: the heat-conductive power of a substance where the quantity of heat flows through a unit area of a unit thickness of the material in a unit time and has a unit difference in temperature.

The British thermal unit (Btu) is the quantity of heat required to raise the temperature of one pound of pure water one degree F. The Btu is the standard of comparison for conduction and general heat transfer.

Convection

This is heat transfer by an automatic circulation that happens at a nonuniform temperature. It occurs because of variations in density and the action of gravity; together they are referred to as *convection* (Fig. 4-3).

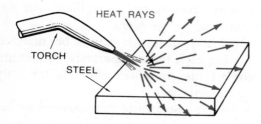

Fig. 4-3. The area beyond the heated steel plate has been warmed by *convection*.

The welder must realize that transferred heat from one body to another can affect his welding.

Expansion and Contraction

When a metal is heated and its dimensions increased, the welding term used is *expansion* (Fig. 4-4). When the metal cools, and has resumed its original size or possibly shrunk, *contraction* has taken place. A rise in temperature determines *expansion*. Conversely, *contraction* is determined by the amount of temperature fall. Variations in the temperature of the parent metal create expansion and contraction problems when welding. Some metals with low ductility, such

as cast iron, could crack or fracture beyond, as well as within, the weld area. When welding these metals, then, a welder has to account for expansion and contraction in the whole piece he is welding.

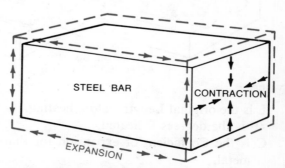

Fig. 4-4. The solid line illustrates the metal bar before heating and after complete cooling. The broken line represents the bar's expansion due to heating.

Table 4-1. Coefficients of Thermal Expansion.

Metal	Coefficient
Aluminum	0.00001234
Copper	0.00000887
Iron, Wrought	0.00000648
Iron, Cast	0.00000556
Nickel	0.00000750
Steel	0.00000636
Zinc	0.00001407

How to Figure Expansion and Contraction

In Table 4-1, the coefficients of expansion for several metals are listed. This coefficient means that if 1 inch of the metal is raised 1° F, the metal will expand that much of an inch.

For example, the coefficient of expansion for cast iron is 0.00000556. This means that if 1 inch square of cast iron (open only on the ends) was heated 1° F, it would become 0.00000556 inch longer.

If there was over 1 inch of material, or if the material was heated over 1 degree, or both, the formula for the new size material would be:

$$L \times °F \times C = N$$

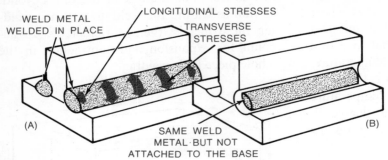

Fig. 4-5. Weld metal shrinkage causing longitudinal and transverse *stress.*

where,

L is the original Length *before* heating,
°F is the degrees F heated,
C is the Coefficient of expansion for that metal,
N is the New length *after* heating.

If 12 inches of *aluminum* was heated 1000° F, the problem would be solved like this:
12 × 1000 × 0.00001234=0.148″ longer.

During welding, some metal shrinkage may be created through heating and cooling. Shrinkage and other problems related to expansion and contraction can be corrected or eliminated. The following information and suggestions, plus practice, will help the welder be able to control these problems.

Analyzing Distortion

Understanding distortion is a complex problem. The weld area is the point of major concern. The enormous temperature difference develops a difficult heat flow problem, created by a nonuniform heat distribution in the part and by heat's ability to change the physical properties of the material. As the temperature increases, so does the specific heat and coefficient of thermal expansion, however, the yield strength and thermal conductivity decrease. Virtually every property of the metal changes.

A consideration of distortion must also include the effect of *restraints*. Restraint from external clamping, internal restraint due to mass, and the stiffness of the steel plate itself must be taken into account. All these factors definitely influence the amount of movement.

Finally, it is necessary to consider *time* as it affects the rapidly changing conditions. The period of time when a specific condition is in effect controls the importance of that particular condition.

These variable conditions are further influenced by the welding process itself. Different procedures, type and size of welding equipment, speed of travel, joint design, preheating and cooling rates—all these bear significantly on the problem.

Obviously, distortion cannot be analyzed by viewing each factor separately. A solution to correct the *combined* effect is the only workable approach.

Shrinkage—Structural fabrication may develop such conditions as *longitudinal shrinkage* and *transverse shrinkage* (Fig. 4-5). These, in turn, can create what is known as bowing, or cambering, as well as angular distortion. When distortion is a factor, it appears

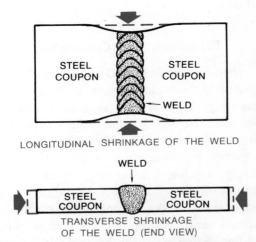

Fig. 4-6. Weld shrinkage due to *distortion.*

as a shortening of the weld area, known as longitudinal or transverse shrinkage (Fig. 4-6). When the transverse shrinkage is not uniform throughout the weld's thickness, angular distortion will result, as in Fig. 4-7.

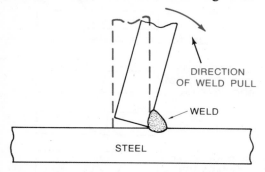

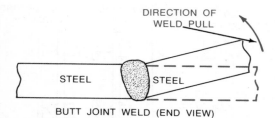

Fig. 4-7. Angular distortion caused by a weld pull effect while heating.

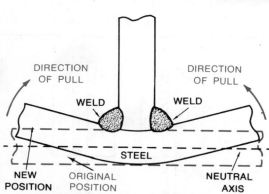

Fig. 4-9. The pulling effect of welds that are not on the neutral axis line. (Courtesy of Lincoln Electric Co.)

Before considering the effects of longitudinal shrinkage on welding, the welder should try to visualize the neutral axis (center) of a structural member as an imaginary line. The shrinkage force around that line acts as in Fig. 4-8. When longitudinal shrinkage acts in a direction that is *not* along the neutral axis of the member, the result is bowing or cambering, as shown in Fig. 4-9.

Shrinkage of the weld metal alone is not the only shrinkage actually encountered. As the weld metal solidifies and fuses with the base metal, it is in its *maximum expanded state,* actually occupying the greatest volume it can occupy as a solid. When it cools, it tries to contract (shrink) to the volume it would normally occupy at the lower temperature, but it cannot because of the adjacent base metal. Stresses develop within the weld, finally reaching the yield strength of the weld metal. At this point, the weld "stretches," or yields, and thins out, adjusting to the volume requirements of the joint being welded. However, only those stresses that *exceed* the yield strength of the weld metal are relieved by this stretching. When the weld reaches room

Fig. 4-8. Shrinkage force acts around the imaginary dotted line.

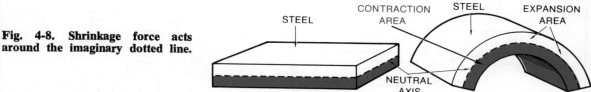

temperature and clamping restraints are removed, the locked-in stresses are partly relieved by causing the base metal to move. This movement deforms or distorts the weld.

Another way of looking at the weld's internal stresses is found in Fig. 4-5. Fig. 4-5(A) shows two heavy steel plates joined by fillet welds. Residual (leftover) longitudinal and transverse stresses are shown as arrows in the fillet welds. To picture *how* these stresses got into the welds, imagine the situation seen in the same figure at (B). Here, the fillets have been separated from the base

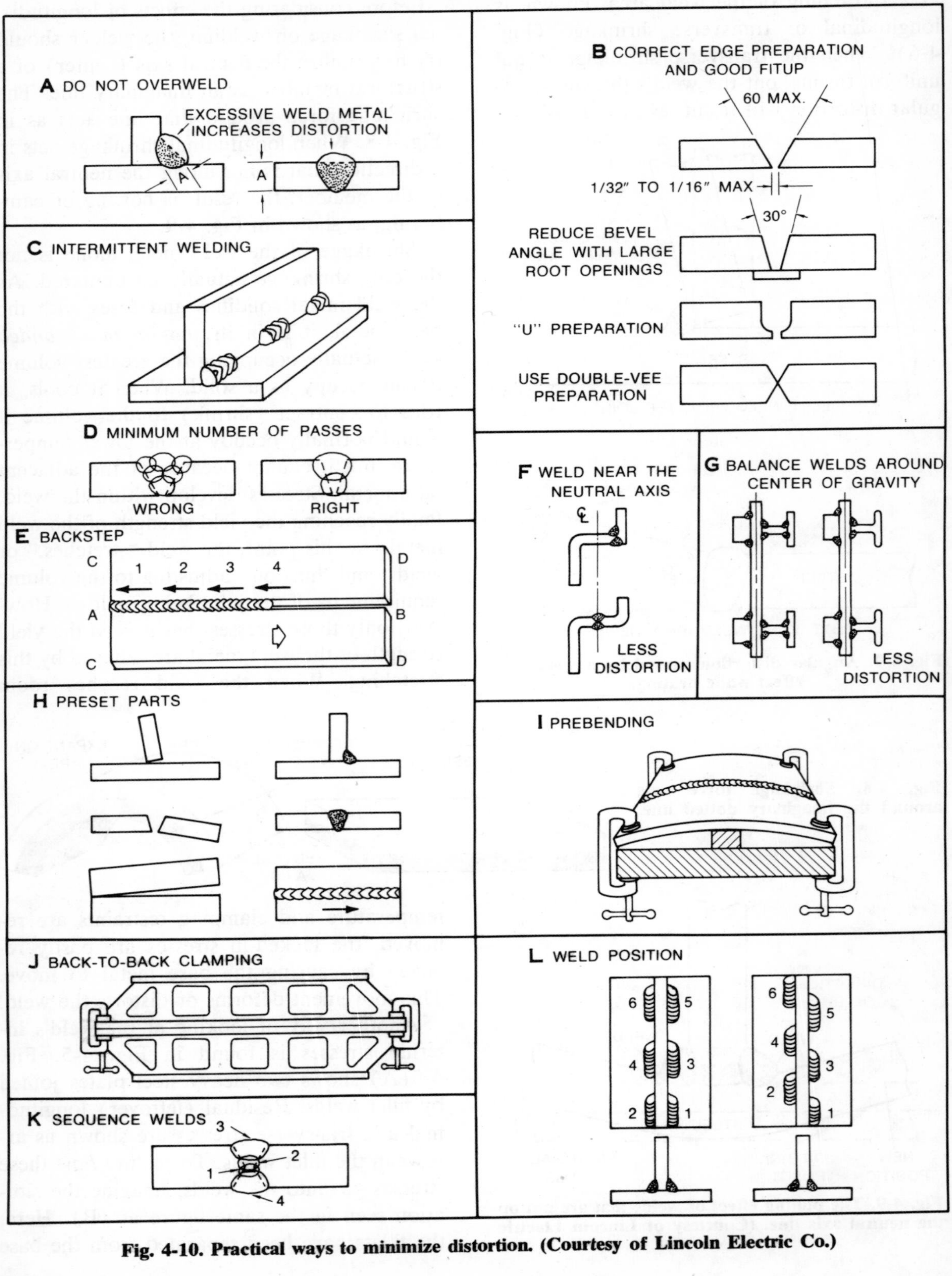

A DO NOT OVERWELD

EXCESSIVE WELD METAL INCREASES DISTORTION

B CORRECT EDGE PREPARATION AND GOOD FITUP

60 MAX.

1/32" TO 1/16" MAX

30°

REDUCE BEVEL ANGLE WITH LARGE ROOT OPENINGS

"U" PREPARATION

USE DOUBLE-VEE PREPARATION

C INTERMITTENT WELDING

D MINIMUM NUMBER OF PASSES

WRONG   RIGHT

E BACKSTEP

C   1   2   3   4   D

A   B

C   D

F WELD NEAR THE NEUTRAL AXIS

LESS DISTORTION

G BALANCE WELDS AROUND CENTER OF GRAVITY

LESS DISTORTION

H PRESET PARTS

I PREBENDING

J BACK-TO-BACK CLAMPING

L WELD POSITION

6   5

4   3

2   1

6   5

4   3

2   1

K SEQUENCE WELDS

3

1   2

4

Fig. 4-10. Practical ways to minimize distortion. (Courtesy of Lincoln Electric Co.)

plates. The same amount of weld metal exists. In its *unattached* condition, the weld metal has shrunk to the volume it would normally occupy at room temperature. It is under no restraint and is stress-free. Because it has no restraints or stresses, it looks much smaller. It is not stretched as in (A).

To get this unattached weld back to the condition in Fig. 4-5(A), it would be necessary to pull it lengthwise, or impose *longitudinal* forces. It would also be necessary to stretch it transversely, or impose *transverse* forces. These forces would *stretch* the weld. The weld metal must "give," or yield, in order to stretch. However, at the time it reaches the *needed* dimensions, it will *still* be under stress equivalent to its yield strength. This residual (leftover) stress will try to deform the weld. In this specific case, as Fig. 4-5(A), it is unlikely that the plates would be deformed much because they are very stiff, compared to the weld size. When the first fillet is laid, however, unless the plates are rigidly clamped, angular distortion may happen.

Preventing Expansion and Contraction

When distortion, also known as *warpage,* is created through expansion and contraction in ductile metals, it can be eased by one or more of the methods in Fig. 4-10.

Prebending and Presetting

One way to control weld shrinkage involves prebending the member, or presetting the joint before welding. In this way, the heat of weld shrinkage pulls the member or connection back into proper alignment.

By looking at point A in Fig. 4-11, a welder can see that a ductile material, such as an iron shaft, will bend when heated, with the point of applied heat forming the bottom of the bend. To overcome this problem, proper preheating before welding should be considered. An experienced welder can cure the problem in many cases by prebending and/or presetting before welding.

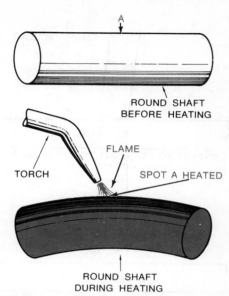

Fig. 4-11. What happens when point A on a cold, straight shaft is heated.

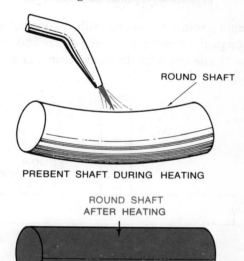

Fig. 4-12. How prebending produces a straight shaft *after* heating.

Therefore, prebending before welding, as in Fig. 4-12, and presetting, as in Fig. 4-13, will help overcome dimensional defects due to distortion and warpage caused by welding heat.

Sequence of Bead Placement

When equal welds are placed on equal sides of the neutral axis of a member (part of a

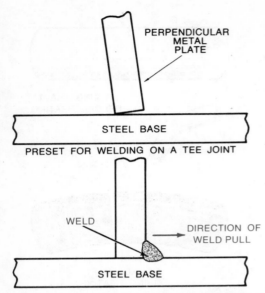

Fig. 4-13. How presetting uses the weld pull to produce a perpendicular welded rod.

weld), some distortion still happens, even though the heating pulls are equal and opposite. Some upset in the compressive area next

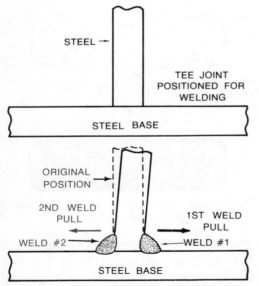

Fig. 4-14. Weld #2 does not completely offset the distortion first created by weld #1, due to the offset of weld #1.

to the weld area occurs after the first weld is made. Because of this upset, the initial distortion from the first weld is not quite offset

by the second weld on the opposite side, as shown in Fig. 4-14.

Where multiple-pass welding is involved, this condition can be corrected, as illustrated in the butt-weld procedure sequence of Fig. 4-15. Here, the first pass is on the top side.

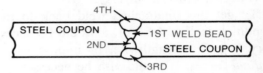

Fig. 4-15. How multiple-pass welding can help offset distortion due to upset.

The second pass is welded on the opposite side. This pass, however, will *not* pull the plates completely back into flat alignment; therefore, the third bead pass is added to the same side. The net result will usually pull the plate slightly beyond the flat position. The fourth pass, then applied to the top side, should bring the plate back into flat alignment.

Frequently, this problem is not of major importance, since the sections to be welded are usually large enough, compared with the size of the weld, to prevent the occurrence of

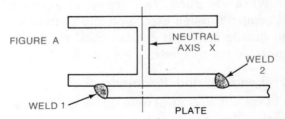

(A) Weld 1 is done first because it is closer to the neutral axis.

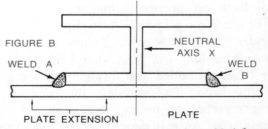

(B) Either weld 1 or weld 2 can be welded first.

Fig. 4-16. Welds not balanced around the neutral axis.

this upsetting. As a result, on large sections, the second weld on the opposite side is just as effective as the first weld.

A welder should also be familiar with welds which would not be symetrically balanced about the neutral axis. It may be possible to take advantage of this difference in distortion by first completing the joint *nearest* the neutral axis, followed by welding the joint on the side *farthest* from the neutral axis (Fig. 4-16).

The I-beam welded to a flat plate in Fig. 4-16A, has welds that are not symmetrical. For this reason, weld 1 is made first, followed by weld 2. Since the I-beam in Fig. 4-16B has the iron plate extending slightly on the left, both welds can be made at the same time.

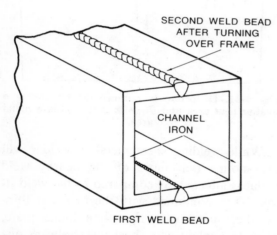

Fig. 4-17. Two channel irons welded together to make a box-type frame.

Box type frames (Fig. 4-17) made from light gauge material such as channel iron can also create distortion problems. Waiting until after the first weld has cooled before making the second weld on the opposite side will usually result in some final bowing, since the second weld cannot pull the member completely straight.

Heating the top side of the member by the first weld causes some expansion and bowing upwards. See Fig. 4-18A. Turning the member over quickly while it is still hot and

bowed and depositing the second weld will help the pulling effect of this less effective weld deposit, Fig. 4-18B. The member will then usually be straighter after cooling to room temperature, as in Fig. 4-18C.

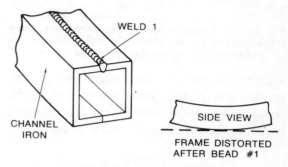

(A) Distortion after the first weld.

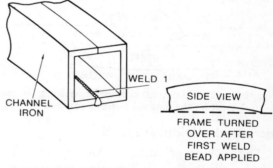

(B) Frame turned over for the second weld.

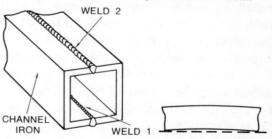

(C) After the second weld, the frame is still slightly bent from the first weld.

Fig. 4-18. Distortion created by welding channel iron together to make box frames.

Proper Welding Process

To reduce shrinkage force in Tee welds or fillet welds, the welder should prepare the joints with a U or double V preparation. Fig. 4-19 illustrates groove Tee welds or fillet welds made with the submerged arc automatic

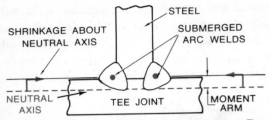

Fig. 4-19. Tee joint welded by submerged-arc. Deep penetration of the arc places the weld closer to the neutral axis and reduces the moment arm.

welding process. The deep penetration of this process lowers the center of gravity of the welds and reduces the moment arm, thereby reducing the shrinkage moment (Fig. 4-20).

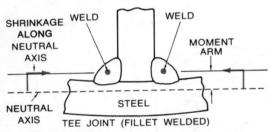

Fig. 4-20. The moment arm after welding a tee joint by the manual welding process.

Increased Welding Speed

The volume of adjacent base metal (Fig. 4-21) contributing to distortion can be controlled by increased welding speeds. Higher welding speeds, using powdered-iron type manual electrodes, semi-automatic and fully automatic submerged-arc welding equipment, or vapor-shielded automatic welding equipment, reduce the amount of metal affected by the arc's heat. They also progressively decrease distortion.

Single Rather Than Multiple Passes

A weld made with more passes will have greater distortion than one made with fewer passes. A welder, by using a single-bevel groove joint welded in the flat position, will immediately notice the difference in distortion, once fewer passes are compared with multiple passes. Any transverse shrinkage can be reduced by using fewer passes or larger electrodes.

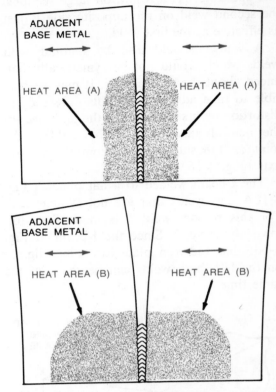

Fig. 4-21. Distortion is reduced (a) by having a smaller heat area and (b) with a larger base metal area.

While welding sheet metal, a welder will also notice that more of the nearby weld metal is affected, as compared to the weld itself. This, combined with the fact that thinner sheet metal is less rigid than thicker plate, helps to explain why sheet metal always distorts easier.

Proper Alignment

Various methods are used for pulling plate edges into alignment and maintaining this alignment during welding. One such method is by using various types of jigs. The most widely used technique, in Fig. 4-22, uses small clips which are welded to the edge of one of the plates. Wedges are then driven between the clips and the second plate, in order to bring both plate edges into alignment. After the plates have been welded together, the wedges are removed and the clips are

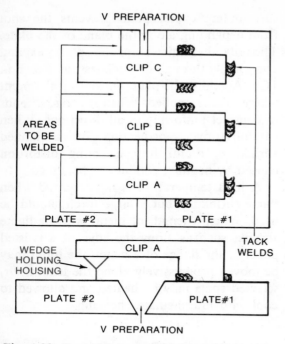

Fig. 4-22. Two plates prepared for welding with a single-Vee preparation. Clips are welded at intervals along the length of plate #1. Wedges are placed in the opening Y, to keep the plates aligned.

knocked off with a hammer. Since the clips can be removed faster by using a hammer rather than a cutting torch, the welder should tack weld the clips just enough to support the wedging procedure.

Bolts welded to the plates (Fig. 4-23) allow pressuring the parts to be welded with a properly designed *strong-back*.

Strong backs can be made from I-beams, channel iron, angle iron, or even railroad iron. The welder should, however, use a

strong back just rigid enough to hold the weldment in the proper position.

Peening

Another method only occasionally used to control distortion is *peening*. Since the weld area contracts, properly applied peening (light hammering) tends to expand it. However, this expansion occurs only near the surface.

Upsetting, or expanding the weld metal by peening, is most effective at higher temperatures where the metal's yield strength is lower.

Unfortunately, most distortion problems occur at lower temperatures, after the yield strength has been restored to its higher value. For this reason, peening does not work as well. Another disadvantage of peening is that it work-hardens the metal's surface.

Flame Shrinkage

Fig. 4-24 shows *flame shrinkage* as a method to correct weld distortion. By directing a gas torch on an area where there is too much metal because of earlier welding, metal displacement in the desired direction occurs, either removing or reducing the distortion. A welder must, of course, be able to recognize the areas of excess metal and must use judgment as to how much heat to apply.

Fig. 4-25 illustrates how controlled expansion and contraction are applied usefully in flame shrinking, often referred to as flame straightening. In Fig. 4-25(A), a gas torch

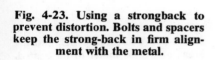

Fig. 4-23. Using a strongback to prevent distortion. Bolts and spacers keep the strong-back in firm alignment with the metal.

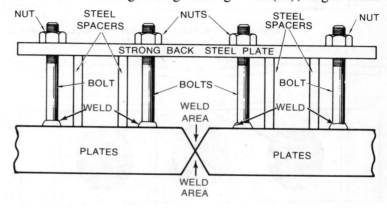

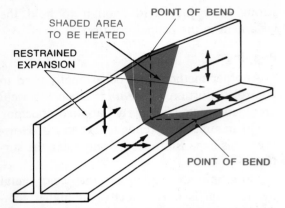

Fig. 4-24. A bent "T" shaped iron marked for flame shrinkage to produce restrained expansion. This will help straighten the "T" iron.

flame is heating a center spot on a distorted steel plate. The spot heats up rapidly and must expand. But, as indicated by (B), the surrounding cooler plate prevents the spot from expanding along the plane of the plate. The only alternative is for the spot to expand abnormally through its thickness, as shown in (C). Actually, the plate *thickens* at the point where heat is applied. When it cools, it tends to contract uniformly in all directions. When carefully done, spot heating gives a planned shrinkage, useful for correcting distortion caused by heating, cooling, or bending.

A high temperature is not required when flame shrinking, but a large torch should be used. When a length of metal is to be flame shrunk, such as along the edge of a twisted panel or the flange of a beam, the torch may be moved progressively along the length. Or, selected spots may be heated and allowed to cool with intermediate checking.

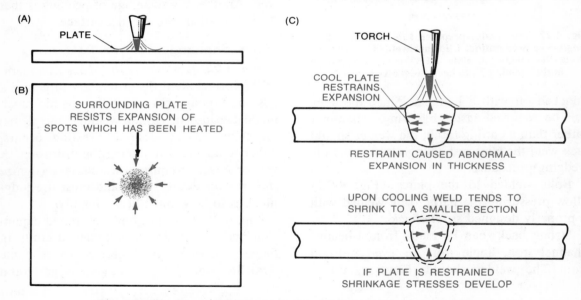

Fig. 4-25. The flame shrinking procedure. (Courtesy of Lincoln Electric Co.)

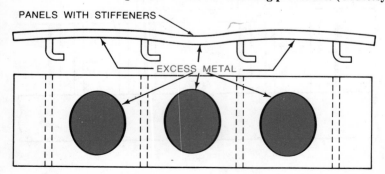

Fig. 4-26. A buckled panel, and the areas of excess metal to be heated. (Courtesy of Lincoln Electric Co.)

A buckled panel, as shown in Fig. 4-26, has too much material in the central area, and flame shrinking will give the shrinkage needed to straighten the panel. If the part of an assembly to be shrunk is restrained, which will be the case in many welds, too much heating could cause the development of locked-in stresses. It is good practice to proceed cautiously when flame shrinking, allowing the metal to cool and checking the results.

To save time, a water spray is sometimes used to speed the cooling. Water is best applied in an atomized spray. This may be accomplished, as shown in Fig. 4-27, by in-

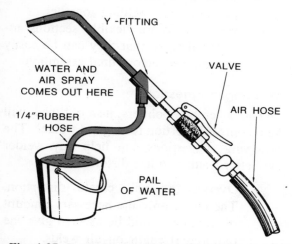

Fig. 4-27. Equipping an air hose to spray an atomized spray for metal shrinkage. (Courtesy of Lincoln Electric Co.)

serting a Y fitting into a compressed air line and running a rubber hose from the fitting to a pail of water. When the valve is opened, the rush of air past the opening in the fitting draws some water into the air stream, creating an atomized spray. When the spray strikes the hot plate, it turns into steam and absorbs a substantial amount of heat. One pound of water will absorb 142 Btus of heat as it is raised from 70° F to 212° F and another 970 Btus in vaporizing into steam. Because of the vaporizing heat required, cooling is rapid, even with only a small amount of water. Since all of the sprayed water is vaporized, the work remains dry.

Suggested Procedures to Cope With Distortion

The following information is presented to help alleviate the distortion problems that welders face during almost all welding.

For instance, angular distortion can be reduced by:

A. Using a Double-Bevel, Vee, J, or U preparation for butt joints. (See Fig. 4-10B.)
B. Using alternating welds from side to side.
C. Beveling the web of a Tee joint to reduce the moment arm of the weld and reduce the angular movement. (Fig. 4-28).

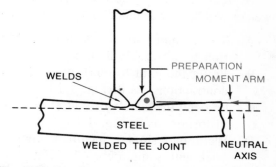

Fig. 4-28. Beveling the web of a Tee joint to reduce the moment arm.

In another case, bending long members by longitudinal welds can be partly controlled in the following ways:

A. Balancing welds about the neutral axis of the member (Fig. 4-29).
B. Making welds of the same size, at the same distance, on the opposite side of the neutral axis of the member.
C. When making welds of different sizes at different distances from the neutral axis of the member, the welds that are further away should be smaller.
D. Where the welding is not *symmetrical* (centered equally around the

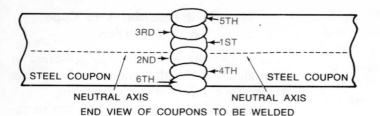

END VIEW OF COUPONS TO BE WELDED

Fig. 4-29. Welding in a staggered order around the neutral axis to reduce distortion.

axis), bending members through welding can be eliminated by:

1. Prebending the member.
2. Breaking the member into sub-assemblies so that each part is welded about its own neutral axis.

Assembly Procedures for Distortion Control

By following these assembly directions, a welder will have less difficulty with distortion:

A. The member should be held in position by a suitable fixture.
B. Preset the joint to offset any expected contractions, as in Fig. 4-10.
C. Prebend the member to offset expected distortion, as in Fig. 4-10.
D. Weld two similar members back-to-back with some prebending, Fig. 4-30.
E. If stress relief is required, weld two similar members back-to-back straight and keep them fastened until after the stress is relieved.

F. Wherever possible, a form of strong-back should be used.
G. Fixtures should be used to insure a proper fit-up and alignment during welding.
H. Use subassemblies and complete the welding on each of them before final assembly and welding together.
I. Weld the more flexible sections together first, so that they can be easily straightened before final assembly.

Welding Suggestions

The assembly pointers just outlined will help control distortion for a good welder. The following suggestions can help any welder get better welds and less distortion:

A. *Over*welding will create distortion. Therefore, only the necessary amount of welding should be used to give the required strength on all welds.
B. A minimum root opening and a minimum included angle should be used whenever possible.

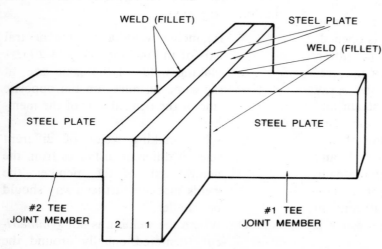

Fig. 4-30. Two similar parts of a weld being welded back-to-back for distortion control.

C. A Double-Vee rather than a Single-Vee joint preparation is preferred. Welding should be applied alternately on both sides, especially for multipass welding.

D. When available, automatic welding is preferred over manual, to take advantage of its deeper penetration.

E. Use shorter and intermittent welds, instead of continuous welds, if the leg sizes of the joint are equal.

F. Use iron-powdered electrodes instead of conventional electrodes, for increased strength.

G. Faster welding speed gives less distortion.

H. Downhand welding should be considered wherever possible, to increase the amount of welding done in a shorter period of time.

I. The first welds should be made on those joints which could cause the most distortion problems.

Carefully Using Expansion and Contraction

In most cases, expansion and contraction are welding *problems*. However, expansion and contraction can be useful for removing broken studs and bolts from castings or housings. This use is covered under upsetting.

Residual stress (internal stress) in a metal due to expansion and contraction may happen when distortion and fracture have *not* taken place. Failure, then, could result from a slight, dynamic stress. Dynamic stress, unlike static stress, is caused by movement usually created under a load. Failure will

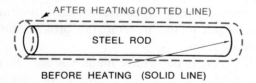

Fig. 4-31. Linear expansion of a steel rod due to heating.

also result where expansion and contraction of metals under extreme temperature changes have been restrained by mechanical means such as clamps or vises. Metal expansion is usually one of two types.

Linear Expansion—A noticeable increase in the *length* of a metal. It is the amount that a unit length of a metal rod expands when it is raised through one degree of temperature. Linear expansion due to heating is shown in Fig. 4-31.

Cubical Expansion—Sometimes called the *volumetric expansion,* where the total *volume*

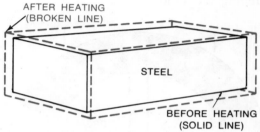

Fig. 4-32. Cubical expansion is where the block expands in all directions, resulting in a net volume increase due to heat.

of the metal is expanded (Fig. 4-32). Generally, three times the linear expansion is a guide to the cubical expansion.

Free Volumetric Expansion and Contraction

When a metal bar is heated, it will have both linear and cubical expansion, according to the previous terminology. If unrestrained, it will expand and contract throughout its volume (the amount depends on the temperature). When cooled, the metal bar will have contracted to its regular size. (See Fig. 4-32.) This type of expansion and contraction is often referred to as *free volumetric.*

Fig. 4-32 illustrates that the bar, before heating and cooling, would take the form shown by the solid lines. On heating, the bar would become larger (the size depending on the temperature) and would take the form shown by the broken lines.

Restrained Expansion and Restrained Contraction

Restrained expansion and contraction creates major problems for an inexperienced welder. When the expansion is restrained, especially in all directions, residual stress is a factor to overcome, especially in more ductile metals. However, if warpage or distortion is prominent, the residual stress is somewhat relieved. In brittle materials, cracking will take place in either the weld or other adjacent areas. Restricted contraction can also cause enough stress to distort areas surrounding the weld.

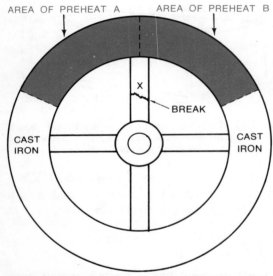

Fig. 4-33. Preheating to avoid stresses. Areas A and B of the cast iron wheel are preheated before welding the break at X. This prevents excessive stress in the wheel.

If the wheel shown in Fig. 4-33 is cast iron, points A and B must be preheated before welding the prepared break at point X. This preheating will compensate for the expansion and contraction that create stresses.

Restrained Expansion and Free Contraction

Fig. 4-34 represents a bar where only one dimension is restrained. The bar, when heated, will expand only on the free dimensional sides. The cubical expansion, in this

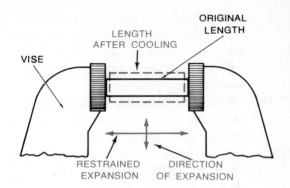

Fig. 4-34. Steel bar held in a vise while heated. Because expansion was restrained while cooling was not restrained, the bar contracts and falls out of the vise when cool.

case, will be equal to the cubical expansion in a free expansion bar of the same size, heated at the same or equal temperatures. Logically, the *free* dimensions of the partially restrained bar will expand a great deal *more* to allow for the restraint on its other dimension.

On cooling, though, *all* dimensions will contract, since contraction is free. The restrained expanded dimension is now smaller than it was originally. Therefore, the other dimensions must now be somewhat larger, since the volume is not changed.

Therefore, the bar in Fig. 4-34, held in the vise and heated, would drop free on cooling, since the restrained expansion dimension was free to contract and is now smaller.

Upsetting

Some of the problems encountered due to heat input, such as expansion and contraction, can be put to effective use with practice and experience. A condition known as *upsetting* is created by heated metal trying to expand in all directions and being restricted by certain areas of colder metal. Upsetting can be put to use in removing broken studs from castings.

If a kerf is cut in the metal coupon, as in Fig. 4-35, and spots are heated at various intervals along the kerf, the heated spots try to expand in all directions. The cooler parts

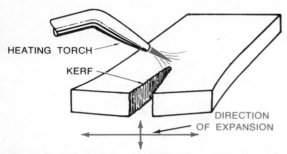

Fig. 4-35. Expansion and distortion by heating a partially cut coupon.

of the metal naturally restrict the expansion and cause the heated portion to expand in the direction of the least resistance.

The metal becomes softer as the temperature increases, and the expansion piling up the hot metal is somewhat relieved. The relieved expansion creates a thickening of the heated area known as the *distortion area*. (Fig. 4-36).

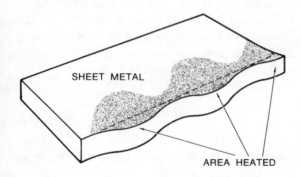

Fig. 4-36. Sheet metal distortion due to cooling.

When metal parts are unevenly heated, some upsetting always occurs. This upsetting is caused by restraining the expansion in one or two directions, at the cost of an expansion increase in the remaining direction.

Removal of Broken Studs or Bolts—Studs, bolts, or other screw-type adapters that have broken off in a casting or housing can be removed by using the free and restraining action of expansion and contraction. See Fig. 4-37.

A nut is welded to the broken stud or screw, as shown in Fig. 4-37. This creates

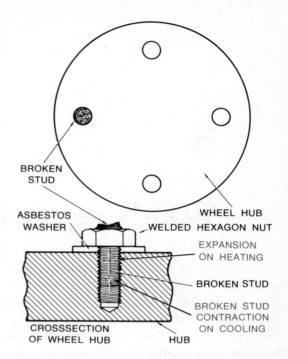

Fig. 4-37. Using the free and restraining action of expansion and contraction to the welder's advantage. Although the stud threads are *restrained* on heating, they are *unrestrained* on cooling. This makes the stud slightly longer and slightly thinner when cool, making it easy to remove.

expansion through the 6000° F temperature of the arc welding process. The internal threaded section restrains the expansion created by the surface of the stud. The *linear* expansion, being unrestrained, expands to a greater degree, to compensate for the overall cubical expansion which would take place *if* total dimensional expansion were free.

When the stud cools, unrestrained volumetric or cubical contraction releases the stud threads from the threaded section of the casting. By turning the nut, the threaded stud can be removed, now that it is longer, but thinner.

Quenching after welding would cause a hard, brittle material to be formed. This could break when it was turned to be removed. For this reason, the heated stud (and welded-on nut) should be allowed to cool normally to room temperature before they are removed.

5

Welding Designs

Welding is the process of heating metallic parts and allowing the metals to flow together, or be forced together, by hammering or compressing after heating. In arc welding, heat is generated at the contact point between the work to be welded and the electrode. The metal parts to be welded and the electrode fuse together in a finished weld. In oxy-acetylene welding, the high tip temperature melts the parent metals, and, with the possible addition of a welding rod, fusion is completed. Several types of joints and types of welds may be used with either of these basic processes.

Types of Joints

Since it is necessary for the welder to prepare materials before welding, it is important to know the different joint styles and the basic, named welds for these joints. The types of joints used during arc or oxy-acetylene welding can be classified as *lap, edge, Tee, butt,* and *corner*, as shown in Fig. 5-1.

Types of Welds

The basic types of welds for the above processes are *bead, groove, fillet, weave* and *plug*. Many variations in the design of the joints and the style of the welds are necessary to keep the weld's cost low and practical.

The welder should be familiar with the following sketches, showing the type of joints and welds and the various preparations that make better welds possible.

Joint Styles

Joint styles may be divided into five basic classifications, outlined as follows: lap, edge, Tee, butt, and corner.

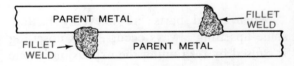

Fig. 5-2. Lap joint welded with double fillets.

(A) LAP JOINT   (B) EDGE JOINT   (C) TEE JOINT   (D) BUTT JOINT   (E) CORNER JOINT

Fig. 5-1. The five basic welding joints.

Lap

The *lap joint* is where two overlapping pieces of metal (Fig. 5-2) are fused together by such welding methods as spot, plug, slot or fillet welding.

Edge

The *edge joint,* shown in Fig. 5-3, is made by placing the surface of one piece to be welded against or on the surface of the piece to be joined, so that the weld fusing the parts

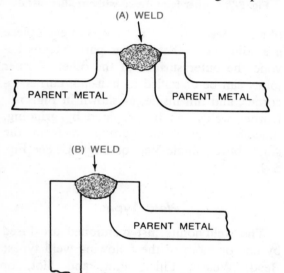

Fig. 5-3. The two basic edge welds. Double flanged edge at (a) and single flanged edge at (b).

is within the outer surface planes of both parts. The two common types of edge joints are the *single-flanged edge* and the *double-flanged edge.*

Tee

The *Tee joint* is welded by placing the edge of one piece on the surface of the other piece, at approximately a 90° angle. This forms a shape similar to the letter "T." The Tee joint can be classified as a plain T, a single-bevel, double-bevel, single-J and double-J, as shown in Figure 5-4.

Butt

In the *butt joint,* the weld is between the surface planes of both fused parts, as shown in Fig. 5-5. The basic butt joint is the square, or plain (Fig. 5-5). The welding operator may, by machining, grinding, or cutting, change the plain joint to a *bevel*-butt joint, a *Vee* joint, a *single-J groove,* or a *single-U groove* (Fig. 5-6). When joints are prepared and welded from both sides, they are classed as double joints.

Corner Joint

In the *corner joint,* the parts are placed at approximately 90° to each other in the form

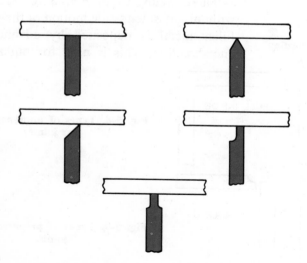

Fig. 5-4. Tee joint weld preparations. (a) Plain Tee, (b) Double bevel, (c) Single bevel, (d) Single-J, (e) Double-J.

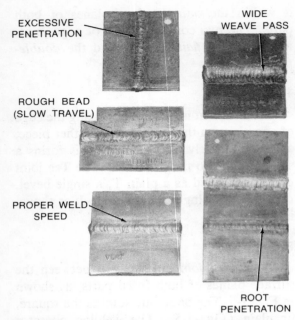

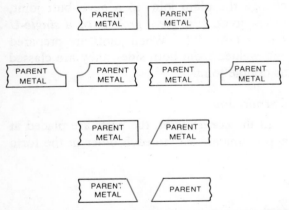

Fig. 5-5. Flat welds made on square butt joints.

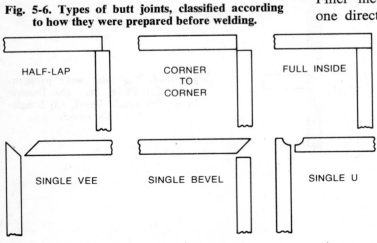

Fig. 5-6. Types of butt joints, classified according to how they were prepared before welding.

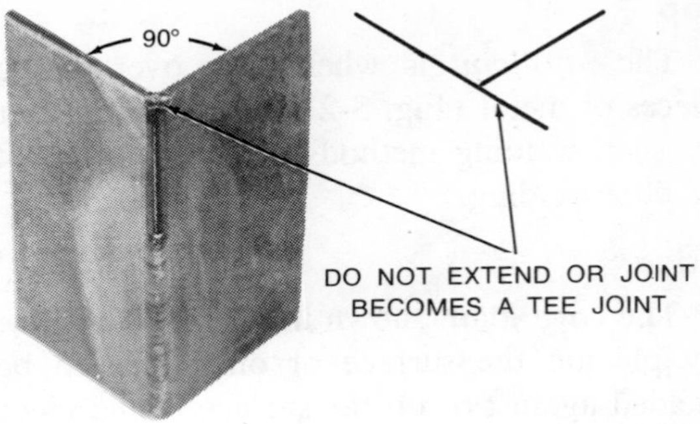

HALF-LAP

CORNER TO CORNER

FULL INSIDE

SINGLE VEE

SINGLE BEVEL

SINGLE U

Fig. 5-8. Types of unprepared corner joints.

Fig. 5-9. Types of prepared corner joints.

Fig. 5-7. Corner joint, fused without filler metal.

of an L. See Fig. 5-7. The edges of each piece are adjusted so that neither part extends beyond the outer surface of the other. Corner joints can be classified as half-lap, corner-to-corner, and full inside, as shown in Fig. 5-8. Corner welds can be prepared by grinding, machining, or flame cutting to form the single-bevel, single-Vee, or single-U. See Fig. 5-9.

Weld Types

The weld joints just mentioned are fused by one or more of the following weld types: Bead, Weave, Fillet, Plug and Slot, or Groove.

Bead Welds

These welds, as in Fig. 5-10, are made without any side-to-side motion of the welder. Filler metal is deposited by advancing in one direction. This is used for building up

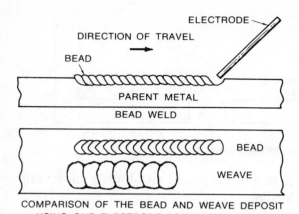

Fig. 5-10. The bead weld is made without the welder moving from side to side.

worn surfaces on round shafts and plates, as shown in Fig. 5-11. *String bead welding,* or welding on material ¾″ thick or more, will be covered under groove welds.

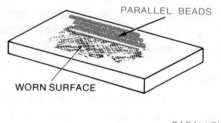

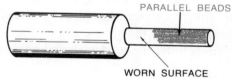

Fig. 5-11. Parallel beads are used to build up a worn surface.

Weave Welds

Running a bead with a side-to-side motion while welding along a given line is *weave welding.* The many variations of weave welding are shown in Fig. 5-12. Enlarged samples of weave welding are shown in Fig. 5-13.

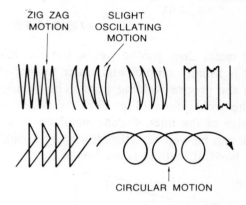

Fig. 5-12. Variations of weave welding.

Differences in the weaving help increase the width of the deposited metal. More metal is deposited at a single pass by weave welding. Welding a Vee groove on heavy materials or building up a pad on flat material makes weave welding more practical than a single bead deposit.

Fillet Welds

Tee, lap and corner joints can all use a *fillet weld* (Fig. 5-14). Fillet welds require more

Fig. 5-13. Practice types of weave welding, enlarged for demonstration purposes only.

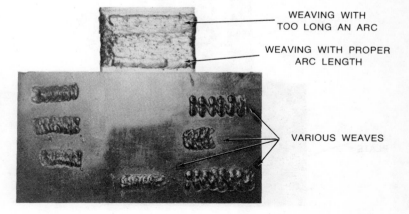

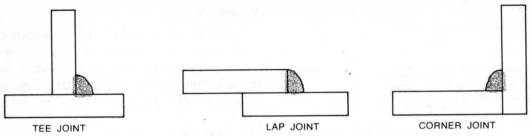

Fig. 5-14. Fillet welds on different joints.

filler metal than groove welds, but the fillet joint is usually simpler and faster to prepare than the groove welding joint. The strength of the weld joints (Fig. 5-14) depends on the size of the fillet. Parent metal up to ½″ thick can be welded satisfactorily by a single fillet weld, if loading is not severe.

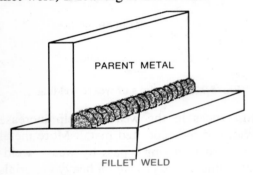

Fig. 5-15. Double fillet weld, used for strength.

Double fillet welding (Fig. 5-15) is used where the strength of the weld must compare to that of the parent metal while under static loading.

Plug and Slot Welds

These welds take the place of rivets. Where two pieces of material have been lapped and fusion must not take place on the edges, a hole or slot, as shown in Fig. 5-16, is made in one of the pieces. It is then placed over the other piece and welded through the opening. This fuses both pieces into one and leaves the outside edges intact. The recommended proportions for plug and slot welds, as shown in Fig. 5-17, will produce strong welds.

Groove Welds

A weld made in the groove between two pieces of metal is known as a *groove weld*. The groove between the pieces to be welded consists of: (1) the groove angle (Fig. 5-18), which includes the total included angle of the

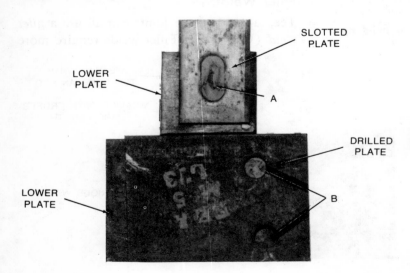

Fig. 5-16. Preparation for plug and slot welds. (a) The slot formed for a slot weld. (b) The holes drilled for a plug weld.

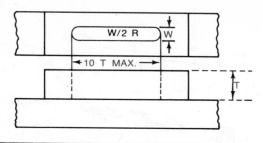

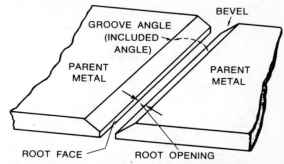

Fig. 5-18. The groove angle is the included angle of the groove.

Dimension T	Dimension W
Under ¼ inch	At least 2 times Dimension T
¼ inch and over	1½ times Dimension T

(A) Slot Weld.

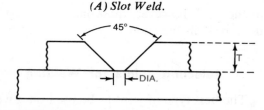

Dimension T	Diameter
Under ⅛ inch	At least ¼ inch
⅛ to ½ inch	At least 2 times Dimension T
over ½ inch	Dimension T + ½ inch

(B) Plug Weld.

Fig. 5-17. Recommended joint proportions for plug and slot welds.

groove, (2) the groove face, and (3) the groove radius.

The types of groove welds are square, single-Vee, single-bevel, single-U, single-J, double-Vee, double-bevel, double-U and double-J. See Fig. 5-19.

Tack Welds

A small weld, shown in Fig. 5-20, used to hold parts in the desired position before starting the weld is called a *tack weld*. Fig. 5-21 shows two plates tack welded before they are actually welded together.

Intermittent Welds

Intermittent welds are beads or weaves placed along a straight line. They are broken intermittently to leave an unwelded part before proceeding with the weld along the welding line or path (Fig. 5-22). Short fillet welds spaced at uniform distances are sometimes preferred to eliminate possible distortion in the parts to be welded because of continuous and extensive welding heat.

Joint Preparation

Most on-the-job preparation is done by oxy-acetylene cutting. Fig. 5-23 shows the carefully controlled oxy-acetylene equipment used to prepare joints. This equipment is called a *radiograph*. In Fig. 5-24, the equipment is in use.

Fig. 5-19. Types of groove welds and intended penetration.

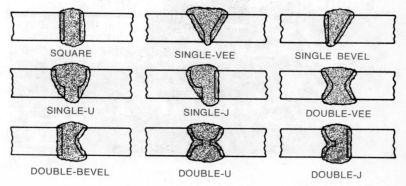

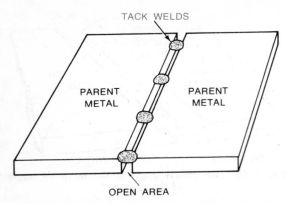

Fig. 5-20. Tack welding. Tack welds are placed between plates to prevent distortion or warping while actually welding.

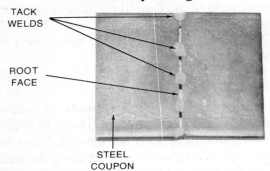

Fig. 5-21. Tack welds made along a single-Vee butt weld, before actually welding.

Grinding, machining, and the arc-air process are other methods used to prepare parts before welding. Unless portable equipment and the power for the equipment is available

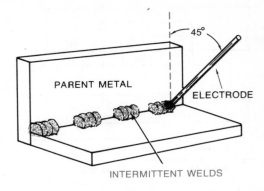

Fig. 5-22. A corner joint with intermittent fillet welds. Intermittent fillet welds are longer and more thorough than simpler tack welds.

in the field, however, these methods must be done only in the shop.

Selecting Joints for Welding

The best type of joint to be used on a weld depends on factors other than the shape of the finished product. Two factors to be considered include the *load intensity* and *cost* of the finished weld.

Load Intensity

This must always be considered before deciding on which joint and equipment to use. The *load intensity* is considered part of one or more of the following four factors.

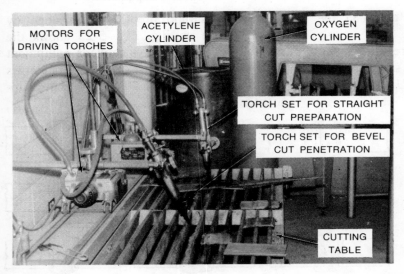

Fig. 5-23. Radiograph. The oxyacetylene cutting equipment used to prepare joints for welding.

Fig. 5-24. A Radiograph being used to prepare metal for welding.

Impact Stress—Various types of pressures, including pounding, bring about *impact stress*. Shock, created by certain applied loads, can cause failure in welds if the joint and welding method are poorly chosen. Welds made on structures subject to severe shock must resist such impact. Proper joint preparation is very important in such cases.

Compression and Tension—The effect of pressures created by applied loads or other stress factors on various kinds of welds can create either *compression* or *tension,* or a combination of both (Fig. 5-25). Preparing joints must be done carefully in such cases to offset any weld failure from compression or tension.

Fatigue—Materials under continuous stress from vibration fail more quickly and under less load, than from other stresses. Preparing joints well can reduce the number of failures on a weld over a given period where *fatigue,* due to stress reversals, must be considered. In

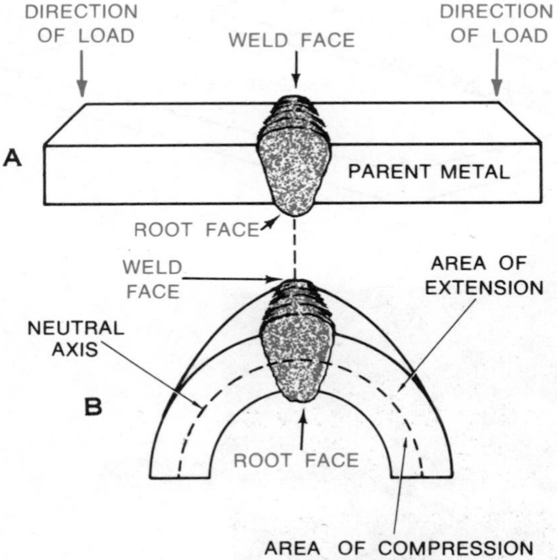

Fig. 5-25. Testing for compression and tension. *(A) Material before severe loading. (B) Material after severe loading. Note the neutral axis and areas of compression and tension.*

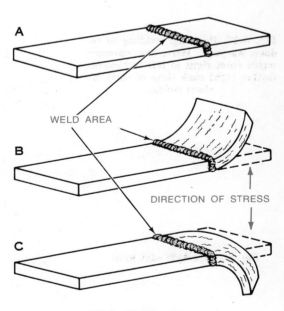

(A) Bar before stress.
(B) Bar under upward stress.
(C) Bar under downward stress.

Fig. 5-26. A bar under fatigue stress.

Fig. 5-26, stress on the bar is constantly reversed. This will lead to *fatigue failure* in the weld.

Warping—During welding, controlling the heat input on the parts being welded is done to offset distortion, commonly called *warping*. Reducing warpage can be done by using good joint preparation and proper welding techniques, such as the back-step method shown in Fig. 5-27. Fixtures and chill bars, or plates, can be used to absorb the excess heat. This could reduce warpage.

Cost of the Finished Weld

Preparing the joints on any weld is important where the *cost* of the finished product must be estimated. The lowest-priced joint will require the least bead fill, but it will give maximum strength and satisfactory characteristics to the finished weld. For this reason, knowing all about the different welds and joints will help keep estimating and production costs low and will produce good welding results.

Advantages and Disadvantages of Various Joints

Edge Joints

This joint (Fig. 5-28) is not practical for structures to be subjected to high intensity loading. Where impact or stress-causing fatigue is possible, a weld other than the *edge joint* should be considered. For simpler loads, the edge joint is fairly easy to weld.

Butt Joints

A major disadvantage of *butt joints* is that they are more costly than lap joints. Moreover, filling the groove in a butt joint using arc welding requires smaller electrodes with less amperage, especially where penetration at the root of the weld is important. (See Fig. 5-29.) Butt welds will present a greater problem to the welder than will fillet welds used with the lap joint, especially where shrinkage and residual stresses are to be minimized. Large electrodes with high current can be used in U-grooves on butt joints, but the time

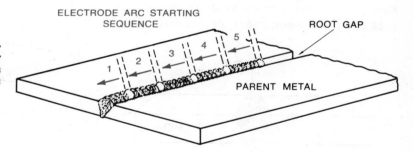

Fig. 5-27. Backstep welding to reduce warpage. Here, the operator welds from right to left but moves farther right each time to start his short welds.

Fig. 5-28. Sample edge joint.

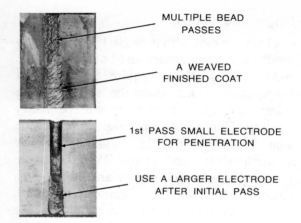

Fig. 5-29. Single-Vee butt joints showing different bead passes for full penetration and high strength.

saved during welding is offset by the required preparation time. Joints which will be required to withstand fatigue stress, shock, or bending should be prepared as butt joints, rather than single- or double-fillet lap joints (Fig. 5-30). Butt joints are found to be more

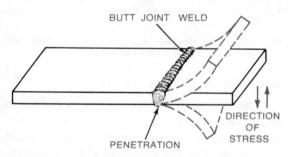

Fig. 5-30. A butt joint welded to withstand fatigue from stress reversals.

satisfactory than lap joints in manufacturing containers or vessels to be filled temporarily with explosive materials. The overlapping surfaces of a lap weld could trap gases, making the empty vessel potentially dangerous should welding ever become necessary on the seams or near the vessel.

Groove-Weld Joints

The root opening, root face, and groove angles, shown in Fig. 5-18, should be designed to give complete penetration and maximum strength, with the least amount of elec-

trode metal. *Groove-weld joints* take in the groove angle, root face, and root opening for considering the weld's entire penetration.

Regardless of the type of preparation on the joint, weld strength depends on whether complete joint preparation has been accomplished during welding.

Square-Groove Joints

These joints are used with the butt, Tee, edge, and corner, as shown in Fig. 5-31. In

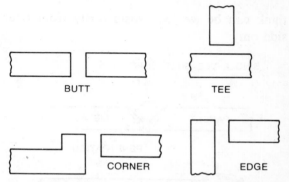

Fig. 5-31. The simplest joint, square groove. Used with many welding positions.

materials up to ⅛″ thick, where welding is to be done from both sides, the *square-butt* joint does not require a root opening for full strength. However, materials up to ¼″ thick may require not only welding from both sides, but an adequate root opening with complete joint preparation, in order to be welded to base metal strength, Fig. 5-32. Where root

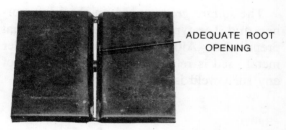

Fig. 5-32. Square butt joint, gapped for welding from both sides.

openings are ample for complete joint penetration and where backing strips are to be used, as in Fig. 5-33, materials up to ³⁄₁₆″

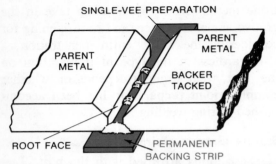

Fig. 5-33. Materials up to ³⁄₁₆″ thick can be welded from one side by using a permanent backing strip.

thick can be welded satisfactorily from one side only.

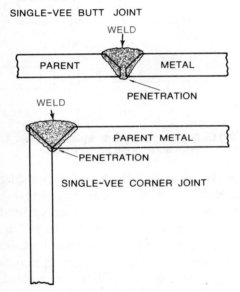

Fig. 5-34. Vee joint preparation for heavy loading.

The *square groove* used on the edge, corner, and Tee type joints requires no special preparation. Although the amount of filler metal used is relatively high, the strength of any such weld is limited to light loading.

Single-Vee Joints

Groove welds with a *single Vee* are mostly used for welding corner and butt joints (Fig. 5-34). Thick plates require the Vee-joint preparation rather than the square-groove joint, especially where there will be heavy loading. Single-Vee joints welded from one side, as shown in Fig. 5-34, have good strength while standing still. When they are placed under impact load or are subjected to stress, created fatigue may cause them to fail.

To offset failure under such conditions, the welder should weld from both sides, making certain of complete joint penetration. Even though the double-Vee groove may be used under a similar condition, only welding from both sides can assure full base metal strength. (Fig. 5-35.)

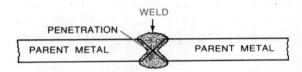

Fig. 5-35. The double-Vee groove weld gives full base metal strength.

Single-Bevel Groove Joints

If welded with complete joint penetration, *single-bevel groove joints* can be used on materials from ¼″ to ¾″ thick. If welded from both sides, they are economical for materials up to ¾″ thick. If welding can only be done from one side, a backing strip and complete weld penetration will usually be strong enough and economical. Fig. 5-33 shows a backing strip application. These single-bevel preparations are used with the corner, butt, and Tee, as shown in Fig. 5-36.

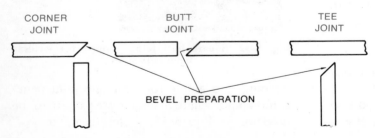

Fig. 5-36. Using a single-bevel preparation on the corner, butt, and Tee joints.

Single-U Groove Joints

This groove-joint preparation is used for butt and corner joints, as shown in Fig. 5-37.

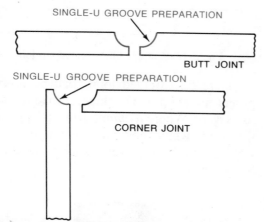

Fig. 5-37. The single-groove preparation is used on materials at least ¾″ thick for complete penetration.

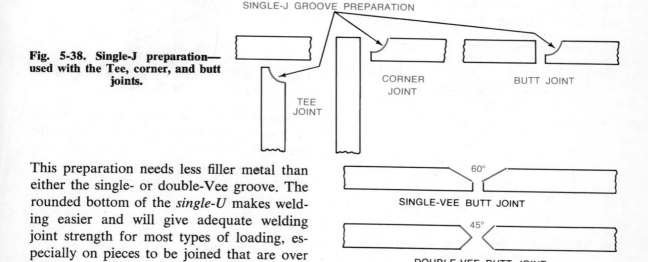

Fig. 5-38. Single-J preparation—used with the Tee, corner, and butt joints.

This preparation needs less filler metal than either the single- or double-Vee groove. The rounded bottom of the *single-U* makes welding easier and will give adequate welding joint strength for most types of loading, especially on pieces to be joined that are over ¾″ thick. Joint preparation takes more time than for Vee grooves; this is a factor where cost and time are concerned.

Single-J Groove Joints

The Tee, corner, and butt joints can be welded after a *single-J* preparation. (See Fig. 5-38.) This type of joint, like the single-U joint, costs more to prepare. However, due to the minimum filler metal required and the

finished weld's strength, it is also suitable for materials over ¾″ thick.

Double-Vee Groove Joints

On plates up to 1½″ thick, or in a situation where a single-Vee groove preparation is not strong enough, a *double-Vee groove joint* is required. The angle of the double-Vee groove can be less than that used on the single-Vee groove, which takes less filler metal but gives greater overall strength. (See Fig. 5-39.)

Double-Bevel Groove Joints

If properly prepared, and if full welding penetration is accomplished, the *double-bevel groove joint* can be used satisfactorily for welding materials as thick as 1½″. It can be used not only in the Tee and butt position, but in the corner-joint position as well. (See Fig. 5-40.)

Fig. 5-39. A double-Vee groove uses less filler metal but gives greater overall strength than the single-Vee groove.

Double-U Groove Joints

When welded with full penetration, the *double-U groove joint* can be used economically on materials over 1½″ thick, if they are welded in the butt position as shown in Fig. 5-41.

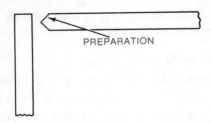

PREPARATION

Fig. 5-40. Corner joint prepared with a double bevel. This can be used for materials up to 1½" thick.

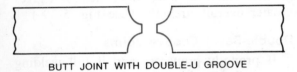

BUTT JOINT WITH DOUBLE-U GROOVE

Fig. 5-41. Double-U groove joint prepared for welding. Used for materials over 1½" thick.

Double-J Groove Joints

Plates at least 1½" thick, prepared with the *double-J groove* and welded from both sides, can withstand all types of severe loading. See Fig. 5-42. The double-J groove can be used satisfactorily on butt, Tee, and corner joints.

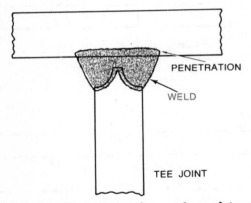

PENETRATION

WELD

TEE JOINT

Fig. 5-42. Double-J preparation used on plates at least 1½" thick.

Fillet-Welded Joint

These joints are easier to prepare and fit than the groove-welded joints, and they sometimes save time. However, the fillet-welded joint will usually require more filler rod than it would if prepared for groove-type welding. The Tee, lap, and corner joints can be classified as single fillet-welded joints. (See Figs. 5-43 and 5-44.) In Fig. 5-44, the extra

metal needed for fillet buildup can be easily seen.

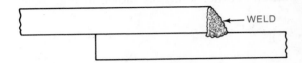

WELD

Fig. 5-43. A single fillet-welded lap joint.

Single fillet-welded joints can be used on parent metal up to ½" thick, if loading is not too severe. Where *dynamic stress* is applied to the joint, which could cause fatigue in the weld, a single fillet-welded joint should not be considered.

PLATE B

PLATE A

Fig. 5-44. A fillet-welded lap joint, showing necessary build-up for strength.

Double fillet-welded joints are also used on the Tee, lap, and corner joints for special applications. While under static loading, the Tee and lap joints are comparable to the parent metal in strength, especially if the weld is large enough.

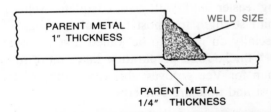

PARENT METAL 1" THICKNESS

WELD SIZE

PARENT METAL 1/4" THICKNESS

Fig. 5-45. Lap-welded joint for maximum strength. This is at least 5 times as thick as the thinner metal.

The lap-welded joint will be of maximum strength if the weld size equals at least five times the thickness of the thinner member, as in Fig. 5-45.

When a corner joint will have various types of loading, the double-fillet-welded joint, Fig. 5-46, should be used.

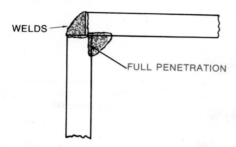

Fig. 5-46. A double-fillet welded corner joint. Welded for maximum strength.

Various types of groove welds have been successfully used with fillet welds, as shown in Fig. 5-47. This combination, used mostly

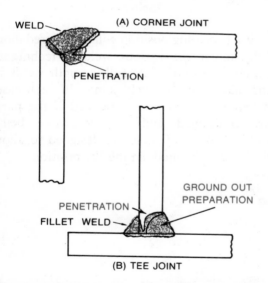

Fig. 5-47. Groove welds used with fillet welds. (a) Single-Vee corner joint, using fillet welds (b) Single-J Tee joint, using fillet welds.

on Tee or corner joints, ensures additional strength while improving the stress distribution in the joints.

Weld Positions

There are four defined welding positions, as in Fig. 5-48. It should be noted that not all welding methods are suitable for all positions. For this reason, the welder must be-

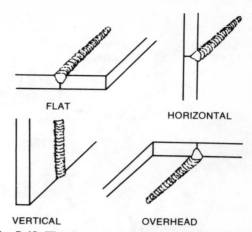

Fig. 5-48. The four most common welding positions.

come acquainted with *all* welding equipment, and the techniques necessary to operate it successfully.

Flat Welding

Wherever possible, and especially in industry, welding should be done in a *flat* position, since it is fastest and cheapest. Fixtures can be used to turn parts so that flat position welding becomes possible, as in Fig. 5-49.

Overhead Welding

The molten metal from the material and rod has a tendency to fall down from the bead position in *overhead welding*. Arc current or flame temperature must never superheat the parent metal to the extent that it falls away due to gravity. Overhead welding is usually the most difficult position to learn and may require a great deal of practice.

Vertical Welding

Usually the welder completes all *vertical welds* from the bottom to the top. With arc welding, lighter materials can be welded down-hand, from the top to the bottom. Again, gravity may try to pull the molten metal down, requiring extra skill and practice.

Horizontal Welding

Horizontal welding creates the same problem as vertical and overhead welding, be-

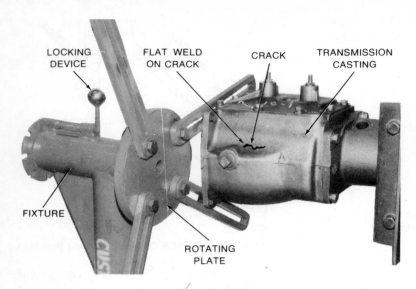

LOCKING
DEVICE

FLAT WELD
ON CRACK

CRACK

TRANSMISSION
CASTING

FIXTURE

ROTATING
PLATE

Fig. 5-49. A rotating fixture used to weld castings and odd-shaped parts more easily.

cause experience and practice are necessary to overcome gravity's pull on the molten metals. This helps to explain how flat welding became so popular.

Flat welding is the easiest welding position. Flat welding production costs are cheaper and some of the distortion caused by slower welding processes is eliminated. Welding can usually be done faster in the flat welding position.

Positioners

Each welding position will be covered more fully under the various welding techniques necessary to complete full strength welds in the many types of weld joints. In each case, a better weld may be produced if the parts are positioned and/or held while being welded. Many of these are designed to allow the weld to be made in the *flat* position.

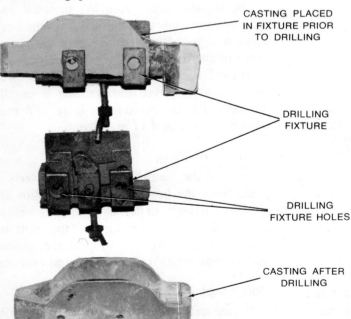

CASTING PLACED
IN FIXTURE PRIOR
TO DRILLING

DRILLING
FIXTURE

DRILLING
FIXTURE HOLES

CASTING AFTER
DRILLING

Fig. 5-50. A drilling fixture for putting accurately spaced holes in an aluminum casting.

Various types of both manual and electric revolving positioners, or weldment holders, are either available through manufacturers or can be produced in the shop. Regardless of the type of positioner, it should be constructed to hold the desired weld firmly, without a long, complicated setup time. It should also be constructed so that most weld joints can be easily approached by the welding operator without any tiring motions.

Fixtures

Devices used to speed up production and lower the cost per unit are called *fixtures*. Where many parts are to be drilled with accuracy and where dimension uniformity is required, the parts can be stacked and drilled while held by a fixture, as in Fig. 5-50.

When parts are placed separately in the *same* fixture for cutting, drilling, welding, or other purposes, they will be uniform. Fixtures should be constructed to insure quick and accurate placement of parts in the fixture, whether for welding, drilling, cutting, or other purposes. Ease of removing the finished part from the fixture is also very important. Moreover, the welder must be aware of the possibility of distortion, and he must make such allowances when designing and constructing a fixture.

Skeleton Frames

If many items of uniform shape and size are going to be made, both fixtures and units known as *skeleton frames* may be considered. Parts of various products can be clamped in place by different type holders fastened to the outside of the skeleton frame. Once the parts are held in place around the frame, they can be tacked to each other. Then, they can be carefully removed and finished being welded, with or without using positioners.

Backers

Pieces of material placed at the root side of a joint before starting the weld proper are called *backers*. The backer can be tack welded in place, as was shown in Fig. 5-33, and, as the parent metal pieces are welded together, the backer becomes a permanent part of the weld proper.

Materials such as flux, copper, asbestos, and carbon can be formed into various shapes at the base (or *root*) of weld joints. (See Fig. 5-51.) These keep the molten metal from falling through the joint due to gravity.

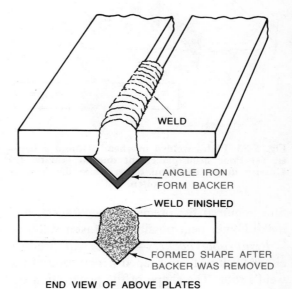

Fig. 5-51. Angle-iron backer is used on the back side of the weld to form a penetration shape.

Permanent backers—These become an integral part of the weld and are often used for bridging a widely spaced gap in the overhead position. (See Fig. 5-52.) Without using a

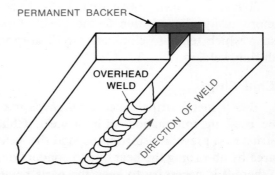

Fig. 5-52. Permanent backer. Used to make overhead welding easier.

backer, an overhead weld joint with poor fit-up cannot be welded to maximum strength, due to the effect of gravity on the bead. (See Fig. 5-53.) Permanent backers eliminate the

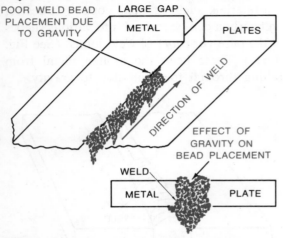

Fig. 5-53. Plates welded overhead without a backer. (a) Poor bead placement due to gravity. (b) Gravity draws the bead down while the bead is molten.

time required for precision preparation, and make higher heat possible for faster welding.

Removable backers—These will produce either flush or form-shaped penetration at the weld's root. The shape of the penetration depends on the shape of the backer holding the molten metal. Removable carbon backers can be easily carved to form a specifically-shaped penetration at the weld root, while metal backers are either ground or machined to shape for the same purpose. Such shaped backers hold the molten metal in a manner similar to die casting.

Joints, as shown above in Fig. 5-52, are more easily aligned, due to the backer-type strip which tends to eliminate any distortion during welding.

Chill Strips

Heavy metal strips called *chill strips* may be used in conjunction with backers. Chill strips help prevent distortion beyond the weld area by absorbing the heat due to conduction. Where it is necessary to keep the parts being repaired as cool as possible while welding,

chill strips can be placed as shown in Fig. 5-54, to help absorb some of the heat.

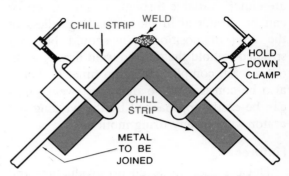

Fig. 5-54. Chill strips being used to absorb some of the heat while welding thin strips of flat iron.

Since chill strips are held in place by clamps, any weld with chill strips will have less opportunity to warp. Alignment of all the pieces while welding is also assured.

Fixtures have been made with chill strips for special production purposes. Production of tanks and containers, using either light aluminum, stainless steel, or other materials, usually uses chill strips.

Welding Terms

These terms are presented to help a welder or a learner review the words and phrases used in the welding trade. These terms will help make the phrases about joints and welding much clearer.

Bevel—A prepared edge cut at an angle for welding. Fig. 5-55 is a *bevel*.

Bevel Angle—The angle formed between the prepared edge of a piece of the parent

Fig. 5-55. A 30° angle beveled on parts to be welded.

metal and the place perpendicular to the piece is the *bevel angle* (Fig. 5-56).

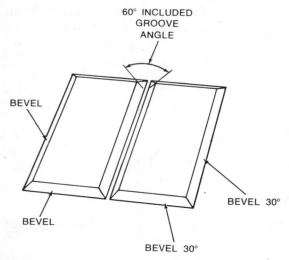

Fig. 5-56.   **A 30° angle beveled on parts to be welded.**

Blowhole—Cavities formed during welding due to trapped flux or gas are known as *blowholes* (Fig. 5-57).

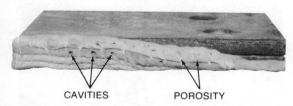

Fig. 5-57.   *Blowholes* **(cavities and porosity) formed in welding due to trapped gases.**

Cascade Sequence—In a *cascade sequence*, weld beads are deposited in overlapping layers (Fig. 5-58).

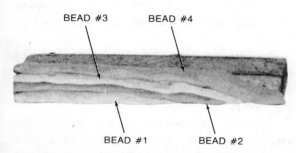

Fig. 5-58.   **A cascade sequence—beads deposited in overlapping layers.**

Coupons—Sample metal pieces used for experimental or welding demonstrations are called *coupons* (Fig. 5-56).

Crater—A *crater* is a natural depression at the end of a weld bead (Fig. 5-59).

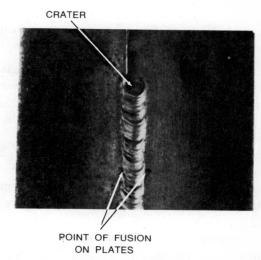

Fig. 5-59. **Crater formation at the end of a weld bead.**

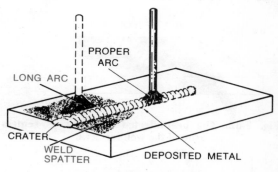

Fig. 5-60. **Deposited metal.**

Deposited Metal—The metal added to fill the joint preparation during welding is *deposited metal*. Fig. 5-60 shows both good deposited metal to fill the joint and poorly deposited metal, called *spatter*.

Depth of Fusion—The distance that fusion extends *into* the base metal from the surface melted during welding is the *depth of fusion* (Fig. 5-61).

Fusion—Joining together molten metals on the surface exposed for welding is simple *fusion* (Fig. 5-59).

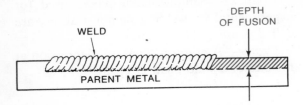

Fig. 5-61. Depth of Fusion—how far the parent metal is cut into.

Groove Angle—The total included angle of the groove between parts to be joined by a groove weld is the *groove angle*. In Fig. 5-55, the groove angle includes both beveled angles together.

Heat-Affected Zone—In the *heat-affected zone*, the parent metal has a change in mechanical properties due to the heat from welding, but has not been melted (Fig. 5-62).

Incomplete Fusion—Poor tensile strength results from *incomplete fusion*. This is a serious weld defect and should be looked for carefully (Fig. 5-63).

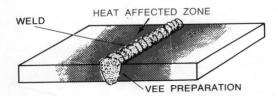

Fig. 5-62. The heat-affected zone always centers around the welding area.

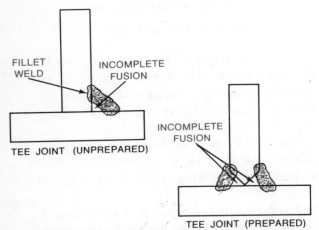

Fig. 5-63. Incomplete fusion on both prepared and unprepared Tee joints.

Joint Penetration—Complete penetration has taken place on the underside of metal if fusion has penetrated the prepared joint, called *joint penetration* (Fig. 5-64).

Kerf—The channel or groove left by removing metal through cutting is a *kerf* (Fig. 5-65).

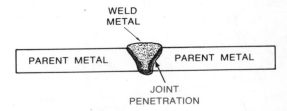

Fig. 5-64. Complete fusion, resulting from joint penetration.

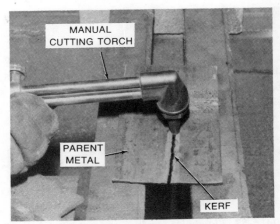

Fig. 5-65. The cut made in the parent metal is the kerf.

Parent Metal—The base metal, or the metal to be prepared, welded, or cut, is the *parent metal* (Fig. 5-65).

Peening—Stress is relieved by pounding the metal with a hammer in *peening* (Fig. 5-66).

Porosity—A welded metal containing holes throughout, due to gas pockets created by improper welding, is said to have *porosity*. See Fig. 5-57.

Residual Stress—Locked-up stress, also known as *residual stress*, remains in a weld because of mechanical and/or heat treatment (Fig. 5-67).

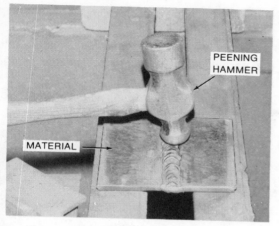

Fig. 5-66. Relieving stress by pounding—*peening*.

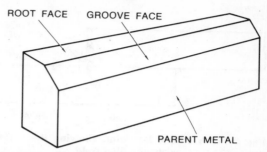

Fig. 5-68. The root face will be partially or completely welded.

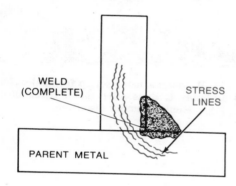

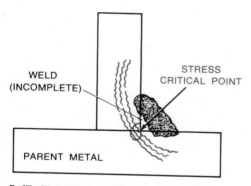

Fig. 5-67. Tee joints welded having stress. (a) Residual stress is less critical due to complete welding. (b) The stress is more critical because of incomplete fusion.

Root Face—The portion of the groove face adjacent to the root of the joint is known as the *root face*. It is on top in Fig. 5-68.

Root Joint—The bottom of two beveled pieces, faced together for welding, is called the *root joint* (Fig. 5-69).

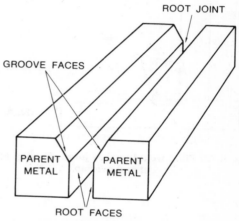

Fig. 5-69. When welded, the root faces are placed together at the root joint.

Slag Inclusion—Nonmetallic materials trapped between the weld metal and base metal are known as *slag*. Slag *inclusion* happens when these are trapped, Fig. 5-70.

Spatter—Metal particles which do not become a part of the weld but adhere to the parent metal at random are together known as *spatter*. See Fig. 5-60.

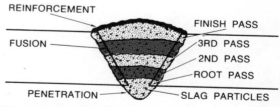

Fig. 5-70. Slag inclusion resulting from not cleaning the weld after each pass.

Toe (of a weld)—The *junction* between the face of a weld and the base metal is the *toe* of a weld. An example would be on the

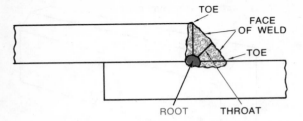

Fig. 5-71. A simple lap joint, showing the various parts of a fillet weld.

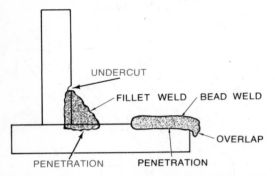

Fig. 5-72. Fillet weld in a Tee joint, showing undercut and overlap.

ends of the hypotenuse (longest side) of a fillet weld (Fig. 5-71).

Undercut—An unfilled groove, melted into the parent metal adjacent to the toe of the weld, is an *undercut* (Fig. 5-72).

Weldment—A *weldment* is an assembly of welded components (Fig. 5-73).

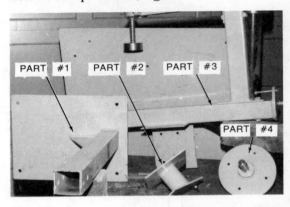

Fig. 5-73. Weldments being set up for final welding. Final assembly is done after minor assemblies to eliminate major distortion.

6

Safety in Welding

A workman with pride in his work considers not only his own safety, but the safety of others. If knowledge and foresight were the first requirements of all workmanship, most serious accidents would be avoided. It is a good idea to remember, before starting *any* project, that "accidents don't just happen;" they are caused by ignorance and carelessness.

A safe workman knows the safety regulations important to his employment. He uses the safeguards, safety appliances, personal protective equipment, and other devices required by good common sense in his work area.

Welding Hazards

Some hazards facing a welder are:

A. *Harmful* or *poisonous gasses* encountered in welding some metals.

B. *Explosions* due to improper handling or operation of acetylene generators.

C. *Explosions* due to improper handling of gases or equipment used in welding.

D. *Burns* to the operator or fellow workers, by contacting flame, sparks, hot metal, and/or hot slag.

E. *Eye injury* from harmful rays, metal particles, chips, and/or slag.

F. *Explosions* within closed objects or vessels which contained flammable ma-

terials but were not properly purged.

G. *Fire* caused by flying sparks, slag, and hot metal.

Poisonous Gases

Welding some metals may give off *harmful* or *poisonous* gases, so a ventilation system complying with the ventilation requirements of the workmen's compensation board should be installed. When welding or burning emits harmful fumes, adequate ventilation or approved respirators should be given to all workmen exposed to the fumes. Oxygen-generating, self-contained breathing apparatus and a fresh-air hose mask with blower are two examples of approved *breathing* equipment available for welding purposes, as shown in Fig. 6-1.

Explosions

Explosion hazards come from improper handling or operating acetylene generators and/or improper handling of gases and equipment used in welding. These suggestions are a good way to avoid potential explosions:

A. Acetylene generators should be installed, operated, and maintained according to the manufacturer's recommendations.

B. Cylinders, piping, and fittings of compressed and liquified gas systems should

CLEAR-VIEW
FACE MASK

AIR HOSE FROM COMPRESSOR
OR CARRYING CYLINDER

Fig. 6-1. Self-contained breathing equipment. (Courtesy of Safety Supply Co.)

be located where they are protected from physical damage.

C. Cylinders, regulators, and hoses of compressed gas systems should be protected from *any* source of heat. This includes direct sunlight. Regulators and automatic welding equipment valves should be used only for the *gas* for which they are intended. A welder should never operate the oxygen cylinders or apparatus with greasy or oily gloves, since grease or oil mixed with oxygen may explode. Oxygen should never be used to blow out pipe lines or in pneumatic tools, for the same reason.

D. *Improper cutting* or *welding* on objects or vessels which have not been properly purged (cleaned or ventilated) of flammables such as oil, gasoline, or tar.

Flammable liquids should be stored in containers complying with safety specifications.

When emptied, they should be destroyed or made safe by steaming and then burning rubbish in the container, whenever possible. A good safety container for gasoline is shown in Fig. 6-2.

Fig. 6-2. A *Protectoseal* can used to handle and store flammable liquids. (Courtesy of Safety Supply Co.)

Welding or other hot work should never be done on any vessel, tank, pipe, or similar structure until tests indicate that the vessel is free of explosive fumes. An explosive meter, as shown in Fig. 6-3, can be used for this purpose.

An inert gas such as nitrogen can be used to purge flammable vapors from vessels or tanks. As an added precaution, adequate and approved fire extinguishers should be readily available where welding, cutting, or heating is taking place.

Burns

Burns caused by hot metal, flying sparks, and direct flames will injure a careless welder. Such burns are considered to be the primary welding hazard. Eye injuries from harmful rays encountered in both arc and oxy-acetylene welding are a second hazard. These and other safety problems will be presented to help make the welder aware of preventive measures.

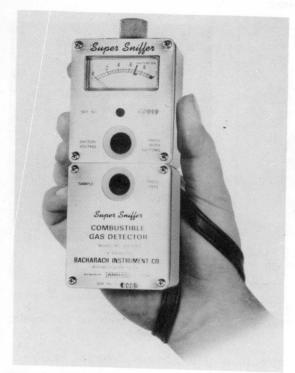

Fig. 6-3. A combustible gas detector. (Courtesy of Safety Supply Co.)

Most burns can be prevented if a welder remembers that metal and torch equipment stays hot long after it has lost its reddish-orange glow.

Clothing

A welder's clothing should not be made of flammable material and should not have any oil or grease on it. Flammable clothing has not only caused painful burns, but it has also caused fires after the welder has gone home.

Basic Clothing—Apparently, denims and cotton drills are the most popular clothing around shops today, probably because they are cheap and shed scattered sparks fairly well. Although they will smolder, there is little danger of this material bursting into flame if it is kept free of oil and grease.

Either tanned leather or asbestos is the best type of material for clothing, but the workman will find them more expensive and quite hot and heavy compared with denims and wool. Some welders prefer wool clothing,

since excessive cleaning does not shorten the material's life to the extent that it does denims. Wool has the added feature of being quite nonflammable.

The welder should not expose any part of his body to substances or rays that would injure his skin, and he should wear clothing suitable for his working conditions. Where a hazard exists, the welder should use not only suitable protective devices, but also suitable clothing.

Head and Neck Covering—It is also a good idea to wear shoulder and neck covering, especially for overhead welding or cutting. Rubber ear plugs could prevent eardrum damage during overhead welding, where hot metal sparks fly down due to gravity, as in Fig. 6-4.

Fig. 6-4. Overhead welding produces more problems than usual because the force of gravity can pull sparks, slag, and molten metal down on the welder.

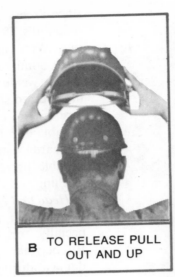

A | TO FIT ON CAP PRESS DOWN

B | TO RELEASE PULL OUT AND UP

C

(A) The parts of a hard hat-helmet combination. (B) The hard hat and helmet parts can be used together or separately. Installing and removing the helmet. (Courtesy, Fibre-Metal of Canada, Inc.)

(C) Using the had hat—helmet for welding. (Courtesy of Safety Supply Co.)

Fig. 6-5. A hard hat used with an arc welding helmet.

Protective head gear, such as caps or beanies, should be worn under the helmet. Under certain welding conditions, a hard hat fastened to the helmet is advisable, as shown in Fig. 6-5.

able in a soft asbestos material, makes it easy and safe to handle the electrode holder. Pants with *cuffs* and large, horizontal-type pockets should never be worn because molten

Fig. 6-6. Asbestos leggings and heavy duty shoes cover and protect the welder's legs and feet. (Courtesy of Safety Supply Co.)

Gloves and Cuffs—A shrink-proof and flameproof gauntlet type *glove*, usually avail-

Fig. 6-7. The welder's arms, hands, face, and eyes are well protected from the welding heat, brightness, and sparks.

metal could get caught in them and burn. Sparks can also lodge in shirt pockets, rolled-up sleeves, and low oxford shoes; for that

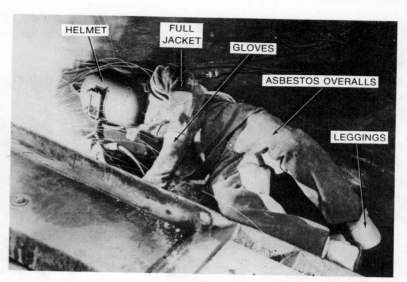

Fig. 6-8. A very well dressed welder. Note the helmet, jacket, gloves, overalls, and leggings. (Courtesy of American Optical Corp.)

reason, a welder should use a cape-like chest covering and high, well-laced boots.

Well Dressed Welders—Excellent welding shoes and leggings are shown in Fig. 6-6. In Fig. 6-7, the welder's eyes, hands and arms are very close to the welding. They are not in danger since he is well dressed for the job.

In Fig. 6-8, a welder safely lies on his side for difficut arc welding. For this job, he needs the complete protection discussed, from head to foot. No part of his body is in danger, since he is well protected.

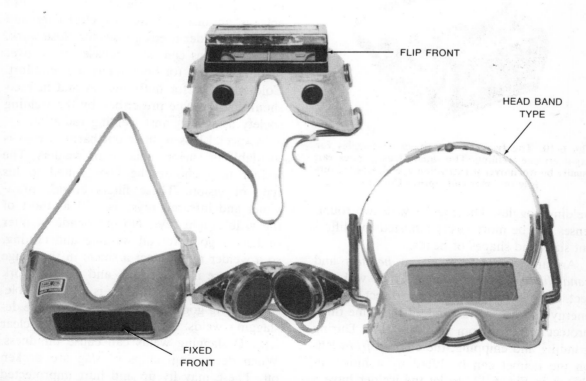

Fig. 6-9. Goggles used for oxy-acetylene welding. They may have a fixed-front or a flip-front. The flip-front lets the welder flip up the dark lens to see his weld *after* the torch is shut off.

Eyes

A welder's *eyes* should always be protected—not only from the injurious rays created during welding or cutting, but also from sparks and hot slag.

Suitable *goggles,* as shown in Fig. 6-9, are available to protect oxy-acetylene welders from rays and sparks. They also protect against flying particles, which are a possible hazard during welding, chipping, or grinding. Fig. 6-10 shows two pairs of oxy-acetylene

Fig. 6-10. The two common types of goggles for oxy-acetylene welding. The square-lens goggles can usually be worn over prescription eye glasses. (Courtesy of American Optical Corp.)

welding goggles. The goggles with two round lenses can be more easily adjusted for different sizes and shapes of heads.

An arc welder has several *helmets* and *hand shields* to choose from. These will protect his eyes and neck from direct radiant energy and hot sparks and, at the same time, protect his eyes from the arc's rays. During grinding and chipping, the colored *front* lens on the helmet can be lifted by a hinge, as shown in Fig. 6-11, to let the welder have a

better view. A clear glass underneath still protects his eyes from slag chips or grinding dust.

FLIP FRONT CLOSED

Fig. 6-11. A welding helmet with a flip-up front lens. Only the clear lens underneath protects the welder's eyes from slag chips and grinding dust.

Goggles, hand shields, and helmets are made of an inflammable material that also acts as an insulator against electricity and heat. If a welder needs glasses for close work, special goggles can be purchased to fit over reading glasses, for both safety and comfort. Colored lenses for both goggles and helmets should be the type prescribed by the welding society as safe against welding radiation.

A special lens to filter out harmful rays is available in various shades for welders. The welder may choose the filter suited to his type of vision. These filters absorb ultraviolet and infrared rays, as well as most of the visible light rays. No one shade of filter or lens is good for all welding and cutting, so a welder must select a shade in the range where glare is eliminated, and yet the work under the arc or welding tip is clearly visible.

Grinding goggles should be worn while deslagging welds. These goggles have a clear lens. Deslagging welds can cause blindness. When deslagging, chips of slag are broken off. These may fly up and hurt unprotected eyes. Using a flip-front helmet, approved grinding shields, or goggles, eliminates danger to eyes when deslagging. Fig. 6-12 shows a

Fig. 6-12. Good goggles for deslagging *only*. **These should not be worn alone for welding.**

welder wearing good goggles for deslagging *only*. These should *not* be used alone for actual welding.

Welding Radiation

Welding and cutting radiation falls into three classifications:

1. Ultraviolet Rays
2. Visible Light Rays
3. Infrared Rays

Ultraviolet Rays—These rays lie beyond the violet end of the visible spectrum. They have wave lengths up to the upper limits of X rays. Strong ultraviolet rays can cause severe inflammation, similar to sunburn, on the eyes and surrounding membranes. The face, arms, neck, and other parts of the body should be kept covered when welding because the intensity of the ultraviolet rays can cause a burn to exposed skin in a very short time.

Visible Light Rays—These rays can be seen by the eye when an arc is struck. They reflect off walls or other objects, causing eye problems to anyone nearby who is not wear-

ing a good lens. Intense visible light rays may cause eyestrain and, possibly, temporary blindness.

Infrared Rays—These rays, situated beyond the red end of the visible spectrum, are certain heat rays with wave length radiations less than radio waves. Infrared rays, also known as ultrared, may cause cumulative effects that may lead to cataracts or retina injuries. Infrared rays are considered to be the most dangerous to the eye of all welding rays.

Lens Shades for Filtering Rays

Choosing the proper lens is very important. The best lens is the darkest shade that will still show a clear outline of the work without eyestrain.

Recommendations have been made for the shades of lenses to be used in various welding operations. These shades vary from No. 5 for light gas welding to No. 14 for arc welding or cutting. Lenses are available in green shades or amber shades, with green being the most popular. In arc welding, the shade numbers work out much the same as for oxyacetylene, except that they must be considerably darker to filter out the higher-intensity rays.

Since all lenses have markings to distinguish the shade, a welder should become familiar with the type of lens and shade best suited for his job and his eyes. Table 6-1 is a guide showing the lens shade recommended by the American Optical Corporation for each type of welding. Using a lighter lens that what is recommended would likely result in some eye damage.

Injury From Ultraviolet Ray Burns

Exposure to ultraviolet rays can create deep and serious burns to the welder's exposed skin. These burns heal very slowly, and they can leave scar tissue which continues to be sensitive to both heat *and* cold. The temperature of the carbon arc, ranging from 6850° to 9550° F, compared with the temperature given off by the rays of the sun, 7650° F,

Table 6-1. AO Filter Glass Shade Numbers.

GAS TORCH CUTTING AND WELDING		ELECTRIC WELDING AND CUTTING	
	Shade Number	Rod Dia.—Inches	Shade Number
Soldering	1.7*	Metal Arc $\frac{1}{16}$ to $\frac{3}{32}$	8
Brazing	3 or 4	Helium Arc $\frac{1}{8}$ to $\frac{5}{32}$	10
Cutting		Metal Arc $\frac{3}{16}$ to $\frac{1}{4}$	12
Light—up to one inch	3 or 4	$\frac{5}{16}$ to $\frac{3}{8}$	14
Medium—one to six inches	4 or 5	Atomic Hydrogen	10 to 14
Heavy—six inches and up	5 or 6	Inert Gas Shielded Arc	10 to 14
Welding		Carbon Arc	14
Light—up to one-eighth inch	4 or 5	Plasma	
Medium—one-eighth to one-half inch	5 or 6	Spraying	5 to 8
Heavy—one-half inch and up	6 or 8	Cutting	8 to 12
		Welding	8 to 12

Note: In oxy-acetylene cutting or welding where torch produces a high yellow light, use AO Noviweld-Didymium glass.
*Medium Calobar
Courtesy of American Optical Corp.

makes it very easy to see the danger from welding ray exposure.

Mild ultraviolet ray burns should be treated immediately by the procedures listed in a good first aid book. Many commercial drug products are adequate for minor exposure. Major burns could be serious and medical attention should be sought immediately.

Arc Flash

When a welding flash causes arc rays to come into contact with unprotected eyes, the injury is called *arc flash*. This usually happens if the helmet is raised and an arc is struck during arc welding. If the flash is frequent enough or severe enough, the eyeballs become covered with many small water blisters. The eyelids moving against the eyeballs cause irritation and pain. The eyes are also hurt by bright light and will water profusely. In extreme cases, blindness will occur for two or three days. If exposed to arc flash, a welder should wear dark glasses and avoid any welding for several days.

Protecting Others From Rays

It is important that a welder has the same consideration for other people's eyes that he does for his own. Striking an arc without warning when others are around is both dangerous and thoughtless. It should be remembered that *reflected* arc rays *will* damage the eyes. Whenever it is necessary to weld outside of a welding enclosure, fireproof portable screens should be set up to protect others from the harmful rays.

Burns

A *burn* is damage to body tissue resulting from extreme heat. Burns, resulting in injury from contacting hot metal, flame, hot slag, or sparks, are of two basic groups: superficial (contact) burns and deep burns.

Contact Burns

Superficial, or *contact,* burns do not penetrate deeply and will generally heal quickly. However, because they are painful, they should receive first aid treatment. There are several products suitable for this purpose on the market. In minor burns, only the outer layers of the skin are damaged; the burned area may be fiery red in color and may blister.

Treating Contact Burns — With minor burns and scalds where the skin is not broken, putting the affected part in cold water will often relieve much of the pain. Because there is the danger of infection, careful washing in soap and water is a good idea. Too, the patient should be kept as comfortable as possible. Hot drinks and blankets will help offset chills brought on by shock.

Deep Burns

These burns, which penetrate beyond the skin and into the flesh, are most often caused by metal at temperatures below the red hot point. These burns take longer to heal, and care should be taken to prevent infection. Major burns can be so deep as to destroy not only the skin, but the underlying muscle and fat as well. Nerves may also be destroyed, and any burn involving over 50% of the body surface is usually fatal.

Treating Deep Burns—On all burns serious enough to require hospitalization, only a dry, sterile dressing should be used. Applying salves or greases severely complicates medical treatment. The burned area should be carefully covered (clothing included where necessary) with bandages or sheeting, to protect the area from infection. The injury should be handled as little as possible, and then only with clean hands.

All persons involved in welding should be aware of these important procedures in case of an accident creating major burns:

A. Handling should be done only if absolutely necessary.
B. Salves or lotions must not be used on the burn.
C. Burned clothing must be left *on* the burn.
D. Blisters should remain unbroken.
E. Cover all the affected areas with dry dressings.
F. Use shock preventive treatment.
G. Get medical help as soon as possible.

Explosions Within Enclosed Objects or Vessels

Every year, newspapers report either injury or death to workmen from exploding vessels. The most serious of these accidents are explosions in tanks containing gasoline or other volatile fuels. (Volatile means *able to explode.*) Accidents of this type, experienced by the welding trade, are usually due to improper

purging (removing by cleaning), which is necessary before welding on tanks or other fuel containers.

Gasoline and other light fuels give off highly volatile vapors. These may accumulate and remain in the weld seams of tanks or containers for *years* after the liquid has been used up. The welder must realize that, during cutting or welding, heat may ignite these volatile vapors. It may also be great enough to vaporize oils, greases, rubber, paint, tar, or other materials, and make *them* potential killers!

Prerequisites for an Explosion

Three "things" are necessary to have an explosion. Basically, the equation for an explosion looks like this:

$$Combustibles + Oxygen + Ignition = Explosion$$

Before an explosion can take place, therefore, there must be:

A. Explosive *combustibles* (material capable of burning).
B. Some method of igniting the combustibles' gases, commonly known as *ignition*.
C. *Oxygen,* or air, necessary to support the combustion.

If any one of the above requirements is eliminated, the welder is safe from an explosion. To safeguard the life of the welder, here are some precautions and procedures for welding on containers. If the container was once filled with *unknown*, or even *known*, contents, serious accidents from explosions can be prevented by following these steps:

Remove all the combustibles—Purging, by steaming or boiling in caustic solution over a long period of time, will eliminate the trapped gases remaining in the seams or open roots of welds. This method is costly and does not lend itself to large commercial jobs.

Remove all the air or oxygen—Removing air or oxygen needed to support combustion can be done by using *inert gas* (gas deficient in active properties, such as nitrogen, argon, or helium). The gas is injected into the vessel before welding; it is especially satisfactory in removing acetylene from manifolds before doing repair work.

All vents should be left open in the container to be purged. Should the combustibles be *lighter* than air, it will be necessary to inject the inert gas into the *top* of the container, as shown in Fig. 6-13.

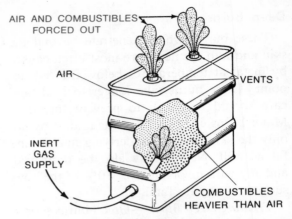

Fig. 6-14. Inert gas is injected into the *bottom* of the container if the combustibles are heavier than the air.

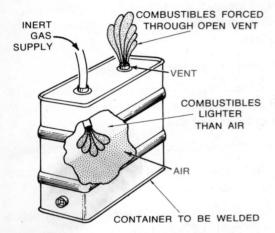

Fig. 6-13. Inert gas is injected into the *top* of the container if the combustibles are lighter than the air.

The purging gas should be delivered to the *bottom* of the vessel if the combustible is *heavier* than air, as shown in Fig. 6-14. Enough purging gas must be used to remove *all* the air.

Displacing air with water—Welding may be necessary near the top of a large, immov-

able vessel or in any area on a small vessel which can be readily maneuvered. In either case, the required repair area will be placed upward or toward the top. Water placed in

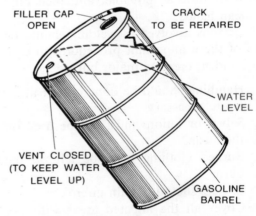

Fig. 6-15. Preparing a storage barrel for welding by filling it with water.

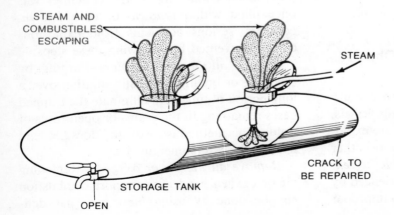

Fig. 6-16. Using live steam to partially purge a tank.

the vessel or container, up to the cracked area, would displace the air up to that point, and if the welder has some knowledge of purging, he can readily weld the vessel without danger, as shown in Fig. 6-15.

Partly removing combustibles and removing air by steaming—Live steam, blown into containers at low pressure, is the most commonly accepted method of purging combustibles from small vessels, as shown in Fig. 6-16. All vents should be open in the container, and welding should be done while the vessel is hot and contains steam. If possible, the steam should continue to purge the vessel while welding is done. This insures maximum safety.

Welding Gas Tanks

This is a topic important to all welders if they plan to stay alive. There are various ways of preparing gas tanks for welding. Some methods are both unsatisfactory and unsafe, unless the workman has a thorough understanding of the methods and a knowledge of the hazards involved. Several methods have been found satisfactory.

Water Method—As mentioned previously, using *water* is limited to jobs where welding is high up on the vessel and where openings are located to let the air escape freely. The volume of air space *above* the water must be small or the danger will still be present.

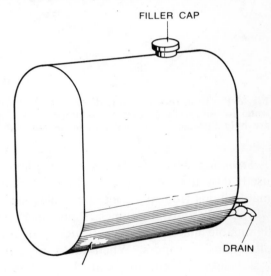

Fig. 6-18. Crack in a lower portion of an oil storage tank.

Fig. 6-17. A gas-detecting device for testing possibly explosive containers. (Courtesy of Safety Supply Co.)

If a positive test with an explosive meter indicates further steaming is needed, the tank should be allowed to cool first. It has been thought that static electricity in the steam from purging a hot tank could create enough ignition to cause an explosion. All newly steamed tanks should be tested with an explosive meter, as shown in Fig. 6-17.

If, as in Fig. 6-18, the vent in the vessel does not allow the combustible vapors to escape, it may be necessary to fill the container completely with water. With a hammer and chisel, an opening may be cut to allow the gases to escape when welding is started (Fig. 6-19). After the crack is repaired, the barrel or vessel can be placed upright and almost filled with water. Then, a torch is revolved around the chiselled opening, while the torch operator stands behind a barricade, in case an explosion occurs. When the operator is assured that no combustibles lie above the water line, the chiselled opening can be safely patched.

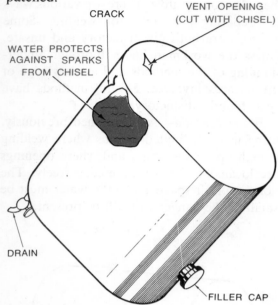

CRACK

VENT OPENING
(CUT WITH CHISEL)

WATER PROTECTS
AGAINST SPARKS
FROM CHISEL

DRAIN

FILLER CAP

Fig. 6-19. Tank turned upside-down and filled with water for safe welding. Because the normal openings are holding water in, a vent had to be cut with a chisel.

Steaming Method—When this method is used, it is first necessary to drain all flammable fluids from the container, since steam will remove only the fumes. Remember that flammable materials, when heated, can give off explosive vapors. When tanks or other vessels are steamed, there must be open vents in the upper part of these containers to allow the steam entering through the bottom of the vessel to escape through the top, carrying with it the explosive fumes. This is shown in Fig. 6-16.

Use the same precautions mentioned previously for purging the tanks carefully. Few people standing near an exploding tank ever live, so be very careful.

Safety Tips on Steaming

The following tips on the most common purging process, steaming, will help to eliminate the dangers of an explosion whenever storage tanks must be repaired:

A. Remove all fluids.
B. Steam at least one hour for every 200 gallons of tank size—about 15 minutes for a 45-gallon drum.
C. Always test after steaming with an explosive meter. (Note: Most natural gas employees keep a meter available.)
D. Use a lighted torch from behind an explosion-proof shield to test around the vents of freshly steamed gas tanks.
E. Use low-pressure steam.
F. Weld as soon after steaming as possible.
G. If further welding becomes necessary after testing the finished job, repeat the entire welding safety procedure.

Proper ventilation helps purging, because the air moving slowly during welding does not permit any vapors formed by the welding heat to accumulate and become dangerously explosive. Circulating air discharges these mixtures through the open vents.

Fumes as Welding Hazards

Fumes are small particles of oxides produced when welding. They can be alleviated, to some extent, by ventilation and/or a larger work area. In arc welding, fumes are produced from the oxides of the parent metal, the electrode, and the flux covering the rod. In oxy-acetylene welding, the tinned flux used for lowering the melting temperatures of the oxides results in dangerous fumes.

Fumes can exist in several ways: as gases that are noxious and injurious to health; as odorous gases rising in smoke; or as a visible or invisible vapor. Knowing about various fumes encountered in welding and the effects of these fumes on the welder's health is very important. These different fumes are listed to help the welder use safe practices in ventilation and ventilating equipment.

Iron Fumes

Since most welding takes place on plain steels, *iron* welding fumes are usually not dangerous in normal concentrations. Iron welding produces the ferric oxides, known also as *hematitie,* which is one of the most important iron ores. (Any natural substance containing any metal is known as an *ore.*)

Copper Fumes

The oxides of *copper* (a reddish, ductile, metallic element) are not considered toxic (poisonous). When welding copper in an enclosed area without ventilation, the welder could become ill. The symptoms, although usually vague, could be described as similar to those developed by breathing zinc fumes.

Zinc Fumes

Metal Fume Fever comes from welding galvanized metal (low-carbon steel dipped in hot *zinc*) in enclosed places without good ventilation. There are usually no serious complications, although the victim may experience a severe upset stomach. The symptoms include a headache, chills (with a rise and fall in temperature over a period of up to six or eight hours), and a tightness in the chest. Zinc fumes rising from the galvanized metal are the cause of metal fume fever.

Lead Fumes

All forms of *lead* (a heavy, soft, inelastic, bluish-gray metallic element) are toxic. Lead oxides are the most *soluble* (capable of being dissolved) in human tissue fluid. Melting, cutting, or welding lead could create lead poisoning, because lead fumes enter the blood stream if inhaled. Lead will accumulate in the various body organs and in the bones. In the bones, it lies dormant but may be set free to circulate again. The symptoms of lead poisoning are a lead line in the gums and a metallic taste in the mouth; constipation, vomiting, and nausea are other symptoms.

Manganese Fumes

Poisoning due to welding *manganese* and its alloys causes respiratory trouble and various changes in the welder's nervous system.

This condition, however, is quite rare, except in welders and workers in poorly ventilated manganese refineries. Manganese is a non-magnetic, metallic element.

Cadmium Fumes

Metals that are *cadmium* plated (or painted) give off deadly fumes when heated. The welder should ventilate the area in which cadmium is being welded. He should know that cadmium, a bluish-white, metallic element, and galvanized materials are similar in appearance. Some materials are also coated with mercury, and these fumes are *also* poisonous.

Fumes of Other Metals

The welder will experience no harmful fumes when welding:

A. Aluminum—A lighter-than-glass, bluish-white, malleable, ductile, metallic element.
B. Titanium—A dark gray, metallic, element resembling tin.
C. Chromium—A hard, grayish-white, metallic element.
D. Nickel—A hard, magnetic, ductile, silver-white, metallic element.
E. Vanadium—A rare, silver-white, metallic element from the phosphorus group.

Nevertheless, in welding of *any* kind, a welder should be sure to have plenty of fresh air for his own comfort and health. Also, some fumes or overlooked conditions might have bad effects.

Ventilation Tips

The regulations set down by most workmen's compensation boards state that all welding areas should have ample, functioning ventilation systems. The following tips should aid both employer and employee:

A. Booths or welding screens should not hinder air movement, which is needed for ventilation. See Fig. 6-20.

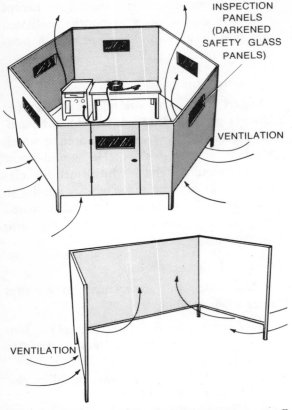

INSPECTION PANELS (DARKENED SAFETY GLASS PANELS)

VENTILATION

VENTILATION

Fig. 6-20. When screens and booths are set up off the floor, they will not hinder the constant and needed flow of ventilating air.

B. Welding on brass or bronze creates toxic fumes which should be eliminated through proper ventilation.

C. Proper ventilation is necessary when welding galvanized iron.

D. Respirators, as shown in Fig. 6-21, should be used in enclosed areas where toxic fumes are prevalent.

E. Air lines and fans can be used to eliminate fumes and toxic gases.

Summary of Protective Measures

This summary of protective measures is written with the hope that repetition may make all welders aware of possible dangers. Using safe practices may be responsible for saving a life, so use safety precautions in all welding and be aware of the following hazards:

A. Welding near flammables.

B. Welding on vessels of unknown content, even though they are "empty."

C. Welding on painted or coated materials.

D. Welding in unsafe positions.

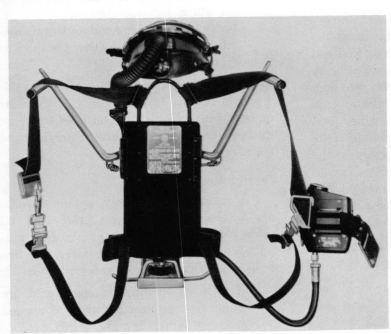

Fig. 6-21. A respirator used by welders when welding in toxic fumes. (Courtesy of Safety Supply Co.)

Adequate ventilation is important in all welding areas, and, in poorly ventilated areas, a fresh air supply should be used continuously when welding. *Respirators* should be used when the ventilation is doubtful.

All types of available safety equipment should be understood by all employees in the area. Knowing where the equipment is and how to use it will insure an immediate and safe rescue when required. In certain welding operations, standards have been given for *goggles* and *lenses,* as was shown in Table 6-1. Protective clothing, essential for the safety and welfare of a welder, has been listed. When the welder must work in close quarters with other workmen, screens or protective partitions, Fig. 6-20, should be used so that welding rays will not create a working hazard.

Fires

Fires, of course, are a hazard anywhere there is heat, and welding is no exception. Fire concerns the welder because it can destroy his shop and equipment, while he himself may be badly burned.

Fires are categorized into one of three distinct classes, depending on the fuel consumed and what is necessary to fight the fire. In order to have a fire there must be a *fuel,* the temperature of which must be raised by *heat* in order for it to burn. As mentioned in the section on explosions, *oxygen* is necessary for combustion; thus, air must be present to have a fire, or the fire will smother. Therefore, the three essentials for a fire are; fuel, heat, and air. If one of the three essentials is eliminated, there is no chance of a fire. Where a fire is *burning,* it is easy to see that by removing the easiest or most practical essential the fire will be extinguished. When a fire breaks out, a decision must be made almost immediately to do one of the following:

A. *Cool* the material to eliminate heat.
B. *Remove* the fuel supply (which is not possible in most fires).
C. *Smother* the fire to keep out the oxygen.

How Fires Are Classified

The welder should know the types of fires and protective measures against them. He should have a ready fire prevention skill to make the protective measures effective, if it becomes necessary. The following information, if followed closely, will help make the welder a well informed and safe worker.

Any fire needs *oxygen,* or air; in addition, any fire needs *heat,* so all fires are alike as far as oxygen and heat go. However, all fires are *not* alike when it comes to the third essential —*fuel.* So, fires are classified according to their *fuel.* Fires may be grouped into *three* classes, according to the fuel of the fire.

Generally, there are three well-known classes of fires: Class A, Class B, and Class C. A fourth class, Class D, is not well known.

Class A Fires—These fires consist of wood, paper, clothing, and similar materials. *Water* is the most effective method of fighting Class A fires, since it will pentrate deeply and cool. On small Class A fires, some extinguishers not containing water will be effective.

Class B Fires—Oils, greases, paints, and other liquids burn to make Class B fires. Once the *flame* is extinguished in a Class B fire, the fire will go out. This is because, in a Class B fire, only the *vapor* on top of the liquid surface is burning. Gasoline, for example, does not contain oxygen, so it will not burn as a liquid. *However,* when the gasoline evaporates into the air, it picks up oxygen. Once the gasoline mixes with oxygen, the *vapor* burns easily.

To fight a Class B fire, just put out the burning *vapor,* because the liquid itself really is not on fire. Foam, powder, sand, and other types of nonflammable materials can be used effectively because of their smothering action, preventing oxygen from reaching the fuel.

Since most flammable liquids will *float* on water, water must *not* be used on this type of fire. Water hitting this type of fire could float the flames over a large area and perhaps create further fire problems.

Class C Fires—The Class C fire consists of connnected (or live) *electrical equipment*. Nonconductive extinguishing materials with a smothering action should be used on Class C fires. Due to the added danger of electrical shock, water should never be used on a Class C fire.

Class D Fires—A fourth class of fires is known as Class D. Class D fires are burning, flammable *metals*. An example of a Class D fire would be burning magnesium shavings.

Many newer fire extinguishers are marked for Class D fires. *Before* working near flammable metals, a welder should find an extinguisher for Class D fires and have it handy.

Classifying Fire Extinguishers

Several systems have been worked out to label fire extinguishers. *First,* they are labeled with big letters A, B, C, and D to tell what type fire they can be used on.

Second, there is an identifying *shape* around each letter, so it can be easily recognized. A has a *triangle* shape, B has a *square* shape, C has a *round* shape, and D has a *star* shape.

Recommended Extinguisher "Suitability" Identification

 Triangle containing the letter "A" (green when shown in color) —identifies an extinguisher suitable for use on "Class A" fires — ordinary combustibles: wood, cloth, paper, rubber.

 Circle containing the letter "C" (blue when shown in color) — identifies an extinguisher suitable for use on "Class C" fires —electrical fires: motors, switches, appliances, etc.

 Square containing the letter "B" (red when shown in color) — identifies an extinguisher suitable for use on "Class B" fires — burning liquids: gasoline, oil, greases, paint, etc.

 Five-pointed star containing the letter "D" (yellow when shown in color) — identifies an extinguisher suitable for use on "Class D" fires — combustible metals.

Extinguishers suitable for use on more than one class of fire may be identified by two or more of the above described symbols. Above symbols in color are available in press-on decal form for marking older model extinguishers. For price or ordering information write: National Fire Protection Association, 60 Batterymarch St., Boston, Massachusetts 02110.

Fig. 6-22. The identification symbols and colors used on good fire extinguishers.

Fig. 6-23. A well-equipped company "fire station." (A) Water-charged extinguisher for Class A fires. (B) Dry chemical extinguisher for Class B fires. (C) Water hose attached to a *standpipe*, for Class A fires.

A   B   C

PORTABLE FIRE EXTINGUISHER SELECTION GUIDE

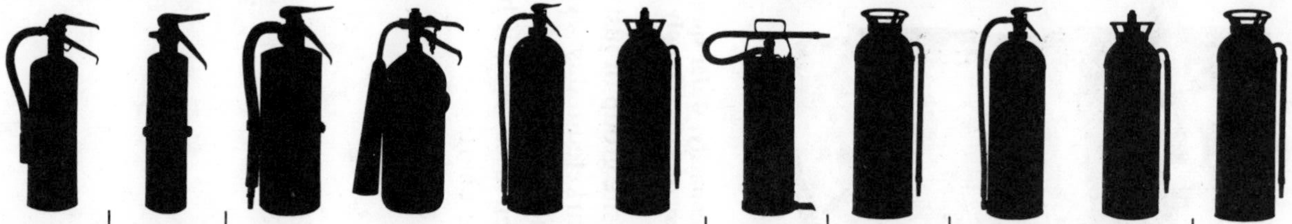

Select the correct Extinguisher for...	DRY CHEMICAL: MULTI-PURPOSE	DRY CHEMICAL: SODIUM BICARB.	DRY CHEMICAL: PURPLE K	CO² : CARBON DIOXIDE	WET CHEMICAL: WATER (Stored Pressure)	WET CHEMICAL: WATER (Cartridge Operated)	WET CHEMICAL: PUMP TANK	WET CHEMICAL: SODA ACID	WET CHEMICAL: LOADED STREAM (Stored Pressure)	WET CHEMICAL: LOADED STREAM (Cartridge Operated)	WET CHEMICAL: FOAM
A	Yes	No	No	No	Yes	Yes	Yes	Yes	Yes	Yes	Yes
B	Yes	Yes	Yes	Yes	No	No	No	No	Yes	Yes	Yes
C	Yes	Yes	Yes	Yes	No	No	No	No	No	No	No
Special Application Information	Dry chemical extinguishers put out flammable liquid fires faster. Recommended for: oil and gas storage facilities, garages and service stations, industrial plants, trucks, busses, cars, and boats —wherever flammable liquids are used or stored.			CO_2 leaves no residue. Protects intricate and expensive electrical equipment . . . also recommended wherever food is handled.	Economical protection. Recommended for heated buildings, factories, office buildings, schools, churches and hospitals. Can be furnished for wall cabinet installation.				An anti-freeze extinguisher with a chemical that tends to retard re-ignition in Class A fires. Ideal protection for unheated buildings. Underwriters' Laboratories-rated to fight Class B fires.		Practical and economical extinguisher for facilities requiring both Class A and B protection.
How to operate	Squeeze handle. Sweep under flames.	Squeeze handle. Sweep under flames.	Squeeze handle. Sweep under flames.	Squeeze handle. Direct at base of flames.	Squeeze handle. Soak burning material.	Invert. Bump. Soak burning material.	Hold hose and pump. Soak burning material.	Invert. Soak burning material.	Squeeze handle. Soak burning material.	Invert. Bump. Soak burning material.	Invert. Blanket fuel surface.
Subject to freezing	No	No	No	No	Yes	Yes	Yes (Unless Polar Crystals are used)	Yes	No	No	Yes

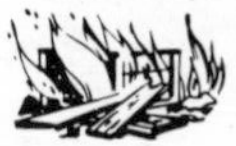

CLASS A FIRES Ordinary combustibles: wood, cloth, paper, rubber.
Triangle containing the letter "A" identifies extinguisher approved for Class A fires.

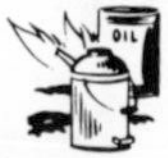

CLASS B FIRES Burning liquids: gasoline, oil, greases, paint, etc.
Square containing the letter "B" identifies extinguisher approved for Class B fires.

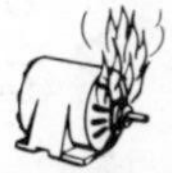

CLASS C FIRES Electrical fires: motors, switches, appliances, etc.
Circle containing the letter "C" identifies extinguisher approved for Class C fires.

COURTESY OF FYR-FYTER COMPANY OF CANADA LIMITED, 19 VICTORIA CRESCENT, BRAMALEA, ONTARIO

Fig. 6-24. Chart showing virtually all the fire extinguishers commonly found in industry, and their applications. (Courtesy, Fyr-Fyter Company of Canada, Ltd.)

Third, the shapes with letters in them have a certain *color* for each shape, if the labels are in color. The triangle-shaped A is colored *green;* the square B is colored *red;* the round C is colored *blue;* and the star-shaped D is colored *yellow.* Good extinguishers (Fig. 6-22) usually have all three:

1. The big *letters.*
2. The special *shapes* around each letter.
3. The special *color* around each letter.

Summary of Fire Extinguishers

A well-equipped shop will have fire extinguishers available for different kinds of fires. In Fig. 6-23, a well-equipped "fire station" is shown. Here, there are portable extinguishers for the three major kinds of fires, plus a long hose attached to a water outlet, available to fight Class A fires. (The water outlet built into a building for fire protection use only is called a *standpipe.*)

Fig. 6-25 shows a large canister used on Class A fires. It contains more water than the smaller extinguisher in Fig. 6-23A. Fig.

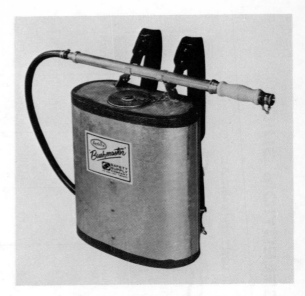

Fig. 6-25. A large-capacity, water spray extinguisher. (Courtesy of Safety Supply Co.)

6-24 is a composite group of just about all the extinguishers marketed today. A good welder will have a general idea of the basic extinguishers and their uses, even if he is not able to memorize the chart.

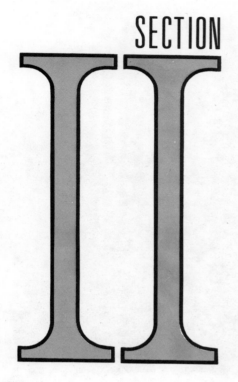

SECTION

II

oxygen-
acetylene
welding

Oxy-acetylene cutting is one of the most widely-used applications of the oxygen-acetylene flame and the oxidation process. Here, a piece is being removed from an automobile rear section being repaired.

7

Oxygen/Gas Welding

A majority of welding done with oxygen and another gas is done with oxygen and acetylene. For this reason, many discussions of oxygen-gas may refer to oxygen-acetylene, or simply oxy-acetylene, welding. Fig. 7-1 shows a group of welding students using oxy-acetylene welding equipment.

Many other gases, however, are used with oxygen to create a welding heat. All of these

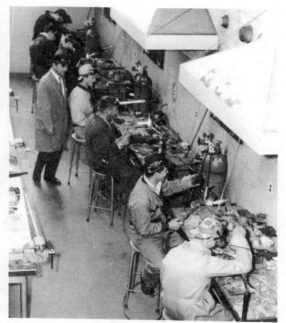

Fig. 7-1. A group of welding students practicing oxy-acetylene welding.

will be reviewed in this chapter, along with the advantages and disadvantages of each.

Recent research on oxy-acetylene welding indicates that the future of metal joining by this method will be either limited or impractical by the year 2000. In other processes, though, its use for cutting metals will continue to be dominant for many decades. It is an excellent process to use as an introduction to welding.

For this reason, the following chapters on oxygen-gas welding will be limited to those procedures and processes still considered suitable for welding various materials. Stainless steel and aluminum, two metals previously welded exclusively by the oxy-acetylene method, are now readily joined by the Tungsten Inert Gas Welding (TIG) process now used in many shops.

The information in the following chapters relating to oxy-acetylene *cutting* will help the welder better prepare joints for weldment and/or repair.

Types of Gases

A welder should have some knowledge of the gases commonly used for welding and cutting. A safety comparison of some gases presently being used for cutting with oxygen is shown in Table 7-1.

153

Table 7-1. Safety Factor For Gases Commonly Used With Oxygen.

SAFETY	Mapp Gas	Acetylene	Natural Gas	Propane
Shock sensitivity	Stable	Unstable	Stable	Stable
Explosive limits in oxygen, %	2½-60	3-93	5-59	2.4-57
Explosive limits in air, %	3.4-10.8	2½-80	5.3-14	2.3-9.5
Maximum allowable regulator pressure, psi.	225 psi at 130° F.	15	None	None
Delivery	Cylinder	Cylinder	Line	Cylinder
Tendency to backfire	Slight	Much	Slight	Slight
Toxicity	Low	Low	Low	Low
Reactions with common materials	Avoid alloys with more than 67% copper		Few restrictions	
Flame temperature in oxygen, ° F.	5,301	5,589	4,600	4,579

Acetylene

This is a colorless hydrocarbon gas which can be produced by combining hydrogen and carbon. It is a member of a large and important group of compounds that contain hydrogen and carbon only. When mixed with minimum or maximum volumes of air, acetylene can become highly explosive. However, air is not a requisite in creating an acetylene explosion. When acetylene is generated or compressed to over 28 lb./sq. in. the gas becomes decomposed into carbon and hydrogen; any shock or ignition could then be considered dangerous. In its free state, acetylene cannot be stored under pressure greater than 15 psi because it becomes unstable. When a slight amount of nonpurified acetylene is mixed with air, it has a strong odor.

Acetylene has always been recognized as a better industrial fuel for welding purposes than propane or natural gas. The Dow Chemical Company, however, has invented a new competitor for acetylene gas. This new industrial fuel is known as *Mapp*® gas. *Mapp* is not as sensitive to shock as is acetylene, and it will no doubt be used in place of acetylene for certain production jobs, especially where cutting is required.

Producing Acetylene—Calcium carbide, a compound made from quicklime and carbon in an electric furnace, is brought into contact with water to form acetylene gas. (Acetylene's chemical composition is designated as C_2H_2.) The reaction of water and calcium carbide is instantaneous, and the gaseous hydrocarbon possesses *endothermic* properties (absorbs heat when it is produced and liberates heat when decomposed). This property, in turn, creates the intense heat of the oxyacetylene flame. Acetylene consists of 92.3% carbon, and the balance is hydrogen.

Argon

This colorless, odorless, and tasteless gas is found in the atmosphere and in some volcanic gases. The symbol for *argon* is A, and its atomic weight is 39.9; its boiling point is −186.1° C. Since argon does not combine with other elements, its *valence* (combining power) is zero, and it is considered an *inert* gas (has no active chemical properties). Argon is also a poor conductor of heat. It is prepared commercially by the *controlled evaporation* of liquid air. Making up 0.94% of air by volume, it is used as a shielding gas for both Tungsten Inert Gas (TIG) and Metal Inert Gas (MIG) welding.

Carbon Dioxide

This well-known compound contains one atom of carbon and two of oxygen. Carbon dioxide is a transparent gas that exists in atmospheric air in small quantities from 0.03% to 0.04%. Carbon dioxide is not combustible, so it is often used as a shielding gas in Metal

Inert Gas welding various types of steel. This heavy gas, designated as CO_2, is formed either by carbon oxidation or by carbonates and acids interacting. Under pressure of 450 psi at $-5°$ F, carbon dioxide changes to a liquid for storage in steel bottles.

Helium

A transparent, nonflammable gas, *helium* is like nitrogen and argon because it has practically no chemical activity. Its chemical symbol is He. Helium is twice as heavy as the lightest known substance, hydrogen. Although generally obtained by distilling liquid air, helium also can be separated from its compounds by using nitrogen or argon. Since helium is an inert gas, it is used in inert gas welding in place of argon, and it is especially beneficial for shielding such hard-to-weld materials as magnesium. The cheaper argon gas, however, satisfactorily shields most metals during welding.

Hydrogen

This is the lightest element known, and it is tasteless, colorless, and odorless in gas form. Hydrogen and oxygen burn together with a colorless, hot flame which can be used as a preheating flame for cutting, especially under water. Hydrogen is also used as a fuel in oxy-hydrogen welding or for low-temperature work, and it is used as a shielding gas in atomic hydrogen welding. Hydrogen is recognized by the symbol H.

Mapp® Gas
(Methylacetylene Propadiene)

This is a liquified acetylenic compound that is not sensitive to shock. Unlike propane and natural gas, it is both fast and hot, and its performance equals that of acetylene. The Btu rating of this gas is 2450 Btu/cu. ft. The 15 psi pressure precaution for acetylene is unnecessary with this new gas. The smooth, clean cuts with *Mapp* gas surpass even those made with propane cutting. When used for welding, little or no backfire or flashback is experienced. These will happen with the oxyacetylene process if precautions are not taken. The work ratio of a cylinder of *Mapp* gas to a cylinder of acetylene is 5 to 1. *Mapp* gas cylinders are easier to handle, since they do not require the stabilizers of porous cement and acetone (necessary for acetylene) which create extra weight. A small *Mapp* cylinder is shown in Fig. 7-2. Other size cylinders are also available.

Fig. 7-2. A smaller *Mapp*® cylinder, lighter in weight than a comparable acetylene cylinder. (Courtesy of Airco Welding Products Division.)

Any comparison of cost between *Mapp* and acetylene gas must take into consideration the nature of the flame cone and the type of work to be done. The *Mapp* gas cone is longer, more bulbous, and has a tendency to spread out; acetylene tends to concentrate the heat near the tip of the cone. Therefore, for cutting jobs 1/4″ to 1″ thick, acetylene would cut costs, especially because it saves oxygen due to its better preheat qualities. *Mapp* gas, however, would provide savings if longer, continuous cuts are made on heavy cross-section steels.

Natural Gas

Distilled under *natural* conditions from the mineral oils stored in the porous layers of the earth, *natural gas* is also a hydrocarbon gas. The Btu rating of natural gas is 1200, as compared to the 1433 Btu rating of acetylene. When these gases are burning under ideal conditions, the acetylene surpasses the temperature of natural gas by about 1700° F. Due to its carbon monoxide, natural gas is poisonous; the welding operator using natural gas must shut off all gas valves after welding as an extra safety precaution.

Oxygen

This is a colorless, tasteless, odorless gas, found in a free state or combined with other elements. *Oxygen* supports combustion, and it combines readily with other substances, usually producing a flame and heat. In cutting and welding, the combustible gas used depends on oxygen for its welding suitability. Oxygen, known by the symbol O, can become a pale, steel-blue liquid under pressure and at a temperature of −119° C.

Propane

This is a liquified petroleum gas, sometimes used as a fuel for flame cutting. When mixed in a 1:5 ratio with oxygen, the temperature of the flame is only about 1000° F. less than an oxy-acetylene flame. To obtain the same volume of heat as in the oxy-acetylene flame, it is necessary to use at least *three times* as much oxygen. The Btu value of this gas is 2537 Btu/cu. ft.

Cylinder Gas Containers

The different welding gases are handled by storing them in carefully built cylinders. These cylinders are built, filled, and transported according to carefully controlled Interstate Commerce Commission (ICC) regulations.

The cylinders must be handled carefully because the gas in them is under high pressure. To get the gas out of the cylinder, a valve fitted on the cylinder's top can be opened to let gases out. *Regulators* are tightly screwed into the valve, to let the gas out in a smooth flow that the welder can use. Even when empty, the cylinders must be handled carefully, stored upright, and marked "MT", a standard abbreviation for "empty."

Most gases are used and delivered from small-to-medium size portable tanks, such as those in Fig. 7-3. For big industrial users, gases can be stored in large, permanent tanks mounted outside, as in Fig. 7-4. The large

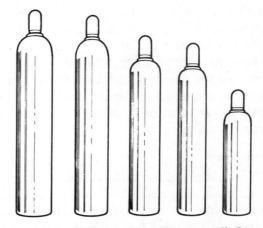

Fig. 7-3. Typical portable welding gas cylinders.

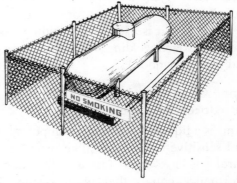

Fig. 7-4. Large, permanent, outside storage tank.

outside tanks are often used to store propane gas for home and farm use.

Natural gas is not usually stored in tanks or cylinders, because it is piped underground to users in home and industry.

Cylinders for different gases vary in color, shape, and, sometimes, in internal construction. Each tank will be discussed separately.

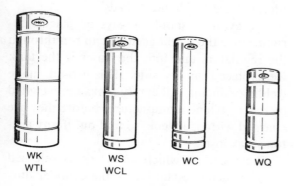

WK
WTL

WS
WCL

WC

WQ

Fig. 7-5. Various sizes of acetylene cylinders.

Acetylene Cylinders

Cylinders of varying sizes, as shown in Fig. 7-5, are used to transport *acetylene*. The acetylene cylinder is made of steel with a concave bottom, which is protected from wear by a welded steel band. Fig. 7-6 shows a cutaway view of a standard acetylene cylinder.

The acetylene cubic content is arrived at by *weighing,* rather than by a regulator dial reading. The acetylene cylinder is weighed before charging. After charging with acetylene, the tank is again weighed and the difference (in pounds) is multiplied by $14\frac{1}{2}$ to give the cu. ft. content of the acetylene. The weight of the empty cylinder, before charging, is always stamped on the cylinder.

After the acetylene cylinder is built, a wet cement mixture of various materials, such as asbestos or charcoal, is packed into the cylinder. The cylinder is then baked, and the mixture forms a porous, sponge-like mass. Under pressure, dissolved acetylene is safe, but free acetylene over 15 psi pressure is unsafe. For safety, then, it becomes necessary to dissolve the acetylene.

Acetone initially dissolves about 25 times its own volume, and as the temperature increases, so does the amount of absorption. About 40% of the cylinder volume is filled with acetone, and, as the cylinder is charged

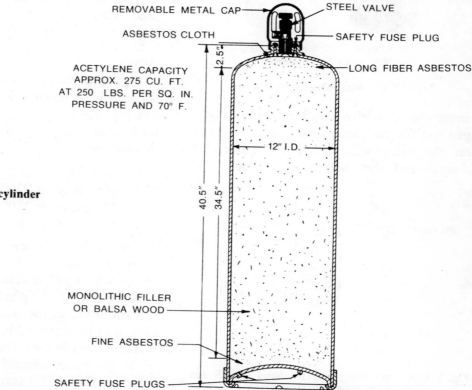

REMOVABLE METAL CAP — STEEL VALVE

ASBESTOS CLOTH — SAFETY FUSE PLUG

2.5"

ACETYLENE CAPACITY APPROX. 275 CU. FT. AT 250 LBS. PER SQ. IN. PRESSURE AND 70° F. — LONG FIBER ASBESTOS

12" I.D.

40.5"   34.5"

Fig. 7-6. Acetylene cylinder cutaway.

MONOLITHIC FILLER OR BALSA WOOD

FINE ASBESTOS

SAFETY FUSE PLUGS

with acetylene, the acetone absorbs the acetylene and swells up in the porous material.

The acetylene cylinder can be charged to 250 psi. In each cylinder are fuse plugs, consisting of steel bolts tapped into the cylinders. These are drilled and filled with a low melting alloy, Fig. 7-6, that melts between 203° and 207° F. In case of fire, the gas escapes *through* the melted fuse plugs, so a torch flame should be kept *away* from all cylinders. Most fuse plugs are on the top and/or bottom of the cylinder.

On the top of the cylinder is a valve, usually covered with a screw-on cap to protect the valve from accidental damage. Fig. 7-7

OXYGEN   ACETYLENE
CAP   CAP

Fig. 7-7. Cylinder valve protection cap.

shows the protective caps used on both oxygen and acetylene cylinders. The connections on the valve will always be left-handed threads and they will have a groove cut

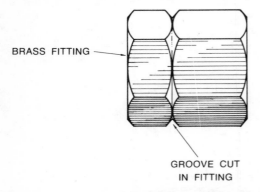

BRASS FITTING

GROOVE CUT
IN FITTING

Fig. 7-8. A brass fitting with left-hand threads has a groove cut around its circumference.

around the circumference of the neck, similar to the groove cut on the left-hand fitting shown in Fig. 7-8. All standard acetylene fittings have left-hand threads.

The cylinder should always be placed upright, since the liquid acetone can run through the regulators and into the hoses if the tank were placed on its side. This would make welding difficult. All empty tanks are weighed by the acetylene manufacturer before they are charged with acetylene, to find out if any acetone was lost.

The speed at which acetylene is released from the acetone and leaves the cylinder during welding is called the *withdrawal rate*. If acetylene is withdrawn faster than $\frac{1}{7}$ the cylinder capacity per hour, acetone will *also* be released through the regulator, hose, and torch. Since acetone deteriorates all the rubber and plastic parts in the system, the regulator diaphragms and hoses might be weakened and ruined. If the acetylene withdrawal is greater than $\frac{1}{7}$ cylinder capacity per hour, it is recommended that a manifold system be used. The oxygen or acetylene manifold system is shown in Fig. 7-9.

Argon

Fig. 7-10 shows an *argon* cylinder, which is constructed much like an oxygen cylinder. The construction allows a high filling pressure for increased capacity. Argon cylinders come in four sizes:

1. 78 cu. ft. at 2200 psi
2. 150 cu. ft. at 2200 psi
3. 275 cu. ft. at 2200 psi
4. 330 cu. ft. at 2640 psi

All of the above cylinders are brown in color, except the 150 cu. ft. cylinder, which is brown with a yellow collar.

Another type of cylinder (Fig. 7-11) can contain argon in the liquid state, as well as oxygen or nitrogen. Although these gases are in liquid form *in* the cylinder, they are drawn off as gas when released from the cylinder. Since the liquified gases require less space

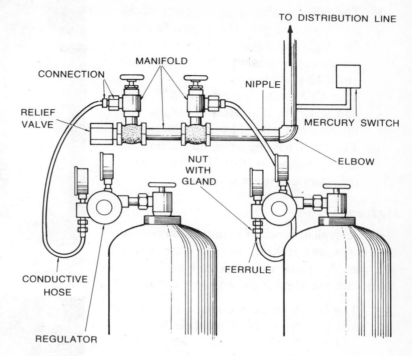

Fig. 7-9. A manifold system used when too much acetylene would be removed from one single tank.

than their gaseous equivalents, fewer cylinder changes are needed. This reduces both transportation and handling costs. This type of cylinder is 58″ high, and it has an outside

cylinder is known as an LC-3 cylinder, manufactured by the Linde Division of Union Carbide Corp.

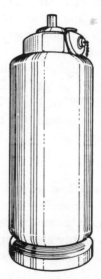

Fig. 7-11. A cylinder used to store *liquid* argon, oxygen, or nitrogen. These are dispensed as a gas when the valve is opened.

Fig. 7-10. The top portion of an argon cylinder.

diameter of 20″; it weighs 215 pounds empty and 514 pounds full of argon. This special

Carbon Dioxide

The two common types of cylinders used for this gas are:

1. The *HK*—having a gas content of 50 cu. ft. at 852 psi. These cylinders are 56″ high and 9¹⁄₁₆″ in diameter. They weigh approximately 145 pounds full and 142 pounds empty.
2. The *HLK*—this cylinder is 57″ high and 9¹⁄₁₆″ in diameter; it weighs 164 pounds full and 114 pounds empty. The HLK has the same cu. ft. content and pressure as the HK type. Both cylinders are silver in color.

Hydrogen

Three typical types of cylinders, known as the H, HS, and HQ, are used to store *hydrogen* gas. All three cylinders can be blue in color with the HS having a yellow collar.

1. The *H* cylinder can contain 191 cu. ft. at a pressure of 2000 psi. It is 56″ high and 9¹⁄₁₆″ in diameter. It weighs 135 pounds full and 132 pounds empty.
2. The *HS*, or middle-size, cylinder holds 118 cu. ft. of gas at the same pressure as the HQ cylinder. The HS cylinder is 51″ high and 9⅞″ in diameter; it weighs 79 pounds empty and only one more pound full.
3. The *HQ* cylinder, when full, consists of 62 cu. ft of gas at 2000 psi. This cylinder is 35″ high and 7″ in diameter. It weighs 64 pounds full and one pound less empty.

Mapp

Cylinders for *Mapp* gas (Fig. 7-12) can be supplied in capacities from 7½ to 115 pounds. The most common cylinder weighs only 50 pounds empty and approximately 115 pounds full. If a cylinder designed for a minimum pressure of 200 psi *Mapp* gas is dropped or struck, it will not explode from the shock. This fuel and its mixtures are very stable to shock, much like common substances such as water and benzene.

Mapp gas is also available in bulk tanks conforming to ASME standards (Fig. 7-13).

Fig. 7-12. Various cylinders used to store *Mapp* gas. (Courtesy of Airco Welding Products Division.)

Fig. 7-13. Permanent outside storage tanks for *Mapp* gas. (Courtesy of Airco Welding Products Division.)

Customers using large volumes of gas for heating and cutting can use the outdoor bulk storage system. With a large *Mapp* tank outside, a manifold system inside the plant distributes the gas to different work stations, as shown in Fig. 7-14.

In Table 7-2, the filling properties for *Mapp* gas are outlined for safety.

Oxygen

Unlike the acetylene cylinder, the *oxygen* cylinder is made of seamless drawn steel, as shown in Fig. 7-15. Oxygen cylinders are

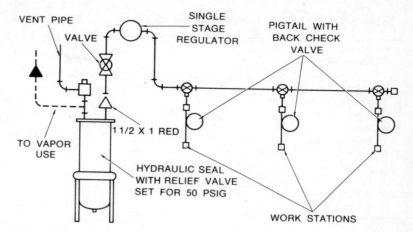

Fig. 7-14. *Mapp* **gas typical manifold installation.**

usually painted red or green, with green being the most common.

The large *Linde*® LC-3 liquid cylinder, mentioned previously, contains the gaseous equivalent of 3000 cu. ft. of oxygen and weighs 463 pounds. This is too large for easy handling, so there are more common cylinders available, containing 20, 80, 122, and 244 cu. ft. of oxygen at a pressure of 2200 psi.

The most common oxygen tank used for shop purposes is the 244 cu. ft., or *full size,* cylinder. This cylinder is approximately 56″ high and 9¼₆″ in diameter. When the cylinder is full of oxygen, it weighs approximately 152 pounds, while it weighs 132 pounds empty. The *half size,* or 122 cu. ft. cylinder, is about 4′ high and 7½″ in diameter. The

half size weighs about 124 pounds full and 114 pounds empty.

Welders should know that oxygen combined with oil is highly explosive; therefore, precautions should be taken to keep the cylinders away from all types of oily or greasy machinery.

Whether an oxygen cylinder is the standard 244 cu. ft. size or the smallest size, the pressure will be the same when filled with oxygen. All oxygen, as well as acetylene, cylinders usually belong to the supplier. Cylinders are often leased to customers on a 25-year basis or are exchanged monthly on a no-charge arrangement; nonleased cylinders usually have a monthly *demurrage* charge if kept over one month. This is a kind of rent on the cylinder itself.

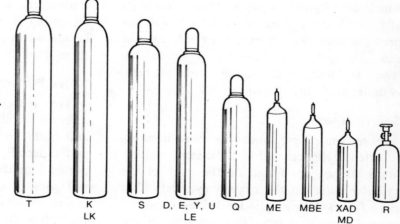

Fig. 7-15. The many sizes of seamless oxygen tanks.

Table 7-2. Mapp Gas Tank Filling Capacities.

Basis:

1. Weight of a gallon of
   Mapp® gas at 60° F. ....4.802 lbs.
2. Weight of a gallon of water at 60° F. ....8.33 lbs.
3. Specific gravity of liquid
   Mapp® gas at 60° F. 0.576
4. Specific gravity of Mapp gas at 60° F. 1.48
5. Standard cubic feet/pound gas at 60° F. 8.85
6. The National Fire Protection Association Standards for
   above-ground storage states that the vessels may be
   filled to the following filling densities with a
   product having a specific gravity of 0.576 at a temperature of 60° F.

Size of Tank in Water Gallons	Pounds of Water Tank Will Hold	Filling Density	Max. Pounds of Mapp® Gas Tank Will Hold	Max. Gallons of Mapp® Gas at 60° F. Tank Will Hold
500	4,164 ×	50% =	2,082 =	433
1,000	8,328 ×	50% =	4,164 =	867
2,000	16,657 ×	53% =	8,828 =	1,838
18,000	149,909 ×	53% =	79,451 =	16,545

50% = 86.60% of actual tank volume capacity.
53% = 91.92% of actual tank volume capacity.

Common Safety Around Gas Cylinders

Because gas is so concentrated in cylinders and under such high pressure, some "common sense" safety ideas are needed to keep the tank from exploding in the shop.

Arc Welding

Electric welding should never be done anywhere near cylinders, since an arc accidentally struck on a cylinder, especially a high-pressure tank like the oxygen cylinder, could cause the tank to rupture. This is particularly true if it is pentrated by the arc while the tank is under such high pressure.

Safety Fuses

Fuse-type plugs, similar to those used on acetylene cylinders and with the same melting temperatures, are found on the back of the *cylinder valve* on the oxygen cylinder. (See Fig. 7-16.) Many cylinder valves on other gas cylinders now offer this same protection

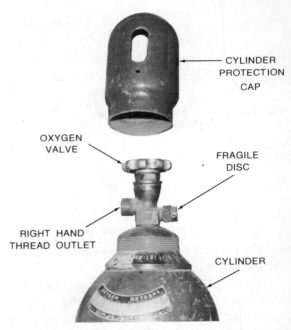

Fig. 7-16. Valve on an oxygen tank. The safety fuse-type plug is located on the right side of the valve.

in case of fire. During a fire, the fragile copper discs heat up and release the gas from the cylinder, as expansion takes place within the cylinder because of the fire's heat.

Cylinder Testing

Gas cylinders are tested and retested for flaws and leaks. For example, oxygen cylinders are factory tested to 3360 psi, and the date of each test is stamped on the cylinder below the valve. Five years from the date of testing, the cylinder is rechecked for flaws which may have developed. Such flaws could be unsafe in any high-pressure cylinder like those used for oxygen.

Valve Protection

The *protective cap* should be securely screwed on all cylinders once the regulators and hoses have been removed (Fig. 7-16), and especially if the tanks are to be transported long distances. Accidentally dropping full cylinders could be dangerous if the valves become damaged or are broken off. When the valves are broken off, the gas can all escape at once, pushing the tank and turning it into a jet-like missle.

Manifold Systems

Supplies of oxygen, acetylene, and other gases, piped throughout a shop to various work areas for better welding efficiency, is made possible through *manifold systems*. A simple three-tank manifold system is shown in Fig. 7-17.

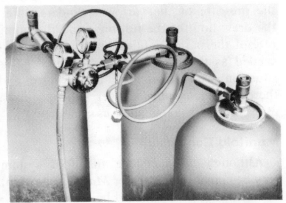

Fig. 7-17. A simple three-tank manifold system. (Courtesy of Airco Welding Products Division.)

If the amount of welding being done requires an extra supply of oxygen, acetylene, *Mapp,* or other gas, a more complex manifold system can be used to connect a number of cylinders, as in Fig. 7-18. This not only increases the supply, but it also allows welding to take place at a number of different stations through an in-shop piping system, as was shown in Fig. 7-14. Manifolds for cylinders in sets of 6, 10, 20, or 30 can be used to distribute oxygen or other gases to various outlets within a certain shop area. These manifolds are usually constructed in two halves; one half can work independently so that the supply to the individual stations is not interrupted during a cylinder change.

Manifold Design

All manifolds must be designed for the type of gas (oxygen, *Mapp,* etc.) being considered; also, all leads, in-plant piping, and connections must be able to withstand the required gas pressures.

Portable manifolds allow the gas to pass directly from the cylinders, through individual leads, to a common coupler block. The block is attached to a common pressure-reducing regulator, which allows a constant, regulated pressure within the plant lines. The single regulator and connecting lines can be seen in Fig. 7-19. The coils of tubing, in Fig.

Fig. 7-18. A complex ten-tank manifold system. (Courtesy of Airco Welding Products Division.)

Fig. 7-19. The regulator and tubing coils used with a manifold system. (Courtesy of Airco Welding Products Division.)

7-19, allow the cylinders to move slightly without breaking the line.

Some types of manifolds are designed to allow cylinders to be connected together in a sequence. The gas content from each cylinder passes through a coupler tee fastened to each cylinder's valve and flows into a common line. One or more regulators, which reduce and regulate the gas flowing from the cylinders to each work station in the shop, become a part of the *shop* pipeline supply system. At each work station, the welder may have a single regulator, as shown in Fig. 7-20.

Fig. 7-20. Single regulator as used at each work station of a manifold-piping supplied shop. (Courtesy of Airco Welding Products Division.)

Special Oxygen Precautions—The welder will find that oxygen manifolds are rarely located in the same area as acetylene manifolds or other combustible gas cylinders. Most fire laws require that manifolds with over 6000 cu. ft. of oxygen be located in special fireproof buildings free of oil, grease, or any other flammable material.

Special Acetylene Precautions—Acetylene can be piped to various shop work areas not only from cylinders, but also after it is produced in various size generators. Acetylene cylinders coupled in a manifold system must have approved flashback arrestors installed

between each cylinder and the coupler block. However, if the manifold system is outside and has only two or three cylinders, only one flash arrester is required—between the coupler block and the regulator. Acetylene and liquified fuel gas must always be manifold-connected in a vertical position.

Manifold Cylinder Pressure

Fuel gas manifolds, such as acetylene, must have only equal pressure cylinders. Should the pressure of the various cylinders differ, the draw limit of the full ones and the filling rate of the partially filled ones would be exceeded as the pressures are equalized. This situation would raise the temperature and thereby increase the danger of explosions, especially with gases such as acetylene.

Manifold Conversion

Although an acetylene manifold system may be converted to a safer *Mapp* gas system, the following points should be considered:

A. Regulators and relief valves should be approved by the *Mapp* gas manufacturer.
B. The line sizes used with acetylene will be ample. Pressure drop through lines will be lower for the same pound/hour flow.
C. Acetylene may be flushed from the system with *Mapp* gas, if the pressure is kept lower than 15 psi at any point in the system. All acetylene should be purged by the *Mapp* gas from all dead-end pockets, usually with the pipe system open at one end. Purging with pressure greater than 15 psi could detonate the acetylene, which is in a free state and ready to explode.

Acetylene Generators

There are two types of generators used to produce acetylene. The first type is known as the *water-to-calcium carbide generator*.

Water-to-Calcium Carbide Generator—
The original type of *water-to-calcium carbide generator* allowed water to drip on calcium carbide. Calcium carbide is obtained by heating limestone in a furnace. The calcium carbide used in this type of generator is usually in the form of bricks or cakes, placed to present only a limited surface area to the water. This surface area, when contacted by water, forms acetylene gas. Today, few welding shops use this kind of generator.

The more modern *carbic acetylene generator* uses calcium carbide in the form of cakes or bars. This generator is sometimes called a *water-recession* acetylene generator, as in Fig. 7-21. When the acetylene is released through

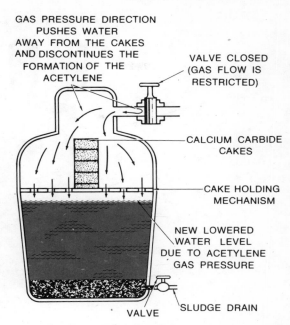

Fig. 7-22. Water position due to acetylene pressure when the water-recession generator has stopped producing acetylene gas.

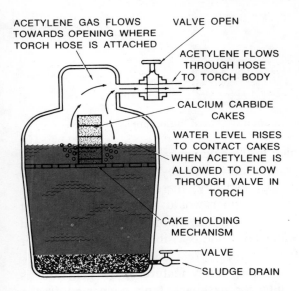

Fig. 7-21. Carbic acetylene generator, also known as a water-recession acetylene generator.

the welding torch, usually during ignition and burning, the loss of gas pressure allows the water to rise and come in contact with the calcium carbide bricks. As long as the water is in contact with the calcium, some acetylene is generated through the gas outlet to the torch, allowing the torch to remain lit.

If the valve on the blowpipe is closed, or if the main gas valve on the generator is shut off, gas pressure continues to build up to a point where the water is forced *down* in the

generator, and away from the calcium carbide. Gas generation stops shortly after the water is forced away from the calcium cakes due to such pressure, and only by allowing the acetylene to escape will the water be able to rise and start generating gas. (See Fig. 7-22.)

For welding purposes, this kind of acetylene generator is not highly recommended since gas is generated as long as the calcium carbide is saturated with the water. Miner's lamps, as well as other former lighting devices using this technique for generating illuminating gas, have been replaced by other, safer lights.

Calcium-to-Water Generator—The second type of generator, known as the *calcium-to-water generator*, is used in both small and large shops. (See Fig. 7-23.) These generators, although usually stationary, are available as portables. Or, shop areas can be supplied from stationary acetylene generators through an in-plant piping system.

This method of producing acetylene gas is instantaneous. Calcium carbide is allowed

HOPPER

PRESSURE
GAUGE

ACETYLENE
PIPE LINE

PRESSURE
RELEASE VALVE

FLASHBACK
TANK

Fig. 7-23. Calcium-to-water generator. The most common type used in the shop.

to drop into pure water, which produces acetylene gas until the calcium carbide supply is discontinued. The control over the gas generated does not present the problem that the water-to-calcium carbide generator does.

The generator, shown in Fig. 7-24, is constructed so it feeds calcium carbide into the water as soon as the torch is lit. In turn, this releases the gas pressure, closing the calcium carbide feeder. When the torch is turned off, the unused calcium carbide continues to produce gas. This gas, in turn, places pressure on the calcium carbide feed control at the bottom of the calcium carbide hopper, thus stopping gas generation.

Safety requirements set forth by Underwriters' Laboratories state that there must not be less than ten inches of space for the generated gas, from the calcium carbide stopper (feeder plate) to the top of the water, as shown in Fig. 7-24. One gallon of water is required for every pound of calcium car-

bide in the calcium carbide hopper. The generators are rated by the calcium carbide hopper capacity; therefore, a 25-pound capacity hopper would require 25 gallons of water in the generator vessel. A maximum water level line is visible for filling the water tank to the correct level.

Another safety feature of the generator is a filter or flashback tank connected to the main container. This filter is filled with water or anti-freeze through which the gas must filter before reaching the mixing chamber in the torch. The filter tank eliminates any ammonia from the acetylene and prevents ignition from reaching the main gas area through flashback (which could result in an explosion).

A 15-pound preset safety valve is attached to the generator to release any gas pressure over 15 psi in the generator. As mentioned earlier, free acetylene is unstable at pressures over 28 psi at 70° F.

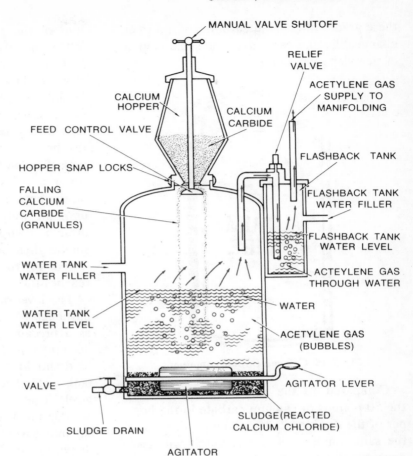

MANUAL VALVE SHUTOFF

RELIEF VALVE

ACETYLENE GAS SUPPLY TO MANIFOLDING

CALCIUM HOPPER

CALCIUM CARBIDE

FEED CONTROL VALVE

FLASHBACK TANK

HOPPER SNAP LOCKS

FLASHBACK TANK WATER FILLER

FALLING CALCIUM CARBIDE (GRANULES)

FLASHBACK TANK WATER LEVEL

WATER TANK WATER FILLER

ACTEYLENE GAS THROUGH WATER

WATER

WATER TANK WATER LEVEL

ACETYLENE GAS (BUBBLES)

VALVE

AGITATOR LEVER

SLUDGE (REACTED CALCIUM CHLORIDE)

SLUDGE DRAIN

AGITATOR

Fig. 7-24. Illustrated cutaway view of an acetylene generator.

Generator Ratings

As mentioned earlier, the rating of the generator is based on the size of the hopper that (when filled) will produce a determined number of cubic feet of acetylene. Regardless of the amount of water and procedures used, one pound of calcium carbide will generate 4½ cu. ft. of acetylene gas.

Generators are also rated by their capacity to develop gas volume over a period of one hour. They are classified as *A*, Single-Rated, and; *B*, Double-Rated. Except for the size of carbide release mechanism in the hopper, they are both similar to the type illustrated in Fig. 7-24.

Single-Rated Generators—This generator uses a coarse calcium carbide such as pea, nut, or lump. Since these grades give the same amount of acetylene per pound, the type used depends on the generators' hopper

mechanism. They cannot be interchanged in a common hopper; for example, the pea-type hopper mechanism would not allow the nut or lump carbide to reach the water.

A 50-pound capacity hopper on a single-rated generator would require a 50-gallon water level with a free 10-inch acetylene gas space. The water level will be marked on the generator body. The 50 pounds of calcium carbide, when fully dissolved in the 50 gallons of water, would give 50 cu. ft. of acetylene every hour for 4½ hours. The 50 pounds of calcium carbide could, therefore, produce 225 cu. ft. of acetylene gas.

Double-Rated Generators—These generators use a fine grade of calcium carbide, such as rice or 14 N.D. (depending only on the type manufactured) purchased in containers as shown in Fig. 7-25. Like the calcium carbides used in single-rated generators, these grades of calcium carbide are not inter-changeable, and each gives the same amount of acetylene per pound.

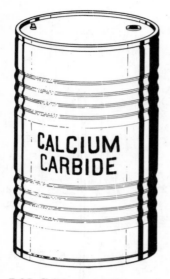

Fig. 7-25. Calcium carbide container.

Compared to the *single*-rated generator, the 50 pounds of calcium carbide in the hopper of the *double*-rated generator will require the same amount of water, but it will produce twice as much gas per hour; in this case, the amount would be 100 cu. ft. Since 50 pounds of calcium carbide can produce only 225 cu. ft. of gas, the 50 pounds of calcium carbide will be consumed in 2¼ hours in this double-rated generator.

These generators, as well as the single-rated generators, can be supplied by various manufacturers in sizes ranging from the small 10-pound hopper type up to the larger 1000-pound stationary units.

Charging Acetylene Generators

To charge the type of generator pictured in Fig. 7-24, the carbide feed valve must first be closed by turning the manual shutoff lever on top of the carbide hopper. The pressure remaining in the generator should be freed through the relief valve on top of the flashback tank. Then, the water should be drained from the sludge drain at the bottom of the generator. The sludge drain is left *open* for the time being.

The calcium carbide hopper on top can now be removed from the generator by releasing the hopper snap locks on the bottom of the hopper. Fresh purging water should now be flushed through the open top where the hopper came off, or through the water tank filler on the tank's side. While the clean water is being flushed in, the agitator lever on the bottom of the tank should be revolved until the water being flushed out the sludge drain is completely clear. The sludge drain should then be shut off and clean water allowed to run into the container until it reaches the marked water tank level line.

The level of the water or antifreeze in the flashback (or filter) tank can now be checked as a precautionary measure. If the water is not up to the flashback tank water level line, fluid should be added through the flashback tank water filler, until the tank has the right level. The calcium carbide hopper should now be filled with the recommended type of calcium carbide, indicated on the container shown in Fig. 7-26.

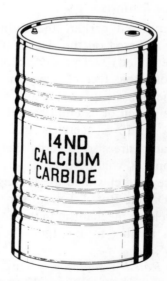

Fig. 7-26. Marking for calcium carbide content.

Most hoppers are turned upside down when removed from the generator, and filled from the bottom with the carbide release mechanism open. After the hopper is filled, the mechanism is closed. The hopper is then placed on the tank top of the generator and held on by screw-down clamps or locks.

A rubber seal is usually installed between the hopper and the tank of the generator, so acetylene will not escape between the two sections. The welder should now check to see that the torch valves are completely closed; after this, the manual feed control valve can be opened to allow the carbide to fall into the water.

As a precautionary measure, the welder should always follow these recharging procedures for *each* new hopper of carbide. Filling with fresh, clean water eliminates overheating which could endanger not only the generator, but the operator as well.

Since acetylene will explode, a welder has to be careful when operating or recharging the generator. Recharging must be done away from flames or people smoking. If the sludge must be stirred, use a wooden paddle. If the water in the generator freezes, it should be thawed with steam or hot water only. When moving a generator, the welder should shut off the carbide feed control valve and lock the manual shutoff lever, so that carbide will not be released into the water. The gas pressure should also be released through the relief valve.

Pipes from generators to shop work areas should be painted red, the color code for acetylene. Copper fittings should not be used in the lines due to the dangerous chemical reaction that would be set up between copper and acetylene.

8

Oxygen/Gas Welding Accessories

Once the fuel and oxygen gases are obtained, other welding tools are needed to get the gases safely out of the tanks and into accurate use as a welding flame.

Regulators are used on the tank (or manifold) valve to release the gas evenly. *Hoses,* or lines, are used to transport the gas to the welder's work. The *torch* itself brings the different gases together for the first time in an accurate mixing chamber, while the many different torch *tips* aim the mixed gases into a usable flame pattern. Each of these pieces of equipment has many variations and must be looked at individually.

In all flame cutting, welding, and heating, oxygen and a fuel gas are combined and burned to provide heat to do a certain type of job. The welder must be able to use these combined heating gases effectively to turn out high-speed, economically finished jobs of high quality.

Gas is controlled by the welder through adjustments made with regulators (Fig. 8-1) and torch valves (Fig. 8-2). The gas and flame size are also controlled by changing the tip sizes. Although there are many tip sizes and shapes, only a few are shown in Fig. 8-3. A group picture of all available torch tips would be almost impossible to assemble! Regulators and torch valves work together to control the amount (volume) of gas flowing through the tip.

Fig. 8-1. Typical regulator used by welders to control gas delivery. (Courtesy of Airco Welding Products Division.)

Usually, complete welding kits are bought for welders. A typical oxygen-acetylene kit is shown in Fig. 8-4.

When a welder finishes welding thin steel sections and has to weld heavier sections of the same type material, he must provide more gas volume for heating. In order to get more gas through the tip, he will need a bigger tip

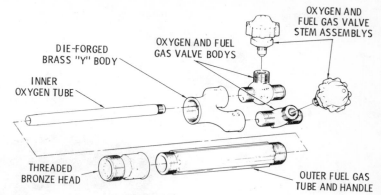

Fig. 8-2. The torch. This tool mixes the gases before welding.

OXYGEN AND FUEL GAS VALVE STEM ASSEMBLYS

OXYGEN AND FUEL GAS VALVE BODYS

DIE-FORGED BRASS "Y" BODY

INNER OXYGEN TUBE

THREADED BRONZE HEAD

OUTER FUEL GAS TUBE AND HANDLE

with its large *orifice*. The orifice is the gas outlet hole (Fig. 8-5). The torch valves will have to be opened wider, and/or the regulators must be set higher, to get more gas through the larger tip.

Regulators

In order to do any flame cutting, welding, or heating, the welder must have enough gas available to burn and generate the amount of heat the job needs.

Gas volume is referred to as *pressure* in the welding industry. The instruments that regulate the volume of gas are known as pressure *regulators,* and the gauges on the regulators are calibrated in pounds per square inch (psi). A typical regulator and its gauge are shown in Fig. 8-6. Welders should refer

to the manufacturers' specification charts for recommended pressure settings for the most efficient use of each tip. Table 8-1 shows a set of recommended pressures to be used for different tips and different metal thicknesses.

A welder should know why the term *pressure* is used in welding terminology rather than *volume*. Though pressure is not the same thing as volume, pressure is a useful indicator of volume. At the same time, pressure keeps the welder aware of specific hazards. Excessive pressure can overstress and rupture regulators or hoses. If pressure is unknown, the stress danger is much higher. Moreover, acetylene requires careful pressure control while being regulated, since free acetylene cannot be safely pressurized to more than 15 psi. Gauges on acetylene regulators indicate such a danger by having a red line extending

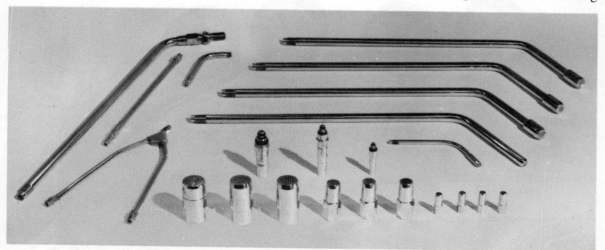

Fig. 8-3. Several of the many torch tips available. (Courtesy of Air Reduction Co., Inc.)

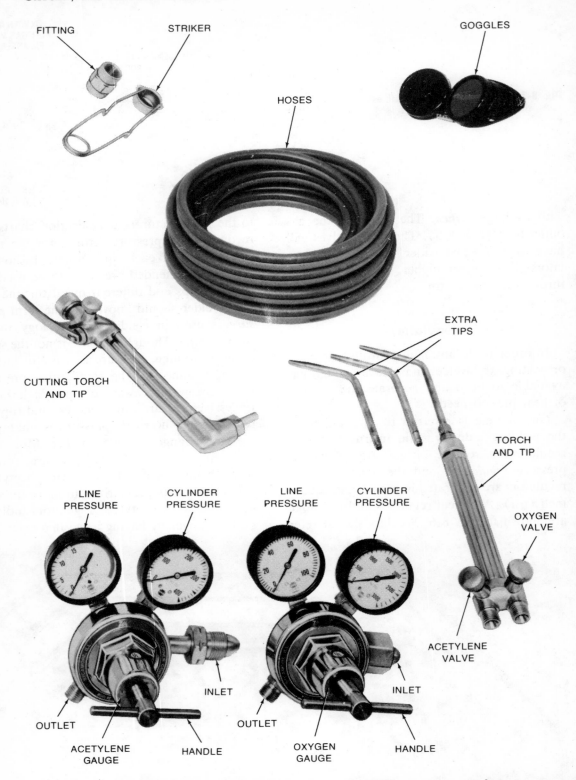

Fig. 8-4. A typical oxy-acetylene welding outfit. This equipment does not usually include the hoses. (Courtesy of Turner Corporation.)

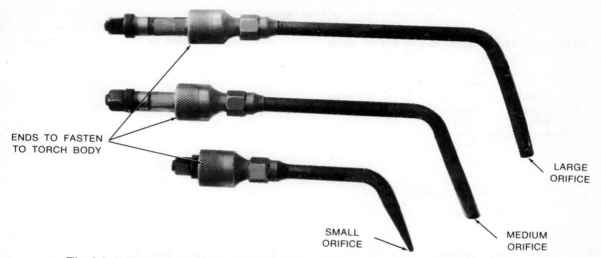

ENDS TO FASTEN
TO TORCH BODY

LARGE
ORIFICE

SMALL
ORIFICE

MEDIUM
ORIFICE

Fig. 8-5. Different welding tips, needed to change orifice sizes for different welding heats.

beyond the number 15 on the face of the gauge.

Fig. 8-6. Typical regulator showing the gauge reading in psi. (Courtesy of Airco Welding Products Division.)

The pressure of a system is usually increased when heavier work requires a greater volume of heat. An increase in gas pressure causes an increase in gas flow, and, within limits, a tip can pass more gas by increasing the pressure. Beyond the pressure limits of a given tip, it will be necessary to install a new tip.

Regulator Qualities

The regulator is the heart of the welding and cutting system. It takes gases at high cylinder pressures and reduces them to easily-controlled working pressures and flows. The four most important attributes of a quality regulator are listed below.

1. It must reduce source pressures to lower working pressures.
2. The operator must be able to set the desired working pressures.
3. It must provide an even, nonfluctuating flow.
4. It must provide high enough flow capacities for the work to be done. It must be able to control the required gas volume.

A nonfluctuating flow is very important when maximum quality is required. A fluctuating flow produces a pulsating, inconsistent flame, which may cause poor job quality. Even when high quality is not required, a relatively stable flow is desirable, to eliminate the need for constantly adjusting the torch.

A fifth, and very important consideration for any gas regulator, is its *flow capacity curve*. This determines how much gas the regulator can handle at given pressures.

Table 8-1. Recommended Pressures for Different Welding Applications.

APPROXIMATE WELDING TIP SIZES AND GAS PRESSURES FOR WELDING TORCHES AND TIPS.

Tip No.	Thickness of Metal - in.	Oxygen & Acetylene Pressure - psi	Acetylene Flow - cfh	Oxygen Flow - cfh
00	1/64	1	0.1	0.1
0	1/32	1	0.4	0.4
1	1/16	1	1	1.1
2	3/32	2	2	2.2
3	1/8	3	3	8.8
4	3/16	4	17	18
5	1/4	5	25	27
6	5/16	6	34	37
7	3/8	7	43	47
8	1/2	8	52	57
9	5/8	9	59	64
10	3/4 up	10	67	73

Pressures and consumptions shown are approximately correct for both separable tips with appropriate mixers and for tip-mixer assemblies.
Gas pressures should be increased slightly for hose lengths greater than 25 feet.

APPROXIMATE CUTTING TIP SIZES AND GAS PRESSURES FOR CUTTING TORCHES AND TIPS.

Tip No.	Thickness of Metal- in.	Oxygen Press - psi	Oxygen Flow - cfh	Acetylene Press - psi	Acetylene Flow - cfh	Hand Cutting Speed - in./min.
0	1/4	30	45	3	9	16 -18
1	3/8	30	75	3	12	14.5-16.5
1	1/2	40	85	3	12	12 -14.5
2	3/4	40	120	3	13.5	12 -14.5
2	1	50	140	3	13.5	8.5-11.5
3	1 1/2	50	180	3	16	7 - 7.5
4	2	50	225	3	16	5.5- 7
5	3	50	290	4	20	5 - 6.5
5	4	60	330	4	20	4 - 5
6	5	50	360	5	27	3.5- 4.5
6	6	55	390	5	27	3 - 4
7	8	60	540	6	36	2.5- 3.5
7	10	70	625	6	36	2 - 3
8	12	70	760	6	36	1.5- 2

Pressures and consumptions shown are approximately correct for tips with light peheating flames when used on clean plate.

Courtesy of Canadian Liquid Air

Flow Capacity Curve

This shows the maximum cubic feet of gas flow per hour possible with various combinations of inlet and outlet pressures. Different regulators have different curves; Fig. 8-7 shows the curve for an H102 *Smith* Oxygen Cutting Regulator. As mentioned previously, flow curves are available with each type of regulator.

In order to determine the maximum flow which could be obtained from using such a regulator, the welder would follow the required lines on the flow chart to a point of intersection. As an example, if a welder had an outlet pressure of 160 psi and an inlet pressure of 800 psi, Fig. 8-7 could be used as follows: A line drawn *up* from the 160 psi outlet pressure would cut the curved 800 psi inlet pressure line at about 9000. Going *over* to the flow figures on the chart's *left* side, the welder would find that, on this specific regulator, a maximum gas flow of 9000 cubic feet per hour is possible at 160 psi.

On the flow charts, inlet and outlet pressures are expressed in pounds per square

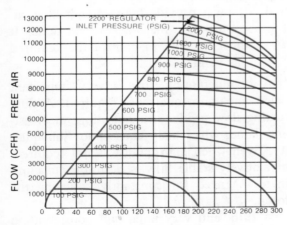

Fig. 8-7. Flow chart for a typical regulator. This chart is for an H102 *Smith* Regulator. (Courtesy of Smith Welding Equipment.)

inch gauge (psig). Flows are expressed in cubic feet per hour (cfh) of free air. Since the curves are based on the results obtained by using *free air*, the flow must be corrected for specific gravity when *other* gases are being used. Correcting is more important when using such gases as hydrogen or helium, especially during atomic-hydrogen or inert-gas welding.

Gases commonly used with regulators, and their specific gravity correction factors, are listed in Table 8-2. Once the maximum flow has been determined, the welder multiplies the cfh by the gas specific gravity to determine the gas flow.

Table 8-2. Specific Gravity Correction Factors for Common Gases Compared With Free Air.

Gas	Correction Factor
Acetylene	1.054
Oxygen	.951
Propane	.810
Butane	.706
Methane	1.342
Hydrogen	3.790
Nitrogen	1.015
Carbon Dioxide	.810
Argon	.852
Helium	2.690

Courtesy of Smith Welding Equipment

Table 8-3. Hose Effect Chart.

Hose Diam. (in.)	Hose Length (ft.)	Cutting Tip Size	Reg. PSI Static	Reg. PSI Flowing	Inlet PSI Torch	PSI Drop In Hose	Gas Flow CFH
3/16	50	3	50	47	37½	9½	169
3/16	100*	3	51	47	26	21	129
3/16	50	5	84½	78	44	34	370
3/16	100*	5	83½	78	22	56	215
3/16	50	7	108	100	24	76	510
3/16	100*	7	106½	100	9	91	270
3/16	50	9	138½	130	19½	110½	735
3/16	100*	9	136½	130	7	123	405
1/4	50	3	50½	47	44½	2½	194
1/4	100*	3	50	47	42½	4½	188
1/4	50	5	86	78	68½	9½	540
1/4	100*	5	85	78	58½	19½	470
1/4	50	7	114	100	68	32	1140
1/4	100*	7	110	100	49	51	870
1/4	50	9	149½	130	56½	73½	1110
1/4	100*	9	144	130	36½	93½	1290
1/4	100**	3	50	47	36	11	164
1/4	100**	5	84½	78	42	36	360
1/4	100**	7	108	100	25	75	560
1/4	100**	9	140	130	18	112	795
3/8	50	3	51	47	46	1	190
3/8	50	5	86	78	74½	3½	580
3/8	50	7	117	100	86	14	1400
3/8	50	9	163½	130	89½	40½	2700
3/8	100*	3	51	47	46	1	198
3/8	100*	5	86	78	72	6	570
3/8	100*	7	115	100	77	23	1280
3/8	100*	9	155	130	70½	49½	2100

*—Two 50 ft. lengths of hose connected together with standard hose unions.

**—Four 25 ft. lengths of hose connected together with standard hose unions.

Courtesy of Smith Welding Equipment

Hose Effects—Welders interested in gas consumption factors must consider the effect of the welding hose *lengths* and inside diameter. Table 8-3 shows how hose length, diameter, and tip size affect pressure and flow.

Hoses are discussed more thoroughly later in the chapter, but this table is necessary to compute gas flow and regulator qualities.

Basic Regulator Types

Many different kinds of regulators are available for gas welding. There are cylinder regulators and line regulators. They may be gauged, gaugeless, preset, or *flowmeter* regulators, as shown in Fig. 8-8.

Basically, all regulators are either a single-stage or two-stage design. Because they are equally capable of reducing pressure, the welder should base his selection on performance and cost differences.

Two-stage Regulators

A two-stage regulator will maintain nearly absolute outlet pressures and flows during long periods of operation. It is best suited for continuous operation and/or jobs requiring high-quality regulation, such as flame hardening, metallizing, and hardsurfacing. Flow charts are usually available for the type

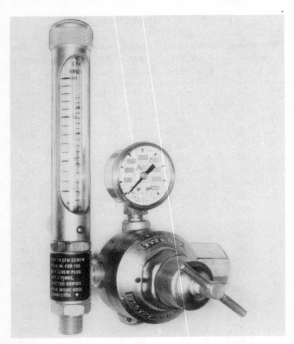

Fig. 8-8. A cylinder-type flowmeter regulator. (Courtesy of Airco Welding Products Division.)

of regulator being used, as in Fig. 8-7. Evaluating these charts gives any welder a better understanding of the differences between flow capacities of two-stage and single-stage regulators.

Since many regulators can be used for more than one type of gas, it is customary for the flow ratings in manufacturers' specifications to be for compressed *air*. Therefore, the welder should use the correction factor described earlier in the chapter for gases other than air. When source and outlet pressures are equal, higher flows come from *line* than from *cylinder* regulators.

One type of two-stage regulator is shown in Fig. 8-9. This is an oxygen regulator. Another type of two-stage regulator is shown in Fig. 8-10. Here, the regulator has a flowmeter attached.

Fig. 8-10. Two-stage regulator with flowmeter attached. (Courtesy of Airco Welding Products Division.)

Single-stage Regulators

The single-stage regulator shows more of a pressure rise or drop as cylinder content is used than does the two-stage regulator. Such a drop is usually not noticeable and doesn't affect the work quality on intermittent, short-term, jobs. The single-stage regulator, as shown in Fig. 8-11, provides both low-cost regulation and accurate enough performance for general welding, cutting, and heating. Several fixed-flow orifices are pictured with the Fig. 8-11 regulator; these regulate the gas flow.

Fig. 8-9. Two-stage regulator.

Fig. 8-11. Single-stage regulator. (Courtesy of Airco Welding Products Division.)

The single-stage regulator provides a greater flow than a comparable two-stage regulator, because the two-stage has more restrictions to the gas flow. For heating, the single-stage regulator will give ample gas flow for multiple torch operation.

Regulator Valve Selection

After a welder has decided if he wants a single- or two-stage regulator, he must decide what kind of regulator *valve* he wants. Regulators, either one- or two-stage, may have either nozzle-type or stem-type valves.

Nozzle-Type Regulators

As gas is used from a cylinder, the pressure in the cylinder goes down. A regulator equipped with a *nozzle* valve will help to decrease the outlet pressure as the inlet pressure goes down. For any outlet pressure setting, the seat is positioned by a balance of three forces. These include the force of the *adjusting spring* (See Fig. 8-12), the force of the *outlet pressure* against the diaphragm, and the force of the *valve spring*. As the cylinder pressure drops, the outlet pressure also drops to maintain the balance of forces.

Stem-Type Regulators

Regulators equipped with *stem-type* valves, as in Fig. 8-13, usually react to a decrease in

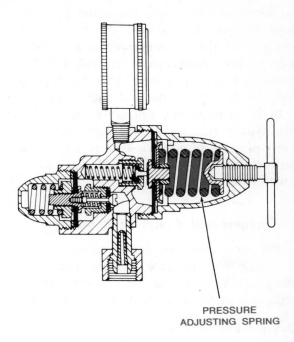

PRESSURE
ADJUSTING SPRING

Fig. 8-12. Positioning the seat of a nozzle-type regulator. The pressure adjusting spring largely helps position the spring. (Courtesy of Union Carbide Canada Ltd.)

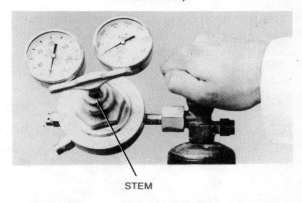

STEM

Fig. 8-13. Regulator equipped with stem-type valve. (Courtesy of Victor Equipment Co.)

source pressure *opposite* the way a nozzle-type valve does. As inlet pressure goes down in a stem-valve regulator, outlet pressure increases. The stem-type regulator acts this way because inlet pressure is used to assist the valve spring in closing the seat. The force of the inlet pressure, plus the force of the outlet pressure against the diaphragm, balances the force of the adjusting spring. As cylinder pressure drops, the outlet pressure must *increase* to maintain the balance of forces. The rise in outlet pressure as inlet pressure goes down can be attributed to the nature of a stem-type valve. This rise in pressure should not be confused with *creep*.

Both the *increasing* delivery pressure of a stem-type regulator and the *decreasing* delivery pressure of a nozzle-type regulator are perfectly normal, and do not indicate a regulator malfunction.

Oxygen and Acetylene Regulators

The most common regulators are *oxygen* and *acetylene*. Because of this, there are many of them on the market—enough to make it worthwhile to study them separately.

Oxygen Regulators

Oxygen regulators are usually shipped from the manufacturer in a protective carton, to protect the delicate gauge from shipping damage. The welder should be just as careful with the regulators once they have been removed from the cartons as the manufacturer was in shipping them. Regulators

should never be abused, or dismantled by welders inexperienced in repairing them.

Oxygen regulators, such as the one in Fig. 8-14, are usually colored green and must be designed so that the high-pressure side of the regulator itself will withstand the pressure of a full oxygen cylinder (2200 psi). To prevent straining the internal mechanism of the gauge, an 800-psi protection margin is built into the gauge. The oxygen regulator shown in Fig. 8-14 has this protection margin, since the maximum pressure indicated is 3000 psi.

Cylinder Pressure Gauge—The contents of the cylinder in cubic feet may be shown on a *second* scale of the *cylinder pressure gauge*. Since this scale indicates the cubic feet content at 70° F, the welder must remember that tank temperatures higher than 70° F would increase the oxygen because of expansion. Should the gas pressure exceed the safe limit, it will be immediately released through a built-in venting system within the working-pressure gauge.

Working-Pressure Gauge—Oxygen regulators are designed to deliver oxygen at pressures from 50 to 75 psi. The regulator shown

CYLINDER
PRESSURE GAUGE

WORKING
PRESSURE
GAUGE

SCREW IN TO
INCREASE WORKING
PRESSURE

REGULATOR

Fig. 8-14. Typical two-stage oxygen regulator. (Courtesy of Union Carbide Canada Ltd.)

in Fig. 8-14 has a 100-psi *working-pressure gauge*. Some regulators used for cutting have working-pressure gauges scaled up to 400 psi. Manifold regulators usually have pressure gauges scaled as high as 400 psi, and they can deliver large volumes of oxygen at pressures as high as 200 psi.

Welders soon realize that the oxygen regulator cannot be accidentally used on the acetylene cylinder. The following safety precaution is taken purposely by tank and regulator manufacturers: The acetylene cylinder and its gauges have left-hand threads and will only accept the *left-hand* thread connecting nut of an acetylene regulator; they will not accept the *right-hand* connecting nut of the oxygen regulator.

Acetylene Regulators

Acetylene regulators, as in Fig. 8-15, have *cylinder pressure* gauges registering up to about 400 psi and usually a 30-psi *working-pressure* gauge, graduated only to 15 psi. Many working-pressure gauges on acetylene regulators only indicate pressures up to 15 psi, as shown on the right in Fig. 8-15. When operating acetylene regulators, *never* increase working pressure above the 15-psi limit, even though all acetylene gauges have vent holes

in the side of the case to relieve excess pressure. Free acetylene can be detonated by pressures over 28 psi at temperatures at or above 70° F. Should the temperature of acetylene be increased above 70°F, the amount of pressure required to explode the gas will *decrease*. For this reason, most pressure gauges on acetylene regulators indicate a *danger zone* on the scale beyond 15 psi.

Regulator Maintenance

A gas regulator is made to control the high pressure of gases which are often extremely dangerous when out of control. Such an important welding tool should be maintained and operated with care.

Oxygen, though not a fuel, can be extremely dangerous if handled or used improperly. Since industrial oxygen is nearly 100% pure, any burning material fed with this type of oxygen would continue to burn out of control. Red hot iron will burn freely when put in direct contact with industrial oxygen.

Lubricating Regulators

Many welders notice that a regulator has several moving parts, so they may feel that a

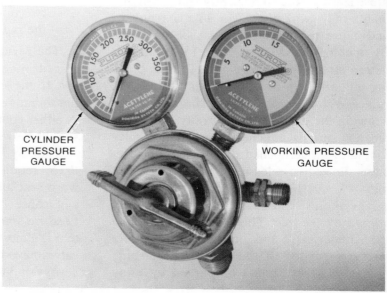

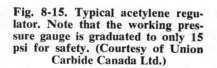

Fig. 8-15. Typical acetylene regulator. Note that the working pressure gauge is graduated to only 15 psi for safety. (Courtesy of Union Carbide Canada Ltd.)

CYLINDER PRESSURE GAUGE

WORKING PRESSURE GAUGE

regulator needs to be lubricated every so often. Actually, this is not necessary.

A regulator can be lubricated correctly whenever it is serviced by an experienced regulator mechanic. At that time, he will use only a small amount of special lubricants on very few parts. For this reason, routine regulator lubrication should *not* be done by the welder himself.

In fact, lubricating regulators may be extremely dangerous for the welder, even though his intentions are good. This holds particularly true for *oxygen regulators.*

As many welders have seen, USE NO OIL is indicated on the face of the oxygen gauge, as in Fig. 8-14. All welders should understand the reason for such a warning. Oils or greases, under the right conditions, are highly explosive when placed in contact with oxygen. In the presence of nearly pure oxygen, common lubricating oil will "flash" at a very low temperature and burn explosively. Lubricating an oxygen gauge with oil could, therefore, start a fire or burn a welder.

Purging Regulators

The valves of all cylinders should be blown out before a regulator is attached, to remove dust or grit. *Dust* or *grit* in the presence of 100% pure oxygen can become a fire hazard. More importantly, such contaminants may work their way into a regulator and cause regulator damage such as *creep.* By opening the cylinder valve for a moment *before* attaching the regulator, the gas blowing out will help to blow out any dirt in the valve. Welders call this *cracking* the valve, allowing the gas to blow out only dirt. After the regulator is attached, the cylinder valve should again be opened slowly, and the regulator allowed to release some gas for cleaning.

The hoses can then be attached and *purged* by the gas. This is done by opening the cylinder valve and increasing the pressure on the regulator. If the welder is attaching *new* hoses, they should be purged by using

pressure from an air hose, to eliminate the talcum-type powder in new hoses.

A process used several places in welding, *purging* is used to blow out or remove dirt, dust, or general impurities from different welding equipment. Purging the cylinder valve or hoses will assure the operator that no dust, dirt, or other foreign matter will enter the small gas passages in either the regulator or torch. These could plug the holes, which could cause *flashback,* creating serious damage to the equipment and possibly injury to the welder.

Each part of the system should be purged as it is attached, and the welder should be sure that no one is in front of the orifice (hole) when the gas is blown out.

Cleaning Regulators

All regulators should receive thorough cleaning and maintenance periodically, which will help eliminate some of the problems mentioned later in the chapter.

Each time regulators are used, the connections should be checked for *leaks.* Bubbles will appear when soapy water is put around a leaking connection. In Fig. 8-16, a leak test is needed around:

—the valve threads into the cylinder.
—the valve-to-regulator connection between the units.
—the regulator-to-hose thread connection.

The regulator inlet threads should not be forced into, or over, the cylinder valve outlet. If the connection seems to need force, the threads should be checked. If the threads in either connector are at fault, the connector should be repaired or replaced.

The regulator valve seat, as shown in Fig. 8-16, requires periodic inspection. A very worn seat allowed to remain in the regulator will cause *creep,* which can be dangerous. Pressure may build up at such a slow rate that it almost can't be detected after the torch valves have been closed. The operator should *always* shut off the cylinder valves and release

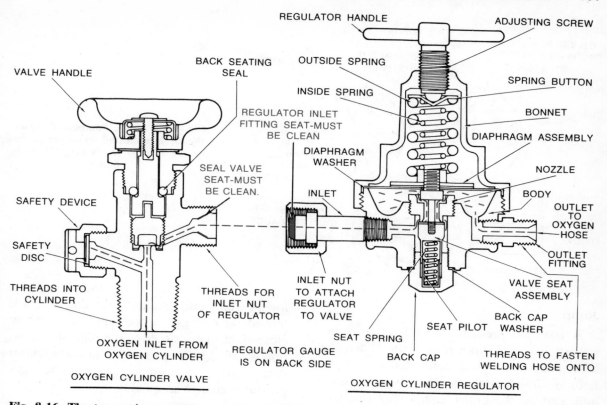

Fig. 8-16. The two major gas delivery parts. On the left is the cylinder valve; on the right, the cylinder regulator.

the pressure in the system by opening the torch valves if the system isn't going to be used for some time. Closing the cylinder valves and opening the torch valves allows both regulator pressure gauges to read 0. If the welder is not acquainted with the regulator's working mechanism, he should only attempt minor maintenance for satisfactory operation and his own safety.

Regulator Problems

A welder should have some knowledge of the things that change a regulator's operation, to insure safer and more efficient regulator operation. Several problems can result from general use, while others come from certain operating conditions.

Frost and Freezing

Any regulator controlling high-pressure gas may get so cold that a layer of frost collects on its outside surface. As the gas ex-

pands, it tends to absorb heat. The heat it absorbs must come from the regulator. As the expanding gas flows through the regulator, it picks up any heat in the regulator. As the regulator loses its heat and gets cooler, the moisture in the air condenses on the cold metal parts of the regulator and freezes. This accounts for the "frozen" regulator appearance.

Although frost on the outside of the regulator does not affect its operation, the gas flow may fluctuate. This causes an unstable flame which, under extreme conditions, will be extinguished. The unstable flame comes from water vapor being present in the gas. This water is *freezing* on the inside of the regulator like frost is freezing on the outside of the regulator. Such freeze-up is not the regulator's fault, but, rather, comes from the welding gas. Frost and freezing are regulator problems that are not caused by the regulator itself.

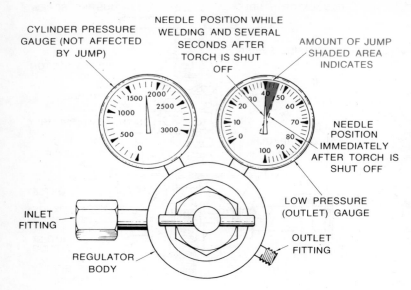

Fig. 8-17. *Jump,* immediately after the torch is shut off. With jump, the needle returns to its original position.

Jump

A common regulator complaint by some welders is a term known as *jump.* When the hose and torch system connected to a regulator is closed at the torch valves, pressure indicated on the outlet gauge may rise momentarily and then stop. This momentary pressure rise is called *jump,* as seen in Fig. 8-17. The amount of jump simply indicates the true difference between gas pressure at the regulator outlet when gas *is* flowing, and pressure when it is *not* flowing. When the torch valve is closed, pressure on the regulator's

low pressure side rises until the diaphragm is moved far enough for the regulator valve to close tight. This closed-valve regulator position is called *lock-up.* Jump is not a regulator defect and is a normal reaction when gas pressures change so quickly.

Creep

Unlike jump, *creep* is a definite regulator malfunction. *Creep* happens when a torch valve is shut off and the pressure on the low-pressure side of the regulator valve increases to an unusually high level. This is shown by

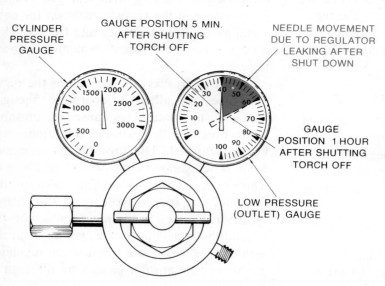

Fig. 8-18. *Creep,* after the torch has been shut off for some time. With creep, the needle is steadily rising.

the slowly-rising needle in Fig. 8-18. Limited creep is hard to distinguish from jump.

When creep is very bad, pressure backs up against a closed torch valve until it is so high that either some part of the system breaks, or pressure in the valve's *low pressure* side equals the pressure from the source.

On low-pressure gas cylinders, such as acetylene, the regulators do not need a safety relief valve. Various high-pressure regulators, however, *do* have automatically-reseating safety relief valves.

Creep always occurs when the regulator is defective and cannot reach lock-up. If a seat or nozzle is worn to the point where no amount of pressure will close it, the valve

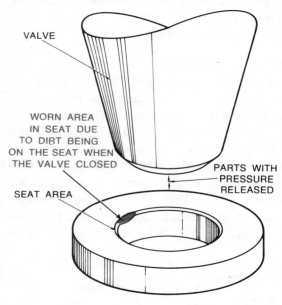

Fig. 8-19. A valve seat in a creeping regulator. The seat has a worn spot that will prevent necessary lock-up.

will leak. Dirt or wear between the seat and valve can create this condition. Creeping in a regulator should be repaired before operating the welding equipment attached to the regulator. See Fig. 8-19.

Regulator Changes From Low Cylinders

Regulators have a tendency to increase or decrease delivery pressures as the source

(tank) pressure drops. This is most obvious in single-stage regulators, but it also affects two-stage regulators at lower cylinder pressures. Although many welders believe that two-stage regulators provide a constant outlet pressure, this is not entirely true.

The outlet pressure of a two-stage regulator *will* change when the cylinder pressure drops below the preset outlet pressure of the regulator's first stage. When a two-stage regulator's *first stage* is preset to deliver a certain pressure to the second stage, then the two-stage regulator will perform as a single-stage regulator whenever the cylinder pressure drops below the first stage preset pressure. With a *single-stage* regulator, the welder can adjust his pressure setting slightly as the cylinder pressure decreases.

Regulator Safety Reminders

The welder should keep in mind a few basic reminders about regulators. This is common sense, because a welder can get hurt easily if a regulator doesn't work right. Also, regulators are fairly expensive and easily damaged if they are misused. They are the tools a welder needs to change sometimes explosive gas under high pressure into a workable, safe, and steady pressure. When using regulators, the welder must remember the following points:

A. Lubricating a regulator should be left to a trained mechanic.
B. Lubricants other than oil should always be used.
C. Clear all passages before applying pressure.
D. Always use the right regulator for the gas being used.
E. Check all connections for leaks.
F. Regulators must connect easily into cylinder valves.
G. Creeping regulators must be repaired before using.
H. Be sure all regulators being used with high-pressure gases are equipped with

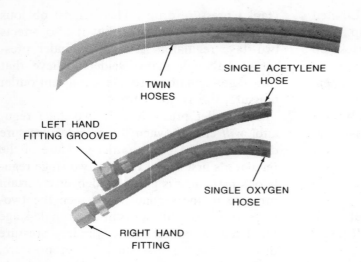

Fig. 8-20. Typical single and twin welding hoses.

clean and correctly set pressure relief valves.

Welding Hose

A major piece of equipment in welding is the high-strength *hose* used to carry the welding gases from the regulator to the torch.

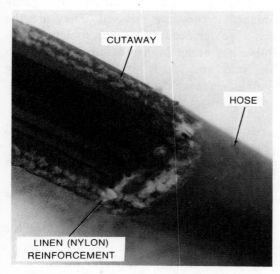

Fig. 8-21. Cutaway of typical welding hose.

Oxy-acetylene hose is factory tested to 400 psi to withstand all normal welding and cutting pressures. Hoses can be purchased as single- or twin-type, as shown in Fig. 8-20.

Twin hoses are made by adhesion joining the two separate lines, which will then not separate under the most severe use.

Hose Construction

Generally, hose is made from a high-quality flame- and oil-resistant synthetic or natural rubber compound, reinforced with a nylon fabric or linen braid. This molded and high-quality rubber hose, vulcanized into cotton or linen fabric, is light, flexible, and durable. It is practically immune to sunlight and weather. (See Fig. 8-21.) Hose is manufactured to eliminate any *coil set*. Coil set, illustrated in Fig. 8-22, could create prob-

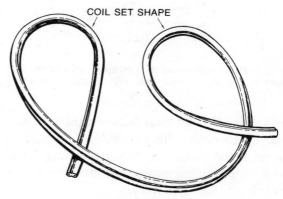

Fig. 8-22. *Coil set* in an oxy-acetylene hose.

lems, especially when welding at various heights.

Hose Coding

Hoses are color-coded according to what gas they are to carry. The common oxy-acetylene twin hose, for example, has the colors red and green in each "twin," one color for each gas.

The red hose has the left-hand connecting fittings for use on acetylene regulators and, of course, is the acetylene hose. The green hose with the right-hand fittings is used for the oxygen regulator and is the oxygen hose. Hoses should *never* be interchangeable; connections having different threads ensures that welders will *not* be able to attach the acetylene hose to the oxygen regulator, and vice-versa. Various *lengths* of hose can be purchased, with diameters ranging from ³⁄₁₆" to ⅜".

Fig. 8-23. Brass connections used with oxy-acetylene welding hoses.

Only the correct brass hose couplers, Y connections, and other hose accessories should be used, such as those in Fig. 8-23.

Before using new hose, the talcum powder should be blown out of the hose by using compressed air. Used hose should be checked periodically for leaks by immersing the hose in water while the hose is under working pressure. Soap and water can be placed around couplings, hose ferrules, and other parts of equipment under pressure; the welder can detect odorless leaks by checking for bubbles.

Welding Torches

While the regulator and hoses give the welder a smooth, even flow of welding gases, he depends on the *torch* to mix the gases for the first time. At the torch he can also adjust the exact flow of gas he needs, and he depends on the torch to hold the welding tip chosen for the flame needed.

Torches come in many sizes, prices, qualities, and styles, but they all look very much alike. A typical torch, with hoses and tip attached, is shown in Fig. 8-24.

Another common oxy-acetylene torch is shown in Fig. 8-25. Hoses are connected to this torch as follows. One end of the green *oxygen* hose is attached to the oxygen regulator with right-hand thread connections, and the other end is fastened to the torch body with right-hand connections.

The red hose has left-hand connections, as indicated by the groove in Fig. 8-26. The red hose connects the acetylene regulator to the opposite side of the torch body shown in

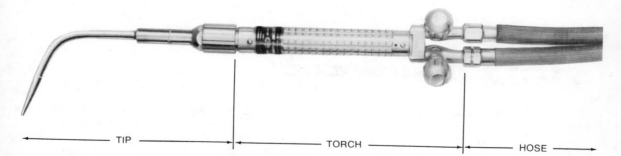

TIP — TORCH — HOSE

Fig. 8-24. Torch body with common tip and hoses. (Courtesy of Airco Welding Products Division.)

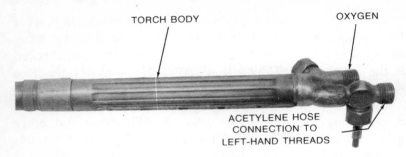

Fig. 8-25. Common torch with no tip or hoses attached.

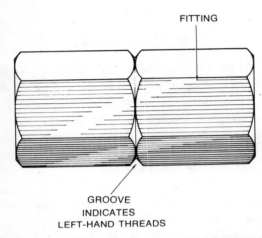

Fig. 8-26. Acetylene fitting. Acetylene fittings have left-hand threads, indicated by the groove cut around the fitting.

Fig. 8-25. The left-hand threads prevent the hoses from getting interchanged, as is the standard practice for oxygen/gas welding equipment.

Welding Blowpipe

The welding *blowpipe,* as in Fig. 8-27, provides a means of mixing the oxygen and fuel gas (usually acetylene). This provides the correct mixture at the tip, which, when ignited, will produce the desired flame for welding or cutting. Blowpipes vary in design, but the various types have similar characteristics.

At one end of the blowpipe (Fig. 8-27) are the gas *inlets* where the oxygen and acetylene hoses are connected. The volume (quantity) of oxygen and acetylene passing from the hoses into their respective inlets is controlled by the inlet *valves.* The desired amount of oxygen and acetylene allowed to flow through the blowpipe can be controlled by these valves, which can be operated independently to allow more of one gas to enter than the other. Welding heads, or *tips,* as they

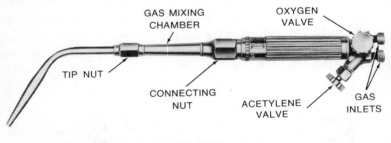

Fig. 8-27. A welding blowpipe. (Courtesy of Union Carbide Canada Ltd.)

Fig. 8-28. A welding tip with a gas mixing chamber.

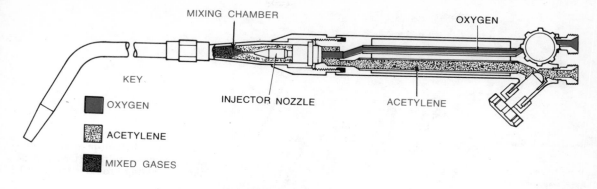

MIXING CHAMBER

OXYGEN

KEY

■ OXYGEN

▨ ACETYLENE

▨ MIXED GASES

INJECTOR NOZZLE

ACETYLENE

Fig. 8-29. Injector-type blowpipe. (Courtesy of Union Carbide Canada Ltd.)

are usually called, are attached directly to the blowpipe with a connection nut that is usually hand tightened. Various-sized tips are available to provide the *mixing chambers* (see Fig. 8-28) and gas orifices for some types of torch bodies. Tips with individual mixing sections provide excellent oxygen-to-acetylene ratios. With individual tips, the welder can select a large number of flame sizes and still use the same blowpipe handle (body) and valves.

Some blowpipes use a mixing chamber adapted between the body and the tip. The blowpipe designed for low-pressure operation, as shown in Fig. 8-27, illustrates the gas-mixing adapter.

Types of Blowpipes

Depending upon the acetylene pressure used, blowpipes are either the *injector-type* (low pressure) or the *medium-pressure* type. This is true for all blowpipes, regardless of what they will be used for.

Injector-Type Blowpipes—Fig. 8-29 shows the oxygen and acetylene hoses connected to their inlets at the rear of an *injector-type* blowpipe. The front of the body forms the terminal for the oxygen and acetylene tubes and holds the oxygen injector. Mixing takes place in the welding tip, and the rear end of the tip is attached to the handle.

As the oxygen passes through the injector nozzle, the oxygen's speed increases until suction is created. In turn, this draws the acetylene in through openings around the oxygen stream, as shown in Fig. 8-30. The gases are properly mixed by the time they emerge from the blowpipe tip to produce a steady flame. The injector-type blowpipe can use acetylene from generators or cylinders at pressures as low as 1 psi. Fig. 8-27 illustrates a blowpipe designed for low-pressure operation.

Medium-Pressure Types—Acetylene from cylinders and medium-pressure generators is used in *medium-pressure* blowpipes. These require that between 1- and 15-psi acetylene

Fig. 8-30. Injector-type blowpipe. Here, the rushing oxygen sucks in the acetylene.

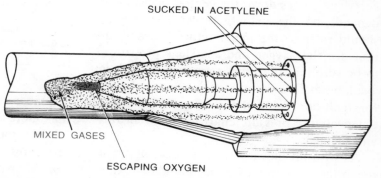

SUCKED IN ACETYLENE

MIXED GASES

ESCAPING OXYGEN

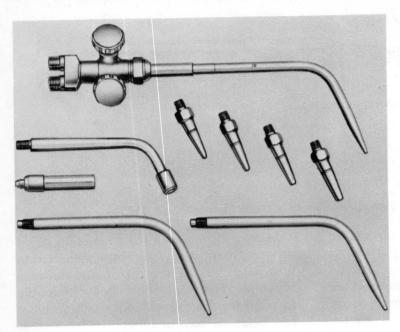

Fig. 8-31. Medium-pressure blow-pipe with assorted tips. (Courtesy of Airco Welding Products Division.)

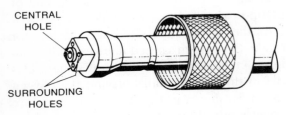

CENTRAL HOLE

SURROUNDING HOLES

Fig. 8-32. Mixer nozzle. The nozzle has a center orifice surrounded by smaller holes.

Table 8-4. Guide to Oxygen and Acetylene Pressures for Different Welding Jobs.

Tip No.	Thickness of Metal - In.	Oxygen & Acetylene Pressure - psi	Acetylene Flow - cfh	Oxygen Flow - cfh
00	1/64	1	0.1	0.1
0	1/32	1	0.4	0.4
1	1/16	1	1	1.1
2	3/32	2	2	2.2
3	1/8	3	8	8.8
4	3/16	4	17	18
5	1/4	5	25	27
6	5/16	6	34	37
7	3/8	7	43	47
8	1/2	8	52	57
9	5/8	9	59	64
10	3/4 up	10	67	73

Pressures and consumptions shown are approximately correct for both separable tips with appropriate mixers and for tip-mixer assemblies.

Gas pressures should be increased slightly for hose lengths greater than 25 feet.

Courtesy of Canadian Liquid Air

pressure be supplied. Medium-pressure type blowpipes, or *balanced pressure,* as they are sometimes called, use a mixer nozzle different from the low-pressure injector nozzle (Fig. 8-31). The mixer nozzle has a central orifice, and around the orifice are several smaller holes. See Fig. 8-32. Whether the oxygen or acetylene comes down the central orifice hole depends on the manufacturer. The welder should refer to the manufacturer's specifications for the oxygen and acetylene pressures, to be sure that the proper amount of each is available before the mixture passes to the tip for ignition. Table 8-4 is a manufacturer's oxy-acetylene pressure chart for different tips and metals.

As with all blowpipes, it is necessary that the orifice size increase beyond the mixing chamber, to allow the gases to expand and mix more thoroughly. Medium-pressure blowpipes usually operate best with almost equal pressures of oxygen and acetylene. The tip size shown on specification Table 8-4 will help determine the necessary oxygen and acetylene pressure.

Cutting Blowpipes

Cutting blowpipes, like welding blowpipes, can be either low-pressure (injector) or medium-pressure types. Cutting blowpipes are designed to allow mixed oxygen and acetylene to be released through *preheat orifices* in interchangeable *heads*. (The cutting tip is often called the *head* of an oxy-acetylene cutting blowpipe.)

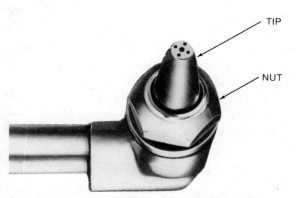

Fig. 8-33. Cutting blowpipe head. The preheat orifices are for the oxygen and acetylene mixture. (Courtesy of Union Carbide Canada Ltd.)

Fig. 8-33 shows the preheat orifices in a cutting blowpipe head. Oxy-acetylene flame burns at these orifices. The center orifice in the cutting tip allows pure oxygen to be released. The pure oxygen does the actual cutting after the metal has been preheated by the oxy-acetylene flame. Cutting with the oxy-acetylene flame is discussed later.

Injector-Type — Low-pressure acetylene generators or cylinders can operate the *injector cutting* blowpipe. The oxygen and acetylene are mixed by the same injector principle as in the injector-type welding blowpipe. Fig. 8-34.

The acetylene hose from the regulator is joined by left-hand connecting nuts to the torch inlet in Fig. 8-34. The oxygen hose is then joined to the right-hand threaded inlet.

Oxygen and acetylene are mixed by the suction created when the oxygen passing through the injector draws the acetylene through holes surrounding the nozzle, as was shown in Fig. 8-32. The oxygen and acetylene mixture is ignited at the end of the tip in a neutral flame to heat the metal red hot. After the metal is red hot, the cutting oxygen valve (Fig. 8-34) is opened by pressing the valve lever down with the thumb. Pure oxygen is blown out through the central orifice in the cutting tip (Fig. 8-33), which removes the hot metal by oxidation.

Medium-Pressure Types—This type of cutting blowpipe utilizes the balanced pressure mixer, like the medium-pressure welding blowpipe. The only difference is that an oxygen valve, controlled by a hand lever, releases pure oxygen onto the preheated metal surface to be cut.

The oxygen and acetylene hoses are connected to the cutting blowpipe through inlets at the back of the torch. The oxygen and acetylene pass through the hose to the valves on the cutting blowpipe. When the valves on

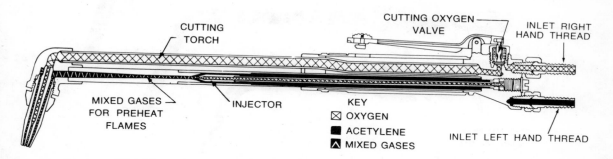

Fig. 8-34. Injector-type welding blowpipe. (Courtesy of Union Carbide Canada Ltd.)

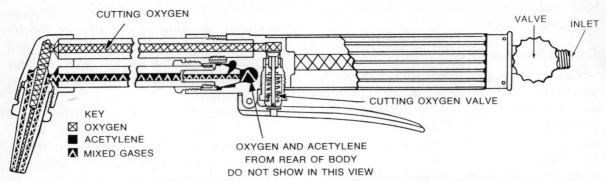

Fig. 8-35. Medium-pressure type welding blowpipe. (Courtesy of Union Carbide Canada Ltd.)

the cutting blowpipe are opened, the oxygen and acetylene flow separately until they join in the mixing chamber shown in Fig. 8-35. Some pure oxygen by-passes the mixing chamber, to be controlled by the cutting oxygen valve. The mixed oxygen and acetylene are ignited and adjusted at the end of the cutting tip for preheating. After preheat-

ing, pure oxygen is released by depressing the oxygen valve lever, as in Fig. 8-35. This allows pure oxygen to pass through the center hole in the cutting tip to oxidize (burn) the preheated metal.

Cutting and welding blowpipes are available from various welding equipment manufacturers to meet specific needs. Any special

Fig. 8-36. *Mapp* **gas or natural gas blowpipe. (Courtesy of Airco Welding Products Division.)**

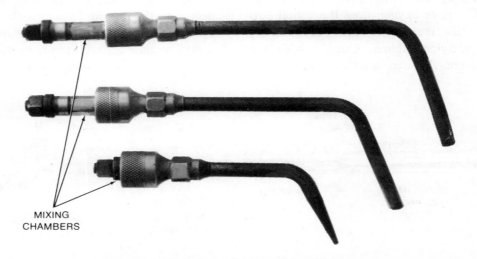

Fig. 8-37. Tips with individual mixing chambers.

equipment, however, is basically the same in design and operation as those mentioned above. Cutting equipment for LP gas, natural gas, or *Mapp* gas is available and should be used when gases other than acetylene are being considered. Fig. 8-36 shows a cutting blowpipe designed to be used with *Mapp* gas or natural gas.

Welding and Cutting Tips

Tips are usually made of heavy-walled copper for greater resistance to reflected heat. Those tips with a *swagged* (bent and shaped cold) construction have a smooth inner passage to eliminate flame turbulence. Tips with individual mixing chambers give the best oxygen-to-acetylene ratio. See Fig. 8-37.

Each manufacturer produces their own type and style of tip and uses their own stock numbering system, as shown in specification Table 8-5. This table, for example, gives specifications for cutting tips made by Smith Welding Equipment. For this reason, most welders refer to the manufacturer's specifications (such as Table 8-6) for the proper tip to use when welding a certain thickness of metal. Various sizes of tips, torches and cutting bodies are available from all welding equipment manufacturers. Fig. 8-38 shows a section from one manufacturer's torch application chart.

Table 8-5. Stock Number Specifications For One Line of Cutting Tips.

MEDIUM PRESSURE			INJECTOR STYLE
Acetylene	LP-Gas	Mapp®	Natural Gas or LP-Gases
SC12 Series	SC21A Series	SC21AM Series	SC31 Series
SC56 Series	SC50A Series	SC50AM Series	SC28 Series
	SC46 Series	SC56 Series	SC36A Series
	SC18A Series		

Courtesy of Smith Welding Equipment

Cutting tips are made from the same material, and in the same way, as welding tips. The holes in cutting tips are uniform, straight, and precision drilled to a specific size, to hold down the turbulence in the gas stream for good cutting. Precision pieces of equipment, welding and cutting tips must be properly cared for, if good work is expected. Turbulence in the gas stream, caused by misuse or

Table 8-6. Welding Tip Size and Application.

Tip Size	Drill Size	Length of Average Inner Flame Cone	Oxygen Pressure (psi)		Acetylene Pressure (psi)		Acetylene Consumption (cfh)*		Metal Thickness
			Min.	Max.	Min.	Max.	Min.	Max.	
000	75	$\frac{7}{32}$"	$\frac{1}{2}$	2	$\frac{1}{2}$	2	$\frac{1}{2}$	3	Up to $\frac{1}{32}$"
00	70	$\frac{7}{32}$"	1	2	1	2	1	4	$\frac{1}{64}$"-$\frac{3}{64}$"
0	65	$\frac{3}{8}$"	1	3	1	3	2	6	$\frac{1}{32}$"-$\frac{5}{64}$"
1	60	$\frac{3}{8}$"	1	4	1	4	4	8	$\frac{3}{64}$"-$\frac{3}{32}$"
2	56	$\frac{3}{8}$"	2	5	2	5	7	13	$\frac{1}{16}$"-$\frac{1}{8}$"
3	53	$\frac{3}{8}$"	3	7	3	7	8	36	$\frac{1}{8}$"-$\frac{3}{16}$"
4	49	$\frac{5}{8}$"	4	10	4	10	10	41	$\frac{3}{16}$"-$\frac{1}{4}$"
5	43	1"	5	12	5	15	15	59	$\frac{1}{4}$"-$\frac{1}{2}$"
6	36	$1\frac{1}{16}$"	6	14	6	15	55	127	$\frac{1}{2}$"-$\frac{3}{4}$"
7	30	$1\frac{1}{4}$"	7	16	7	15	78	152	$\frac{3}{4}$"-$1\frac{1}{4}$"
8	29	$1\frac{1}{4}$"	9	19	8	15	81	160	$1\frac{1}{4}$"-2"
9	28	$1\frac{7}{16}$"	10	20	9	15	90	166	2"-$2\frac{1}{2}$"
10	27	$1\frac{7}{16}$"	11	22	10	15	100	169	$2\frac{1}{2}$"-3"
11	26	$1\frac{7}{16}$"	13	24	11	15	106	175	3"-$3\frac{1}{2}$"
12	25	$1\frac{1}{2}$"	14	28	12	15	111	211	$3\frac{1}{2}$"-4"

*Oxygen consumption is 1.1 times the acetylene under neutral flame conditions.

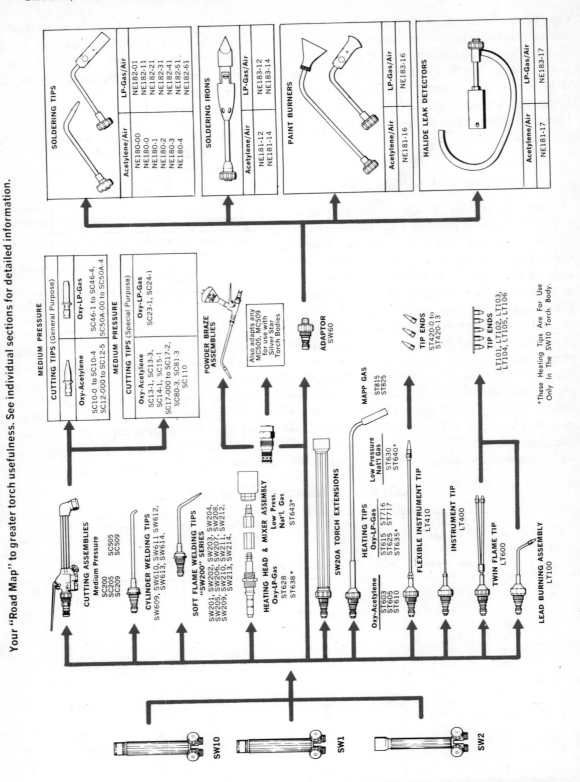

Fig. 8-38. Torch guide chart for various torch tips and applications. (Courtesy of Smith Welding Equipment.)

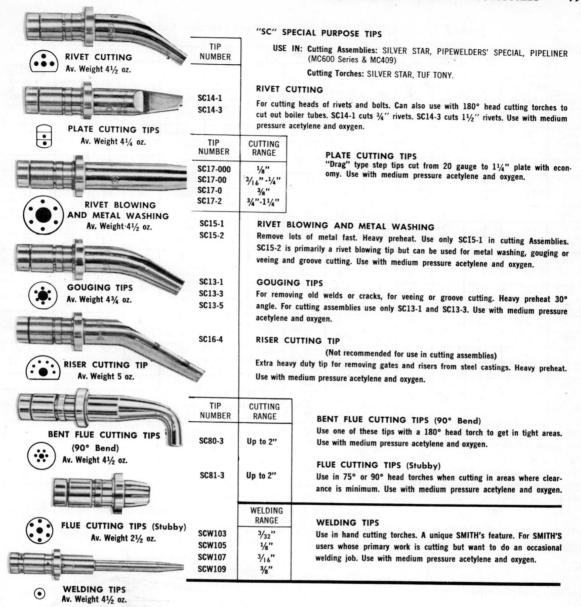

"SC" SPECIAL PURPOSE TIPS

USE IN: Cutting Assemblies: SILVER STAR, PIPEWELDERS' SPECIAL, PIPELINER (MC600 Series & MC409)

Cutting Torches: SILVER STAR, TUF TONY.

TIP NUMBER		

RIVET CUTTING

| SC14-1 | |
| SC14-3 | |

For cutting heads of rivets and bolts. Can also use with 180° head cutting torches to cut out boiler tubes. SC14-1 cuts ¾" rivets. SC14-3 cuts 1½" rivets. Use with medium pressure acetylene and oxygen.

TIP NUMBER	CUTTING RANGE	
SC17-000	⅛"	
SC17-00	³/₁₆"-¼"	
SC17-0	⅜"	
SC17-2	¾"-1¼"	

PLATE CUTTING TIPS
"Drag" type step tips cut from 20 gauge to 1¼" plate with economy. Use with medium pressure acetylene and oxygen.

| SC15-1 | |
| SC15-2 | |

RIVET BLOWING AND METAL WASHING
Remove lots of metal fast. Heavy preheat. Use only SC15-1 in cutting Assemblies. SC15-2 is primarily a rivet blowing tip but can be used for metal washing, gouging or veeing and groove cutting. Use with medium pressure acetylene and oxygen.

SC13-1	
SC13-3	
SC13-5	

GOUGING TIPS
For removing old welds or cracks, for veeing or groove cutting. Heavy preheat 30° angle. For cutting assemblies use only SC13-1 and SC13-3. Use with medium pressure acetylene and oxygen.

| SC16-4 | |

RISER CUTTING TIP
(Not recommended for use in cutting assemblies)
Extra heavy duty tip for removing gates and risers from steel castings. Heavy preheat. Use with medium pressure acetylene and oxygen.

TIP NUMBER	CUTTING RANGE	
SC80-3	Up to 2"	
SC81-3	Up to 2"	

BENT FLUE CUTTING TIPS (90° Bend)
Use one of these tips with a 180° head torch to get in tight areas. Use with medium pressure acetylene and oxygen.

FLUE CUTTING TIPS (Stubby)
Use in 75° or 90° head torches when cutting in areas where clearance is minimum. Use with medium pressure acetylene and oxygen.

	WELDING RANGE	
SCW103	³/₃₂"	
SCW105	⅛"	
SCW107	³/₁₆"	
SCW109	⅜"	

WELDING TIPS
Use in hand cutting torches. A unique SMITH's feature. For SMITH's users whose primary work is cutting but want to do an occasional welding job. Use with medium pressure acetylene and oxygen.

Fig. 8-39. Various types of cutting tips from one manufacturer. (Courtesy of Smith Welding Equipment.)

poor hole cleaning, may produce roughness on the face of a cut.

There is a wide selection of cutting tips available today for different cutting jobs, as the manufacturer's chart in Fig. 8-39 shows. Varying types and sizes are produced by different manufacturers. For this reason, cutting and welding tips cannot be used on a torch other than the one they are designed to fit. Many factors, which will be mentioned

in the oxy-acetylene cutting chapter, control tip selection. Good welders refer to the specifications outlined by the welding manufacturer before either adjusting the oxygen and acetylene pressures or choosing a welding or cutting tip. Good welders also learn much from studying the many charts made up by the welding industry. One of these charts is Table 8-7. While they have little advertising value, most welding equipment companies

Table 8-7. Acetylene and Oxygen Medium Pressure Cutting Tips.

Metal Thickness	SC10 and SC12 Series Tip Number	OXYGEN PRESSURE P.S.I.			Fuel Gas Pressure	CONSUMPTION C.F.H.			Cutting Speed	Kerf Width	DRILL SIZES			Recommended Number of Cylinders (Single or Manifold)
		Cutting Pressure		Preheat Pressure At Reg.[2]		Cutting Oxygen	Preheat Oxygen	Fuel			Cutting	Preheat SC10 Series	Preheat SC12 Series	
		At Regulator	At Torch											
⅛″	000	20	20	3	3	18	7	6.5	28	.035	72	71	75	1
³⁄₁₆″	00	20	20	3	3	24	7	6.5	26	.050	68	71	75	1
¼″	0	30	30	4	4	40	7.5	7	22	.055	62	70	74	1
⅜″	0	35	35	4	4	50	7.5	7	20	.055	62	70	74	1
½″	1	35	35	4	4	75	11	9.5	19	.080	56	68	71	1
⅝″	1	40	40	4	4	85	11	9.5	17	.080	56	68	71	1
¾″	2	36	35	4	4	105	12	10.5	16	.095	54	65	70	1
1″	2	41	40	4	4	115	12	10.5	14	.095	54	65	70	1
1¼″	2	51	50	4	4	135	12	10.5	13	.095	54	65	70	1
1½″	3	42	40	5	5	170	14	12	12	.100	51	65	68	1
2″	3	47	45	5	5	180	14	12	10	.100	51	65	68	1
2½″	4	38	35	5	5	240	15	13	9	.125	45	60	62	1
3″	4	44	40	5	5	265	15	13	8	.125	45	60	62	1
4″	4	54	50	5	5	315	16	14	7	.125	45	60	62	1
5″	5	56	50	6	6	420	29	25	7	.150	41	60	60	1
6″	5	67	60	6	6	485	29	25	6	.150	41	60	60	1
8″	5	78	70	6	6	550	30	26	5.5	.150	41	60	60	1
10″	6[1]	83	70	6	6	750	32	28	5	.203	32		60	1
12″	6[1]	125	90	6	6	975	32	28	4.5	.230	32		60	1
14″	7[1]	100	82	6	7	1250	34	30	4	.250	28		56	1

SC12 Series: Medium preheat for general hand and machine cutting (6 and 8 preheats).
SC10 Series: Medium preheat for general hand cutting.

[1]SC12 series only.

[2]For 3-hose machine cutting torches only.

The figures shown here were gathered under average machine cutting conditions, using clean steel. Pressures shown are for 25 feet or less of ¼″ I.D. hose. If longer hose is used, pressures should be increased.

spend a great deal of time and money on these charts to advance the science of welding.

Tip Reconditioning

When tips become *bent,* the welder should be able to straighten them without damaging the orifice, especially if the bend is not too severe. Placing the tip on a block of wood and striking it gently near the bend with a hammer wrapped in leather will usually straighten the tip without compressing the copper. See Fig. 8-40.

Slag can be removed from the tips by tapping the base of the nozzle on a piece of hardwood. Then, the orifice is cleaned with the proper-sized drill cleaner. When using a

Courtesy of Smith Welding Equipment

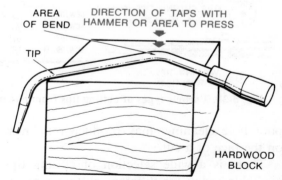

Fig. 8-40. Straightening a bent tip.

cleaner, make sure that the motion is up and down in the hole, so the hole doesn't become "bell-mouthed." Fig. 8-41 shows the right way to clean a tip so the hole stays straight.

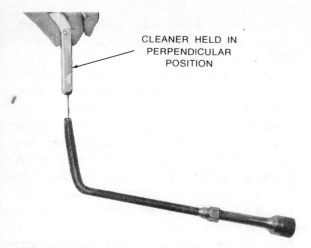

Fig. 8-41. Correct way to clean tips without damaging the orifice (hole).

In Fig. 8-42, two tips have been cut in half to show how they were cleaned. The left tip was cleaned with an up-and-down motion,

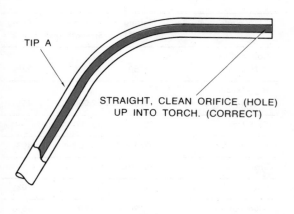

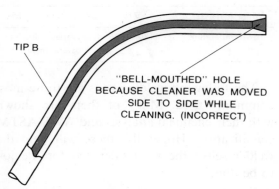

Fig. 8-42. Tip cleaning effects. Tip A was cleaned the right way, so it has a clean, straight orifice. Tip B has a "bell-mouth" because the cleaner was moved side-to-side while cleaning.

but the right tip was cleaned with the cleaner leaning to one side. This made the tip bell-mouthed. A bell-mouthed tip makes it hard to weld with a good flame. After cleaning the orifice with a cleaning tool, the welder should move the end of the tip across a length of emery cloth. This will square the face and help eliminate flame turbulence.

Tip Cleaners—These can be either the *serrated* or *twist drill* type. Fig. 8-43 shows

Fig. 8-43. Common tip cleaner package in an assortment of sizes. (Courtesy of Airco Welding Products Division.)

the common serrated tip cleaner with an end file. Both types of cleaners should be used to eliminate the bell-mouth effect caused by improper cleaning. When using the twist drill cleaner, the proper drill must be used for the orifice being cleaned. Although this is also true when using serrated cleaners, damage is seldom done to the tip because of using an improper size serrated cleaner. Some welders find that the twist drill type cleaner may break off in the orifice if the cleaner is moved side to side, instead of up and down. For this reason, many welders use the softer serrated cleaners, which are more difficult to break.

Torch Lighters

Because of danger to the welder, it is not a good idea to light the torch with matches. Matches need two hands for safety, which doesn't leave any hands for holding the torch! The safer way to light the torch is by using a spark-type torch lighter. Most welders use these because they are quick and fast, and they can be operated with one hand. These cause a spark by rubbing a flint across a rough striker.

Flint lighters, containing from one to three flints, are available for igniting the acetylene as it leaves the end of the welding or cutting tip. (See Fig. 8-44.) As the flints wear out

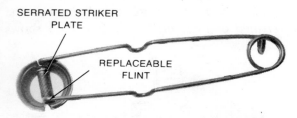

Fig. 8-44. Flint type spark lighter. Squeezing the handles together pulls the flint across the striker plate, causing a spark. (Courtesy of Airco Welding Products Division.)

from scratching the serrated striker piece, new flints can be easily installed, usually by screwing them onto the striker handle.

Welding Rod

During most fuel-gas-with-oxygen welding, more filler metal is needed than the metal melted off the base metal. To add more metal for a stronger weld, welders add metal from a welding *rod.*

A welding rod usually has the same diameter as heavy wire. Welding rods come in several diameters with different *coatings* to give the weld good properties and to protect the rod while it is being stored. After the welding rod is melted into the weld, it is known as *filler metal.*

Filler metal should match the parent metal as close as possible for good, strong fusion welding. Any filler metal must "weld in" properly to give the weld both strength and ductility. Before choosing a filler rod, the welder must consider what properties are necessary in the parent metal, such as its ability to accept various heat treatments. Bronze and other low melting temperature filler rods may be necessary to overcome stress distor-

tion where the heat used to melt high-temperature rods could change the parent metal's properties.

The welder should know where to find information covering various types of filler rods. Table 8-8 shows such information. Many manufacturers sell welding rods with similar characteristics. Since each manufacturer has their own rod classification number/name, however, welders must follow the manufacturer's *conformance* (data information) for the AWS-ASTM classification. One manufacturer's data chart is shown in Fig. 8-45. Most of the rods used with oxy-acetylene welding are 36 inches long, except cast iron rods, which are usually 24 inches long.

Table 8-8. Properties of Common Welding Rods.

Rods	Melting Point(°F)	Tensile Strength (psi)	Elong. in 2"
Copper-Coated Mild Steel	2750	52,000	23%
High-Tensile, Low-Alloy Steel	2750	62,000	20%
Cast Iron	2200	40,000 Max.	—
Stainless Steel	2550	80,000	30%
Bronze	1598—1625	55,000±	—
Everdur®	1866	50,000	20%
Aluminum	1190±	16,000	25%
White Metal	715	52,000 Max.	8%
Low-Temperature Brazing Rod	1170—1185	(Varies with parent metal)	

The following welding rods are those most commonly used. Many of them are shown with their chemical analysis and AWS-ASTM classification. Hopefully, these will help the welder select the right filler rod for the job to be done.

Mild Steel Rod

Most *mild steel* rods are available with copper coating to protect the rod from rust

Welding Rod

WELDING AND BRAZING ROD CONFORMANCES

Listing of the welding and brazing rods in the tables below only indicates conformance with the chemical requirements of appropriate specification and does not imply approval by that agency.

STEEL, CAST IRON, AND COPPER BASE RODS

Airco Number ▶		1	4	7	9	10	20	21	22	23A	27	92	1010
AWS-ASTM Classification ▶					RCI	RCI-A	RB Cu Zn-A	RB Cu Zn-D	R Cu Zn-B	R Cu	R Cu Zn-C	R Cu Si-A	R Cu Si-A
Specification	Type of Welding												
AWS A5.2 ASTM A251	Oxyacet.	RG-60	RG-60	RG-45									
AWS A5.7 ASTM B259	Oxyacet. & TIG						●	●	●	●	●	●	●
AWS A5.8 ASTM B260	Brazing						●	●					
AWS A5.13 ASTM B399	Oxyacet.											●	●
AWS A5.15 ASTM B398	Oxyacet.				●	●	●	●	●		●		
ASME SA-251	Oxyacet.	GX-60	GX-60	GX-45									
ASME SB-259	Oxyacet. & TIG						F-35			F-31	F-35	F-32	F-32
ASME SB-260	Brazing						F-106	F-106					
QQ-R 571B	Oxyacet. & TIG						●	●	●	●	●		
MIL-R 908A	Oxyacet.		Class 1		Class 2*								
MIL-R 18818A	Surfacing by TIG												
MIL-R 19631B	Oxyacet. & TIG						●	●	●	●	●	●	●

*Phosphorus and sulfur must be checked before shipment.

Fig. 8-45. A *conformance chart* **for one manufacturer's welding rods. (Courtesy of Air Reduction Co., Inc.)**

and corrosion while it is in storage. These are low-priced, general purpose rods; they are used for welding sheets, pipe, and plate with a low carbon content. The ultimate tensile strength of this type of filler metal is 52,000 psi. It has a 23% elongation in two inches, which indicates ductility high enough to stretch considerably before breaking. These rods are available in diameters from 1/16 inch to 1/4 inch. The chemical analysis of mild steel welding rod is:

Carbon	0.06% Max.
Manganese	0.25% Max.
Phosphorus	0.025% Max.
Sulfur	0.035% Max.
Silicon	0.03% Max.

Alloy Steel Rod

These rods are designed for welding low-carbon and low-alloy steels, including pipe. They are good for welding pressure systems that comply with the codes set down by the American Standards Association (ASA). This rod gives the welded joint a 22% elongation in two inches, with an ultimate tensile strength of 62,000 psi. The chemical analysis of an *alloy steel rod* is:

Carbon	0.15% Max.
Manganese	0.30-0.60%
Phosphorus	0.035% Max.
Sulfur	0.040% Max.
Silicon	0.10-0.30%
Nickel	1.00-1.50%
Chromium	0.30% Max.

Cast Iron Rod

This rod is available with or without alloying materials added, such as molybdenum and nickel. The straight *cast iron rod* consists of approximately 3% silicon, for soft machinable deposits. Gray cast iron welded with this type of rod will be machinable if the proper precautions have been taken during pre- and postheating procedures. A flux, unnecessary with steel welding, is needed for cast iron. The type and reason for using flux is covered later in this chapter.

A moly-cast iron rod, with small amounts of molybdenum and nickel added, gives strong cast iron deposits on alloy cast iron. Both the above rods can be obtained from various manufacturers. They are 24 inches

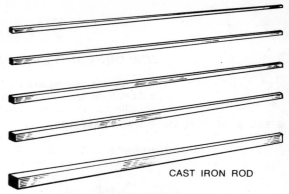

CAST IRON ROD

Fig. 8-46. Cast iron welding rod.

long and square-shaped, as in Fig. 8-46. The chemical analysis of cast iron rods is shown in Table 8-9.

Table 8-9. Chemical Analysis of Common Cast Iron Rods.

Alloying Element	Cast Iron Rod	Moly-Cast Iron Rod
Carbon	3.00-3.50%	3.25-3.50%
Manganese	0.60-0.75	0.50-0.70
Phosphorus	0.50-0.75	0.20-0.40
Sulfur	0.10 Max.	0.10 Max.
Silicon	2.75-3.00	2.00-2.50
Nickel	–––––	1.20-1.60
Molybdenum	–––––	0.25-0.45

Brazing Rod

In brazing or bronze welding, the added metal becomes a part of the joint, but not a part of the base metal. This is because the base metal is not actually melted.

Rods used for brazing, of course, are not made out of steel, but, basically, of brass and copper.

Standard Brazing Rods

Bare or coated bronze rods are available for braze welding steel, cast iron, brass, and bronze. Most bronze rods are free-flowing but have enough flow resistance to allow the welder to control the molten puddle. The melting point of these rods is about 1625° F. When used on steel, the tensile strength is about 50,000 psi. The chemical analysis of brazing rods is usually:

Copper	58.50-59.90%
Iron	0.40- 0.80%
Manganese	0.01- 0.09%
Silicon	0.04- 0.14%
Tin	0.75- 1.10%
Lead	0.05% Max.
Aluminum	0.01% Max.
Chromium	0.05% Max.
Zinc	Balance

Manganese Bronze Rods

These rods are used when a higher-strength bronze filler metal is required. Manganese bronze rods give better bonding qualities than most other types of bronze rods used for brazing malleable iron or steel. The chemical analysis of this rod, which has a tensile strength of about 50,000 psi, is:

Copper	58.50-59.90%
Iron	0.35- 0.50%
Tin	0.75- 1.10%
Manganese	0.13- 0.33%
Nickel	0.25- 0.40%
Silicon	0.05- 0.10%
Aluminum	0.01% Max.
Lead	0.05% Max.
Zinc	Balance

Deoxidized Copper Rods (Bare)

Bare rods are available for welding deoxidized copper. They produce a weld with a tensile strength of 25,000 psi. The chemical analysis of this rod, also known as a *deoxidized copper* rod, is:

Tin	0.65-0.90%
Silicon	0.10-0.30%
Manganese	0.10-0.25%
Phosphorus	0.15% Max.
Aluminum	0.01% Max.
Lead	0.05% Max.
Copper plus Silver	98.00% Min.

Nickel-Silver Brazing Rods

This rod is used primarily on maintenance and production jobs. It replaces the conventional bronze, cast iron, and steel brazing rods. The tensile strength of *nickel-silver* brazing rods can run up to 80,000 psi. This rod has a low melting temperature of 1725° F, with a melting point of 1650° F. The chemical analysis of the rod is:

Copper	46.00-50.00%
Nickel	9.00-11.00%
Silicon	0.15% Max.
Iron	0.25% Max.
Lead	0.05% Max.
Aluminum	0.005% Max.
Others	0.50% Total
Zinc	Balance

Low-Temperature Brazing Rods

Brazing alloy filler metals and fluxes to meet the low-temperature AWS and ASTM classifications are sold by many manufacturers. These filler metals are used for joining all types of metal alloys with one or more joints to close tolerances. Some filler metals can be used without a flux, especially when joining copper to copper. During most brazing, however, the easiest and best job is done with a flux.

The welder who is going to use this type brazing rod should use a rod manufacturer's chart for the type of rods available, and the brazing temperature range required to do a specific job. Fig. 8-47 is a chart from one manufacturer.

Special Welding Rods

Certain metals need to have welding rods that are not normally on hand, such as aluminum and stainless steel. Since a weld must have the same composition as the base metal, special welding rods are needed for each kind of metal.

These welding rods and the special properties of each may be discussed separately.

SILVER BRAZING ALLOY SPECIFICATIONS

AIRCO Alloys	AWS-ASTM Class	% Silver	% Copper	% Zinc	% Cadmium	% Others	Solidus °F	Liquidus °F	Brazing Temp. Range °F
AIRCOSIL 50	BAg-1a	50	15.5	16.5	18	—	1160	1175	1175-1400
AIRCOSIL 3	BAg-3	50	15.5	15.5	16	Ni 3	1195	1270	1270-1500
AIRCOSIL 45	BAg-1	45	15	16	24	—	1125	1145	1145-1400
AIRCOSIL 35	BAg-2	35	26	21	18	—	1125	1295	1295-1550
AIRCOSIL 15	BCuP-5	15	80	—	—	P 5	1185	1500	1300-1500
AIRCOSIL A	—	9	53	38	—	—	1510	1600	1600-1750
AIRCOSIL B	—	20	45	35	—	—	1430	1500	1500-1700
AIRCOSIL C	—	20	45	30	5	—	1140	1500	1500-1700
AIRCOSIL D	—	30	38	32	—	—	1370	1410	1410-1650
AIRCOSIL E	BAg-4	40	30	28	—	Ni 2	1240	1435	1435-1650
AIRCOSIL F	—	40	36	24	—	—	1330	1445	1445-1650
AIRCOSIL G	BAg 5	45	30	25	—	—	1250	1370	1370-1550
AIRCOSIL H	BAg-6	50	34	16	—	—	1270	1425	1425-1600
AIRCOSIL J	BAg-7	56	22	17	—	Sn 5	1145	1205	1205-1400
AIRCOSIL K	—	60	25	15	—	—	1260	1325	1325-1550
AIRCOSIL L	—	54	40	5	—	Ni 1	1325	1575	1575-1750
AIRCOSIL M	BAg-8	72	28	—	—	—	1435	1435	1435-1600
AIRCOSIL N	—	80	16	4	—	—	1360	1490	1490-1650
AIRCOSIL P	BAg-Mn	85	—	—	—	Mn 15	1760	1778	1780-2100
AIRCOSIL Q	—	50	28	22	—	—	1250	1340	1340-1550
AIRCOSIL R	—	40	30	25	—	Ni 5	1240	1560	1560-1750
AIRCOSIL 5	BCuP-3	5	88.75	—	—	P 6.25	1190	1480	1300-1550
AIRCOSIL S	—	25	52.5	22.5	—	—	1500	1575	1575-1750
AIRCOSIL 60	—	60	30	—	—	Sn 10	1095	1325	1325-1550
AIRCOSIL AE-100	—	92.5	7.3	—	—	Li 0.2	1435	1635	1635-1850
AIRCOSIL 105	—	45	30	12	—	Mn 13	1298	1298	1298-1500

Fig. 8-47. Typical manufacturer's brazing rod chart. (Courtesy of Air Reduction Co., Inc.)

Stainless Steel Rods

These rods are available for oxy-acetylene welding most stainless steels. Today, oxy-acetylene welding, when used, is confined mainly to welding light gauge, chrome-nickel stainless steels. Welders planning to weld stainless steel with oxy-acetylene welding should contact the rod manufacturer for information on these rods, and then follow the specification charts available. (See Fig. 8-48.)

Aluminum Welding Rods

These rods are available for welding all grades of aluminum alloys, cast aluminum alloys, and sheet aluminum. Aluminum welding rods are made with various chemical differences, for welding base metals with like differences. These rods have melting points from about 1060°F to about 1165°F. Many aluminum rods available will have a chemical analysis *similar* to, but varying slightly from, the following *general purpose* rod analysis:

Silicon	11.00-13.00%
Copper	0.30% Max.
Iron	0.80% Max.
Zinc	0.20% Max.
Magnesium	0.10% Max.
Manganese	0.15% Max.
Other elements (each)	0.05% Max.
Other elements (total)	0.15% Max.
Aluminum	Balance

STAINLESS STEEL SPECIFICATIONS

AIRCO	AWS A5.9-62 ASTM 371-62	Class	SA-371 F-No.	A-No.	MIL-E-19933D
A308	ER 308	ER 308	F-7	A-7	MIL-308
A308ELC	ER 308L	ER 308	F-7	A-7	MIL-308L
A309	ER 309	ER 309	F-7	A-7	MIL-309
A310	ER 310	ER 310	F-7	A-8	MIL-310
A316 ·	ER 316	ER 316	F-7	A-7	MIL-316
A316ELC	ER 316L	ER 316	F-7	A-7	MIL-316L
A347	ER 347	ER 347	F-7	A-7	MIL-347

Fig. 8.48. Stainless steel welding rod specification chart. (Courtesy of Air Reduction Co., Inc.)

Before welding, the welder should check with the rod manufacturer about the type rod required to meet the AWS-ASTM classification. Fig. 8-49 is an aluminum welding rod specification chart showing the properties of aluminum welding rods.

ALUMINUM SPECIFICATIONS

AIRCO	AWS A.5.10-61 ASTM B285-61	SB-285 Class	SB-285 F-No.	QQ-R-566A
No. 1100	R 1100	R 1100	F-21	1100
No. 4043	R 4043	R 4043	F-23	4043
No. 5183	R 5183	R 5183	F-22	
No. 5356	R 5356	R 5356	F-22	5356
No. 5554	R 5554	R 5554	F-22	5554
No. 5556	R 5556	R 5556	F-22	5556

Fig. 8-49. Aluminum welding rod specifications. (Courtesy of Air Reduction Co., Inc.)

Fluxes

Impurities in the air and dirt on the parent metals can cause poor welds. The best welds would result if the metal was perfectly clean and welded in an inert atmosphere or in a vacuum. Because those conditions are not too common, welders must try to eliminate dirt and air as much as possible while welding. This is usually done by a mixture of chemicals called a *flux*. The flux may be a coating on the rod or a can of flux into which the rod is dipped before welding.

While oxy-acetylene welding steel, welders usually find that either a mild steel rod or a high-tensile, low-alloy rod can be welded easily without adding flux. Since mild steel melts at about 2750°F, the heat is enough to melt the iron oxides in the material and float them to the top, thereby keeping the weld's insides clean and strong.

Since cast iron is also an iron compound, it also has this same iron oxide. However, due to cast iron's lower melting temperature of 2200°F, the iron oxides cannot be dissolved or melted during oxy-acetylene welding. When this happens, it is necessary to use a flux which will lower the melting temperature of the oxides. This will allow them to float to the surface where they can be removed.

While welding, therefore, one of the flux's purposes is to *lower* the melting temperature of the oxides, so they can be removed for a stronger weld. Fluxes used with brazing or soft soldering also prevent oxidation of the base metal by a protective coating. Too, they promote *capillarity* (interaction between contacting surfaces of a liquid and a solid) of

the filler metal. Among others, a cast iron welding flux is shown in Fig. 8-50.

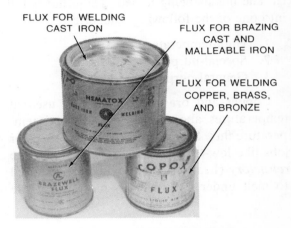

Fig. 8-50. Common welding fluxes.

The oxides of commercial metals other than steel have melting points *higher* than the metals themselves. Therefore, when these metals are fluid at welding temperature, the oxides still cannot be fused or melted. The flux used must be able to float these high-temperature oxides away from the weld in the form of *slag*.

Due to differences in the chemical make-up and melting point of fluxes, the welder must

Fig. 8-51. A good assortment of welding and brazing fluxes. (Courtesy of Airco Welding Products Division.)

use the right flux for: A. The filler metal being used; B. The method being used, and; C. The metals being joined. These fluxes fall into one of the following categories:

1. High-temperature.
2. Special-purpose.
3. General-purpose.

Copper and brass brazing alloys used at temperatures above 1500° F use high-temperature flux. Special purpose flux is used for jobs like low-temperature brazing, where the *refractory* (heat-resistant) oxides are required to melt under 1100° F. In most cases, the

general-purpose flux, which usually melts below 1500° F, can be used with copper-phosphorus alloys and the more common silver-brazing alloys. Joining copper to copper-phosphorus alloys requires no flux during silver brazing, since the phosphorus in the brazing material deoxidizes the copper surface.

Due to the many variations in the types of rods and fluxes being produced by many manufacturers, the welder should have information on fluxes for the type of rods he intends to use. Fig. 8-51 shows most of the many fluxes made by one company.

9

Preparing For
Oxy-Acetylene Practice

When a welder has the equipment needed for oxy-acetylene welding and cutting, he has to assemble the equipment into a safe working unit. This is called *setting up* for welding and cutting.

The needed equipment begins with a pair of *cylinders,* one containing acetylene and the other oxygen. A *regulator* to control the gas flow is attached to each tank. These control the cylinder gas pressure, to make controlling the welding flame possible.

Next, a hose, built to hold the gas pressure and resist oil and general shop abuse, is attached. It carries the gas from each cylinder regulator to the welding or cutting torch assembly. One end of the hose is attached to the regulator, the other end of the hose is attached to the torch body.

The torch mixes the gases and passes them on to the tip. Or, the torch may be a *cutting* torch, as shown with the other welding equipment in Fig. 9-1.

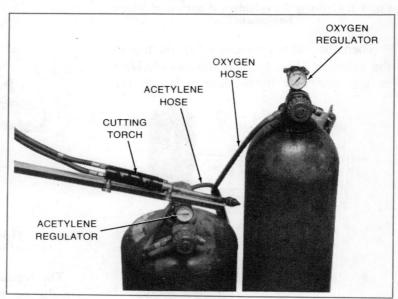

Fig. 9-1. Cutting torch setup, ready to use with common gas cutting equipment.

OXYGEN REGULATOR

OXYGEN HOSE

ACETYLENE HOSE

CUTTING TORCH

ACETYLENE REGULATOR

Setting Up Tanks

The first job is to get the tanks set up, held in place, and made ready to work with the oxygen and acetylene cylinders side by side in an *upright* position. For safety, of course, the welder should never lay the acetylene tank down on its side. The tanks should then be chained or otherwise held in place, as in Fig. 9-2.

Fig. 9-2. Chaining the cylinders. (Courtesy of Victor Equipment Co.)

Then the valve protection cap on top of the cylinder (Fig. 9-3) is unscrewed. Dust or grit that may have accumulated in the

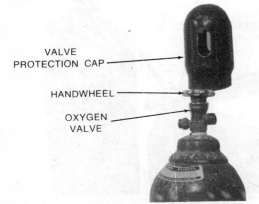

VALVE
PROTECTION CAP

HANDWHEEL

OXYGEN
VALVE

Fig. 9-3. Valve protection cap, unscrewed from the cylinder. During shipping, the cap protects the valve.

valve opening is cleared out by *cracking* the valve. Cracking the valve means opening the valve and blowing dirt or grit from the valve outlet with the released gas pressure. Caution should be used when cracking the valve. Stand to one side of the outlet and keep the gas flow away from flames. When cracking the *oxygen* valve, the welder must be sure that the tank is away from grease or oil that could cause an explosion. The tanks are now set up and chained, and they have clean valves.

Attaching Regulators and Hoses

The welder now has to hook up the right regulator and hoses to each cylinder. The regulators cannot be interchanged, since there is the *right-hand* thread connection for the oxygen and the *left-hand* thread connection for the acetylene. Once the regulators have been screwed on to the tank outlets by hand, they can then be tightened by the regulator nut (or gland) wrench (Fig. 9-4).

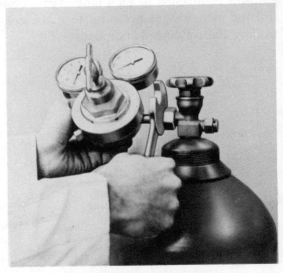

Fig. 9-4. Tightening the regulator onto the tank valve. (Courtesy of Victor Equipment Co.)

The regulator adjusting handle (Fig. 9-5) should be released on each regulator by screwing *counter-clockwise*. Usually, two turns will release the regulator mechanism

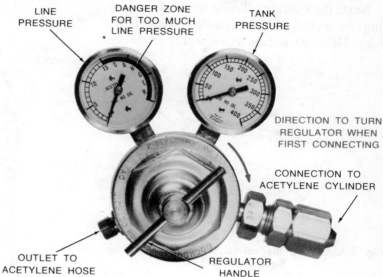

LINE PRESSURE

DANGER ZONE FOR TOO MUCH LINE PRESSURE

TANK PRESSURE

DIRECTION TO TURN REGULATOR WHEN FIRST CONNECTING

CONNECTION TO ACETYLENE CYLINDER

REGULATOR HANDLE

OUTLET TO ACETYLENE HOSE

Fig. 9-5. Acetylene regulator, showing how to turn the regulator handle when first setting up. (Courtesy of Dockson Corporation.)

enough to stop the gas from entering the hose when the cylinder valve is opened. When the welder turns the regulator handle, he is actually compressing a large spring in the regulator, as shown in Fig. 9-6.

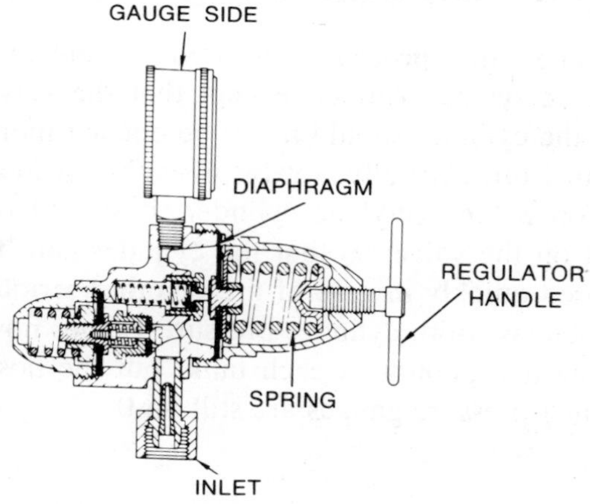

GAUGE SIDE

DIAPHRAGM

REGULATOR HANDLE

SPRING

INLET

Fig. 9-6. How the regulator handle affects the spring. (Courtesy of Union Carbide Canada Ltd.)

Now, one end of the hose is attached to the regulator. The hoses on the other end are attached to the cutting or welding torch. See Figs. 9-1 and 9-7. The *green* hose goes to the *oxygen* cylinder, but the welder can't mistakenly put this hose on the wrong cylinder, because of the different threads used on each cylinder. To review, the *green* hose has *right*-

hand threads for *oxygen*, while the *red* hose has *left*-hand threads for acetylene.

If the hose is new, of course, it should be *purged* (blown out) to get rid of the talcum powder inside a new hose. Purging a new hose

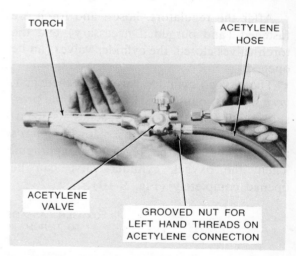

TORCH

ACETYLENE HOSE

ACETYLENE VALVE

GROOVED NUT FOR LEFT HAND THREADS ON ACETYLENE CONNECTION

Fig. 9-7. Connecting the hoses to the torch. (Courtesy of Victor Equipment Co.)

is usually done by using compressed air to blow out the powder under pressure. If purging is done with oxygen, the welder must carefully remove any explosive materials. Once the hoses are purged, the welder can finish connecting the *other* end of the hoses to the torch, as in Fig. 9-7.

Next, the operator should turn off the cutting or welding torch needle valves (Fig. 9-8). How the needle valve works is shown

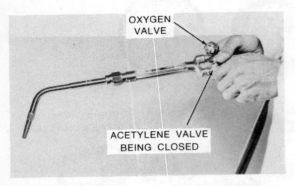

Fig. 9-8. Closing the valves. (Courtesy of Victor Equipment Co.)

in Fig. 9-9. Here, the passage from the hose is blocked off when the valve is screwed all the way in.

Opening Valves

After the regulators, hoses, and torch are fastened (and purged if necessary) and the torch valves closed, the cylinder valves can be opened.

The oxygen cylinder valve is opened first, *very slowly,* counter-clockwise, until the pressure needle on the gauge reaches its maximum. Opening the valve slowly guards against rupturing the regulator diaphragm. Then, the oxygen cylinder valve can be opened completely (Fig. 9-10).

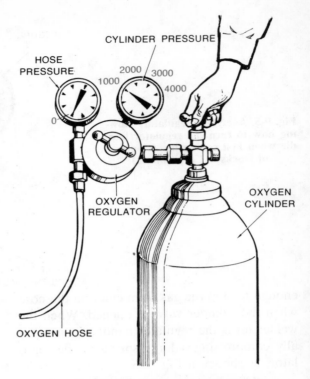

Fig. 9-10. Opening the oxygen cylinder valve.

The same procedure is used for opening the acetylene cylinder, except that the valve on the cylinder should *never* be opened more than 1 turn. Usually, a valve *wrench* is needed to open the acetylene cylinder; it should be left on the valve, so that the cylinder can be closed quickly in case of fire or other serious problems. Both cylinder pressure gauges now show the amount in each tank, but the hose (line) pressure gauges are still on 0.

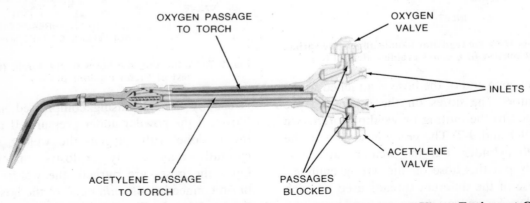

Fig. 9-9. How the valves work, as illustrated by a torch cutaway. (Courtesy of Victor Equipment Co.)

The required hose pressure is set by turning *in* the adjusting pressure screw on the regulator. The exact pressure will depend on the type of tip being used for each welding job. Hose pressure adjustments, called the *torch balancing* process, will be explained later.

It may be necessary to readjust the regulator pressures after the oxygen and acetylene mixture has escaped through the tip, to compensate for a gauge drop. Even so, the welder must *always* remember to keep the acetylene pressure under 15 psi at all times. Even while cutting heavy 12″ thick material, acetylene gas pressure does not need to exceed 6 psi.

Each tip used on a welding torch has an orifice (opening) designed to use a certain amount of acetylene. Therefore, a welder has to know the maximum amount of acetylene that will go through the tip he wants to use. He must "figure in" all the factors that tell him which tip size to use for the job. Most of the factors are listed below, but experience will help any welder learn factors most important to him.

Type of Weld—The tip size and the amount of acetylene to be used will depend on whether the welder is going to do low-temperature brazing, braze welding, or fusion welding.

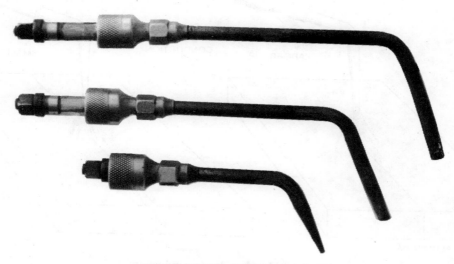

Fig. 9-11. Common torch tips.

Since the hose and all connections are now under gas pressure, a welder can check all possible areas for gas leaks by using soapy water as he looks for bubbles at connections and along the hose. Then, any leaking connections can be tightened before lighting the torch.

Tip Selection

Torches need to have *tips* put on the end, so the mixed gasses can be sent out in a certain direction for a good flame. Fig. 9-11 shows two tips ready for a torch.

Welding Technique—Welders who prefer a *forehand* welding technique instead of a *backhand* technique need a smaller welding tip to produce good welds. These techniques are explained in a later chapter.

Welding Position—Less heat is used in *vertical* and *overhead* positions (Fig. 9-12) than in the flat position. Sometimes, it may be necessary to use a smaller tip size if adjusting the flame doesn't reduce the heat enough.

Type of material—The type of material to be welded also helps tell what tip size to use. For example, when welding large aluminum sheets where the melting tempera-

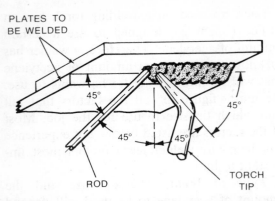

Fig. 9-12. Overhead position for the welding torch and rod.

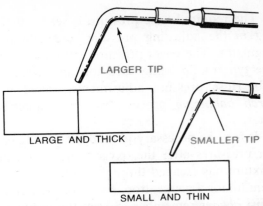

Fig. 9-14. Generally, the larger welding job needs a larger tip, if the metals are the same kind.

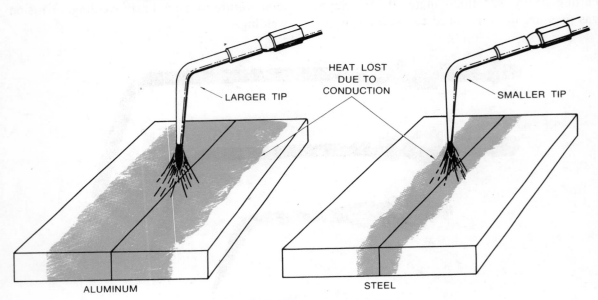

Fig. 9-13. Heat loss due to conduction.

ture is lower than steel, heat radiation is so great that a *larger* tip size is needed than when welding steel of the same size! See Fig. 9-13.

Thickness and Size—The *thicker* the material and the larger its *size,* the larger is the tip needed to do a good job (Fig. 9-14).

When a welder knows about all the factors affecting the tip size, he will be able to understand and use fine flame adjustments for quality welds. To do these adjustments, the welder must be able to *balance* the hose delivery pressure of the oxygen and acetylene after lighting the torch.

After the tip is chosen, it is screwed into the end of the torch body, as in Fig. 9-15. Although some tips may need to be tightened with a wrench, many torch and tip seals and seats can be damaged by overtightening. So, *if* they are tightened with a wrench, they should only be tightened lightly.

Lighting the Torch

Before lighting the torch, purge both hoses by opening the torch needle valves *one at a time,* and increase the gas pressure by turning the regulator handle (Fig. 9-16). Pres-

TIP   TIP NUT   TORCH

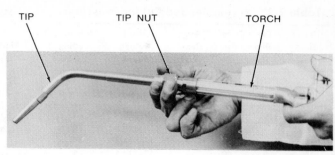

Fig. 9-15. **Tightening the torch tip by hand. (Courtesy of Victor Equipment Co.)**

sure used to purge the acetylene hose should *remain under* the 15-psi safety limit, but pressure used on the oxygen hose can be set at 30 psi for purging. After purging the hoses, which takes just a couple of seconds of releasing gas, the oxygen and acetylene pressures should be shut off at the regulators.

Then, each valve is opened on the torch, and the regulator is adjusted to give the right pressure called for in tables such as Table 9-1 for cutting and Table 9-2 for welding. Each valve is done separately and one valve must be shut off before adjusting the other one.

Basic Method

To actually light the torch, the oxygen valve is left *closed*. The acetylene valve is "cracked" (opened slightly), about $\frac{1}{6}$ of a turn. Then, holding the torch away from anything that burns, squeeze the spark lighter under the tip, and the acetylene will light. Goggles, of course, must always be worn.

The acetylene valve is then opened slightly, but not so far that the flame jumps away from the tip, as in Fig. 9-17. Then, oxygen is added by opening the oxygen valve slightly. The two valves are then adjusted to provide the desired flame.

Alternate Method— Balancing the Delivery Pressure

A more accurate method of adjusting the torch is by using the regulators to balance the delivery pressures. Here, the hoses are purged and the valves are closed on the torch.

Fig. 9-16. **Purging the hoses. The acetylene valve is done first. [Note: In this picture, no tip has been installed on the torch. Purging can be done with or without a tip installed.] (Courtesy of Smith Welding Equipment.)**

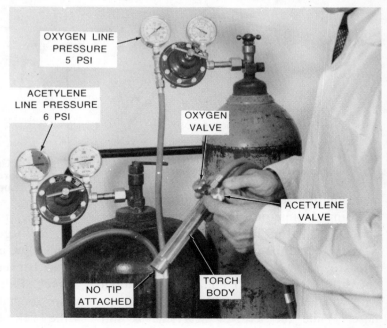

OXYGEN LINE
PRESSURE
5 PSI

ACETYLENE
LINE PRESSURE
6 PSI

OXYGEN
VALVE

ACETYLENE
VALVE

TORCH
BODY

NO TIP
ATTACHED

Table 9-1.   Recommended Cutting Pressures.

Metal Thick-ness	Cut-ting Tip	Oxygen* Pres-sure	Acety-lene* Pres.†	Hand Cutting Speed	Machine Cutting Speed
Inches	Size No.	lbs. sq. in.	lbs. sq. in.	in./min.	in./min.
¼	0	30	3	16 -18	20
⅜	1	30	3	14.5-16.5	19
½	1	40	3	12 -14.5	17
¾	2	40	3	12 -14.5	15
1	2	50	3	8.5-11.5	14
1½	3	45	3	6.0- 7.5	12
2	4	50	3	5.5- 7.0	10
3	5	45	4	5.0- 6.5	8
4	5	60	4	4.0- 5.0	7
5	6	50	5	3.5- 4.5	6
6	6	55	5	3.0- 4.0	5
8	7	60	6	2.5- 3.5	4
10	7	70	6	2 - 3	3½
12	8	70	6	1.5- 2	3

Courtesy of Airco Welding Products Division.

*Gas pressures are for hose lengths up to 25'. Increase pressure for longer lengths of hose.

†Acetylene pressures are for tips with light preheating flames for use on clean plate. On rusty, scaly or painted surfaces use tips with larger preheat flames. These tips require higher acetylene pressures than shown; oxygen pressures remain the same.

Table 9-2.   Recommended Welding Pressures.

Metal Thickness (in.)	Tip Number	Oxygen* Pressure (psi) approx.	Acetylene* Pressure (psi) approx.	Acetylene† Usage (cfh) approx.
¹⁄₆₄	00	1	1	0.5
¹⁄₃₂	0	1	1	1
¹⁄₁₆	1	1	1	2
³⁄₃₂	2	2	2	5
⅛	3	3	3	9
³⁄₁₆	4	4	4	16
¼	5	5	5	25
⁵⁄₁₆	6	6	6	30
⅜	7	7	7	40
½	8	7	7	60
⅝	9	7½	7½	70
¾ up	10	9	9	90

Courtesy of Airco Welding Products Division.

*Pressures and consumptions are for separable tips with appropriate mixers. Operating pressures for tip-mixer assemblies to normal flow will be somewhat higher for the smaller sizes. Gas pressures are for hose lengths up to 25'. Increase pressure for longer lengths of hose.

†Oxygen consumption calculated at 1.1 times acetylene consumption.

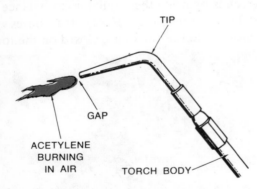

Fig. 9-17. Excess acetylene causes the flame to form a gap between itself and the torch tip.

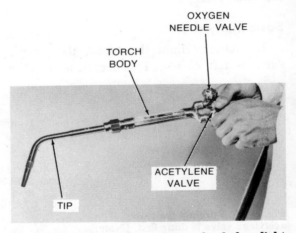

Fig. 9-18. Opening the acetylene valve before light-ing the torch to balance the delivery pressures. (Courtesy of Victor Equipment Co.)

Then the regulator handles are screwed out, releasing the line pressure. However, there is enough gas left in the hose to light the torch.

With goggles on and the regulator pressure released, the welder opens up the acetylene valve fully (Fig. 9-18) and lights the torch with what little acetylene there is left in the hose. While the leftover acetylene is still

burning, the acetylene pressure is increased at the regulator. Then, the acetylene flame will blow away from the tip, leaving a slight gap between the flame and the tip of the torch. Balancing should be done with a #3 tip, so that about a ⅜" gap exists between the back of the flame and the end of the tip, as shown in Fig. 9-17. A tip twice that size would require a gap of about ¾″, to give maximum pressure and flow for that tip size.

After adjusting the regulator for the maximum acetylene pressure that the tip can support, the next step in balancing is to *close* the acetylene needle valve on the torch very slowly, until the gap between the end of the flame and the tip is completely gone, as in Fig. 9-19.

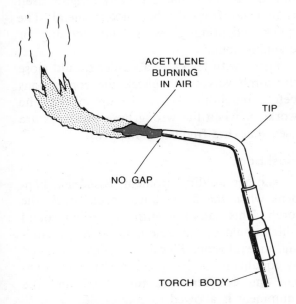

Fig. 9-19. Closing the acetylene needle valve slightly will bring the flame back to the torch tip from where it was in Fig. 9-17.

Since oxygen is going to be added, the gap must be gone completely so that the oxygen pressure will not blow out the flame. If the acetylene flame is smoky when the gap is closed, the acetylene flow must be increased just enough so that the flame quits smoking. This adjustment assures the operator of minimum acetylene gas flow from the torch tip, and eliminates the danger of *backfire,* to be explained later.

Now, the torch oxygen needle valve is opened all the way, so a cone will form inside the flame, as shown in Fig. 9-20. The oxygen regulator can now be adjusted for the pressure called for in the chart or table being

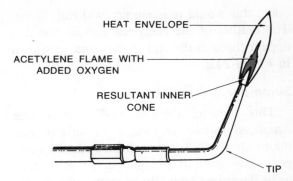

Fig. 9-20. A smaller interior flame appears as oxygen is added.

used. Each needle valve on the torch should now be opened gradually and alternately, to get the right flame.

By using this oxygen and acetylene flow adjustment, the flame is not likely to blow out. Once the acetylene needle valve is fully opened, the operator is assured of maximum acetylene flow. With *both* valves fully opened, the regulator adjusting screw on the oxygen valve is used to form a neutral flame, as shown in Fig. 9-21.

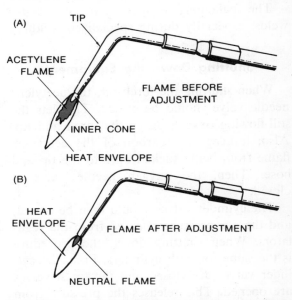

Fig. 9-21. As oxygen is increased (or acetylene decreased), the acetylene flame completely disappears.

The oxygen pressure should not be increased to a point where it *shortens* the inner cone; this would indicate an oxidizing flame. The feather of the acetylene should just disappear through the end of the cone, as shown in Fig. 9-21B.

Summary

This lighting method has included the *equalized pressure technique.* Because of this, the torch needle valves can be operated independently without the danger of *flashback,* to be described later. By operating the needle valves, it is easy to set a flame size suitable for the work at hand.

All the preceding steps must be followed if it becomes necessary to use a different sized tip. If the flame is reduced to a point where the acetylene would be smoky when the oxygen is shut off, a "popping" at the tip could happen. This "popping" is called *backfire,* and comes from low acetylene gas feed. A smaller tip will overcome this problem. For this reason, it is a good idea to work more in the mid-range of a tip's capacity, rather than at the extreme maximum or minimum.

The balancing method can give better welds, especially during specialized welding.

Shutting Down the Equipment

When shutting the torch off, the acetylene needle valve should be closed. This lets the still-flowing oxygen purge the torch and tip. Also, it keeps the carbon of the acetylene flame from being sucked up into the tip and hose. Then, the oxygen needle valve is closed.

The cylinder valves should then be closed, and the gas pressure released from the regulators. When shutting down, the procedure is the same for both cylinders. First, the cylinder valves are closed and the torch valves are opened. This releases the pressure from the hose and regulators. Then, the adjusting screws on the regulators are released (turned out), and the torch valves are again closed. Finally, if the equipment is to be left for long periods, the welder should turn in the pressure

adjusting screw slightly (Fig. 9-5) which releases the spring pressure on the valve seat.

Torch Problems

Backfire

Sometimes, a welder may experience *backfire,* due to trouble with the torch and tip. Backfire happens when the flame backs up into the tip. The flame usually relights itself right away from the hot molten metal. The quick relighting gives a loud popping or cracking sound.

Faced with backfire, a welder should close the torch valves and check the connections before relighting. When the tip touches the work, or when the wrong hose pressures are used, backfire frequently happens.

Flashback

Another welding problem, *flashback,* happens when the flame burns back *inside* the torch. This causes a shrill squealing sound which could create serious problems, including an explosion. Flashback can be avoided by equalizing the gas pressures after the oxygen and acetylene are turned off and the equipment is allowed to cool.

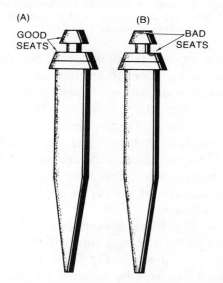

Fig. 9-22. Two seat cutting tips with (a) good seats and (b) bad seats.

Explosions due to flashback will happen in the line carrying the least pressure. Since flashback happens *beyond* the mixer, it can destroy the hose and even the regulator. Faulty seats (Fig. 9-22) can cause flashback, as well as poorly operating the torch valves.

Torch Flames

Different kinds of welding need different flames—for example, different amounts of either oxygen or acetylene or different flame lengths and widths. The differences come from how the metal is made up (how much carbon or other alloy it contains) and how thick or rusty the metal is. Fig. 9-23 shows the more common welding flames and their characteristics.

Several common flames will be covered completely. These are the ones most often used in the welding industry.

Acetylene Burning in Air

When acetylene is first ignited and has only the air's oxygen to support it, the flame is known as *acetylene burning in air* (Fig. 9-24). This flame's temperature is only about

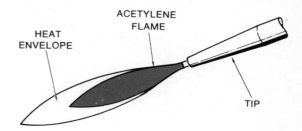

Fig. 9-24. Acetylene burning in air.

WELDING FLAME (APPROX. TEMP.)

EFFECT ON METAL

Not suitable for welding.

a. Acetylene burning in air (1,500° F.)

Metal boils and is not clear.

b. Strongly Carburizing Flame (5,700° F.)

Similar to neutral flame—little or no puddling is necessary.

c. Slight Excess Acetylene Flame (5,800° F.)

Metal is clean and clear, flowing easily.

d. Neutral Flame (5,900° F.)

Excessive foaming and sparking of metal.

e. Oxidizing Flame (6,300° F.)

CUTTING FLAME

EFFECT ON METAL

Not suitable for cutting.

f. Acetylene burning In air

Excess acetylene helps to get heat down to the bottom of material being cut, this is especially suitable for cutting cast iron.

g. Strongly Carburizing Flame—Preheat only

h. Strongly Carburizing Flame—Cutting Oxygen flowing

Standard adjustment for cutting steel.

i. Neutral Flame—Preheat only

j. Neutral Flame—Cutting Oxygen flowing

Fig. 9-23. Oxygen-acetylene flames. Temperatures, characteristics, and applications.

1500° F, and is *not* suitable for welding. Burning pure acetylene in air produces thick, black strings of carbon that can be found in any oxy-acetylene welding area.

Carburizing Flames

Any flame producing a feather *beyond* the cone, as in Fig. 9-25, is known as a carbon-

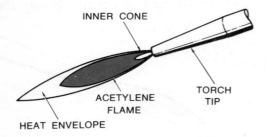

Fig. 9-25. Carburizing flame. This contains too much acetylene for normal welding.

izing or a *carburizing* flame. The carburizing flame shown in Fig. 9-26 has a temperature of about 5700° F and is known in the welding trade as a *2X* flame.

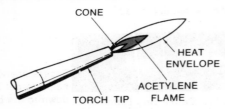

Fig. 9-26. The 2X welding flame, a carburizing flame.

A flame with too much acetylene is called a carburizing flame because it adds carbon to the metal. As a matter of fact, acetylene *is* almost all carbon. The acetylene part of a flame is called the acetylene *feather*.

There are a variety of tip sizes for welding, yet tip size alone has no bearing on the temperature. To compare tips, comparisons are made in relation to the length of the *neutral cone*. Too much acetylene in the flame produces a feather at the end of the cone, as in Fig. 9-27A. *In*creasing the amount of acetylene (or *de*creasing the amount of oxygen) causes the acetylene feather to get

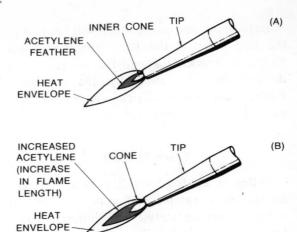

Fig. 9-27. As the acetylene gas is increased, the acetylene feather increases.

larger, as in Fig. 9-27B. In both figures (A and B) the heat envelope and inner cone parts of the flame stay about the same size.

The cone itself is said to be X long (Fig. 9-28), and the feather is twice that long, or 2X. For this reason, the flame in Fig. 9-28 is said to be a *2X flame*. This would be a carburizing flame. Fig. 9-29 shows a *3X*

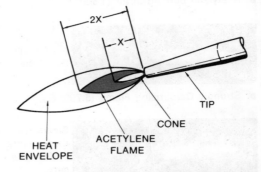

Fig. 9-28. A 2X flame. The acetylene feather is twice as long (2X) as the inner cone (X).

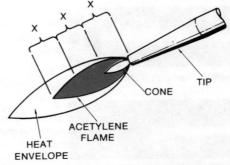

Fig. 9-29. A 3X flame, strongly carburizing.

flame, where the acetylene feather is three times as long as the inner cone. This is called a strongly carburizing flame.

A carburizing flame with less than a 2X feather is called a reducing flame, since the amount of acetylene has been reduced to shorten the feather and bring it closer to the end of the cone. Fig. 9-30 shows a 1⅛ re-

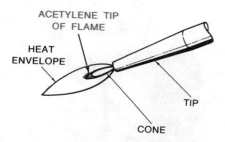

Fig. 9-30. A 1⅛ X, or reducing, flame.

ducing flame, with the feather only ⅛ longer than the cone.

Carbonizing (carburizing) flames *add* carbon to the weld. In reasonable amounts, this carbon increases the tensile strength of the weld, but it also causes a loss of ductility and shock resistance. Carbonizing flames are used to give certain conditions in the weld. For example, the reducing flame (1⅛ X) is used when welding mild steel with a low-alloy, high tensile strength welding rod. This type of flame is usually used with backhand welding and will be discussed in a later chapter. The 2½ X or 3X flame is used for hardsurfacing and aluminum brazing. When using low melting temperature rods, such as Easy Flo®, the carbonizing flame is a must.

Neutral Flame

This flame has a temperature of about 5850° F. As shown in Fig. 9-31, it is neither oxidizing nor carburizing. Since the feather does *not* show beyond the cone, it is considered to be an equal distance with the cone. These is no excess oxygen or acetylene in the flame, so nothing is added or taken away from the weld metal. When welding on mild steel, the welder should keep the molten

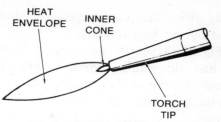

Fig. 9-31. A neutral flame.

puddle in a quiet, mirrorlike condition; this indicates a *neutral* flame.

The neutral flame is used for welding low-carbon steels and cast iron. It is also used in *brazing,* mentioned later. In flame cutting, the outside holes in the cutting tip (Fig. 9-32) have a neutral flame for preheating. Since the neutral flame will not add or take anything

Fig. 9-32. A neutral flame is produced from the preheat holes of the cutting torch.

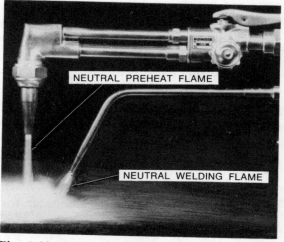

Fig. 9-33. Common welding and cutting flames. (Courtesy of Victor Equipment Co.)

away from the parent metal, it should be used during heat treating.

Fig. 9-33 shows two welding flames, including the cutting torch neutral preheat flame and the normal neutral flame. These are the most common flames for welding and cutting.

Oxidizing Flame

Once a welder has a neutral flame, he can add still *more* oxygen to get an *oxidizing* flame. Cutting *down* on the acetylene would have the same effect. In either case, the inner cone becomes shorter than a neutral flame's cone, and the flame gives off a harsh, hissing sound. The oxidizing flame burns at about 6300° F. Fig. 9-34 shows an oxidizing flame with a short, stubby cone.

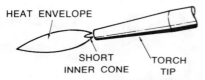

HEAT ENVELOPE
SHORT INNER CONE
TORCH TIP

Fig. 9-34. The oxidizing flame.

In an oxidizing flame, there is more oxygen than acetylene. When this flame is used to weld mild steel, the weld is oxidized; this makes it porous and weak.

The oxidizing flame is used for welding brass and bronze, as well as for brazing galvanized iron. The prepared edges of cast iron are sometimes seared with an oxidizing flame before being brazed.

The oxidizing flame is the least used welding flame. It has the fewest practical uses in welding, but it is easily identified by its short cone and harsh, hissing sound.

Oxy-Acetylene Setup Review

For quick reference, here is a step-by-step sequence to quickly review the safe setup for oxy-acetylene welding.

1. Put the oxygen and acetylene cylinders together near where they are to

be used and secure them from falling (Fig. 9-35).

Fig. 9-35. Chaining the tanks so they can't fall. (Courtesy of Smith Welding Equipment.)

2. Unscrew the valve protection caps. Put the caps away safely, since they must be put back on the empty cylinders before they are returned (Fig. 9-36).

3. Examine the outlet valve threads, shown in Fig. 9-36. Wipe off any dirt with a clean cloth so that absolutely no dirt will enter the regulator. The wiping cloth must not have *any* oil or grease on it.

4. Quickly open and close the oxygen and acetylene cylinder valves to make sure they do not stick and to blow out any dirt or dust in the valve. This is called *cracking* the valve.

    a. The oxygen cylinder valve has a hand wheel for operating (Fig. 9-37).

    b. The acetylene valve may need a valve *wrench*, which is fur-

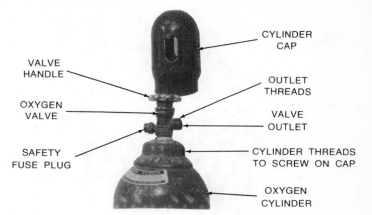

Fig. 9-36. Cylinder valve and protection cap.

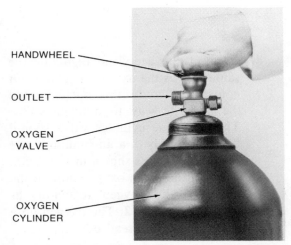

Fig. 9-37. Cracking the oxygen valve. (Courtesy of Victor Equipment Co.)

nished by the acetylene supplier (Fig. 9-38).

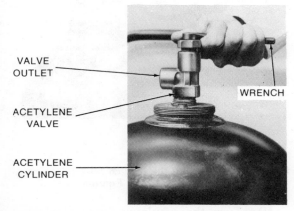

Fig. 9-38. Cracking the acetylene valve. (Courtesy of Victor Equipment Co.)

5. Be sure that the regulator *inlet* connection is absolutely clean and free of any dirt, grit, or grease. Attach the oxygen regulator to the oxygen cylinder valve, using a 4-way open end wrench often furnished with welding outfits. Do not use too much force, but tighten the regulator nut firmly (Figs. 9-39 and 9-40).

Fig. 9-39. Tightening the oxygen regulator in place. (Courtesy of Smith Welding Equipment.)

6. Be sure the regulator adjusting screw is in the "out" position, releasing tension on the regulator. The adjusting screw is turned *counter-clockwise* to release the tension, as in Fig. 9-40.

7. Open the oxygen cylinder valve *slowly,* so the high-pressure regulator

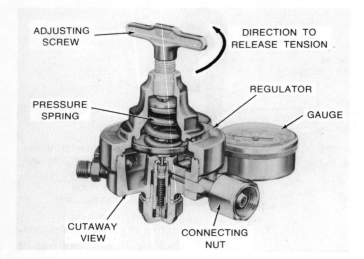

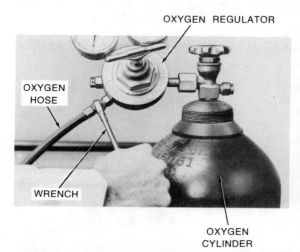

Fig. 9-40. Components of a regulator and adjusting screw. (Courtesy of Victor Equipment Co.)

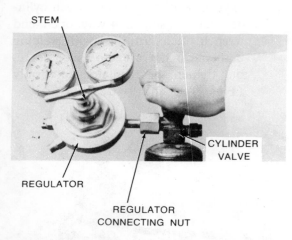

Fig. 9-41. Opening the oxygen cylinder. (Courtesy of Victor Equipment Co.)

gauge needle will move up slowly until as much as 2200 psi is registered, if the cylinder is full (Fig. 9-41). Stand to one side of the regulator when the oxygen cylinder valve is opened. When full pressure has been reached, the valve can be opened all the way. If the valve leaks, close it immediately and call the supplier. A cylinder valve shouldn't need force to be opened or closed. If the valve will not open or close by hand, call the supplier and have the cylinder replaced.

8. Connect the acetylene regulator the same way as the oxygen regulator. Of course, the connecting nut is turned the opposite way to tighten. NOTE: The acetylene cylinder valve should be opened only to a maximum of *one* complete turn, as shown in Fig. 9-38.

9. Connect the green oxygen hose to the outlet of the oxygen regulator, as shown in Fig. 9-42. The oxygen hose has *right*-hand thread connections.

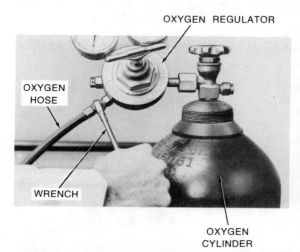

Fig. 9-42. Connecting the oxygen hose to the regulator. (Courtesy of Victor Equipment Co.)

10. Turn the regulator screw clockwise, until a reading of 5 psi shows on the

regulator low-pressure gauge. Purge the hose by letting the oxygen blow through the hose until you are sure the hose is clean inside, as in Fig. 9-43. *New* hose, of course, has tal-

Fig. 9-43. Purging the oxygen hose. (Courtesy of Smith Welding Equipment.)

cum powder in it to protect the hose lining while in storage. Left in the hose, dust, dirt, and especially talcum powder, may enter the small gas passages in the torch and plug them, possibly causing flashback.

11. Connect the red acetylene hose to the acetylene regulator outlet. The acetylene hose has *left*-hand thread connections. Purge the acetylene hose like the oxygen hose. Remember, though, the acetylene *will* burn. Keep it away from open flames while you are blowing out the hose. The method is similar to the oxygen purging in Fig. 9-43.

12. Unscrew the regulator screws until gas stops coming out of the hoses.

13. Connect the free ends of the hoses to the welding or cutting torch. Remember that green oxygen hose connections are *right*-hand thread, while the red acetylene hose connections are

left-hand thread (Fig. 9-44). The torch valves are usually marked OXY and ACET.

Fig. 9-44. Connecting the hoses to the torch. The grooved nut on the acetylene hose means left-hand threads. (Courtesy of Smith Welding Equipment.)

14. Select the tip or nozzle suitable for the job you are going to do. Refer to a welding and cutting tip selection chart.

15. Install the tip. Tips are held onto the torch body by a wrench-type connection nut. In Fig. 9-45, a welding tip is being installed on the torch; in Fig. 9-46, a cutting tip is being installed. Many welding tips should not be tightened into the torch with a wrench. The cone end of many tips has two Hypalon (synthetic rubber) sealing rings. A *hand tight* fit is all that is necessary for a gas tight joint, as in Fig. 9-47.

16. Partly open the torch oxygen valve and adjust the oxygen regulator until the pressure is right for the tip being used (Fig. 9-48). Then close the valve.

17. Partly open the torch acetylene valve and adjust the regulator pressure for

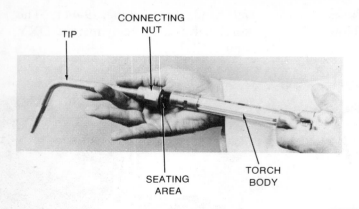

Fig. 9-45. Installing a welding tip. (Courtesy of Victor Equipment Co.)

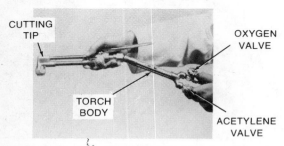

Fig. 9-46. Installing a cutting tip. (Courtesy of Victor Equipment Co.)

the tip being used (Fig. 9-49). Then close the valve.

NOTE: The regulator pressure will rise slightly when the torch valves are closed, as in Fig. 9-50. This is why the regulator pressures are set with the valve open. All pressures in weld-

Fig. 9-47. Firmly hand tightening a new style tip onto the welding torch body. (Courtesy of Smith Welding Equipment.)

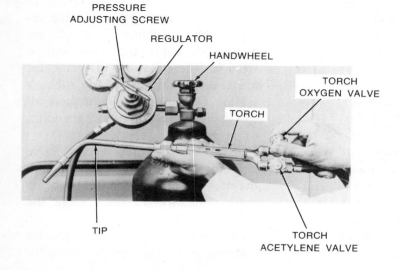

Fig. 9-48. Opening the torch oxygen valve to adjust the delivery pressure at the regulator. (Courtesy of Victor Equipment Co.)

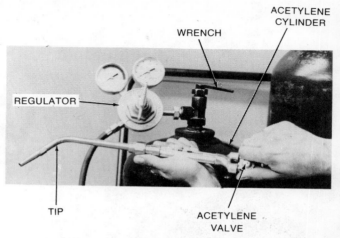

Fig. 9-49. Opening the torch acetylene valve to adjust the delivery
pressure at the regulator. (Courtesy
of Victor Equipment Co.)

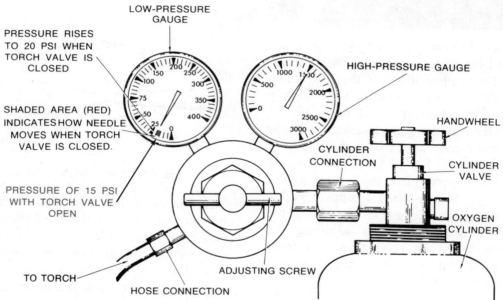

Fig. 9-50. After the pressure is set, the low-pressure gauge rises slightly when the torch valve is closed.

ing and cutting charts are *flowing*
pressures with the torch valves *open*.

18. Put goggles and protective clothing in
place.

19. Hold the torch in one hand and the
spark lighter in the other, as shown
in Fig. 9-51. Point the torch away
from people, cylinders, or anything
that might burn.

20. Open the torch acetylene valve approximately ⅙ turn ·and light the gas
by squeezing the spark lighter like
a pair of pliers.

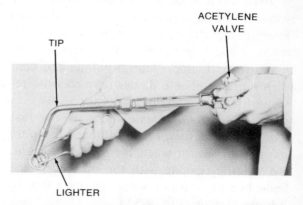

Fig. 9-51. Position for lighting the torch. (Courtesy
of Victor Equipment Co.)

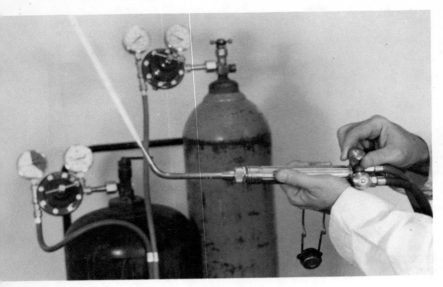

Fig. 9-52. Adjusting the oxygen valve for a neutral flame. (Courtesy of Smith Welding Equipment.)

21. Keep opening the acetylene valve until the flame stops excessive smoking. Then, when it leaves the end of the tip about ⅛″, close the valve slightly to bring the flame back to the tip.

22. Open the oxygen valve until a bright inner cone appears on the flame, as shown in Fig. 9-52. Keep opening the oxygen valve until the feathery edges of the flame disappear and a sharp inner cone is visible. This is the *neutral flame*. Keep adjusting the torch oxygen valve back and forth until you are sure what a neutral flame looks like.

Torch Flame Review

As a torch is set up, lit, and adjusted for whatever flame is needed, there are four different flames the welder sees as he adjusts the torch. These include: Pure Acetylene Flame, Carburizing Flame, Neutral Flame, and Oxidizing Flame. These flames may each be reviewed in the order they appear.

Pure Acetylene Flame—This is the first flame a welder sees, because there is no pure oxygen turned on when the torch is first lit. No welding is done with the pure acetylene flame because it isn't hot enough, even though it looks violent (Fig. 9-53).

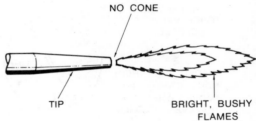

Fig. 9-53. Pure Acetylene Flame.

Carburizing Flame—As soon as oxygen is added to the acetylene, the flame tries to form a cone in the middle. Some acetylene *still* burns in front of the cone in a carburizing flame, however, because there is too much actylene for the oxygen. A carburizing flame is sometimes used on purpose to add carbon to the weld (Fig. 9-54).

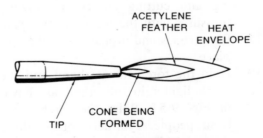

Fig. 9-54. Carburizing Flame.

Neutral Flame—This is the most common flame, used for most general oxygen-acetylene welding. To get the neutral flame, a welder keeps opening the oxygen valve slightly, until the acetylene feather just disappears into the cone. The cone will be long, bright, and clear (Fig. 9-55).

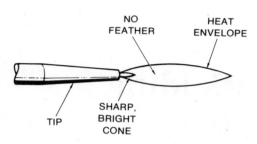

Fig. 9-55. Neutral Flame.

Oxidizing Flame—If the oxygen valve is opened more after the neutral flame appears, the long, sharp cone becomes shorter and fuzzy around the edges. This adds oxygen to the weld metal that causes defects, so an oxidizing flame has only a few good applications (Fig. 9-56).

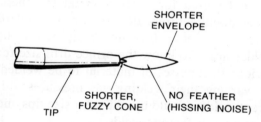

Fig. 9-56. Oxidizing Flame.

Shutting Down Review

When you have finished welding, certain steps should be taken, in order, to quickly and safely get the oxygen-acetylene equipment put up for the day. Here is a quick review of the steps in order:

1. Turn off the acetylene torch valve.
2. Turn off the oxygen torch valve.
3. Close both cylinder valves.
4. *After* the cylinder valves are closed, open the torch valves again. This will relieve the left-over pressure on the hose and regulators.
5. Turn the regulator screws out to release pressure on the regulator mechanism.
6. Close the torch valves.

The above steps are all that is needed for shutting down just for a couple of hours or overnight. However, if the equipment isn't to be used for over a day, the regulator screws should be turned back slightly, to relieve spring pressure on the valve seat.

If the equipment needs to be taken apart for moving, it is disassembled in the reverse order of how it was put together. In case the torch was just shut off, you should be careful not to get burned on any part of the torch or tip that may still be hot. Immediately after the regulators are removed, the protective caps should be screwed on the cylinders.

All welding equipment should be stored and handled carefully, especially the regulators, tips, and torches. The hoses need to be coiled carefully so that they don't get kinked or dirty. When tanks are empty, they are still valuable and dangerous, so they should be marked "MT" and chained or fastened upright until they are picked up to be refilled.

10

Oxy-Acetylene Welding Techniques/Practice

A technique is a way of doing something. Usually, there is more than one technique for anything a person wants to do. This isn't to say that one technique is better than another, but simply different.

When welding with an oxygen and acetylene torch, a welder can choose from two widely used *techniques* for oxy-acetylene welding. These refer to how he holds and moves the torch and welding rod. The two techniques used in oxy-acetylene welding are known as the *forehand* and the *backhand* technique. Either method is influenced by the torch tip that the welder uses.

Tip Size

The welder's experience has a lot to do with the size and type of tip best suited for the job. A welder using the forehand method usually uses a smaller tip than if he were welding with the backhand method. The type and size of tip vary according to the type of material being welded. Welding certain parent metals could require different types and sizes of tips than would be used to braze the same material. The material's thermal conductivity has a bearing on tip size, and for this reason, a new welder should be well acquainted with the heat transfer discussion.

If a tip is too large for the job being performed, the base metal may burn and create a weak weld. Too small a tip could increase the welding time due to lack of heat in the base metal. Each welding equipment manufacturer supplies information similar to Table 10-1, to assure the welder of the proper size tip to be used for a given metal thickness. Since manufacturers do not use a standard number or letter for similar tip sizes, the welder needs to know the orifice drill size for various thicknesses of material to be welded. By varying their welding gas pressures, different welders could use different size tips and perform satisfactory welds.

Table 10-1. Tips Used for Welding Mild Steel.

Base Metal Thickness	Tip Drill Size Number	Oxygen psi	Acetylene psi
24 Ga. to 1/16″	70	3 - 4	3 - 4
18 Ga. to 1/8″	62	4 - 5	4 - 5
1/8″ to 3/16″	54	4 - 5	5 - 7
3/16″ to 3/8″	48	5 - 6	8 -10
3/8″ to 1/2″	42	7 - 9	8 -10
1/2″ to 3/4″	37	7 - 9	8 -10

Forehand Welding

The most popular welding technique is known as *forehand welding*. In forehand

224

welding, the torch is inclined at an angle of about 15° from straight up in the direction of travel, as in Fig. 10-1. At this angle, the

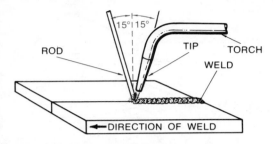

Fig. 10-1. In forehand welding, the torch is tilted about 15°, so that the flame *preheats* the metal to be welded.

heat from the torch preheats the work to be welded. During this technique, the rod is dipped in the puddle at about a 15° angle, as shown in Fig. 10-1. Then the rod is withdrawn.

Experienced welders usually vary the angle of the torch and/or filler rod to suit the job they are doing. If the torch is sloped away from the vertical at an angle over 15°, the puddle can be forced along the joint from the torch pressure. This helps to eliminate the chance of burn-through on light gauge (thin) metal. Holding the torch more vertical allows more penetration in forehand welding. The withdrawn rod end, however, must always be kept within the heat area to help prevent the heated filler metal from oxidizing. See Fig. 10-2.

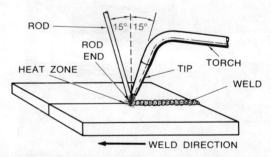

Fig. 10-2. Keeping the filler rod in the heat zone helps keep it from being oxidized.

In forehand welding, the bottom edge of the prepared joint must be kept in a molten

state, in order to accept good fusion with the melting rod (Fig. 10-3). If this does not

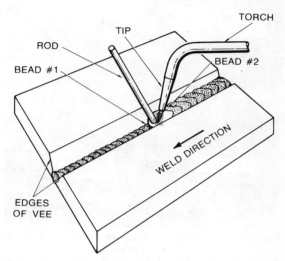

Fig. 10-3. The bottom edge of the prepared joint must be kept molten, so that the filler rod will fuse well.

happen, the melted rod material will form the shape of the groove preparation and look like a die casting. In welding, this is known as *cold shuts* (Fig. 10-4). This condition can

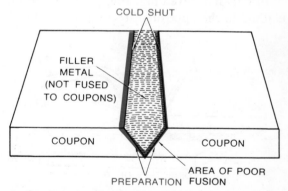

Fig. 10-4. A welder's "cold shut." The filler metal has become solid without fusing very well to the base metal.

usually be avoided by keeping the torch more vertical with the plate being welded and by allowing the flame to *keyhole*. Keyholing at the joint and leading edge of the puddle indicates that the root area is melted and ready to accept the filler rod. A typical keyhole is shown as having been made in Fig. 10-5.

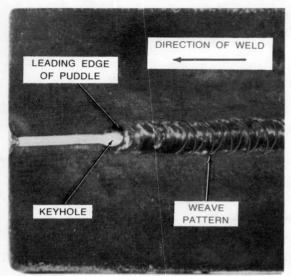

Fig. 10-5. Allowing the flame to *keyhole* for better penetration.

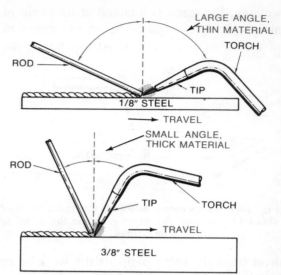

Fig. 10-7. Thicker materials have to be welded with the rod and torch in a more vertical position.

Backhand Welding

The second welding technique is known as backhand welding. It is not as popular as forehand welding, but it has to be used on thick materials for good joint penetration and strength.

In backhand welding, Fig. 10-6, the torch is tilted in the opposite direction to the direc-

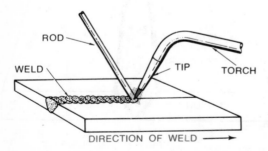

Fig. 10-6. In *backhand* welding, the torch handle is tilted away from the weld.

tion of travel. The right-handed welder starts at the left and goes to the right of the piece being welded. The amount of torch angle used depends on the thickness of the material to obtain good penetrattion. The thicker the material, the more it is necessary for the welder to keep the torch upright, as in Fig. 10-7. By keeping the torch more vertical,

forming the keyhole will be easier, as in Fig. 10-5. In thick metal, a good keyhole is usually necessary for complete penetration.

In most cases though, the torch will be kept upright and within the 15° limits shown in Fig. 10-6. Again, the experienced welder will usually vary the angle to suit the conditions and materials being welded.

Unlike the puddle in forehand welding, the puddle in backhand welding (Fig. 10-8) is

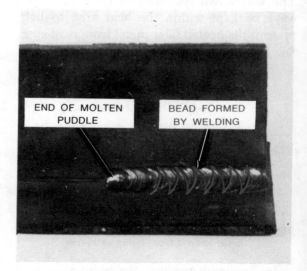

Fig. 10-8. A solid puddle formed in backhand welding. The weave pattern is shown on top of the weld.

held back until the edges of the groove are ready to receive molten filler metal. The rod is held in the molten puddle all the time while it is moved from side to side or in a circular motion. This results in a weld low in stress and with good ductility. The tip used with the backhand technique should always be at least one size larger than the tip used for welding the same material with the forehand technique. The larger tip size compensates for heat being deflected by the filler rod, since the filler rod is always left in the molten puddle. See Fig. 10-9.

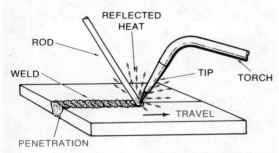

Fig. 10-9. A larger tip is used with backhand welding because some heat is reflected by the filler rod.

When to Backhand—Always use the backhand technique for welding metal over ⅜″ thick. This helps to get complete penetration at the weld's root. Also, this technique is recommended for welding steel with low-alloy, high tensile steel filler rods. When using these filler rods, be certain to use a reducing flame (1⅛X). See Fig. 10-10. The

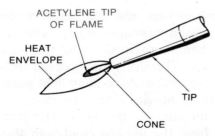

Fig. 10-10. Reducing flame. Used with low alloy, high-tensile strength steel filler rods.

slight feather on the flame helps make a higher tensile strength weld, since the alloys in the rod will not be burned out. A free

type (highly fluid) puddle should not be used with low-alloy, high tensile steel rods.

Welding Positions

Oxy-acetylene welding techniques can be used in one or more of the four common welding positions shown in Fig. 10-11.

1. Flat position (downhand).
2. Vertical position.
3. Horizontal position.
4. Overhead position.

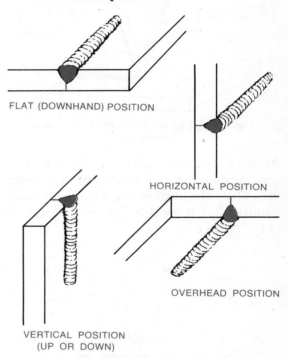

Fig. 10-11. The four common welding positions.

Flat Position

Welding in the flat position can be done with either the forehand or backhand technique. Also called downhand welding, the flat position in Fig. 10-12 is being done with the backhand technique; in Fig. 10-13, the forehand technique is being used.

Flat Position-Forehand Technique—Using the forehand technique in the flat welding position, the torch is inclined opposite to the direction of travel, as shown in Fig. 10-13.

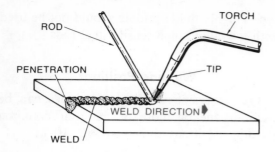

Fig. 10-12. Flat welding with the backhand technique.

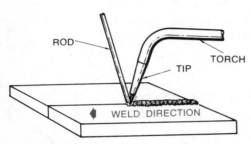

Fig. 10-13. Flat welding with the forehand technique.

As mentioned earlier, the amount of inclination depends on the joint type and metal thickness. The thicker the metal, the more "straight up" the welder has to hold the tip (Fig. 10-14).

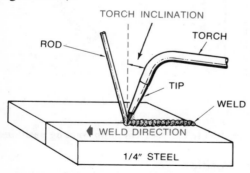

Fig. 10-14. Welding on thicker steel forces the welder to hold the torch straighter.

The heat from the tip is directed ahead of the weld to preheat the work piece and prepare the joint edges for welding. The rod is placed in the puddle and left for two or three seconds before being withdrawn. Of course, the heated end of the rod shouldn't be withdrawn far enough to cool, or oxidation will occur and create a poor weld. As the rod is

allowed to melt in the puddle, a weld bead will develop by adding the filler metal. Once the proper bead height is noticed, the filler rod is carried forward to a new puddle area; this has been melted ahead of time by the advancing torch. The torch and rod work together and another weld is formed, as shown in Fig. 10-15. Continuous crossing of the torch and rod while running the weld bead creates the weave shown in Fig. 10-15.

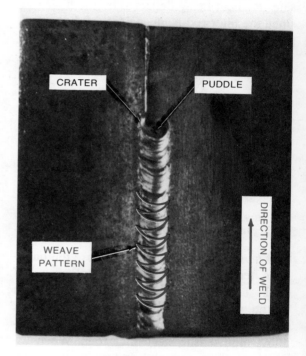

Fig. 10-15. Forehand technique showing the weave pattern.

Most welders find that welds made with the forehand technique have a better appearance than those made with the backhand technique. Therefore, the forehand technique should be used when welding light-gauge iron or steel. In heavier metal, the molten puddle may run into the bottom of the joint and away from the finished bead with the forehand technique. This would fill the gap at the joint bottom with hot metal before the parent metal was hot enough to accept it for complete fusion. The resulting cold shut condition can usually be avoided with practice, guarding against a poor weld at the bottom of the joint. See Fig. 10-16.

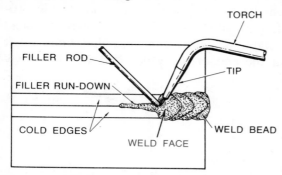

Fig. 10-16. Weld rundown on a thick plate, creating a cold shut.

Flat Position-Backhand Technique—Using the backhand technique in the flat position, as in Fig. 10-12, the torch is inclined in the same direction as the weld travel. A right-handed welder works the weld bead from the left to the right. While using the back-hand technique in the flat position, the flame pressure force keeps the molten puddle from running down to the bottom of the joint to create the cold shut in Fig. 10-16. The puddle, being held back by the flame force, immediately fuses with the edges of the groove once they have heated enough.

During the backhand technique, the welder usually holds the torch steady and only moves the filler rod within the puddle. Since the rod is continuously lowered in the puddle as it melts, there is no danger of having oxides form in the center of the weld from the rod end being exposed to the air. Fig. 10-17 shows how the weld bead extends on one side of the groove because the rod shielded one side of the work from the heat.

It is a good idea to try to keep the molten puddle in the desired position by using both the stirring action of the filler rod and the flame pressure. This will help eliminate "one-sided" welds, as in Fig. 10-17. Molten metal tends to flow toward the hotter area during oxy-acetylene welding. Welds made with the backhand technique are automatically post-heated because the flame is always directed

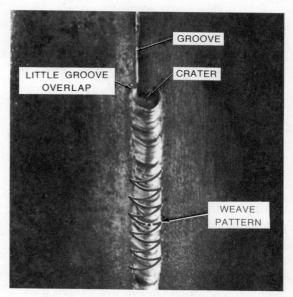

Fig. 10-17. Improper rod and torch positions caused the weld bead to move to one side of the groove.

towards the finished weld. This assures the welder of a weld with good ductility and low internal stress.

Using the backhand technique, it is recommended that the coupon be prepared with narrower Vees. Or, if using joints with a 60° or 70° included angle, a root face or land should be left on the prepared bevels, as shown in Fig. 10-18. The filler rod used in

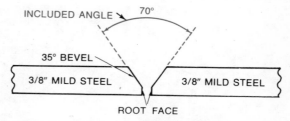

Fig. 10-18. Heavy plate prepared with a 70° included angle, to help eliminate burn-through.

the backhand technique is usually one size smaller than the rod used for forehand welding, to help cut down the filler rod's heat-shielding effect on the bevel edges to be welded.

Tip Position—As mentioned previously, the welding tip angle often depends on the welder's experience. A beginner can start off

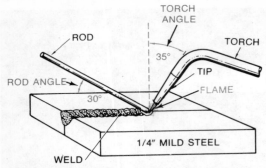

Fig. 10-19. A good angle for a practice weld.

by tipping the torch about 35° from the perpendicular, toward the finished weld, as in Fig. 10-19. By allowing the bevel edges to melt at this angle, a good practice weld can be had on mild steel up to ¼″ thick. The filler rod should be held in the molten puddle at a 30° angle to produce good practice welds. See Fig. 10-19.

The angle of the torch, whether using the forehand or backhand technique, should be an equal distance from the two plates being welded in the flat position, as in Fig. 10-20. If additional heat is required to melt the beveled joint edges, the operator can raise the torch to extend the heat radiation or angle the torch straighter up.

As the puddle is built up to the required height by the melting rod, the torch is drawn

to the right, away from the bead. This allows the puddle to follow and melts a new section on the beveled edges. All *craters* (Fig. 10-15) should be filled to the desired bead height. Some welders find that craters left at the end of the weld bead will fall away

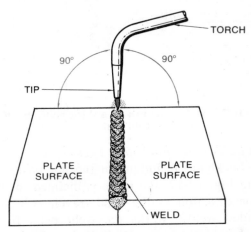

Fig. 10-20. The torch tip is the same distance from both plates for good welding.

or depress more if filling is continued *without* withdrawing the rod and torch from the puddle area. If the rod and torch are withdrawn, a crater will form at the bead's end as the molten puddle solidifies. Then, the welder can reheat the crater until a fluid, mirror-like puddle appears. Filler rod is then added by

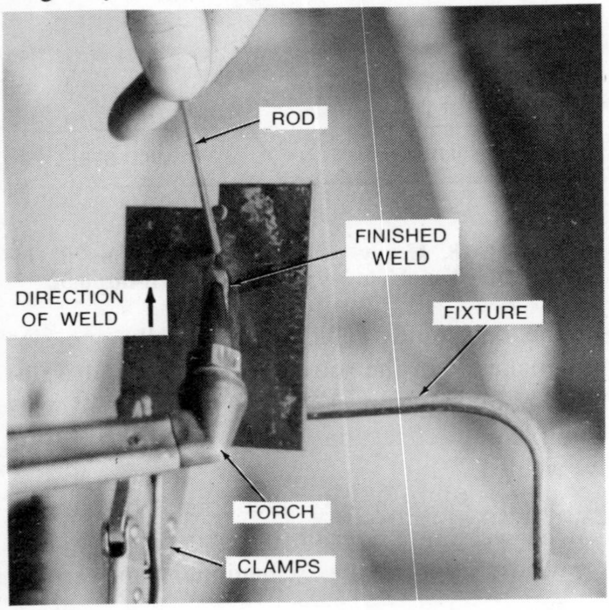

Fig. 10-21. Forehand technique used for vertical welding. (Note: Although the welding torch *looks* like a cutting torch, it is actually a special kind of welding torch.)

dipping and withdrawing in the forehand technique until the puddle builds up to the required height.

Vertical Welding

The forehand technique shown in Fig. 10-21 is the most common method of vertical welding. The rod is dipped into the molten pool at an angle of about 30° to the horizontal, as in Fig. 10-22, and then drawn away.

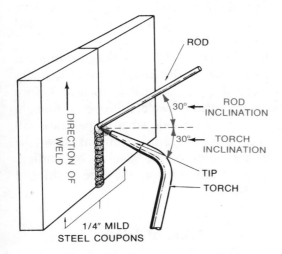

Fig. 10-22. Good torch angle for using the forehand technique on a vertical weld.

The torch, at a similar but opposite angle, weaves the molten metal into shape for a finished bead contour, as shown in Fig. 10-23.

Welding is usually done from the bottom of the joint to the top. This lets the pressure of the flame play an important part in keeping the molten pool in place and usually free from gravitational pull. The procedure for laying the bead is the same as in flat position welding, except that the puddle should be kept more plastic and not as free-flowing as in flat position welding.

Horizontal Welding

During horizontal welding, many welders find that the forehand technique is best for good bead contour with complete fusion and penetration. Horizontal welding is done from the right to the left with the rod and torch

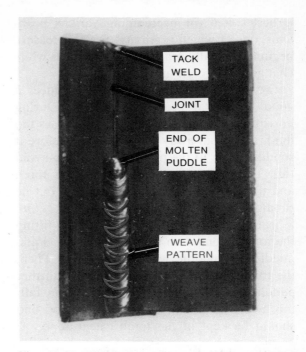

Fig. 10-23. Finished bead contour for a vertical weld, forehand technique.

inclined about the same as in the vertical welding position. See Fig. 10-24.

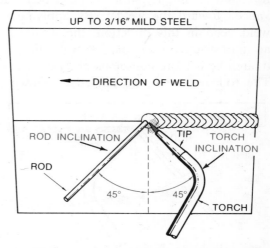

Fig. 10-24. Rod and torch angle for horizontal welding.

Many experienced welders use different rod and torch movements to place the bead where it is needed. By practicing, most welders are able to determine which rod and torch angle is best suited for overcoming various prob-lems, one of which is gravity. For this reason, beginners usually practice various welds while changing the torch and rod angle.

Oxy-acetylene welding usually produces a good horizontal bead in one pass. During horizontal welding, if the puddle becomes too fluid, it will fall away from the joint before it solidifies, due to the pull of gravity. To avoid this, the welder should use the same torch and rod technique as for flat welding, but he should allow the puddle to solidify *before* reheating the next part of the bead. By lifting the torch, along with the rod, from the liquid puddle, the puddle will instantly jell. Then, the torch and rod can be lowered for further puddling without danger of the puddle fall-ing away. As usual, of course, the rod end should be kept in the torch's heat zone.

Overhead Welding

After mastering the horizontal type oxy-acetylene weld bead, most welders will have little trouble welding in the overhead position.

The overhead welding position uses the forehand technique, with the bead being ad-vanced toward the welder. By applying the weld bead in this direction, the welder has an unrestricted view of the weld line, al-though he is more conscious of flying sparks. The rod and torch are angled as shown in

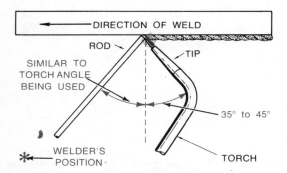

Fig. 10-25. Overhead welding requires different angles, so the welder can see the weld and not get burned by flying sparks.

Fig. 10-25, but they can be moved from this standard position to give the required pene-

tration and to place the molten metal where it is desired at the same time. As with vertical welding, it is a good idea to keep the puddle less fluid than in flat welding.

Position Review

In all welding positions, it may be neces-sary for the welder to change the torch angle to a more "straight up" position, to be sure that the parts being welded are completely penetrated. This change depends on how thick the material is and is usually needed for thicker materials.

More information on welding positions is presented later in this chapter with the meth-ods needed to weld various metals. Before these metals can be welded properly, though, the welder must decide what preparation will be necessary. Preparation means how the joints are going to be shaped by cutting, grinding, or fitting before they are positioned for welding.

Types of Joints

For the best welding jobs possible, most welders will be required to *prepare* materials before welding. This means knowing the dif-ferent joint styles available for welding, to-gether with the basic, named welds used on each joint. In oxy-acetylene welding, of course, the high tip temperature melts the parent metal. Then, after a welding rod is added, fusion is completed in one of these four types of *weld joints:*

1. Butt
2. Fillet
3. Lap
4. Flange

Butt Joint

In the *butt* joint (Fig. 10-26), the weld joining the parts is between the surface of both fused parts. The butt joint, when welded by a skillful welder, is the strongest weld joint. It is not only the most common joint

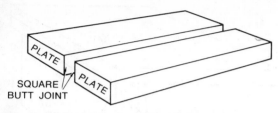

Fig. 10-26. Square butt joint. Simple and strong, if welded properly.

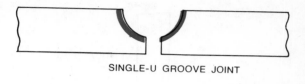

SINGLE-U GROOVE JOINT

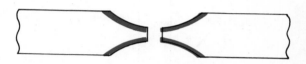

DOUBLE-U GROOVE JOINT

Fig. 10-29. U- and double-U preparations, used where distortion could be a problem.

used for joining metal, but it is the simplest to align for maximum weld strength.

The *square* butt joint, shown in Fig. 10-26, is the most common butt joint used for materials from ⅛″ to ³⁄₁₆″ thick. The *Vee* butt joint is used for metal ³⁄₁₆″ to ⅝″ thick. See Fig. 10-27. The included angle is often

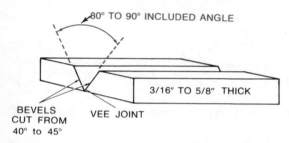

Fig. 10-27. Vee-joint preparation. Used for medium thickness steel.

as much as 90°; this allows for better puddle formation and plate edge fusion. The Vee joint, like other butt joints, is prepared by machining, grinding, or cutting.

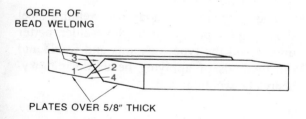

Fig. 10-28. Double-bevel preparation. Used for very thick steel.

The double bevel in Fig. 10-28 is used for material over ⅝″ thick. The U groove and double-U groove (Fig. 10-29) are used in the place of the Vee and double-Vee, respectively, where distortion might be a problem.

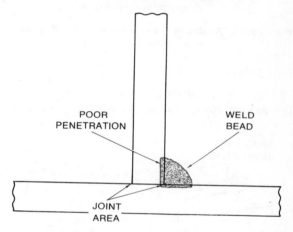

Fig. 10-30. Fillet joint with little root penetration due to poor heating of the base metal.

Fillet Joint

The *fillet* joint is a popular joint for welding metal with the oxy-acetylene torch. This joint, illustrated in Fig. 10-30, requires a wide, molten pool of parent and filler metal, to make sure that the welder has obtained maximum weld strength. When using a fillet weld for fusing two pieces of metal, a weld bead should be placed on each side of the vertical, especially where Tee joints are being used. Fig. 10-31 shows a strong Tee joint. It has been welded on both sides for maximum strength and it has good penetration on both sides.

Information on the techniques used for fillet welding will be given later in this chapter.

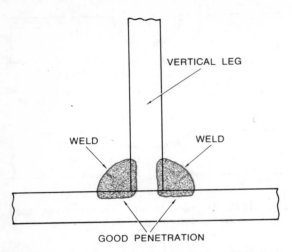

Fig. 10-31. A strong Tee joint with good penetration.

Lap Joints

Fig. 10-32 shows a *lap* joint welded with the oxy-acetylene torch. As with any welded

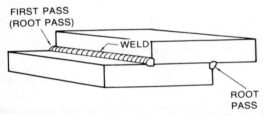

Fig. 10-32. Lap joint welded for maximum strength.

lap joint, both sides of the lap must be welded for maximum joint strength. The lap weld is used to join the edge of one plate to the face of another, where one plate is placed on top of the other, as shown in Fig. 10-32. The lap joint is usually joined by a fillet weld. The procedure mentioned later in the chapter for Tee joint welding will also apply to lap joint welding.

Flange Joint

Little or no filler metal is usually needed with a *flange* joint (Fig. 10-33). This is because the flange, on melting, forms the filler metal necessary to join the two pieces of metal into a strong unit. Such a joint is used only on thin-gauge sheet metal. The welder

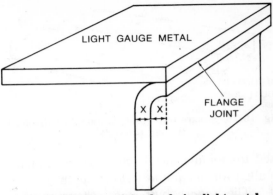

Fig. 10-33. Flange joint for fusing light metal thicknesses.

should bend the flange on each plate so that the flange height is equal to the metal's thickness, as shown in Fig. 10-34.

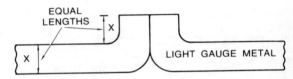

Fig. 10-34. Preparing thin metal for a flange joint.

Joint Review

For more information on joint styles and when considering different types of joints, the oxy-acetylene welder should refer to the chapter on joint styles. The types of joints and welds used in arc welding have many things in common with the joints and welds used in oxy-actylene welding.

The following methods for welding steel sheets and pipe will help the welder better understand the types of joints and welds and the methods of applying beads during oxy-acetylene welding.

Welding Steel

Today, arc welding has taken over most welding jobs on material over ⅜″ thick. However, oxy-acetylene welding equipment may still be used for several reasons:

A. Arc welding equipment may not be available.

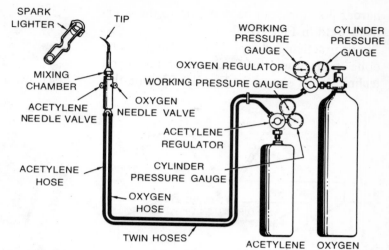

Fig. 10-35. The basic parts of an oxy-acetylene welding outfit.

B. Arc welding equipment may not be easily or economically moved to the welding job location.

C. Arc welding equipment needs a high-voltage power source. That source, or a generator, may not always be available.

D. Welders skilled in arc welding may not be available.

E. The job may require heat-treating or other treatments requiring heat only, but no welding.

For any of these reasons, then, oxy-acetylene equipment and welders may be called in to use oxy-acetylene equipment, as pictured in Fig. 10-35. For these reasons, basic oxy-acetylene welding continues to be a valuable skill.

Practice Welds

Good practice welds can be made on ¼″ thick *coupons* (welding plate samples) for all positions. The simplest practice weld

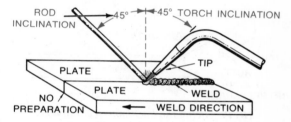

Fig. 10-36. The easiest and simplest practice weld.

would be a flat position weld on ¼″ coupons, using a butt joint and no edge preparation. This setup is being done in Fig. 10-36. Then,

Fig. 10-37. Practicing overhead welding position. [Note: Although the *torch* in this picture is a welding (not a cutting) torch, it is an unusual design.]

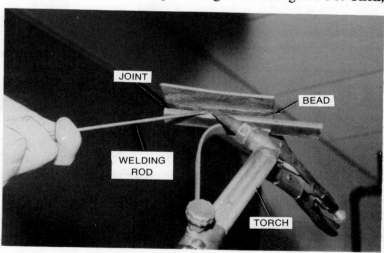

harder positions can be done, up to the over-head weld in Fig. 10-37.

For practicing, flat position welding can be done with either the forehand or backhand technique, as shown in Fig. 10-37. For this

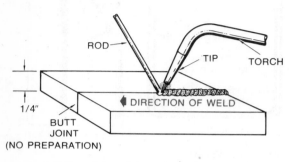

(A) Practicing the forehand technique.

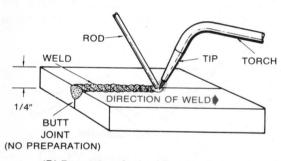

(B) Practicing the backhand technique.

Fig. 10-38. Practicing flat position welding with no butt joint preparation.

reason, both techniques are shown for welding ¼″ thick mild steel. It Fig. 10-39, the

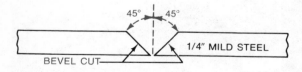

Fig. 10-39. Bevel preparation on what was a simple butt joint. This is now a 90° single-Vee butt joint.

plates are each beveled at a 45° angle along the edge, to form a single-Vee, 90° included angle butt joint. In Fig. 10-40, the welder is welding this joint with the forehand technique. In Fig. 10-41, he is using the backhand technique. The prepared angle for the

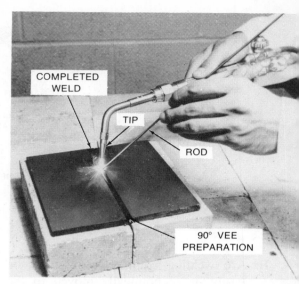

Fig. 10-40. The forehand technique on a 90°-Vee preparation butt weld. (Courtesy of Victor Equipment Co.)

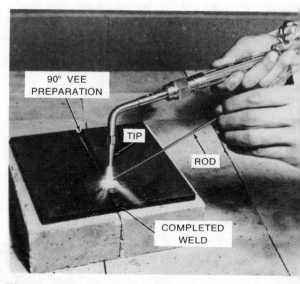

Fig. 10-41. The backhand technique on a 90°-Vee preparation butt weld. (Courtesy of Victor Equipment Co.)

backhand technique, however, is only 60°, not 90°.

To practice, the welder can prepare coupons, (pieces of prepared metal) as in Fig. 10-42. These can be cut by an oxy-acetylene cutting torch, Fig. 10-43, or other mechanical means. Bevel joints are usually prepared by either chipping, machining, grinding, or flame cutting; the exact choice depends on both economy and convenience.

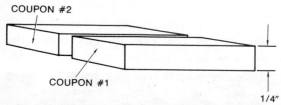

Fig. 10-42. Sample welding *coupons*, ¼" thick.

Forehand Single-Vee Butt Weld

For a forehand, single-Vee butt weld, beveled plates with ¹⁄₁₆" root faces are placed side by side with a gap between them (Fig. 10-44). The gap is from ¹⁄₁₆" on one end to ³⁄₁₆" on the other. This allows for expansion as the weld is made. The coupons to be welded are then *tack welded* first, to help control distortion. See Fig. 10-45. By tack welding before complete welding, most problems are avoided that could happen through the expansion and contraction caused by the heat.

Setting Up—For this practice, the welder would be using a torch with an orifice drill size (tip number) 48, as mentioned previously and shown in the specification Table 10-1. A high-test steel welding rod, ³⁄₁₆" in diameter, is used with the right flame and gas pressure to start the weld. For this par-

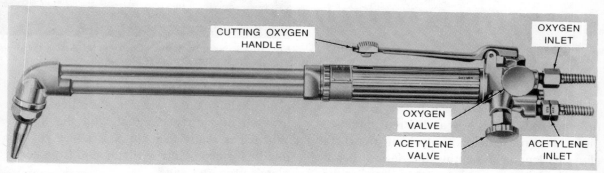

Fig. 10-43. Cutting torch, as used to prepare thicker welding coupons. (Courtesy of Union Carbide Canada Ltd.)

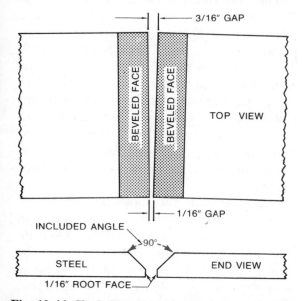

Fig. 10-44. Single-Vee coupons prepared for a butt position weld. Note the ¹⁄₁₆" root face.

ticular weld, the acetylene pressure is set at 5 psi, and the oxygen pressure is set at 16 psi.

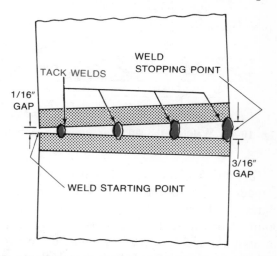

Fig. 10-45. Tack welding the coupons to control distortion.

Torch and Tip Position—The torch is held so that the flame at the tip is angled about 45° from the perpendicular. A neutral or 1⅛ X reducing flame should be used. The filler rod, which can be used either straight or bent, is angled toward the flame from the opposite direction to the torch, which is held at a similar angle. Fig. 10-46 shows the correct torch

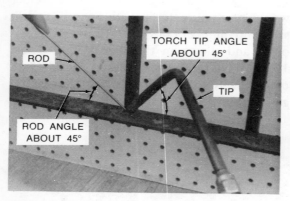

Fig. 10-46. Torch tip and rod held at the right angle for a practice weld.

tip and rod angles on a different practice piece. The rod can be easily bent anywhere along its length by heating the desired bending point with the torch and then pressing the rod against the welding table. Fig. 10-47 shows the setup being welded with a bent rod.

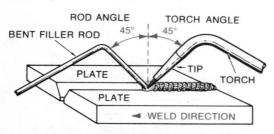

Fig. 10-47. Welding the 90°-Vee preparation butt weld.

Tacking the Coupons—The first tack weld is made at the bottom of the Vee, where the plates are closer together. See Fig. 10-45. The point to be "tacked" should be preheated in an area about 2″ to 2½″ in diameter. When the area starts to turn red, the torch flame is concentrated on the end of the Vee. At this

time, the filler rod tip is allowed to warm up in the flame's outer envelope. By using a neutral flame on steel, the welder will notice a mirror-like puddle of molten metal forming at the end of the Vee, directly under the flame. The neutral flame, as in Fig. 10-48,

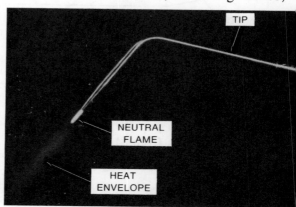

Fig. 10-48. Neutral flame. (Courtesy of Airco Welding Products Division.)

will add no carbon to the weld. When this clear puddle starts to form, the filler rod is lowered into direct contact with the flame. When it is ready to melt, it is added to the puddle formed by the molten metal from each plate's prepared edge.

The filler metal is continuously added to the puddle, and the torch flame pressure forces it to bridge the gap between the two plates, forming a single bead. See Fig. 10-49. The metal from the filler rod should not be added unless the bottom and sides of the Vee are properly melted, or the fusion of parent metal and filler metal will form the type of lamination known as the *cold shut*. A cold shut, of course, is incomplete fusion and causes the weld to be weak and of poor quality.

The tack weld, also shown in Fig. 10-49, was completed by allowing the filler metal to continuously build up the molten puddle inside the joint. When tacking parts to be welded, the welder should direct the flame more toward the rod than the parent metal. This keeps the tack puddle to a minimum size. Too much heat directed toward the puddle could enlarge the fluid area, causing molten metal runoff. This would create an unsatisfactory practice weld.

During the process of dipping the rod into the puddle, practice and experience will help the welder move the filler metal to any desired point by moving the rod and flame pressure.

After the tack is made at the one end, the puddle is allowed to solidify and the same procedure is used to tack the other end of the plates. When the final tack is solid, the unit is placed on a table, usually covered with *fire bricks,* as in Fig. 10-50.

Welding after Tacking—Welding can then be started from the right to the left, with the wide gap of the Vee to the left. See Fig. 10-51. Welding then progresses from the right to the left, with the flame always between the completed weld and the filler rod, as shown in the forehand welding technique of Fig. 10-40.

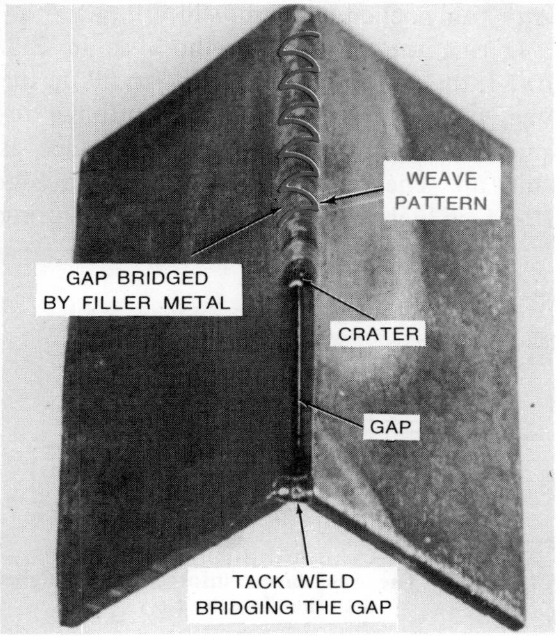

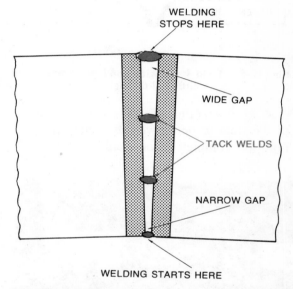

Fig. 10-49. Metal from the filler rod melted to bridge the gap.

Fig. 10-51. Positioning the tacked assembly for welding.

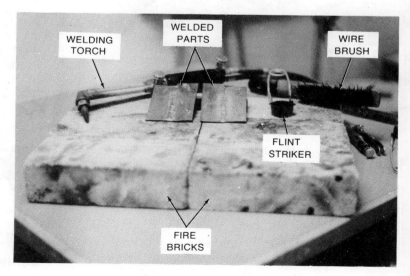

Fig. 10-50. Fire bricks and welding equipment.

The tack on the right is remelted until the bottom of the Vee and the edges of each plate are fluid enough to retake the melted filler rod. Usually, an experienced welder uses the molten filler and parent metals to fill the bottom of the Vee preparation first, for a distance of ¼" to ½". This assures *complete* penetration without excessive fall-through, as in Fig. 10-52.

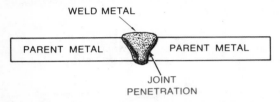

Fig. 10-52. Weld bead with good penetration and little fall-through.

A *keyhole* (Fig. 10-53) is carried ahead of the weld to help give the needed penetration for complete root weld strength. The

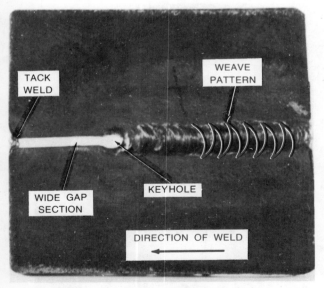

Fig. 10-53. *Keyhole* carried ahead of the weld to insure good penetration.

torch and rod should be moved to cross opposite each other near the center of the Vee, to keep the finished bead smooth and uniform. As the torch melts one side of the Vee and is swung over to melt the opposite side,

the rod fills the finished melted side, since the rod melts more quickly than the heavier parent metal. The filler rod can also be added to the fluid pool by dipping it in and out, as long as the rod's heated end is kept within the torch heat zone. As mentioned earlier, if the heated end of the rod is allowed to cool, it will oxidize. Then, when it is reintroduced to the molten puddle, it will leave impurities in the weld. These impurities might create *porosity* (air pockets) in the weld.

During the one-pass method, as in Fig. 10-53, the filler metal is allowed to fill up the Vee, fusing each side in turn until the required reinforcement on the weld face is built up. The reinforcement, shown in Fig. 10-54, is higher than the plate itself. During

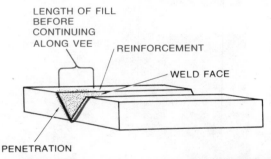

Fig. 10-54. The weld face reinforcement is higher than the plate itself.

this procedure, the weld apears somewhat slanted from the bottom of the Vee to the reinforced weld face, as shown in Fig. 10-55. The bottom of the Vee *ahead* of the finished weld, although filled with metal, is again heated to the mirror-like puddle. Filler metal

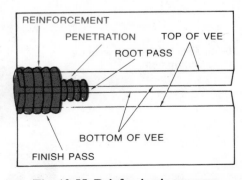

Fig. 10-55. Reinforcing in progress.

is again melted into the open Vee area, as bridging the plates with filler metal continues. The unfilled portion of the previous weld requires both melting the beveled edge of each plate and *re*melting the slanted portion of the previously incompleted weld. See Fig. 10-55.

Although the above application looks like a step formation, the welder with enough practice will continue this method with little or no hesitation. By doing this, he will be able to complete a well-penetrated and completely fused weld.

The welder may want to fuse only the bottom edges of the Vee for complete penetration. If so, each successive cover pass should *not* be applied until the initial pass and the beveled plate edges have been reheated. Then, when they form a molten puddle, the filler rod can be accepted for complete fusion.

Backhand Single-Vee Butt Weld

In the backhand technique, welding progresses from the left to the right, with the torch flame directed toward the finished weld, as in Fig. 10-41. The rod is always between the flame and finished weld, as shown in Fig. 10-56.

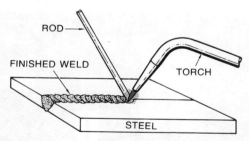

Fig. 10-56. Backhand welding with the torch tip pointed toward the finished weld.

Practicing Backhand—When practicing backhand welding, a larger tip is used than the one in forehand welding, if the metals are equally thick. For welding ¼″ metal, a #50 drill size tip is recommended. Acetylene pressure should be set at 5 psi and oxygen pressure at about 20 psi.

If using a ³⁄₁₆″ diameter high tensile steel rod for ¼″ thick material, it is a good idea

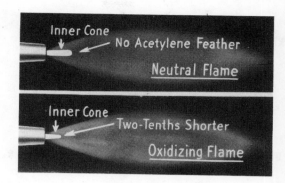

Fig. 10-57. Neutral and oxidizing flames. (Courtesy of Union Carbide Canada Ltd.)

to practice while using a reducing flame, $1\frac{1}{8}$ X, rather than a neutral flame. If a neutral flame were used in the oxidizing flame zone (Fig. 10-57), the alloy in the steel filler rod could be burned out from oxidation. This would create a weak weld.

Preparing the Coupons— The pieces to be welded for practice with the backhand technique should be beveled 30° each, to produce a 60° included angle when placed side by side. See Fig. 10-58. This angle is ade-

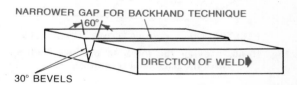

Fig. 10-58. Two 30° bevels produce a 60° included angle for using the backhand technique.

quate for the backhand technique because the welding flame is directed on the edges of the Vee, always ahead of the puddle. This eliminates the need for the welder to crisscross the torch and rod, which is necessary with the forehand technique.

Tacking the Coupons—Backhand welding, especially with alloyed, high tensile steel rod, requires a more plastic (less fluid) puddle than was used with the forehand technique. Preparation and tack welds are completed like those for the forehand technique. Then, the weld bead is ready to be completed from left to right, as shown in Fig. 10-59.

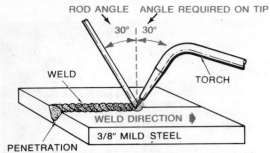

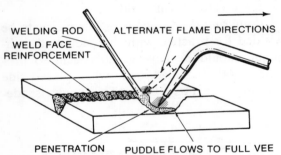

Fig. 10-59. Right-handed welder using the back-hand technique.

Welding after Tacking—After the flame heats the left hand tack weld to welding temperature, the metal from the previously warmed rod is melted into the weld puddle. From this point on, the flame is alternately turned toward the bottom of the Vee and the finished weld face, as in Fig. 10-60. This lets

Fig. 10-60. Moving the torch to produce buildup with the backhand technique.

the puddle move downward to fill the Vee. The filler rod metal, which is constantly being melted, can then be forced *back* toward the weld until the needed weld face reinforcement is complete. Fig. 10-61.

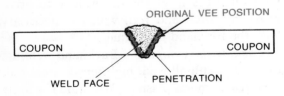

Fig. 10-61. Weld face within the bead joint.

The welder continues welding by first filling a short distance along the bottom of the Vee and then working the puddle backwards.

All the time, he continues adding melted filler rod metal to completely fill the Vee joint. The edges of the Vee in front of the puddle are moved forward along the bottom of the Vee. This process is continued along to the end of the joint. All the time, care must be taken to keep the reinforcement height uniform throughout the weld length.

Single-Vee Butt Weld, Vertical Position

Although it is faster and cheaper to weld in the flat position, there are times when it may be necessary to perform a repair by welding in *other* than the flat position.

In the *vertical* position, for example, the welder might have trouble with the molten puddle dropping down, if some method weren't available to control gravity. A more plastic puddle is easier to control in the presence of gravity. The welder must, therefore, keep the welding puddle smaller and more plastic in any welding position besides flat. Once a drop is allowed to form, the force of gravity has the power to make the drop fall. Note the bead rundown from gravity in Fig. 10-62.

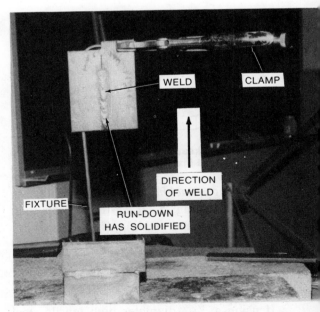

Fig. 10-62. Bead rundown due to gravity in a poor vertical position weld.

The plasticity (pliability) of the puddle and the gas pressure in the flame can help a practicing welder keep the puddle in the required position. This will help overcome gravity's pull on the molten metal.

In flat welding, the plates being welded tend to support the puddle like a dish. The puddle of metal can be placed in any position on the dish by the flame pressure and rod movement. The rod and flame pressure are still used to move the weld puddle as needed in the vertical, overhead, and horizontal positions.

Unfortunately, though, in the vertical and horizontal positions, the dish is turned up on its side. There, it seems to be open on the *side,* instead of the *top.* In the overhead position, of course, the dish is turned upside down! For these reasons, each small and molten puddle, once placed as desired, must be allowed to solidify. By raising the torch briefly or moving it to one side, the puddle will immediately solidify. Then, if the bead is placed properly, the bead will act as a step or a ledge to hold each bead in successive order. The face of each supporting ledge or step, however, must be melted enough before applying the next small molten puddle, to ensure that complete fusion between the beads has taken place. See Fig. 10-63.

Practicing Vertical Welds—Practice welding in this position uses the same size plates mentioned in the other exercises. They should be tack welded the same way, and with the same space allowance between the plates for expansion and contraction that was shown in Fig. 10-51.

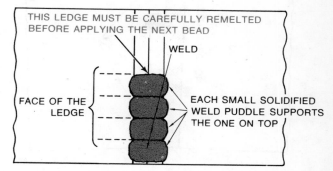

Fig. 10-63. How the beads support each other in vertical position welding.

Setting Up Equipment—The tip used for vertical position practice on ¼″ thick metal should have the same orifice size that is used for flat welding–a #48 drill size tip. The acetylene should be set at 5 psi and the oxygen set at 16 psi. The rods usually used are low-alloy, high tensile strength steel, and it is a good idea to use these rods for all position practicing. The rod size used for *this* exercise should be ³⁄₁₆″ diameter.

Positioning for Practice—After tacking the pieces into a single unit, the unit is placed in a *fixture* (Fig. 10-64) and swiveled to hold the practice piece in the vertical position, with the narrow space at the bottom. See Fig. 10-65.

Fig. 10-64. Welding fixture used to put the practice piece in the right position.

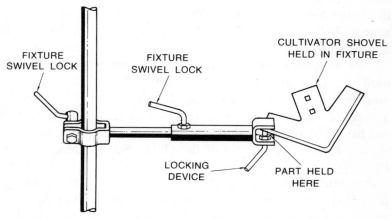

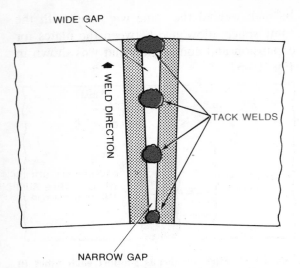

Fig. 10-65. Tack welded parts positioned for vertical welding practice.

The flame and rod inclination will be the same as for the flat position *during melting.* The flame, however, must *change* directions constantly, to keep the puddle small and use the burning gas pressure to place the puddle where it is wanted.

Welding after Positioning—The bottom of the Vee, at the lower end of the plate in Fig. 10-65, is then filled about ¼″ to ⅜″. The weld must be completely fused in the plate edges to give complete penetration. If kept small, the initial puddle can be more fluid than the next puddle used to cover it. In order, then, these puddles will stack up to fill the Vee joint. After the Vee has been well fused and penetrated over a short distance, the flame and rod are again lowered. Then, a new shallow and sticky puddle is formed and fused to the initial one.

This procedure of buildup is carried on until a part of the weld material, about ⅜″ long and slightly wider than the Vee, forms a type of "stair tread." For reinforcement, the height of the weld should form a crown slightly higher than the tip of the plates. See Fig. 10-66.

After the initial "stair tread" has been completed with good fusion, a technique similar to the backhand welding technique can be

used to finish the weld. In this technique, the first puddle is moved forward to the bottom of the Vee and then backward to the top of the Vee. Although each tread is allowed to solidify, welding is continuous. Therefore, the flame must return almost immediately to the face of the ledge to raise it to the required temperature for a new puddle. This procedure is continued until the full length of the Vee has been filled. With continued practice, it is usually possible to make a good weld in the vertical position as easily as in the simpler flat position.

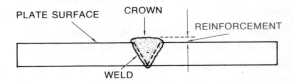

CROSS SECTION OF WELDED VERTICAL JOINT

Fig. 10-66. Cutaway view of a good vertical weld sample.

Single-Vee Butt Weld, Horizontal Position

In the *horizontal* position, gravity is still a problem. Instead of pulling the weld down the Vee, as in the vertical position, gravity tries to pull the weld toward the lower plate.

Fusion to the top plate must be watched carefully, or most of the metal will be deposited on the lower plate. This would obviously create a poor and weak weld.

Preparing, Positioning, and Setting Up the Horizontal Position—In order to practice the horizontal position, the welder should tack weld ¼″ thick prepared plates and place them in a fixture as shown in Fig. 10-67. The plates should be prepared with a 60° included angle. The neutral flame, used with a ³⁄₁₆″ high tensile strength steel rod, can be used for practicing this type of weld. The pressure of the acetylene gas is set at 5 psi and the oxygen is set at 16 psi.

Practicing Horizontal Welds—In horizontal welding, gravity's pull is similar to the pull in vertical and overhead welding. However, in the horizontal position, the bevel of the *lower* plate, together with careful rod movement, will keep the plastic-like bead from falling away.

Fig. 10-67. All-purpose welding fixture.

This welding technique is similar to that used during vertical position welding, but it may be difficult to weld an equal amount of bead buildup on both the top and bottom plates. The metal has a tendency to build up, due to gravity, on the *lower* Vee groove edge. Of course, the lower Vee groove edge is on the lower plate.

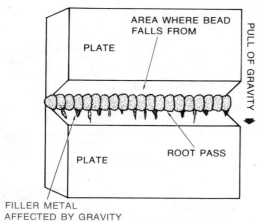

Fig. 10-68. How gravity pulls the weld metal down to the bottom plate.

To overcome this problem (Fig. 10-68) it is necessary to direct the flame toward the upper edge. Directing the flame toward the upper edge and moving the rod toward that edge will help control gravity's pull. Care

must be taken when directing the flame toward the upper plate edge. Too great a temperature produces an overly fluid puddle, which is difficult to keep in position while adding filler metal. For this reason, the puddle formed on the edge of the top plate should be sticky rather than fluid.

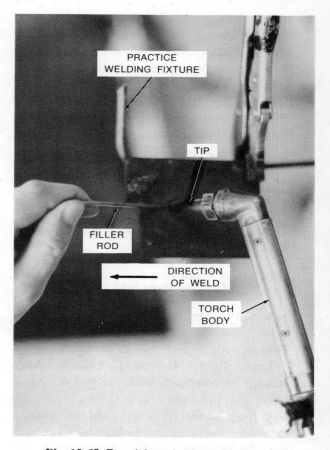

Fig. 10-69. Practicing a horizontal weld.

Welding should be completed in one pass whenever possible. Good penetration, complete plate fusion, and a well-formed bead contour indicate a good horizontal weld. The basic technique is shown in Fig. 10-69.

Single-Vee Butt Weld, Overhead Position

This is probably the most difficult position, since the pull of gravity pulls the molten metal straight down, away from all parts of the weld area. Nevertheless, welding can be done successfully in the overhead position, although it may take more practice.

Preparing, Positioning, and Setting Up— For overhead welding practice, the acetylene and oxygen pressure should be set the same as for vertical position welding—5 and 16 psi respectively. Each of the ¼″ steel plates is beveled at a 30° angle, and they are then tacked and gapped as they were for flat position practice.

After tacking, the plates are fastened in a fixture, so they can be positioned where the welder can weld them in the overhead position. See Fig. 10-70. A neutral flame should

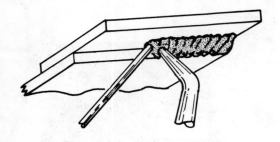

Fig. 10-70. A simple joint positioned for a partially overhead weld.

be used for overhead practice, even when the welder uses a high tensile strength steel welding rod, usually ³⁄₁₆″ in diameter.

Practicing the Overhead Position—When making an overhead weld, the welder should direct the tip almost *into* the weld. In this way, the weld direction is toward the operator for ease of welding and better visibility, as in Fig. 10-71. The welding, however, is still

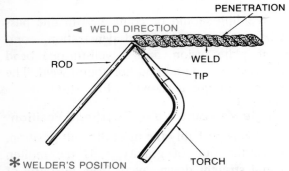

Fig. 10-71. Overhead welding position—forehand technique.

from right to left; the bottom of the Vee is first built up slightly ahead of the welding action. Then the puddle is returned to finish filling the Vee.

During this procedure, the puddle must be kept small. This allows the rod to be used for moving the puddle without any filler metal loss from gravity's pull. As with the other welding positions, the plate edges must first be heated to the proper temperature.

As with the other three weld positions, the Vee must be well penetrated and filled throughout its complete length with a well-crowned, reinforcement type weld. With some practice, many welders experience fewer problems with overhead than with vertical position welding.

Fillet, Lap and Tee Welds

These welds can easily be welded in all positions with some extra practice. Once the welder has successfully mastered the different butt joint techniques used with ¼″ thick metal, he can easily move on to these welds.

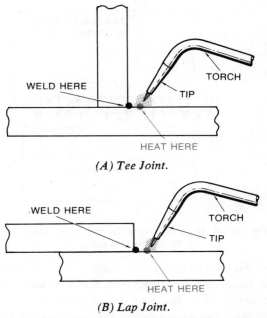

(A) Tee Joint.

(B) Lap Joint.

Fig. 10-72. Directing the flame heat for a strong weld.

The most important point in welding a fillet, lap, or Tee joint is to obtain *complete* fusion at the root or corner of the joint. The flame should not be directed too far vertically into the joint, since the reflected and burned gas from the flame will create an air pocket. This makes the molten puddle difficult to control. Also, as in horizontal position welding, the vertical leg in one of these welds (Fig. 10-72) must be heated hot enough to attract the molten filler metal *without* gravity pulling the molten metal down. Usually, if the flame is directed toward the lower plate, it will reflect enough heat to the upper plate so that very little *direct* heat will be required on the upper plate before welding.

Once a welder has practiced complete penetration of the beveled plates in all positions, especially those with a *root face,* as in Fig. 10-73, he will easily progress to welding

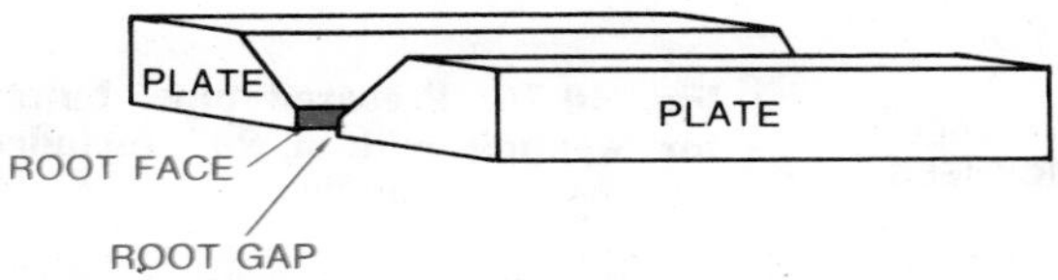

Fig. 10-73. Butt joint, prepared with a *root face.*

metal only 1/8″ thick or less. This requires flame movement for *keyholing,* as was shown in Fig. 10-53, to speed up bead penetration. The size of tip orifice and rod will be somewhat smaller with thinner metal, and tables such as Table 10-2 should be referred to whenever the welder isn't sure what changes need to be made.

Equalizing pressure within the hoses and flame adjustment should be practiced by a welder as he uses different types of torches and tips. By practicing with different equipment and procedures, good control will be possible, and the welder can be assured of good, sound, strong welds.

Pipe Welding

Today, most pipe welding is done by shielded arc or inert gas welding. Some pipe

Table 10-2. Comparison Chart of Tip and Rod Sizes.

*Steel Thickness	Tip Size	Rod Dia. (in.)
20 ga. and lighter	#0	1/32
16-20 ga.	#1	1/16
14 ga.	#2	1/16-3/32
1/8″	#3	3/32
1/8″-5/32″	#4	3/32
1/4″	#5	1/8
	#6	1/8
	#7	1/8 backhand—3/16 forehand
	#8	3/16
	#9	1/4

*Thickness shown above is for mild steel, butt welding, forehand technique. Oxy-acetylene welding of steels thicker than 1/4″ is seldom done today since the metallic-arc process is faster, cheaper, and leaves a narrower heat-affected zone.

Courtesy of Canadian Liquid Air.

welding, such as that done by natural gas companies for gas input lines to buildings, is still being done with oxy-acetylene welding. For this reason, it is a good idea to go over the common pipe welding presently being done with the oxy-acetylene process.

Various sizes of pipe, from 1″ to 6″ in diameter with wall thicknesses up to 1/4″, can be welded satisfactorily with oxy-acetylene welding. Pipe up to 2″ in diameter is usually butted together with or without spacing, although spacing between pieces to be welded is never greater than 1/8″. See Fig. 10-74.

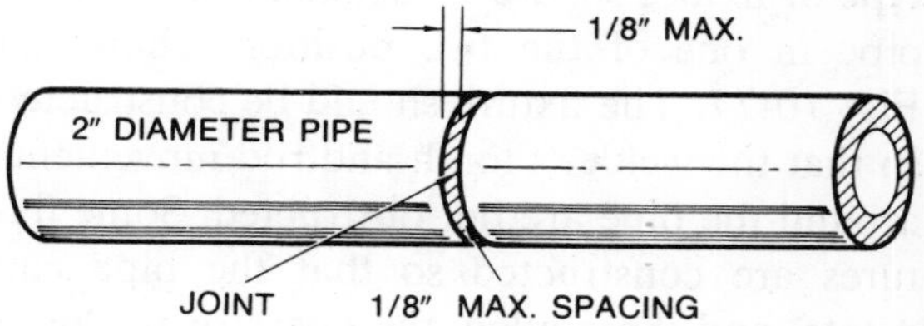

Fig. 10-74. Unprepared pipe, butted for welding with a gap not over 1/8″.

Pipe with a wall thickness of 3/16″ to 1/4″ is always beveled, usually by means of a pipe beveler, such as the one shown in Fig. 10-75. The beveled pipe, when butted together for

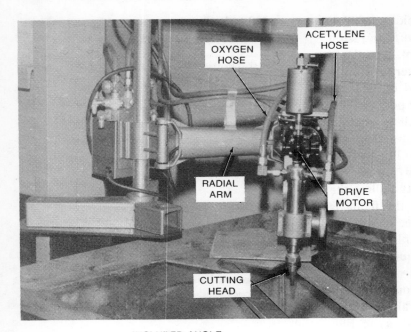

Fig. 10-75. Pipe bevelling machine.

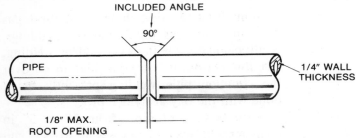

Fig. 10-76. Prepared pipe, butted for welding with a 90° included angle.

welding, usually has a 90° included angle. See Fig. 10-76.

Practicing Pipe Welding

For practice, the welder should use 45° beveled pipe sections, 6″ in diameter. Some type of fixture should be available to hold the pipe in one of the two positions shown in Fig. 10-77. The fixture should be constructed so that the welder's torch and rod movements around the pipe are not obstructed. Some fixtures are constructed so that the pipe can rotate, and these allow the operator to do all the welding in the flat position.

Equipment—The acetylene and oxygen pressures should be set at 5 and 16 psi, respectively, for welding 6″ diameter pipe. The tip orifice size should be a #48 drill. A neutral flame should be used with a ³⁄₁₆″ or ¼″ high tensile strength steel welding rod.

Tack Welding—The pipe should have a ⅛″ root opening all around the joint. A tack weld is then made on the *top* portion of the joint. The tack should bridge the gap, fusing both side of the joint and penetrating completely. It should be about ¾″ long and fill the joint within ¼″ of the top of the required

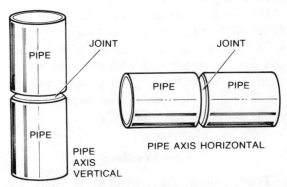

Fig. 10-77. The two positions used to hold pipe for welding.

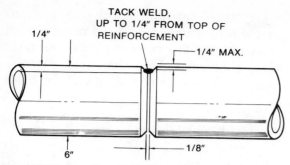

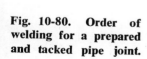

Fig. 10-78. Tack weld on bevelled pipe for a butt joint.

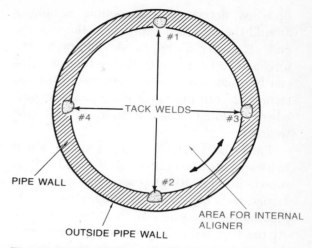

Fig. 10-79. Complete tack welds, viewed from the top (end) of the pipe, cutaway.

reinforcement, as in Fig. 10-78. When the tack weld has cooled, the pipe is turned a half turn in the fixture, and another tack is welded opposite the first. The welder should check the ⅛″ spacing between the pieces to be welded after placing the first two tacks, since expansion or contraction from the heat could have increased the distance. After the second tack has cooled, the pipe is given a ¼ turn for a third tack. Then, a final ½ turn positions the pipe for a final, fourth, evenly-spaced tack.

Final Welding—After all the tacks have been completed as shown in Fig. 10-79, the unit is ready for welding. An area at the bottom part of the pipe between tack welds #2 and #3, as shown in Fig. 10-80, is heated to melting temperature to receive the melted filler metal, thereby welding the bottom of the joint. From this point, the *forehand* technique, *vertical* position is used to fill and penetrate the root of the groove while building up and fusing the sides of the bevel. Using a ledge to support the molten metal, as mentioned earlier with vertical weld posi-

Fig. 10-80. Order of welding for a prepared and tacked pipe joint.

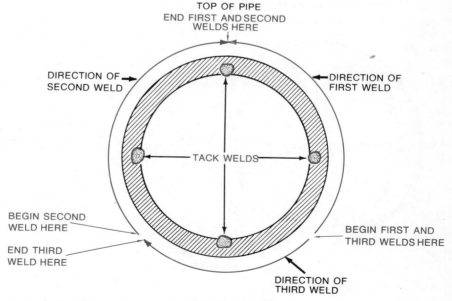

tions, is necessary. As welding continues around the pipe and approaches the top, the pull of gravity decreases. At this point, the forehand technique with *flat* position welding can be used. The welder continues with this technique until the *top* of the pipe is reached, where welding stops temporarily at tack weld #1.

Then the torch is brought down about halfway between tack welds #3 and #4, as shown in Fig. 10-80. From this point, the second weld begins and continues around the pipe on the opposite side from the first weld. At the top, both the first and second welds stop in the same place. The welder must be sure to remelt the crater at the end of weld #1, to have a strong bond where the welds meet.

The final weld (#3) is the most difficult if the pipe cannot be rotated into position. In Fig. 10-81, the first two welds are done, and weld #3 is about to be made.

Weld #3 is an overhead position weld if the pipe cannot be rotated. As an overhead weld, the puddle must not be too fluid, or gravity will pull it down. In each step, the welder must carefully remelt the previous puddle before adding filler rod metal.

As with the first two welds, the ends must be carefully melted, so that weld #3 fuses completely with the beginning ends of welds #1 and #2.

Vertical Pipe Welding

Welders are sometimes called on to weld pipe in the *vertical* position, as in Fig. 10-82. The joint, then, is in a *horizontal* welding position. The material, pressures, and flame setting recommended previously for horizontal pipe welding also work well for vertical pipe with a ¼″ wall thickness. The pipe is spaced and tacked as before. Then, the joint is placed in a suitable fixture, free of obstructions, to be held for welding.

Before practicing *vertical* pipe welding, it is a good idea to learn about and practice horizontal pipe welding and the horizontal joint techniques on simple butt joints.

Welding Practice Review

This chapter has dealt with the first welding practices, once the equipment is set up and ready to weld. Oxygen and acetylene gases are generally used with a ³⁄₁₆″ filler

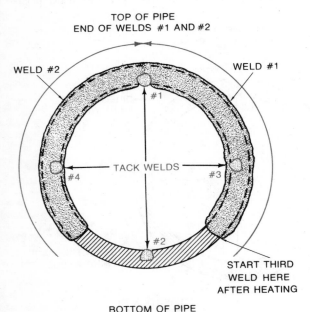

Fig. 10-81. First and second welds completed, ready for weld No. 3.

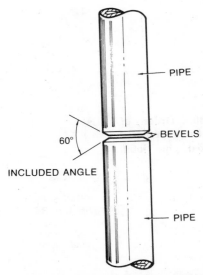

Fig. 10-82. Pipe positioned to be welded in the vertical position, horizontal weld.

rod on ¼″ mild steel sample plates (coupons).

Practice Welding

Most practice welding is done on butt joints, either unprepared or prepared with a 60° or 90° included angle. Pipe welding is a simple addition to the position practices. With pipe welding, though, it is as if the coupons were bent around, like a pipe.

After welding for some time, individual welders usually develop their own particular style, as well as certain favorite habits, equipment, settings, and techniques. This chapter has discussed the processes used by average welders—those methods that seem to work the best for most welders. Hopefully, a new welder will have "good luck" using the most common settings.

Techniques

There are two basic welding techniques, *forehand* and *backhand*. They refer to how the welder holds the torch and rod. The most commonly used method is *forehand*.

In forehand welding, the flame points ahead of the weld. The flame is between the filler rod and the completed weld, as was shown in Fig. 10-1.

In backhand welding, the flame points back to the finished weld. The rod is between the flame and finished weld, as was shown in Fig. 10-6.

While forehand welding is more popular, it is not necessarily any better than backhand; it is just easier for many welders. Others develop the backhand technique skillfully and like to use it because of its automatic postheat qualities.

Weld Positions

Four basic weld positions are used: Flat, Vertical, Horizontal, and Overhead. These are arranged in order of practice and ease, although vertical and horizontal welds are about equal in difficulty. Overhead welding should be tried only after the other positions

are mastered. These positions are shown in Fig. 10-11.

Joint Designs

The four basic joint designs used for practice welding are the butt joint, fillet joint, Tee joint, and lap joint. The most common for practice is the butt joint, either plain (Fig. 10-26) or prepared (Fig. 10-39).

After the butt joint is mastered, practice can move on to the Tee, lap, and fillet joints, covered briefly in the chapter.

General Pointers

Although methods vary, a few general pointers do apply to all welding. Weld metal should never be allowed to fall into the molten pool from beyond the heat shield area of the flame, as in Fig. 10-83. This

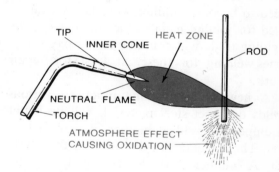

Fig. 10-83. The filler rod must be kept in the protective *heat envelope* at all times.

would cause oxidation, since hot metal is easily oxidized by the air and is harfmul to the weld.

The parent metal should always be heated to the proper melting temperature, and the sides of the joint should be melted the entire length of the weld. This allows the filler metal to be accepted, forming a solid unit weld.

If the base metal and filler metal are puddled together properly, and if the rod is added evenly and steadily within the protective outer envelope of the flame, the welded metal can be tested and found to be as strong or stronger than the parent metal.

11

Oxy-Acetylene Welding
Common Metals

Many metals other than steel can be joined with oxy-acetylene welding. As many different metals come into wider use, welders will need to become familiar with the procedures used for joining metals other than steel. The welding trade will continue to use oxy-acetylene welding for these, at least for repair work.

A new welder, having practiced various welds on sheet steel material, will be better equipped to weld a number of new metals, too. These metals are discussed in this chapter. A number of these new metals may require a *flux* to be welded properly.

Oxy-Acetylene Fluxes

Many of the other metals joined by oxy-acetylene welding require more than flame heat and filler metal to form a successful weld. Various *fluxes,* outlined in Table 11-1, are required when welding some metals. The following information should give welders an idea of flux's value in producing sound welds free of oxides.

If the melting temperature of a metal's *oxide* is close to or above the melting point of the metal itself, a flux is usually necessary. This is more easily understood by the welder

who has done oxy-acetylene welding on mild steel and cast iron.

When welding mild steel, no flux is needed, yet flux *is* necessary when welding cast iron. Both metals contain varying amounts of carbon, and, since they are ferrous materials, they have the same iron oxides. The melting temperature of iron oxide is 1900° F., and the melting temperature of mild steel is 2750° F. Mild steel's melting temperature is high enough to melt the steel's iron oxides and allow them to float to the top. This eliminates blow-holes and oxides trapped in the weld. Due to the higher carbon content of cast iron, the melting temperature of cast iron is approximately 2400° F. Though higher than the melting temperature of the iron oxide, this is not considered high enough to break down the oxides. When flux is added, the melting temperature of the oxides is lowered to the point where the flame temperature for cast iron welding will melt the oxides.

For this reason, it is important to check both the melting temperature of parent metals *and* the melting temperature of their oxides. The flux not only lowers the melting point of the oxides, but also forms a shield over the molten (heated) area. This excludes air and

Table 11-1. Welding Flames and Fluxes for Different Metals.

Base Metal	Filler Metal	Flame	Flux
Carbon Steel	Mild Steel	N	No
	High-Tensile— Low-Alloy	N	No
Wrought Iron	Low-Alloy	N	No
Gray Cast Iron	Cast Iron	N	Yes
	"Moly-Nickel" Cast Iron	N	Yes
Stainless Steel 18.8	Stainless Steel (347)	N-SR	Yes
Aluminum 1S, 2S, 3S	Aluminum	SR	Yes
Phosphorized Copper (deoxidized)	Deoxidized Copper	SO	Yes
Silicon Bronze (Everdur® 1010)	Everdur® (copper-silicon)	SO	Yes
Lead	Strips of Base Metal	N	Yes
Yellow Brass Red Brass-Bronze	Canadian Liquid Air Altem Super Bronze*	O	No
Zinc Diecast	L A Diecast	R	No

N—Neutral
R—Reducing
SO—Slightly Oxidizing
O—Oxidizing
SR—Slightly Reducing

*Where Altem Super Bronze is specified, "FL" Bronze and Tobin Bronze may also be used, using Brazewell Flux or Copox as specified.

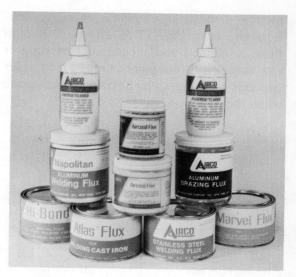

Fig. 11-1. A good assortment of welding fluxes. (Courtesy of Airco Welding Products Division.)

prevents oxidation. It also assists in *flowing* the metal, and for this reason the flux should become fluid before heating causes the flux to rapidly oxidize.

There is no universal flux for oxy-acetylene welding. Flux can be obtained in powder or paste form in containers of various sizes. Fig. 11-1 shows most of the fluxes made by one manufacturer.

Welding Cast Iron

To weld *cast iron* using oxy-acetylene welding, a cast iron filler rod is used with a neutral *flame*. The welding *tip* used should be

the same size as would be used to weld similar-sized, mild steel material. A chart for the tip size would be one such as Fig. 11-2. A cast iron welding *flux* is necessary to break down the iron oxides.

Drill Size	HEAVY DUTY For use in SW1 & SW2	MEDIUM DUTY For use in MW5 & PW1	LIGHT DUTY For use in AW1, AW5, AW6	Range Inches	Pressure Each Gas	Consumption c.f.h.
78			AW2000			.65
76			AW200	Up		1.3
75			AW20	to	3 p.s.i.	1.7
71	SW201	MW201	AW201	1/32		2.3
69	SW202	MW202	AW202	Up		3.0
67	SW203	MW203	AW203	to		3.2
63	SW204	MW204	AW204	3/32	5 p.s.i.	4.3
57	SW205	MW205	AW205	1/8		6.0
56	SW206	MW206	AW206	5/32		9.0
54	SW207	MW207	AW207	3/16		12.0
52	SW208	MW208	AW208	1/4	8 p.s.i.	17.0
49	SW209	MW209	AW209	3/8		23.0
44	SW210	MW210	AW210	1/2		36.0
40	SW211	MW211		5/8		49.0
34	SW212	MW212		3/4-7/8	11 p.s.i.	66.0
30	SW213			1" and		90.0
26	SW214			over		121.0

Fig. 11-2. Welding tip specification chart. (Courtesy of Smith Welding Equipment.)

Preparation

Most breaks in cast iron should be prepared with a 90° included angle and a root face. The root face helps to support the

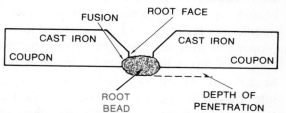

Fig. 11-3. The root face on cast iron helps make good penetration.

molten metal and eliminate excessive fall-through, as in Fig. 11-3. The parts should then be aligned and tacked so that support can be given to the pieces (Fig. 11-4). This prevents sagging, which could take place when the cast iron was heated.

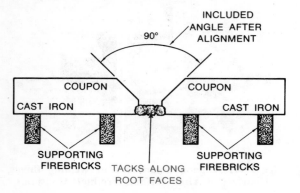

Fig. 11-4. Aligning cast iron for welding.

Preheating

Thorough preheating is very important when welding cast iron. A full red preheat temperature should be used to help eliminate cracking in the weld area during cooling, after the weld is completed. Preheating, using an oxy-acetylene flame, should be distributed evenly over the *entire* area of the piece to be welded. A furnace should be used for some cast iron pieces, especially those considered too large for flame preheating with the torch.

Technique

After preheating, the flame is directed to the bottom of the Vee preparation, at the right end of the piece to be welded. There, a puddle is formed in the same manner as for steel welding (Fig. 11-5). A cast iron rod, heated on the end while the parent metal puddle was being formed, is immersed in the proper cast iron flux about ½″ on the heated end. The flux will melt and adhere to the heated rod. The flux-dipped rod is brought into the molten puddle, melting the rod and helping to fuse and penetrate the edges of the preparation and the bottom of the Vee.

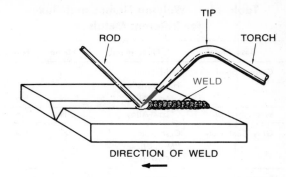

Fig. 11-5. Forehand technique used to weld cast iron.

Stirring with the rod loosens the oxides and allows them to float to the surface. The filler rod should not be allowed to drop into the molten pool; this would create hard spots in the finished weld.

When the bottom of the Vee is fully fused, the flame is allowed to melt the bevel sides by moving the flame from one side to the other. The molten metal from the sides will then run down to mix with the molten pool at the bottom of the Vee. With the flame being moved from side to side, as in flat position steel welding, the rod continues to melt, filling the Vee to the required height. Care must be taken, however, to make sure that any excess molten metal running ahead into the bottom of the Vee (Fig. 11-6), is eventually brought to melting temperature again for good penetration.

Further fluxing of the filler rod becomes necessary when white spots or gas bubbles

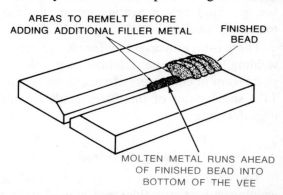

Fig. 11-6. Any molten metal running ahead in the bead must be remelted before welding.

appear in the puddle, indicating that impurities are present. These spots can be skimmed from the molten pool by moving the rod like a paddle while welding. If the impurities adhere to the rod during skimming, they can be removed by hitting the rod *gently* on the welding table. If the rod is struck too hard, it may break, since it is non-ductile cast iron.

Welding cast iron can usually be completed in one pass, especially where about an inch of the Vee is filled to the required height for reinforcing. See Fig. 11-7. Before the

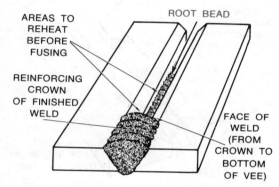

Fig. 11-7. Finished weld on one end. The face of the weld must be remelted.

next inch of welding is started, the face of the exposed finished section must be heated to the right welding temperature. This gives complete fusion at the bottom and sides of the Vee, so they will combine into a strong and nonporous unit. *Porosity* refers to how many air pockets and impurities are trapped

Fig. 11-8. Cross-section of a good gray cast-iron weld. Magnified 500 times. (Courtesy of Bethlehem Steel.)

in the weld. This process is carried on until the full length of the joint is completed. Fig. 11-8 shows a magnified view of a good gray cast-iron weld.

Alternate Method—Some welders like to fuse about three inches of the bottom of the joint, followed by a two-inch second pass. These are finally covered with a one-inch finished pass, as shown in Fig. 11-9. After the

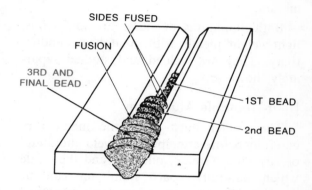

Fig. 11-9. Multiple-pass cast iron welding.

third (final) pass, the first pass is continued for another three inches or more. The second and third passes are alternately extended respectively. This form of *cascade* welding is shown in Fig. 11-10. This procedure is car-

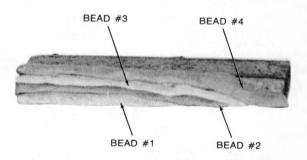

Fig. 11-10. Cross-section view of a cascade weld.

ried on until the total weld is completed. Each time, the exposed faces of each weld must be remelted for complete fusion of the next phase. Also, a welder using this method should be aware of the fact that rapid heating and cooling, as well as excess flux, can create serious problems in the finished weld.

Welding Aluminum

Information on welding *aluminum* will cover the common two aluminum types—Heat-Treatable and Commercially Pure. The 53S Heat-Treatable Aluminum is one of those that can be solution heat-treated and consists of an aluminum-magnesium-silicon alloy. On the other hand, 2S Commercially Pure Aluminum contains a minimum of 99% aluminum.

Both aluminums, of course, are nonferrous, light-weight metals with great heat conductivity. They are welded and studied separately, however.

Heat-Treatable Aluminum

The welder must clean the metal very carefully before attempting to weld aluminum of any kind. Grease, oil, dirt, and the oxide which naturally coats aluminum can be cleaned off with a nonflammable chemical solvent, such as caustic soda mixed with hot water. If a heavy oxide coating covers the area to be welded, the area should be brushed or pickled in acid. Either a cold 10% sulfuric acid solution or a hot 2% nitric acid solution can be used satisfactorily. In many cases, wire brushing with hot water is good enough.

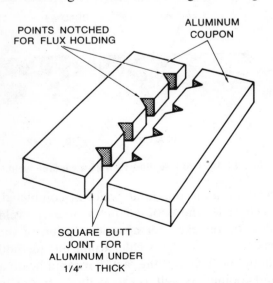

Fig. 11-11. Joint preparation for aluminum up to ¼″ thick.

Joint Preparation—The joint preparation for aluminum is the same as for steel. Thicknesses over ¼″ should be beveled. As an added precaution, the edges of both the square and Vee joint should be notched about ¹⁄₁₆″ to ⅛″ deep and about ³⁄₁₆″ apart, as shown in Figs. 11-11 and 11-12. This will assure that there is full weld penetration and that the flux will penetrate well down into the Vee. The notching can be easily done with a

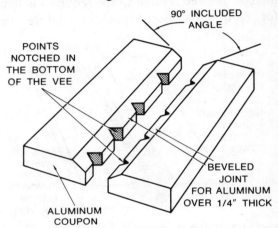

Fig. 11-12. Joint preparation for aluminum ¼″ thick or over.

hammer and chisel, due to the metal's softness. Aluminum up to ¹⁄₁₆″ thick should be tack welded at 2″ intervals. On heavier material, tack welds can be placed up to 10″ apart, since distortion will not be as great on the heavier material.

Preheating—Aluminum over ⅜″ thick should be preheated between 650° and 750° F. Heat-treatable aluminum alloy is sensitive to cracking near the weld during cooling, so it should be free to expand and contract. For welding this metal, the welder should use an aluminum *rod* containing 5% silicon. A powdered *flux*, mixed with water to form a paste, should be applied to the edges of the preparation and to the end of the rod. The acetylene and oxygen pressure should be set the same as for similar thicknesses of steel, but it is advisable to use a *tip* one size larger. This is due to the difference in the heat conductivity of aluminum and steel.

An excess acetylene *flame,* 1½X, is recommended, as in Fig. 11-13. The tip of the

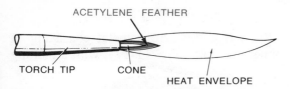

ACETYLENE FEATHER
TORCH TIP   CONE
HEAT ENVELOPE

Fig. 11-13. The 1½X flame recommended for welding aluminum.

inner cone should be kept at least ⅛" away from the molten metal. Since aluminum shows no indication of reaching melting temperature through color change, as steel does, initial overheating may be a minor problem. This can be easily overcome through practice welding.

Technique — The forehand *technique* is preferred, and welding can be done in all positions by using the same technique described for steel. Due to the low melting point of aluminum, 1218° F, the welder may, at first, have trouble recognizing when the molten puddle is ready for the filler metal. Because of this, he may have some difficulty with the metal being too hot and falling through.

Fig. 11-14. Grinding a welded aluminum piece. (Courtesy of The Norton Company.)

The end of the filler rod, unlike that in steel welding, should always be kept in the puddle and stirred vigorously while welding. This allows all the entrapped oxides to float to the surface. Whenever possible, all welds should be made in one pass.

Clean-Up—After welding, all flux and slag should be removed by washing for 15 minutes in a 5% sulfuric acid solution at 150° F. This is followed by a clear water rinse. Grinding can then be done if needed, as in Fig. 11-14.

After cleaning the welded part, which may also be a casting, it should be postheated to 970° F and quenched. This will restore the original strength to the parts.

Commercially Pure Aluminum

Commercially Pure Aluminum, also called 2S Aluminum, is prepared for welding the same way as is Heat-Treatable Aluminum. Preweld cleaning, similar to that used for Heat-Treatable Aluminum, is required. In many cases, however, wire-brushing without added solution will usually remove oxides as well as oil and grease. If preheating thick sections (usually required due to the high heat conductivity of aluminum), the heat should be at least 650° F, but not over 750° F.

Flux, Rod, Flame, and Tip—Powdered *flux*, mixed with water and kept thoroughly mixed by frequent stirring, should be applied to the prepared joint and welding rod. The *rod,* of the same purity as the base metal, can then be welded with an excess acetylene *flame,* as in Fig. 11-13. The *tip* size is the same or one size larger than would be used for welding the same thickness of steel.

Technique—The forehand *technique* (Fig. 11-15) is used with continuous puddling, the same as for welding steel. By using these methods, the oxides will be broken down through the flux action, and a strong, porous-free, weld will result. Again, the welder has to be careful not to overheat any aluminum.

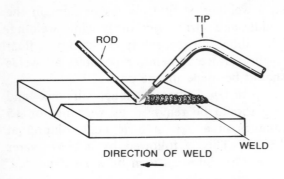

Fig. 11-15. The forehand technique used for welding aluminum. Note the 1½X flame.

Since little or no color change can be observed before the metal reaches its melting temperature, this alloy may also suddenly "fall through."

Clean-Up—The finished weld should be cleaned in a mixture of hot water and 5% sulfuric acid at 150° F for 10 minutes. This will remove the slag and flux that naturally hold onto the weld. Grinding and polishing can then be done on the face of the weld if necessary, as shown in Fig. 11-16.

Fig. 11-16. Grinding a small welded aluminum piece. (Courtesy of The Norton Company.)

After welding, 2S Aluminum will not require postheating to restore the metal to its original strength. This is a major difference

between Commercially Pure Aluminum and Heat-Treatable Aluminum.

Welding Brass

Brass is a nonferrous metal consisting of copper and zinc (Fig. 11-17), with a melting

Fig. 11-17. Annealed brass. Approximately 70% copper and 30% zinc. (Courtesy of Bridgeport Brass.)

temperature of 1650° F. Brass can be joined with oxy-acetylene welding using a bronze welding *rod,* preferably with the same composition as the brass being welded. Since the zinc part of brass has a relatively low boiling point and vaporizes easily, it is necessary to have a controlled flame and a suitable filler rod when welding brass. These help reduce the zinc vaporization during welding. For this reason, the welder should refer to a manufacturer's filler rod specification chart before attempting to weld brass.

Preparation

Brass, like aluminum, should be brushed or washed in solvents to remove grease, dirt,

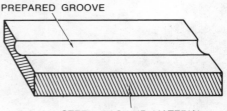

Fig. 11-18. Steel back-up block used in welding brass.

oil, and other foreign matter. It may also be cleaned by grinding. Preparing brass for welding in the same way as for steel of the same thickness will give good joint strength.

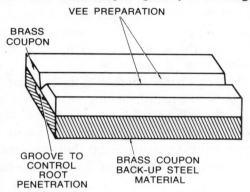

Fig. 11-19. Brass coupons beveled and mounted over the prepared groove.

When brass reaches melting temperature, it has a high fluidity. For this reason, using grooved steel or asbestos back-up material is helpful for uniform root penetration. See Figs. 11-18 and 11-19.

Fixtures are usually needed to hold the molds and parts during welding. The fixtures, however, should be so designed so as *not* to act like chill strips. To eliminate any chill strip effect, asbestos may be placed between the fixture and the metal being welded. This will hold the heat in, which is very necessary during brass welding.

Flux, Flame, and Tip

Since a coating will form on the molten puddle during welding, the welder will need a suitable *flux* to weld brass. Mixed with water, the powdered flux is then painted on both the edges to be fused and on the filler rod. Most welding supply companies have powdered fluxes available to be mixed with water. The mixing forms a paste that should be stirred periodically and then applied as required.

The *flame* used will be slightly oxidizing (Fig. 11-20). This will prevent the zinc from *fuming,* recognized as white puffs of smoke. Since there is a rapid dissipation of heat dur-

ing brass welding, it will be necessary for the operator to use a *tip* about twice the size of that required for the same thickness of steel.

Preheating

Temperatures ranging around 450° F will do a good job of preheating brass, although

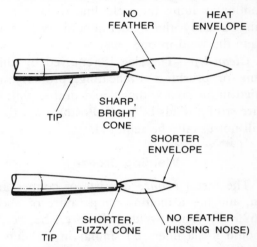

Fig. 11-20. The oxidizing flame is used to weld brass. This is one of the few uses of the oxidizing flame.

overheating can burn out the zinc and weaken the base metal. Due to the rapid dissipation of heat, the operator should weld in an area without any cold drafts.

Technique

Many welders prefer the forehand *technique* for welding brass, although welding heavier sections may require changing to the backhand method.

Most brass welds are completed in the flat position. After some experience, the welder may find that satisfactory welds can also be made in the vertical position, especially if a molding device is used to hold the fluid metal against the surfaces being welded. Using an oxidizing flame, the welder can tack weld the aligned parts in a manner similar to tacking steel. When the unit has been tacked, it is placed in the appropriate fixture.

The metal is then reheated, using an oxidizing flame. Heating starts at the extreme right of the joint with the forehand technique until a puddle is formed. The filler rod is then added, while the welder makes sure that too much heat is not applied in one spot. The puddle must be molten, however, before the filler rod is added, or a poor weld joint will result. Once a puddle has been formed, it is continued along the weld line without stopping. Usually, the welder should try to complete the weld in one pass.

Clean-Up—The slag is removed by chipping or brushing. Preheating, properly applied in an area without cold drafts, will ensure strong welds by itself. Postheating, then, will not usually be required.

Welding Bronze

The metal *bronze* consists of copper and tin, and has a melting temperature of about 1625° F. When welding bronze, the welder should use a filler *rod* similar in makeup to the base metal, especially when a color match is required. Any good bronze welding rod will give an adequate weld if a few "rules of thumb" are followed.

Preparation

To weld bronze, it is essential to remove the grease, oil, dirt, and other contaminants. This can be done by wire brushing, grinding, chipping, or using solvents. As with brass welding, the welder should take advantage of *molds,* especially in the vertical position. The joint preparation is the same as that used for welding steel of the same thickness. However, because of bronze's high fluidity, back-up bars or strips should be considered. These are similar to those used with brass, as illustrated in Figs. 11-18 and 11-19.

The welder should also consider using fixtures to hold the parts to be welded. Insulating the parts from the fixture's chilling effect may be done with asbestos wrappings.

Flux, Rod, Flame, and Tip

A *flux,* similar to the one used for brass welding, is required to help eliminate the

coating that forms on the molten bronze puddle. These fluxes are also mixed with water into a paste before being applied to the edges of the joint preparation and to the end of the filler *rod*. The end of the filler rod can be dipped into the flux every so often to eliminate the scum (oxides) floating on the molten pool, if necessary.

The oxidizing *flame* is usually used because it will retard melting until a red heat is reached. If an oxide coating forms on the weld puddle, the welder should adjust the flame more toward neutral, but still oxidizing. Since the bronze will dissipate the torch

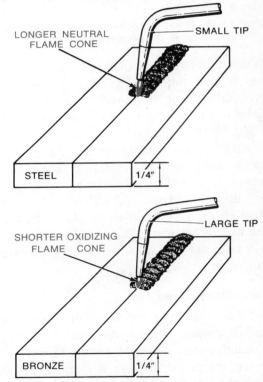

Fig. 11-21. Welding bronze requires a larger tip and a shorter flame than does welding steel.

heat rapidly, it is necessary to use a *tip* about twice as large as the one that would be used for welding steel of the same thickness. See Fig. 11-21.

Technique

After tacking the parts to be welded in a manner similar to tacking steel plates, the bronze is preheated to about 450° F. The preheating and welding of the parts should be in an area free of drafts, since the rapid heat dissipation can create a welding problem if the pieces to be welded are hard to melt.

The welding *technique* used by many welders is forehand, but for heavier bronze it is usually necessary to use the backhand technique. In either case, the puddle must be carefully controlled by moving the rod and flame in the same manner as for steel welding. As with steel welding, the welder must be sure to melt out any tack welds that the flame comes into contact with. He must also be sure that the puddle is molten before attempting to add any filler metal.

Clean-Up—It is not necessary to postheat bronze after welding. Because no postheating is needed, the welder can immediately deslag the weld by chipping, grinding, or other cleaning methods.

Welding Deoxidized Copper

Deoxidized copper (Fig. 11-22) melts at 1980° F. This melting temperature is over

Fig. 11-22. Tough pitch copper, photographed under a microscope. (Courtesy of Bridgeport Brass.)

300° higher than the melting point of brass or bronze. However, like brass and bronze, deoxidized copper must be preheated to about 450° F before welding.

During welding, the operator should make every effort to prevent chilling the metal due to rapid heat dissipation. Copper can be welded in the same position and with the same techniques as brass and bronze. However, due to copper's high fluidity, the welder should be very careful when welding it in the vertical position.

Preparation

Before welding deoxidized copper, all dirt or other foreign matter must be cleaned from the pieces to be welded. This may be done by chipping, grinding, or using cleaning solvents. When cleaning parts by chipping or grinding, the parts should be cleaned at least ½" beyond the point where the weld will finally extend, as shown in Fig. 11-23.

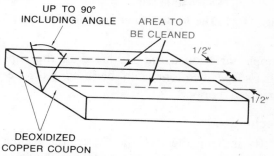

Fig. 11-23. Preparing deoxidized copper coupons for welding.

When welding deoxidized copper, joints are prepared in the same manner as would be used for steel of the same thickness. The copper should be protected by asbestos sheeting when it is placed in a fixture. This will reduce its rapid heat dissipation and the chill bar effect of the fixture. The asbestos sheeting will hold the heat in and speed up the welding process.

After welding deoxidized copper, the welder should loosen the fixture as the metal begins to cool; this relieves some of the stress in the weld area.

Tip, Flux, Flame, and Rod

A *tip* about twice the size as would be used for welding the same thickness of steel will be needed to contain enough heat for welding deoxidized copper. However, the welder will not need a *flux,* as he did with brass or bronze welding. If welding presents some difficulty due to any contaminants in the copper, a *flux* can be used, as mentioned for brass and bronze welding. The flux will clear up and float away the impurities, giving a better and easier weld.

The welder should adjust the torch for a neutral *flame* (Fig. 11-24). A deoxidized

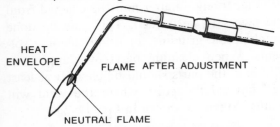

Fig. 11-24. The neutral flame used to weld deoxidized copper.

copper or bronze *rod,* as listed in the supplier's specifications, can be used with deoxidized copper.

Technique

By moving the rod and flame, the welder will be able to control the puddle in the same manner as with steel. The tack welds should be melted out and a molten puddle formed before adding any filler rod. When welding heavy sections of deoxidized copper, the weld should be started at least two inches from one edge, and then welded back to that edge. After this, the starting point of the weld is remelted, and welding is then continued toward the other edge, as shown in Fig. 11-25. The forehand *technique* can be used on thinner sections, while the backhand is used on thicker sections.

Clean-Up

When the weld is finished, usually in one pass, it should be cleaned by wire-brushing. After cleaning, the weld should be peened gently with a peening hammer to reduce the grain size. Following peening, the welded material is postheated between 1100° F and 1200° F. This will restore the copper's duc-

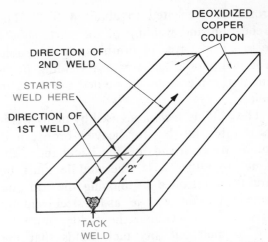

Fig. 11-25. Procedure for welding deoxidized copper.

tility while improving its hardness and strength.

The weld area can be ground, if necessary, to improve the appearance. Grinding would be done by using equipment as shown in Fig. 11-26.

Fig. 11-26. Grinding a finished weld to improve the appearance.

Welding Everdur®

Everdur® is a copper-silicon alloy with lower thermal conductivity than copper. It is not as sensitive to overheating as are the other copper alloys. *Everdur* melts at about 1866° F.

Preparation

Like other copper alloys, *Everdur* requires cleaning to ensure strong and porous-free welds. The parts to be welded are prepared the same way as is steel of similar thickness. Then, the parts should be tacked and mounted in a fixture to prevent distortion.

Flame, Tip, Flux, and Rod

Either a slightly oxidizing or neutral *flame,* concentrated in a small area of the weld, will be good for welding *Everdur*. As with the other copper alloys, it is necessary to use a large *tip* to keep the heat in the parent metal for welding. Molten *Everdur* is naturally protected with a flux-like, liquid glass film during oxy-acetylene welding. However, even with these protective qualities, the welder should use a good *flux,* as was used with the other copper alloys. This will make sure that the oxides are removed and the weld is strong. A copper-silicon filler *rod,* with a composition close to that of the parent metal, is used.

Technique

Everdur is usually welded using the fore-hand *technique*. Although the metal can be difficult to work with, it can be welded in the vertical position, if the precautions described for the other copper alloys are taken.

After preheating the parts to a temperature of about 450° F, the tacks are melted out. Then, the copper-silicon filler rod is added to a small weld puddle after fluxing. With the flame concentrated on a small weld puddle, the welder must weld more rapidly than with the other copper alloys.

Material up to ⅜" thick can be welded in one pass. Heavier sections will require two passes, because the flame has to be concentrated on the small puddles. Welds, whether single or double pass, are applied the same way as for deoxidized copper, Fig. 11-25.

Clean-Up

After welding, the slag can be removed. Since postheating is not required, peening can be done after cleaning, while the material is either hot or cold. The mechanical properties of the metal are improved if peening is done while the material is hot. However, cold peening and annealing will improve the grain structure.

Welding Nickel

The metal *nickel* is used commercially, either by itself or alloyed with copper and other elements. An excellent property of nickel is that it compares with stainless steel in corrosion-resistance. This metal, and its alloying elements *Inconel*™ and *Monel*™, can be joined by oxy-acetylene welding. Nickel melts at a fairly high 2646° F.

Preparation

The surfaces must be thoroughly cleaned to ensure the welder of a strong, porous-free weld. The included angle in a prepared nickel joint should not be more than 75° (Fig.

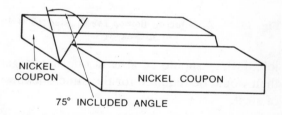

NICKEL COUPON   NICKEL COUPON

75° INCLUDED ANGLE

Fig. 11-27. A 75° included angle is usually used for welding nickel.

11-27), although any joint will be satisfactory if the grooves are kept at this angle.

Only thin sections of nickel require fixtures. These should be designed to accept differently shaped chill blocks and grooved back-up bars near the joints, as in Fig. 11-28. Preheating is not required.

Rod, Flame, and Tip

Flux is not required when welding nickel with the oxy-acetylene method. The filler *rod*

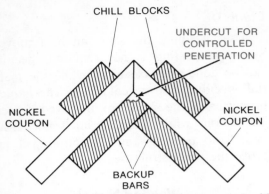

Fig. 11-28. Chill blocks and back-up bars used with welding nickel.

should be nickel, with a composition as close as possible to that of the parent metal. The *flame* is adjusted to be slightly acetylene; this will ensure that elements required for corrosion-resistance and strength will remain intact in the weld area. A reducing flame (Fig. 11-29) may also be adequate for this welding

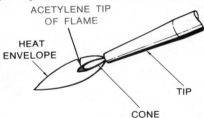

Fig. 11-29. A reducing flame, 1⅛ X, used to weld nickel.

job. The *tip* should be at least one size larger than would be required for the same thickness of steel.

Technique

Nickel can be welded satisfactorily in any position. Although the forehand *technique* is usually preferred by experienced welders, the backhand technique may also be used.

Nickel is very sluggish when molten and should be welded in one pass. Any rewelding would create a poor, porous weld due to oxidation. Unlike steel, the melted area should never be puddled. The rod should be touched gently to the surface of the sluggish pool. This will allow the filler metal to flow smoothly into the pool. Like steel, the rod should al-

ways be kept in the flame's heat zone to avoid oxidization.

After filling the joint to the desired height in one spot, as in welding cast iron, the sluggish puddle can be moved ahead and a new area completed to the desired height until the weld is finished.

Clean-Up

Finishing up a nickel weld is fairly easy. After welding, the welder needs only to wire-brush the finished weld. Postheating and peening are not required with nickel.

Welding *Inconel*

Inconel melts at about 2540° F and, like all the previously mentioned nonferrous metals, needs to be cleaned thoroughly before welding. Like nickel, this metal can also be welded in any position, if the previously mentioned precautions are followed. Any type of joint used for steel can be used when welding *Inconel*.

Preparation

The prepared groove should have an included angle of 75° to assure the welder of a strong, sound weld. Fixtures should be used for holding tacked, thin sections of *Inconel*. Adding chill blocks and grooved back-up bars, as in Fig. 11-28, may also be necessary.

Rod, Flux, Flame, and Tip

Filler *rods* of *Inconel* are available, along with a special *Inconel flux*. The flux, dissolved in an alcohol-shellac solution, is kept in a nonmetallic container to avoid any chemical reaction.

A soft, slightly reducing *flame* is recommended with a welding *tip* at least one size larger than for steel of the same thickness.

Technique

The forehand welding *technique* is best suited for welding this material, which should be finished in one pass. *Inconel* does not need preheating and, like nickel, it is sluggish on melting. The *Inconel rod* should be kept within the outer flame envelope to keep the rod free from oxidation. The flux-dipped rod should be allowed to touch the surface of the weld pool. This allows the filler metal to flow smoothly into the pool, while filling the melted and flux-coated edges of the joint.

Clean-Up

The finished weld joint will be covered with a flux coating. This can be removed by sand-blasting or cleaning in the 50% nitric acid bath previously mentioned. After the weld has been soaked in the bath for approximately 15 minutes, it should be washed thoroughly with clean water.

Welding *Monel*

Monel melts at 2370° F and is more sluggish than nickel or *Inconel* in the molten state.

Preparation

Before attempting to weld *Monel* by the oxy-acetylene process, the welder must thoroughly clean the surfaces to be welded.

Like *Inconel*, *Monel* can be welded in any position. However, it should be held in a fixture with chill blocks set up near the joints and back-up bars supporting the weld material. The joints used for welding mild steel can be used when welding *Monel*. Where a groove joint is used, the 75° included angle is still found to be the most satisfactory.

Flux, Rod, Flame, and Tip

After cleaning, a special *Monel flux* is painted on the edges of the joint. This special flux is similar to *Inconel* flux and is available from most welding supply companies. The *Monel* paste that is usually used is mixed in alcohol or boiling water and is also kept in a nonmetallic container. *Monel* welding *rods* are available, and they should be used whenever welding this material.

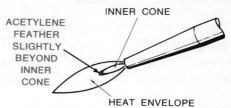

A slightly excess acetylene *flame,* as in Fig. 11-30, should be used. The metal will

Fig. 11-30. Slightly acetylene flame used to weld Monel metal.

require a *tip* one size larger than the size ordinarily used for welding steel of the same thickness.

Technique

Preheating *Monel* metal is not needed. A *Monel* welding rod, with composition similar to the parent metal, should be dipped in the flux periodically. This will remove or break down any contaminants that may be in the weld area.

Monel metal flows quite freely when melted, and the welder should try to avoid puddling. The flame envelope should always be in contact with the fluxed end of the filler rod, to protect it from the harmful effects of the atmosphere. As with *Inconel* welding, the *Monel* rod is gently touched to the weld pool surface, allowing the filler metal to flow smoothly in. Using the forehand welding *technique* with one pass should be practiced for best results.

Clean-Up

Postheating after welding is not required. However, the welder should remove all the flux from the weld bead surface by using a wire brush and hot water.

Welding Magnesium

Magnesium metal is similar to aluminum in many ways. It is lighter than aluminum and is used in many production pieces. It welds much like aluminum and melts at 1240° F. Most of the magnesium castings welded in production are welded with the Tungsten Inert Gas (TIG) method. However, in many cases, the welder is required to join broken pieces of magnesium by using oxy-acetylene welding. The following information will help the welder use oxy-acetylene equipment to repair magnesium.

Preparation

Before welding magnesium, the welder must remove all dirt, oil, and other foreign matter from around the edges of the prepared joint. Usually, this is done by a chemical solvent. Wire-brushing the bevel joint to about one inch from the prepared edges will usually

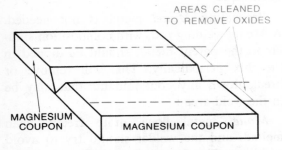

Fig. 11-31. Prepared magnesium coupons.

remove any oxides that may create problems. See Fig. 11-31. If the parts to be welded are heavily oxidized, it will be necessary to immerse them in a 50% nitric acid bath for about ten minutes, after which they can be washed in clean water.

The joint preparation for magnesium is similar to that used for aluminum or steel. Magnesium over ¼″ thick should always be beveled. Notching is desirable for any magnesium material over ⅛″ thick, as was shown in Fig. 11-11 for aluminum. The tack-welding procedure for this metal is also the same as for aluminum. Thinner material is tacked every 2″ along the joint, while thicker material, such as ⅜″ to ½″, should have 10″ between each tack.

Magnesium and its alloys are very sensitive to cracking as they cool, especially around the weld bead. For this reason, fixtures used for tacking and aligning should be loosened to allow the welded parts to expand and contract of their own accord. Preheating the magnesium to between 600° F and 700° F will help alleviate its tendency to crack while cooling.

Flux, Rod, Tip, and Flame

Powdered *fluxes* are available that can be mixed with water to form a smooth paste. This flux should be kept at an even consistency by stirring the mixture periodically. Only a minimum amount of flux should be used when welding magnesium parts. It should be applied to the end of the filler rod and along the edges and surfaces of the prepared joint. The filler *rod*, which should be the same composition as the parent metal, can then be dipped into the flux as required.

The size of the *tip*, as with other types of welding, depends on the thickness of the material to be welded. Generally, many welders find that by using a tip one size larger than would be used for welding the same thickness of steel, they can better control the puddle. Regardless of the tip size being used, a neutral *flame* will do a good job of welding magnesium.

Technique

Magnesium, like aluminum, shows no color change when it reaches its melting temperature. By using the forehand welding *technique*, practice magnesium welds can be made by using a continuous puddle of molten filler rod for adequate penetration. During magnesium welding, one pass proves to be the most satisfactory method for making strong and porous-free welds. The material can be welded in all positions, and the forehand welding technique is preferred where possible. Experienced welders usually try to avoid lap joints, even in the flat welding position.

Clean-Up

The flux on the finished weld should be removed by using a wire brush and hot water. Another good method, where possible, is boiling the part in a 0.5% sodium dichromate solution for approximately one hour and then rinsing it with clean water. Postheating is not required to complete the magnesium welding process.

Welding Stainless Steel

Like magnesium, most of today's *stainless steel* is being welded in production with the Tungsten Inert Gas (TIG) method. Still, many welders will be expected to be able to join or repair stainless steel parts by using oxy-acetylene welding. For this reason, information on welding the common grade 18-8 Stainless Steel is given here. The familiar and popular 18-8 Stainless Steel gets its number name from the percentage content of chromium and nickel, respectively.

Preparation

Stainless steel must be cleaned thoroughly with a solvent, especially around the weld joint. This will remove the oil, grease, dirt, or other contaminants that may be clinging to the surface of the pieces to be welded. Joint preparation for stainless steel is the same as for mild steel in oxy-acetylene welding. Since thin sections of stainless steel may warp during heating, the parts to be welded should be tacked in various places to hold the parts in alignment. Fixtures are often used for this, especially while welding thin sections. Chill blocks and grooved back-up bars may be added when required. These are shown in Fig. 11-28 for welding nickel.

Stainless steel can be welded in all positions. The filler rod and torch are held in the same general positions as were described for mild steel welding. Although preheating 18-8 Stainless Steel is seldom required, some welders prefer a slight torch preheating on metal over 1″ thick. This tends to speed up the welding after a puddle is established.

Rod, Flame, Flux, and Tip

The filler *rod* should consist of about 18% chromium and 8% nickel, together with

stabilizing alloy elements such as titanium, molybdenum, or tungsten. A slightly excess acetylene *flame* is preferred. This flame ensures that these elements remain intact and that they will not be destroyed if the flame becomes more oxidizing than neutral. The added stabilizing elements resist the damage that could happen to the stainless steel's excellent corrosion-resisting properties through heating.

The joint area should be painted with a good stainless steel *flux*, available from most welding supply companies. For welding stainless steel, a *tip* is used that is at least one size smaller than the one used for welding steel of the same thickness.

Technique

The joint and filler rod end are both fluxed. Then, using the reducing flame, the torch is held so that it strikes the joint to be welded

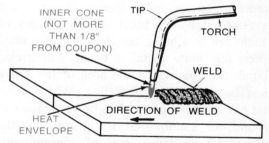

Fig. 11-32. Torch position for welding stainless steel.

from a nearly vertical position. The inner cone is held up to ⅛″ away from the puddle, as shown in Fig. 11-32. Once the puddle is formed, it should be kept small, and the welding rod should be kept near the inner cone of the flame. The welder continues welding, using the forehand *technique* and adjusting his welding speed to compensate for the closing and opening gap between the plates. The changing gap may be caused by expansion and contraction as he welds. If he welds too fast, the gap between the plates will tend to close, and, if he is welding too slowly, the gap will tend to open.

Clean-Up

After finishing a stainless steel weld, the excess flux can be removed by wire-brushing or by using a chemical solution similar to the one used on magnesium. Postheating is not required after welding stainless steel.

Welding White Metal

White metal castings usually fall into one of three categories alcording to their basic composition; i.e., zinc, aluminum, or magnesium. Identifying each of these is usually done by weighing them. Zinc alloy, which melts at approximately 800° F, is the heaviest and most commonly used. It is the type covered in this section. Magnesium alloy is the lightest of the three types.

Preparation

The welder must be especially careful when cleaning white metal before welding. Oil, grease, paint, or other foreign material can be removed by chemical cleaners, many of which are available on the market today. When the white metal is chrome or nickel plated, the coating on the area of the break must be removed by grinding. White metal that has not been plated can be wire-brushed, ground, or sandblasted to insure that the area to be welded is free of contaminants that could affect the finished weld.

Butt joints are usually preferred, with a bevel edge ground or filed on each section to be welded. Experienced welders will often use the torch heat to melt out a groove for preparation. Where possible, some welders fabricate a type of die from metal to hold the ex-

cess filler metal. See Fig. 11-33. Where a fabricated holder is impractical, the parts to be welded should be aligned in a fixture. The fixture will then support the parts so as to prevent sagging from applying heat.

Preheating is recommended *if* it is possible to put the parts to be welded on a ½″ steel plate. The torch then heats the plate for *indirect* preheating. See Fig. 11-34. The steel plate should be heated to approximately 500° F to 600° F for the best results.

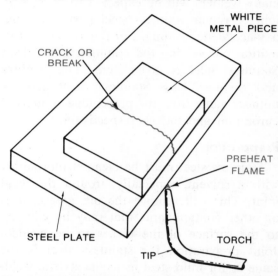

Fig. 11-34. Preheating a broken white metal piece.

The parts should be set up for welding in the flat position, since welding in any other position is almost impossible.

Rod, Tip, Flame

A white metal welding *rod,* usually rubbed with emery cloth to remove the oxides, is

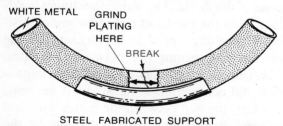

Fig. 11-33. Fabricated support often made to hold white metal while welding.

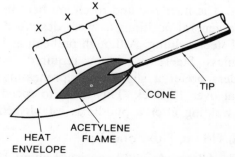

Fig. 11-35. A 3X carburizing flame used to weld white metal.

used to weld and fill a white metal joint. Since the melting temperature of the parent metal is relatively low, very little heat is required for fusion. The *tip* should, therefore, be no larger than a #72 size drill, if a sound repair is to be welded in the broken parts. The torch should be adjusted so that the *flame* is 2½ X or 3X, as shown in Fig. 11-35. Oxy-acetylene welding *flux* is not needed to weld white metal.

Technique

Either the backhand or forehand welding *technique* can be used. Many experienced welders prefer the backhand welding technique, using the filler rod to reflect some of the heat. The tip of the torch should be kept about ¾″ away from the base metal, until the metal begins to flow. The flame should then be turned parallel to the surface (Fig. 11-36) and the heat delivered with the side

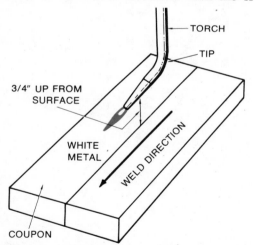

Fig. 11-36. When welding white metal, the torch tip is held parallel to, and ¾″ above, the surface.

of the flame until the rod is heated. When the base metal and filler rod are molten, the rod is touched to the joint where the rod and base metal unite.

Many experienced welders use a small, mild steel oxy-acetylene welding rod, flattened on the end like a spoon, to transfer molten metal to different places on the prepared joint. With practice, many welders find that

a more compact and well-formed bead is possible by using the mild steel rod.

Caution—If a welder forgets to grind off the plating before heating the part, the white metal will become fluid before the plating

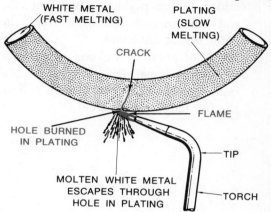

Fig. 11-37. If plating is left on the white metal, the metal underneath will be molten before the plating melts.

breaks down. Then, after the plating melts, the white metal fluid will run away or fall due to gravity. See Fig. 11-37.

Clean-Up

After welding, there is usually excessive bead metal buildup on the weld. This can be ground or filed to the required size with little trouble. If the weld has been well built, it will be just as strong after the clean-up as before.

Other Metal Welding Summary

The welder's skill and the proper type of filler metal together determine a weld joint's strength. For this reason, many welders want to practice welding on metals besides steel, after referring to the manufacturer's rod specification chart.

These principles are involved in most of the welding operations outlined in this chapter:
A. In almost all cases, the welder should not let the welding rod melt above the weld pool. This usually creates oxidation on the rod end, resulting in a poor weld.

B. To make the strongest weld, the welder must keep the metal to be welded at the proper heat for fusion to take place.

C. Once the pool of molten metal is formed, the end of the rod should be added to the puddle. After that, the rod must not be removed from the flame's heat envelope.

D. As in all previous welding operations, the welder must melt the parent metal on both sides of the preparation and on the full length of the weld.

E. The filler rod should be melted by either the flame or molten puddle. Then, the filler rod will fill the joint by being carried along evenly by flame pressure.

F. The filler rod should never be added to the parent metal until the parent metal is at the temperature to accept the rod for complete fusion.

G. Preheat and postheat should be considered only if recommended in any of the welding methods.

12

Oxygen/Gas Flame Cutting

Flame cutting is a process whereby an oxygen-and-gas flame is used to cut metal by removing some of the metal. A common saw does the same thing, and the small amount of metal the saw actually removes is referred to as *chips* or *dust*. Flame cutting is often used in steel fabrication, construction, and demolition work.

The flame-cutting process may be *manual,* where the welder is required to use a hand-operated cutting torch, as in Fig. 12-1. Or, the torch may be machine driven as is the *radiagraph* machine shown in Fig. 12-2. Either way, a very hot flame is used to cut the metal, as in Fig. 12-3.

Chemical Reaction

The process of flame cutting metals containing iron (ferrous metals) is a *chemical reaction*. The reaction works because of the fact that oxygen has a chemical attraction to ferrous metals when they are heated above the *kindling* temperature. (The kindling temperature is the temperature needed to make something burn.) Therefore, during the cutting process, the welder will always use more oxygen than the ignition gas being used, which is usually acetylene.

Red-hot iron, exposed to the extra oxygen used for cutting, will be removed through a

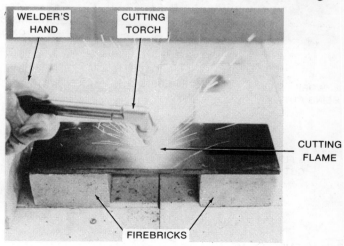

Fig. 12-1. Hand, or manual, flame cutting. (Courtesy of Victor Equipment Co.)

WELDER'S HAND

CUTTING TORCH

CUTTING FLAME

FIREBRICKS

chemical reaction known as *oxidation*. When iron oxidizes, it simply burns up, leaving ashes known as *slag*. Also, the energy of the cutting oxygen stream aids in removing the metal by washing away as much as ⅓ of the total molten metal, still in its metallic form. This can be seen in Fig. 12-4.

In some cases, flame cutting may be done while using a gas *other* than acetylene for heating. Propane, natural gas, and *Mapp* gas are sometimes used in place of acetylene. Construction companies and welding shops may use these other gases for greater safety, more convenience, and better economy. The

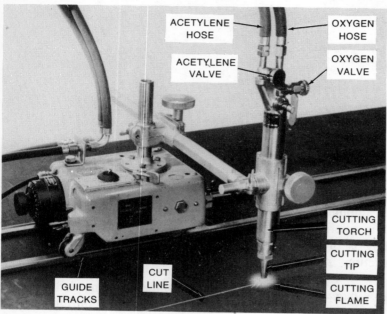

Fig. 12-2. Machine-driven cutting torch, a radiagraph. (Courtesy of Airco Welding Products Division.)

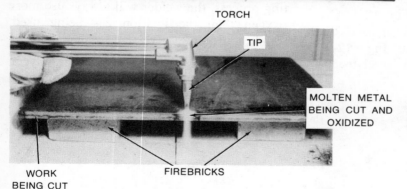

Fig. 12-3. Simple flame cutting process. (Courtesy of Victor Equipment Co.)

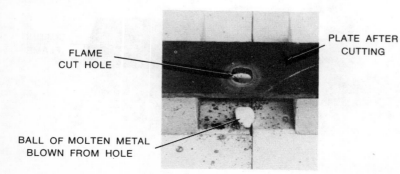

Fig. 12-4. A small ball of molten metal can be seen below the flame-cut hole. (Courtesy of Victor Equipment Co.)

welding temperatures of these gases, under ideal conditions, are shown in Table 12-1.

Table 12-1. Heat Value and Flame Temperature of Fuel Gases.

Gas	Heat Value (Btu per cu. ft.)	Flame Temperature with Oxygen (Degrees F)
Acetylene	1433	6300
Butane	2999	5300
Mapp®	2406	6000
Methane	914	5000
Natural Gas	1200	4600
Propane	2309	5300

Mapp gas (Fig. 12-5) is now being used extensively in many shops in place of acetylene, especially for cutting.

Information about using *Mapp* gas for cutting is outlined later in the chapter.

Regardless of the fuel gas used to preheat the metal, the oxygen creates the chemical reaction with the red-hot metal. In fact, the oxygen does three things in oxygen-gas flame cutting:

1. It combines with the fuel gas to provide a very hot preheat flame.
2. It combines with the iron in the ferrous metal to oxidize the iron. Oxidized iron, like oxidized anything, is burned up.

Fig. 12-5. Medium size Mapp® gas cylinder. (Courtesy of Airco Welding Products Division.)

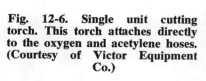

Fig. 12-6. Single unit cutting torch. This torch attaches directly to the oxygen and acetylene hoses. (Courtesy of Victor Equipment Co.)

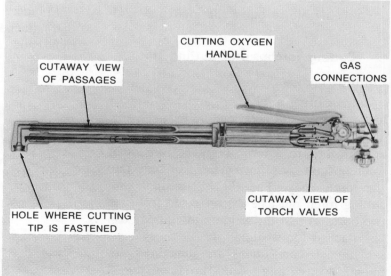

CUTTING OXYGEN HANDLE

GAS CONNECTIONS

CUTAWAY VIEW OF PASSAGES

CUTAWAY VIEW OF TORCH VALVES

HOLE WHERE CUTTING TIP IS FASTENED

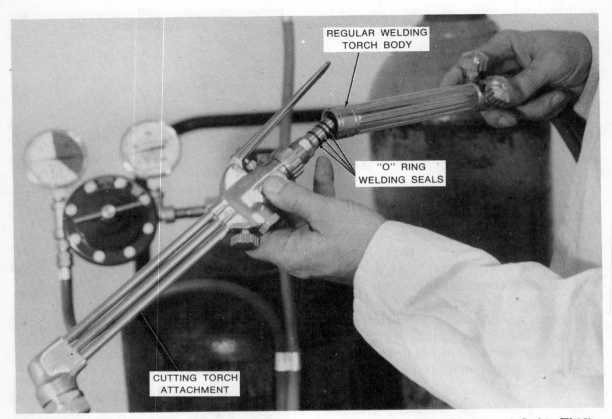

Fig. 12-7. Installing a cutting torch attachment to a regular torch body. (Courtesy of Smith Welding Equipment.)

3. It forces the molten metal out of the preheated area by blowing the metal out. Some of the metal is removed by oxidizing; the rest of it is removed by being blown out while molten.

Cutting Torches and Cutting Tips

Cutting Torches

Some cutting *torches* are attached directly to the oxygen and acetylene hoses, as in Fig. 12-6. Other cutting torches, as in Fig. 12-7, are attached to a torch body (as are other welding tips) by screwing them directly into the torch body. The Fig. 12-6 single-unit cutting torch must be disconnected from the two hoses so that the welding torch body can be attached for welding.

Although connections are sometimes different, most cutting torches in cutaway look like the one in Fig. 12-9. Here, the separate passages and connections for the two gases can be seen, as well as the large passage for a heavy flow of cutting oxygen.

Connecting the Hoses—Any cutting torch that attaches directly to the hoses has the right-hand oxygen outlet threads and the left-

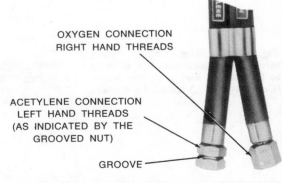

Fig. 12-8. Twin welding hose, as would be directly attached to some cutting torches. (Courtesy of Airco Welding Products Division.)

hand acetylene outlet threads that are the industry standard. As shown by the groove in the nut, the left-hand threads are on the acetylene side of the twin hose in Fig. 12-8. The reverse threads make it impossible to connnect the hose ends to the wrong outlets.

Tightening—The type of torch that attaches to the body of the welding torch, like nozzles or tips, is usually *hand tightened* only. By hand tightening only, the threads cannot become crossed or destroyed. A wrench should not be used on nozzles or cutting attachments because the seals on the end of the tips could be damaged by overtightening. See Fig. 12-10.

Torch Tips

Tips can be mounted into either torch at the torch *head* (Fig. 12-11). The size of the tip will again depend on the thickness of the material to be cut. For this reason, the welder should refer to tip selection charts for choosing a tip. See Fig. 12-12.

Cutting tips may be removable, such as those in Fig. 12-13. They have different numbers of preheating holes. If a welder has to use a tip with a *four hole* preheating system, he will have to turn the holes for the type of cut he intends to make before tightening the tip in the torch head. This will produce a smooth flame-cut surface, as outlined below.

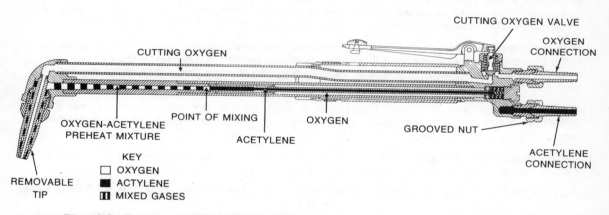

Fig. 12-9. Cutaway of common cutting torch. (Courtesy of Union Carbide Canada Ltd.)

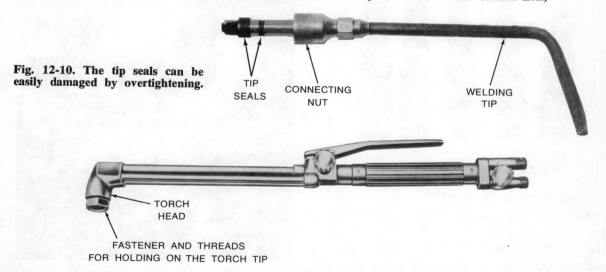

Fig. 12-10. The tip seals can be easily damaged by overtightening.

Fig. 12-11. Single unit welding torch with the torch head ready for a tip. (Courtesy of Airco Welding Products Division.)

RECOMMENDED RANGE OF CUTTING TIP SIZES FOR: (left-side bracket categories) 3 HOSE MACHINE CUTTING TORCH; HEAVY DUTY AND 2 HOSE MACHINE CUTTING TORCHES; "STANDARD" CUTTING TORCHES; "C" CUTTING TORCHES; "STANDARD" CUTTING ATTACHMENTS; "C" CUTTING ATTACHMENTS; 550 CUTTING ATTACHMENTS

THICKNESS OF STEEL INCHES	TIP SIZE NUMBER Small	Medium	Large	OXYGEN PRESSURE P.S.I. Min.	Max.	OXYGEN CONSUMPTION C.F.H. Min.	Max.	ACETYLENE PRESSURE P.S.I.	ACETYLENE CONSUMPTION C.F.H. Min.	Max.	CUTTING SPEED INCHES PER MIN. Min.	Max.
1/64 to 1/8		000		5	20	30	60	5	6	11	20	35
1/8	000	00	0	5	25	50	100	5	7	15	20	35
1/4	00	0	1	7	25	60	110	5	9	18	15	30
3/8	00	0	1	7	30	60	120	5	10	20	15	25
1/2	0	1	2	10	35	65	130	5	12	25	12	25
5/8	0	1	2	13	38	80	150	5	13	27	12	22
3/4	1	2	3	15	40	100	160	5	15	28	12	20
1	1	2	3	20	45	120	180	6	17	30	10	20
1 1/2	2	3	4	25	50	150	225	6	18	35	8.0	17
2	2	3	4	30	55	175	275	7	20	45	6:0	15
3	3	3	5	35	60	225	375	7	22	47	5.0	12
4	3	4	5	40	65	325	475	8	25	50	4.0	10
5	4	5	6	45	70	375	550	8	27	58	4.0	8.0
6	4	5	6	45	75	425	650	9	30	6C	3.0	6.5
8	5	6	7	50	80	625	750	9	33	63	2.5	5.0
10	6	7	8	55	85	625	900	10	35	65	2.0	4.0
12	6	7	8	60	90	725	1050	10	37	68	2.0	3.5
14	7	8	9	55	100	825	1200	11	40	70	2.0	3.0
16	7	8	9	60	110	950	1300	12	42	73	1.5	3.0
18	8	9	10	65	120	1050	1450	12	45	75	1.5	3.0
20	9	10	12	80	130	1150	1550	13	50	80	1.5	3.0
24	12	14		90	140	1350	1750	14	50	80	1.5	3.0
28	14	16		100	150	1650	2400	15	60	85	1.0	2.5

SELECTION OF TIP SIZE: This table is intended to be used as a guide for selection of the proper size cutting tip and for approximate operational data. Under the heading "Tip Size" there are generally three tips listed that can be used for a particular thickness of steel. The "Medium" size tip will give the most economical results when used by the average operator.

SPEED AND ECONOMY FACTORS: Maximum cutting speeds and economy are affected by a number of variable factors; the highest speed and lowest gas consumptions will be made by experienced operators on good, clean metal and long cuts, using the correct tip and operating pressures. Dirty metal, short cuts, poor material, wrong tip size and operating pressures will result in the lowest speed and cutting economy.

HEAVY CUTTING SUGGESTIONS: When cutting heavy sections, the oxygen pressures can be reduced from those shown in the table if large I.D. hose (3/8" to 1/2"), large capacity regulators, large capacity three hose torches and an adequate manifolded oxygen supply are used.

PRESSURE CONNECTIONS: The pressures in the above table are at the regulator and approximate when using 25 feet of 1/4" I. D. hose. Add about one pound for each additional 25 feet of hose.

Fig. 12-12. Specification chart for cutting tip sizes and gas recommendations. (Courtesy of Victor Equipment Co.)

Fig. 12-13. Removable cutting tips.

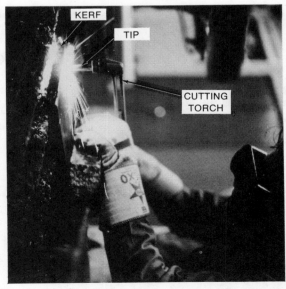

Fig. 12-14. Correct welding position for a straight perpendicular cut. (Courtesy of Airco Welding Products Division.)

Positioning the Preheat Holes—If the welder is required to make a straight perpendicular cut, as in Fig. 12-14, he must position the preheat holes so that two holes follow each other *along* the line of cut. The line of cut is also known as the *kerf*. The desired preheat hole position is shown in Fig. 12-15.

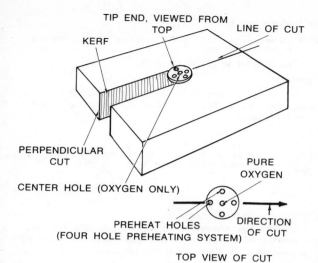

Fig. 12-15. **Positioning the four preheat holes for a straight vertical cut.**

When a bevel is to be cut, the preheat holes in the tip must be changed to follow the line of cut shown in Fig. 12-16. By setting the preheating holes in this position, the top and side of the metal to be cut (Fig. 12-17) will receive good preheating where the metal is to be cut. To understand the preheat hole positions, look at the torch tip end, as in Fig. 12-18.

Where tips have more than four preheating holes, as in Fig. 12-19, positioning is unnecessary. These cutting tips vary in size and shape for different cutting jobs.

Installing the Tip—Once a tip is selected, the welder must make sure that the *seat* (Fig. 12-20) is clean and undamaged before installing it in the torch head. When the tip

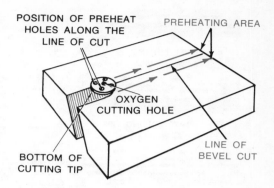

Fig. 12-17. **How the edges of the bevel cut are preheated for cutting.**

has been installed and is seated in the torch head, the nut holding the tip is tightened with a wrench, as shown in Fig. 12-21. Excessive tightening could damage the seat of the tip. This, in turn, would result in poor cutting caused by flame fluctuations.

If necessary, the tip can then be cleaned by using a tip cleaner, as shown in Fig. 12-22. The correct size cleaner is chosen from a chart such as shown in Fig. 12-23.

Operating the Cutting Torch

Fig. 12-24 shows a complete cutting torch setup. As can be seen, this cutting attachment connects directly to a common torch body, converting the common torch to a cutting torch assembly.

A new welder may notice two new controls on the cutting attachment—the preheat oxygen valve and the cutting oxygen valve. The

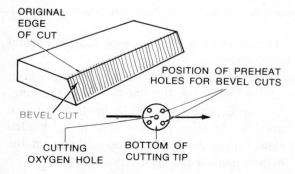

Fig. 12-16. **Positioning the four preheat holes for a bevel cut.**

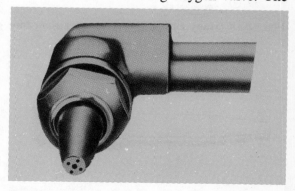

Fig. 12-18. **Cutting tip with four preheat holes. (Courtesy of Union Carbide Canada Ltd.)**

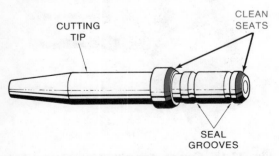

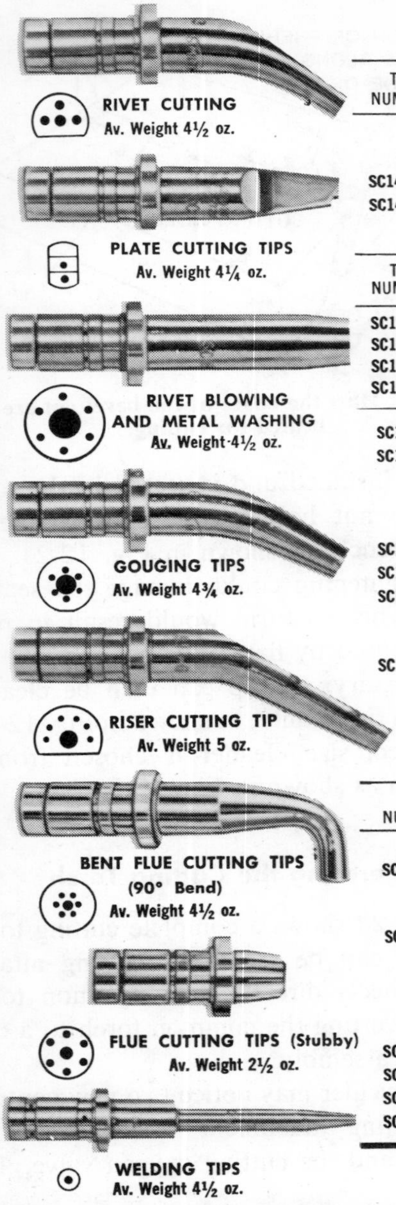

RIVET CUTTING
Av. Weight 4½ oz.

PLATE CUTTING TIPS
Av. Weight 4¼ oz.

RIVET BLOWING AND METAL WASHING
Av. Weight 4½ oz.

GOUGING TIPS
Av. Weight 4¾ oz.

RISER CUTTING TIP
Av. Weight 5 oz.

BENT FLUE CUTTING TIPS (90° Bend)
Av. Weight 4½ oz.

FLUE CUTTING TIPS (Stubby)
Av. Weight 2½ oz.

WELDING TIPS
Av. Weight 4½ oz.

"SC" SPECIAL PURPOSE TIPS

USE IN: Cutting Assemblies: SILVER STAR, PIPEWELDERS' SPECIAL, PIPELINER (MC600 Series & MC409)

Cutting Torches: SILVER STAR, TUF TONY.

TIP NUMBER		
SC14-1 SC14-3		

RIVET CUTTING
For cutting heads of rivets and bolts. Can also use with 180° head cutting torches to cut out boiler tubes. SC14-1 cuts ¾″ rivets. SC14-3 cuts 1½″ rivets. Use with medium pressure acetylene and oxygen.

TIP NUMBER	CUTTING RANGE	
SC17-000	⅛″	
SC17-00	³/₁₆″ -¼″	
SC17-0	⅜″	
SC17-2	¾″-1¼″	

PLATE CUTTING TIPS
"Drag" type step tips cut from 20 gauge to 1¼″ plate with economy. Use with medium pressure acetylene and oxygen.

SC15-1 SC15-2		

RIVET BLOWING AND METAL WASHING
Remove lots of metal fast. Heavy preheat. Use only SC15-1 in cutting Assemblies. SC15-2 is primarily a rivet blowing tip but can be used for metal washing, gouging or veeing and groove cutting. Use with medium pressure acetylene and oxygen.

SC13-1 SC13-3 SC13-5		

GOUGING TIPS
For removing old welds or cracks, for veeing or groove cutting. Heavy preheat 30° angle. For cutting assemblies use only SC13-1 and SC13-3. Use with medium pressure acetylene and oxygen.

SC16-4		

RISER CUTTING TIP
(Not recommended for use in cutting assemblies)
Extra heavy duty tip for removing gates and risers from steel castings. Heavy preheat. Use with medium pressure acetylene and oxygen.

TIP NUMBER	CUTTING RANGE	
SC80-3	Up to 2″	
SC81-3	Up to 2″	

BENT FLUE CUTTING TIPS (90° Bend)
Use one of these tips with a 180° head torch to get in tight areas. Use with medium pressure acetylene and oxygen.

FLUE CUTTING TIPS (Stubby)
Use in 75° or 90° head torches when cutting in areas where clearance is minimum. Use with medium pressure acetylene and oxygen.

	WELDING RANGE	
SCW103	³/₃₂″	
SCW105	⅛″	
SCW107	³/₁₆″	
SCW109	⅜″	

WELDING TIPS
Use in hand cutting torches. A unique SMITH's feature. For SMITH'S users whose primary work is cutting but want to do an occasional welding job. Use with medium pressure acetylene and oxygen.

Fig. 12-19. Many cutting tips have more than four preheat holes. (Courtesy of Smith Welding Equipment.)

CLEAN SEATS

CUTTING TIP

SEAL GROOVES

Fig. 12-20. Cutting tip with seals and seats, showing the area to be cleaned before installation.

preheat oxygen valve looks like a normal torch valve.

The cutting oxygen valve, though, is controlled by a long, flat handle. With the long, flat handle, the valve can be opened and shut quickly. In fact, the valve is spring-loaded *shut,* unless the welder presses down on the handle *against* the spring pressure.

To operate the torch, the regulators must first be adjusted for their new job.

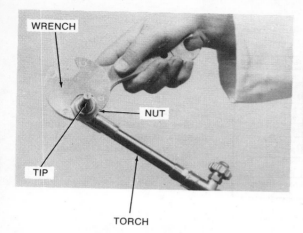

Fig. 12-21. Tightening the torch tip. (Courtesy of Victor Equipment Co.)

Fig. 12-22. Tip cleaners. (Courtesy of Airco Welding Products Division.)

TIP CLEANING DRILL SIZES [AIRCO TIPS]

Tip Size	Cutting Holes All Styles Except 81	FLAME HOLES — STYLES							STYLE 81		Welding Tips All Styles
		124	138	144	164	195	198	199	Cutting	Preheat	
00	69	74	77	66	68	—					77
0	64	71	77	64	66	—		61			74
1	57	69	74	62	63	57	59	59			69
2	55	68	71	60	62	57	57	57			64
3	53	66	69	59	60	56	56	56			57
4	50	65	68	57	59	55	56	56			55
5	47	63	68	57	57	55	55	55			52
6	42	61	65	56	57	54			36	49	49
7	36	59	63	56	56	54			1/8″	49	45
8	1/8″	57	61	55	56	53			26*	49	42
9	27*			54	55	52			21*	49	36
10	19*			54	55	52			16*	49	1/8″
11	12*			56	54						—
12	3*			57	56						—
13	C*			57							—
14	K*			56							—

*Cleaning drills for these sizes are too large for practical use.

INSTRUCTIONS FOR USE

Cleaners are provided for the purpose of removing foreign matter from tip orifices. Insert cleaner in flame end of tip, keeping central and in line with hole in order to avoid scratching or deforming hole. After removal of cleaner blow out holes with a jet of oxygen or air.

Fig. 12-23. Cutting tip cleaning recommendations. (Courtesy of Airco Welding Products Division.)

Adjusting the Regulators

Since a large amount of oxygen is required for cutting, the oxygen torch valve is fully opened. A second, *preheat* oxygen valve controls the flow of oxygen for the neutral preheat flame. See Fig. 12-24.

When the torch oxygen valve is fully opened, no oxygen will escape until either

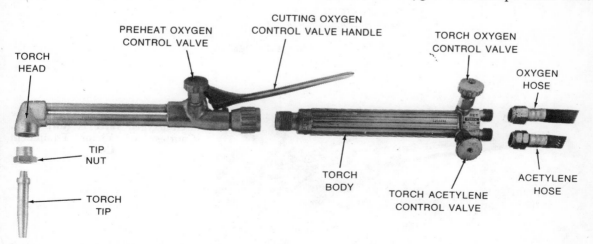

Fig. 12-24. Parts of the complete cutting torch assembly. (Courtesy of Union Carbide Canada Ltd.)

AIRCO "45" High Speed **MACHINE CUTTING TIP**
Approximate Guide
HIGH SPEED CUTS — Straight Line Cutting*

Thickness—Inches	¼	⅜	½	¾	1	1¼	1½	2	2½	3	4	5	6	7	8
Tip Size	0	1	1	2	2	3	3	4	4	6	8	8	8	10	10
Oxygen Pressure Lb. per sq. in.	120	85	100	100	110	85	100	100	110	85	85	110	125	80	100
Acetylene Pressure Lb. per sq. in.	3	3	3	3	3	4	4	4	4	4	4	4	4	4	4
Speed—Inches per Min.	26	24	22	20	18	17	16	13	12	10	9	8	7	6	5½
Cutting Oxygen Cleaning Rod Number ■	0	1	1	2	2	3	3	4	4	6	8	8	8	10	10
Preheat Orifices Cleaning Drill Number ■	70	68	68	68	68	63	63	63	63	61	59	59	59	58	58
Exit End of Cutting Orifices—Cleaning ■	Use ONLY the Cleaning Fibre provided by the tip manufacturer to clean the exit end of tip.														

*For shape cutting and higher quality cutting use slower speeds.
■ See "Cleaning Instructions" furnished.
Not recommended for use with less than 80 P.S.I. or more than 125 P.S.I. cutting oxygen pressure.

Fig. 12-25. Specification chart for cutting tip sizes and gas recommendations. (Courtesy of Airco Welding Products Division.)

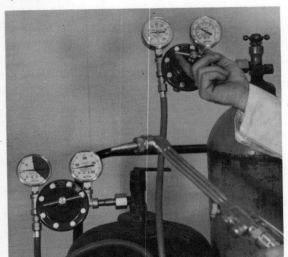

Fig. 12-26. Adjusting the regulators for cutting. (Courtesy of Smith Welding Equipment.)

the *preheat* oxygen valve or the *cutting* oxygen valve (Fig. 12-24) is opened. Opening the torch valve alone will not release oxygen.

The oxygen and acetylene regulators should be set according to the specification chart, as was shown in Fig. 12-12. According to the thickness of the material to be cut, very high oxygen settings may be necessary. Because different manufacturers have different tip sizes, the chart being used must be from the same manufacturer as the tip being used. Comparing Figs. 12-12 and 12-25 shows that different tips are recommended for cutting the same thickness of steel. This is because the tips and charts are from different manufacturers. See Fig. 12-25.

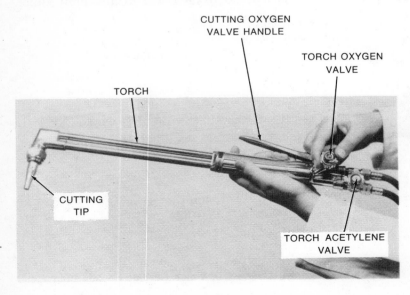

CUTTING OXYGEN
VALVE HANDLE

TORCH OXYGEN
VALVE

TORCH

CUTTING
TIP

TORCH ACETYLENE
VALVE

Fig. 12-27. Being sure that the torch oxygen valve is fully open. (Courtesy of Victor Equipment Co.)

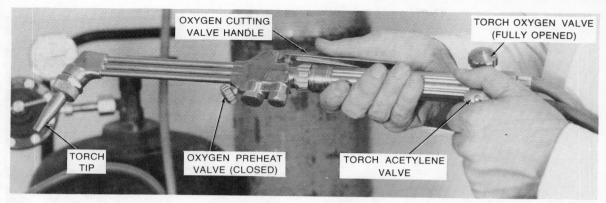

Fig. 12-28. Opening the acetylene valve to light the preheat flame. (Courtesy of Smith Welding Equipment.)

Once the correct chart has been found, the regulators can then be set, as in Fig. 12-26. In Fig. 12-26, the acetylene regulator is already adjusted for 6 psi, and the oxygen is being set to a much higher 35 psi needed for cutting.

Lighting the Preheat Flame

After setting the regulators, the *preheat flame* can be ignited. The first step is to be sure that the torch oxygen valve is fully *open*, as in Fig. 12-27. The oxygen preheat valve, identified in Figs. 12-24 and 12-28, should be fully closed. The cutting oxygen lever can then be pressed down, as in Fig. 12-28, to purge the large center orifice of the cutting tip. The acetylene (or other ignition gas) valve is then opened about 1/6 turn. See Fig. 12-28. The gas coming out of the preheat holes of the tip can then be ignited with a *flint lighter* (Fig. 12-29). A flint lighter is the only safe and recommended method of igniting welding gases. Match or cigarette lighter flames are unsafe for igniting welding gases. A flint lighter is a cheap investment for safe welding.

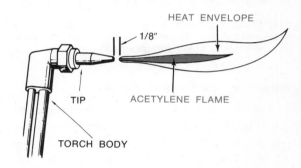

Fig. 12-30. Initial pure acetylene flame adjustment.

Fig. 12-31. Bringing the flame back to the tip by reducing the acetylene. (Courtesy of Smith Welding Equipment.)

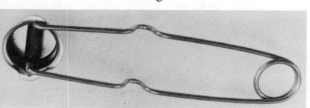

Fig. 12-29. Flint lighter. (Courtesy of Airco Welding Products Division.)

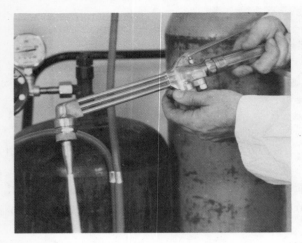

Fig. 12-32. Adjusting the preheat oxygen valve for a neutral preheat flame. (Courtesy of Smith Welding Equipment.)

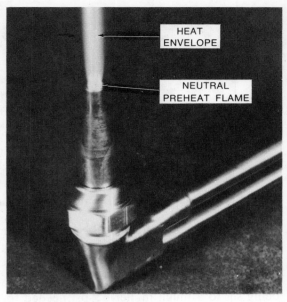

Fig. 12-33. Neutral preheat flame. (Courtesy of Victor Equipment Co.)

Adjusting the Preheat Flame

The flame pressure is then increased by opening the acetylene valve until the flame jumps about ⅛″ from the tip end, as in Fig. 12-30. The welder should then return the flame to the tip by reducing the pressure slightly, as in Fig. 12-31.

Now the preheat oxygen can be gradually released by opening the preheat oxygen valve until a neutral flame is established, as in Fig. 12-32. Then, each preheat hole will have a neutral flame, Fig. 12-33.

After adjusting the preheat flame through the preheat orifices, the cutting oxygen lever is again depressed, as in Fig. 12-28. This will show the welder whether or not the cutting passage (Fig. 12-34) is restricted. If the oxygen is not restricted by dirt or slag in the center hole, the oxygen will make a dark path through the center of the flame and change the neutral flame to a slightly carburizing flame. *While the oxygen cutting lever is de-*

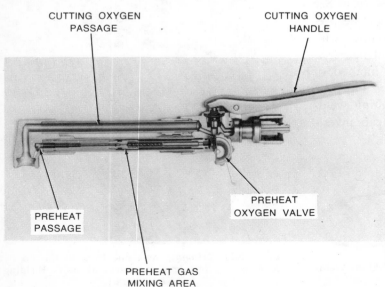

Fig. 12-34. Torch cutaway, showing the cutting oxygen passage. (Courtesy of Victor Equipment Co.)

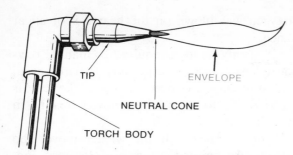

Fig. 12-35. Neutral flame, adjusted with the oxygen cutting valve open.

TIP
ENVELOPE
NEUTRAL CONE
TORCH BODY

pressed, the welder should readjust the preheat oxygen valve until the preheat flame becomes neutral again, as in Fig. 12-35. The welder can then close the cutting oxygen valve. The torch is now set up and adjusted, ready to cut the metal into the desired shape.

Caution

Before actually doing any flame cutting, *be sure that the hoses are not under the metal being removed!* Hot metal and slag from heavy plate could not only damage the hose, but could cause a serious fire or even explosion if the hoses were penetrated by

Fig. 12-36. Easily-handled fire extinguisher. To be kept near the cutting area.

sharp or hot metal falling from the cutting table.

Flammable material should be removed from the area where cutting is to be done. Nevertheless, a *fire extinguisher,* such as the one shown in Fig. 12-36, should be readily available in case of fire.

The welder must always turn *off* the gases at the cylinder valves if a fire should break out, since the gases would support combustion and make the fire extinguisher useless. (In a piping system, the gases should be shut off at the main station valve.)

Of course, good oxy-acetylene welding *goggles* (Fig. 12-37) must be worn at all

Fig. 12-37. Oxy-acetylene welding goggles. (Courtesy of Safety Supply Co.)

times while cutting. The goggles should be darker than for normal welding, because the oxidized metal burns very brightly when it is cut. A *hard hat* (Fig. 12-38) should be worn when cutting has to be done overhead, or in any position where the hot metal and slag might fall or bounce on the welder's head.

Fig. 12-38. Hard hat. (Courtesy of Safety Supply Co.)

Flame-Cutting Procedure

When the preheat flame is properly adjusted, and the welder has taken in all the safety precautions, he is set up to do actual cutting. As a new welder starts cutting, he may be surprised at how fast the process takes place when metal is cut. Of course, practice will build confidence.

Preheating

By using the neutral preheat flame, the welder heats a spot on the metal to a cherry red color, where he intends to start the cut. The preheat flames are held just above the metal's surface, as in Fig. 12-39. This not only protects the preheat holes against metal blow-back from the pure cutting oxygen, but it also keeps the torch end away from the heat reflected by the metal's surface.

Fig. 12-39. Preheating the metal. (Courtesy of Victor Equipment Co.)

Straight Cutting

Once the preheated spot is cherry red in color, the cutting oxygen lever can be slowly depressed. This allows the pure cutting oxygen to attack the metal, removing the metal by flame pressure and oxidation. By allowing the cutting oxygen to flow out *slowly,* the spot won't be cooled by the fresh oxygen stream. This cooling could happen if a large gush of

oxygen was blown on the heated spot all at once.

A steady hand is required in manual oxyacetylene cutting, especially where even, straight cuts are needed. Once the cut is started, the torch must be moved steadily in the direction of the cut.

Cutting Speed—If the welder moves too slowly, the finished cut groove (the *kerf,* Fig. 12-40) will close up behind the torch with molten metal. This will give a poorly fused welding effect.

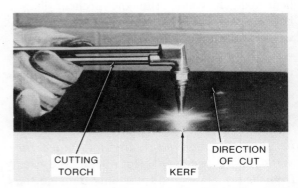

CUTTING TORCH

KERF

DIRECTION OF CUT

Fig. 12-40. The kerf is the cut made with a cutting torch.

If the welder moves too fast, the metal at the cutting point will not be properly preheated, and probably will not oxidize. When this happens, the welder must preheat a new starting point at the edge of the kerf where cutting stopped. Then, the cutting action can again be started when the preheated area becomes cherry red. See Fig. 12-41.

Marking the Cutting Line—Whether cutting along a straight line or a curved line, the welder should mark the line of cut with *soapstone* (a type of talc used for marking

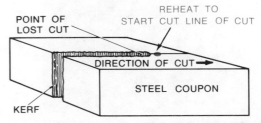

POINT OF LOST CUT

REHEAT TO START CUT   LINE OF CUT

DIRECTION OF CUT →

STEEL COUPON

KERF

Fig. 12-41. Reheating the kerf where the cut was "lost."

steel). If this is not possible, he could use a center punch to mark the line of cut every ¼″ to ⅜″, as shown in Fig. 12-42.

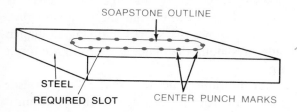

SOAPSTONE OUTLINE

STEEL

REQUIRED SLOT   CENTER PUNCH MARKS

Fig. 12-42. A slot marked with a soapstone and center punch around the line of cut.

find that this cutting technique gives an even, smooth, and straight cutting edge.

Cutting Thin Metal—If the torch head is inclined even as much as 15° to the horizontal and is moved in the direction of the cut, material thinner than ¼″ can be cut without leaving large holes along the kerf, Fig. 12-44.

Bevel Cutting

When *bevel* cutting is required, the torch head must be inclined in a position to allow

Fig. 12-43. Correct position for cutting metal over ¼″ thick. (Courtesy of Victor Equipment Co.)

With the preheat flame just above the surface of the metal to be cut, and the head held at right angles to the work (Fig. 12-43), the welder can work along the line of cut at the proper speed. With practice, many welders

the cutting oxygen to act on the preheated metal along the top edge of parts to be welded. This allows the metal to be removed, forming the required bevel angle, such as the one shown in Fig. 12-45. The tip should be

Fig. 12-44. Angling the cutting torch to cut thinner metals. (Courtesy of Victor Equipment Co.)

kept high enough so that the tip outlet holes are away from the splashing effect of the molten metal being struck by the cutting oxygen. Bevel cuts in production work are

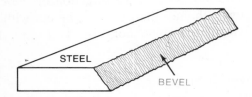

Fig. 12-45. Bevel cut made with oxy-acetylene cutting.

usually done by a *radiagraph,* such as the one being adjusted in Fig. 12-46. This produces a more accurate cut with less gas.

Fig. 12-46. Adjusting a radiagraph for a smooth bevel cut.

Piercing

When a hole is cut in metal (*piercing*), the preheat flame is held so that it just clears the

point where the piercing is to be done. When the area below the cutting tip is a bright cherry red, the oxygen cutting lever is pressed very slowly and the torch head is tipped slightly away from the perpendicular. Tipping allows the sparks and slag to glance sideways along the plate, away from the torch tip.

At the same time, the welder slowly raises the torch from the surface of the plate until a hole is pierced. The hole, in turn, will allow the sparks to drop through, away from the tip; the torch can then be lowered to its normal cutting position. After the torch is lowered to its normal cutting position, it can be moved around to cut the hole to the desired size. With practice, either perpendicular or bevel sides can be cut as required.

If the plate cannot be completely pierced by this technique, either more oxygen pressure or a larger tip size is needed.

Slot Cutting

Holes or shapes in the middle of plates, as in Fig. 12-47, can be cut out by piercing a hole in the center of the area to be removed. Then, using a straight line cut from the pierced hole to the marked cut line, the desired portion can be removed.

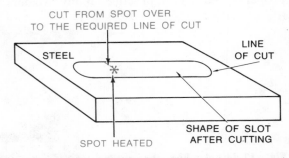

Fig. 12-47. Procedure for cutting a hole or slot in steel plate.

Once any cutting is started, the torch must be moved across the metal fast enough to produce a continuous cutting action. The cutting may either be done manually (by hand) or automatically, using the production radiagraph.

Hand and Automatic Cutting Methods

Through practice, many welders obtain satisfactory cuts with the *manual* method. *Automatic* torches are set the same for preheating as are manual torches. However, a motor drives the automatic torch along the line of cut at a speed easily regulated by a variable-control mechanism.

When the starting point is cherry red, the cutting oxygen lever is engaged for a continuous flow of oxygen. Then, the torch travel

their cutting arm for circle and shape cutting. Radius rods are also available on automatic cutting torches for circle cutting. Multiple cutting heads are commonly used on automatic cutting machines, especially where many pieces of the same shape are required.

Flame Cutting With *Mapp* Gas

Until recently, the major fuel gas in common use with oxygen has been acetylene. Today, a new fuel gas is available. *Stabilized*

Fig. 12-48. A three head radiagraph, used to make three oxyacetylene cuts at the same time. (Courtesy of Airco Welding Products Division.)

running mechanism is engaged at a speed necessary for a smooth kerf edge on the thickness of material being cut.

The torch head on most automatic cutting torches can be tilted to several angles and then fastened for making different bevel cuts. Using these machines, many different and complex cuts can be made evenly. Often, several can be made at one time, as in Fig. 12-48.

Using manual cutting torches, some welders can compete with automatic cutting equipment for quality cuts in straight and simple designs. These welders often use a straight-edge *guide* for line cutting and "tip-to-work" distance gauging. Others may use a *radius rod* or other simple guide to steady

Fig. 12-49. Small, portable Mapp® gas cylinder. (Courtesy of Airco Welding Products Division.)

methylacetylene-propadiene or *Mapp* gas, as it is known, offers greater safety in handling and transporting, with a versatility that many welders feel is equal to acetylene. *Mapp* gas was originally developed by the Dow Chemical Company. *Mapp* gas is available in light-weight portable cylinders, as in Fig. 12-49, as well as larger conventional cylinders (Fig. 12-50).

Fig. 12-50. A large, stationary Mapp® gas cylinder. This type of cylinder is often used with commercial installations. (Courtesy of Airco Welding Products Division.)

Natural gas and propane, although not as versatile as acetylene, have been used in place of acetylene for cutting and heating. Although they are low in cost and have other advantages, they are inferior to acetylene for many applications, and cannot be used for many welding purposes. Adding stabilizers to mixtures of methylacetylene-propadiene has made them safe while maintaining their good heating characteristics. As *Mapp* gas becomes more widely available, propane and natural gas will probably decline in use for heating and cutting.

The following information will acquaint the welder with the cutting techniques used in oxygen-and-*Mapp* gas cutting. The comparison between *Mapp* and other industrial fuels is shown in Fig. 12-51.

	MAPP Gas	Acetylene	Natural Gas	Propane
SAFETY				
Shock sensitivity	Stable	Unstable	Stable	Stable
Explosive limits in oxygen, %	2.5-60	3.0-93	5.0-59	2.4-57
Explosive limits in air, %	3.4-10.8	2.5-80	5.3-14	2.3-95
Maximum allowable regulator pressure, psi	Cylinder (94 psig at 70°F)	15	Line	Cylinder
Burning velocity in oxygen, ft/sec	15.4	22.7	15.2	12.2
Tendency to backfire	Slight	Considerable	Slight	Slight
Toxicity	Low	Low	Low	Low
Reactions with common materials	Avoid alloys with more than 67% copper	Avoid alloys with more than 67% copper	Few restrictions	Few restrictions
PHYSICAL PROPERTIES				
Specific gravity of liquid (60/60°F)	0.576	—	—	0.507
Pounds per gallon liquid at 60°F	4.80	—	—	4.28
Cubic feet per pound of gas at 60°F	8.85	14.6	23.6	8.66
Specific gravity of gas (air=1) at 60°F	1.48	0.906	0.62	1.52
Vapor pressure at 70°F., psig	94	—	—	120
Boiling range, °F, 760 mm Hg	-36 to -4	-84	-161	-50
Flame temperature in oxygen, °F	5,301	5,589	4,600	4,579
Latent heat of vaporization at 25° C, BTU/lb	227	—	—	184
Total heating value (after vaporization) BTU/lb	21,100	21,500	23,900	21,800

Fig. 12-51. Chart comparing Mapp® gas with other common fuel gases. (Courtesy of Airco Welding Products Division.)

Mapp Gas Flames

Mapp gas is similar to acetylene and other fuel gases in that the welder can, by varying the gas pressures, produce a neutral, a carburizing, or an oxidizing flame.

Lighting and Adjusting the Torch—The cutting torch is lit the same way as for acetylene. When the *Mapp* gas valve is opened slightly and the *Mapp* gas is lit, the preheat oxygen valve can then be gradually opened. This slowly increases the preheat oxygen flow, until the yellow *Mapp* flame becomes blue with an added yellow feather on the end of the preheat cones. This *Mapp* gas flame is known as the *carburizing flame* (Fig. 12-52).

If a little more oxygen is mixed with the *Mapp* gas, the yellow feather will disappear. At this point, the preheat cones will be a dark blue color and more sharply defined. These *neutral flames* (Fig. 12-53) will remain stable even when a small, additional amount of preheat oxygen mixes with the flame gases.

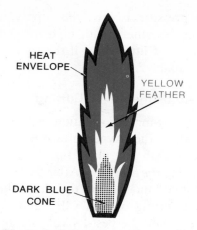

Fig. 12-52. Carburizing Mapp® gas preheat flame.

After adjusting the torch for a neutral flame, the welder can change the dark blue flame to a lighter blue by increasing the flow of preheat oxygen. This is done, as usual, by further opening the preheat oxygen valve on the torch. As the oxygen is increased, the pressure sound of the torch will increase and

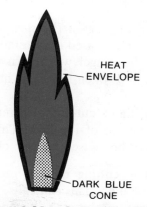

Fig. 12-53. Neutral Mapp® gas preheat flame.

the cones will become shorter. This indicates that the flame is changing from neutral to oxidizing. See Fig. 12-54.

Flame Characteristics—*Mapp* gas preheat flame cones are at least 1½ times longer than acetylene cones produced by the same tip. Therefore, a welder accustomed to oxy-acetylene cutting may accidentally adjust the flame to be similar to an acetylene flame, when using a one-piece *Mapp* tip (Fig. 12-55). If a *Mapp* flame is adjusted to look like acety-

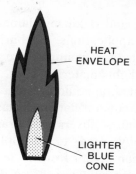

Fig. 12-54. Oxidizing Mapp® gas preheat flame.

ONE PIECE (TIP MIX)

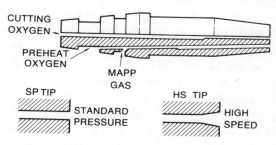

Fig. 12-55. One-piece Mapp® cutting tip.

lene, the flame will be too cold for cutting, since this adjustment will not have enough fuel to produce the needed cutting heat.

Using the Neutral Flame—A neutral flame is used for most cutting jobs, although car-

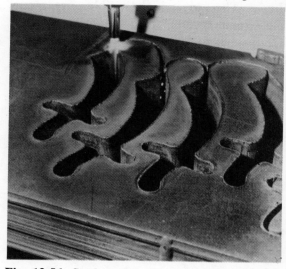

Fig. 12-56. Stack-cutting shapes in light material. Many thin plates are fastened together so they will be cut equally. (Courtesy of Airco Welding Products Division.)

burizing and oxidizing flames are useful for special applications. Carburizing flame adjustments, for example, are used for *stack cutting* light material (Fig. 12-56) where slag formation could create a problem.

Stack cuts on flat stock have square edges free of burrs, slivers, and drag. Stack cutting is usually used to multiply productivity. Stacks can be cut from 2″ to about 6″ thick. If a strong oxidizing flame is used for stack cutting, slag is produced in the kerf; this tends to weld the plates together, making them difficult to separate after cutting.

Using the Oxidizing Flame—A moderately oxidizing flame is used for faster cutting starts and piercing, since the flame is slightly hotter (and the burning velocity is higher) than the neutral flame. Oxidizing flames are also used for beveling joint preparations. At an angle to the plate surface, the tip does not use all of its preheat flames to make the bevel cut. Therefore, an oxidizing flame may be used to get maximum heat from the small portion of the preheat flame that is actually preheating the critical area.

The Stinger—In order for the welder to make good quality cuts using *Mapp* gas with oxygen, he will need what is known as a good *stinger*. The stinger is the visible oxygen cutting stream coming from the center hole of the tip when the cutting oxygen valve is opened, allowing the cutting oxygen to escape (Fig. 12-57).

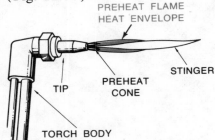

Fig. 12-57. The stinger is pure oxygen, blown out of the center hole of the cutting tip.

In order to have a good stinger, it is necessary to have the right volume and pressure of oxygen. If the oxygen volume is ample but the pressure is low, the supply will be poor. On the other hand, high oxygen pressure will be of little value if the right volume is not available.

Flame Adjustments for Cutting

The following information will help the welder adjust the oxygen pressure to be used for cutting with any given tip.

Adjusting the Stinger—After putting a soft, low-volume flame on the tip, the welder can open the cutting oxygen valve and vary the pressure to find the best-looking stinger. Low cutting-oxygen pressure will give very short stingers, about three inches long, that appear to break up at the end. By increasing the oxygen pressure, the stinger will become longer and more defined at the end. This long, unbroken stinger indicates the correct cutting oxygen pressure for the given *Mapp* gas tip. The stinger will then remain constant over a fairly wide range of cutting oxygen adjustments.

Effects of Oxygen Pressure—If the oxygen pressure is increased *after* establishing a good stinger, the stinger will return to the original short and broken form it had under low pressure.

When flame cutting, slag may try to stick to the bottom of the kerf. This tendency to stick may be caused by any of three things:

1. The cutting oxygen pressure may be too high. This can cause high iron content in the slag, because the oxygen speed through the kerf is great enough to blow out the molten iron before the metal becomes oxidized.
2. The welder's travel (cutting) speed may be too fast. If the cutting speed is too fast, not enough time is allowed to thoroughly oxidize the molten iron. Then, the effect is the same as if the oxygen pressure was too high.
3. The slag may have too much metallic iron in it. This is caused by the iron not being completely oxidized. Either

of the two causes above, or both, can cause the slag to have too much metallic iron in it, causing the slag to stick to the bottom of the kerf.

As with any flame cutting, then, cutting oxygen pressure and cutting speed are very important for successful work.

If the surface of a cut shows a concave or a notching effect on the surface (Fig. 12-58),

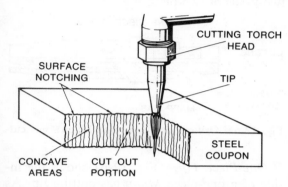

Fig. 12-58. Effects of too high an oxygen cutting pressure.

the welder is using too high an oxygen pressure. If the oxygen pressure is too low, a wide kerf is often produced at the bottom of the cut, together with excessive slag formation. The correct oxygen cutting pressure leaves a clean, straight edge.

Coupling Distance

The distance between the end of the preheat cones and the surface of the material being cut is known as the *coupling distance.*

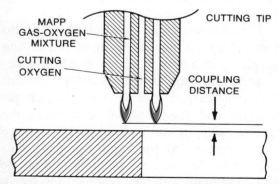

Fig. 12-59. The coupling distance is between the ends of the preheat cones and the work being cut.

Shown in Fig. 12-59, the coupling distance is important to the quality of the cut. When cutting plates up to 3″ thick with *Mapp* gas, the coupling distance must be about ⅛″. When piercing, or for very fast starts, the preheat cones should touch the surface of the plate for faster preheating.

For cutting plates thicker than 6″, the coupling distance should be increased to get more heat from the secondary flame cones. Material over 12″ thick should be cut using a ¾″ to 1¼″ coupling distance.

Torch Angle

The torch angle is sometimes called the *lead angle* (Fig. 12-60). It is the angle be-

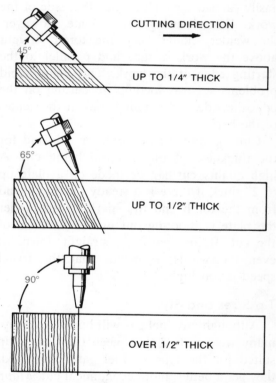

Fig. 12-60. The torch angle used with Mapp® gas cutting on various thicknesses of metal.

tween the *torch center* and the *work piece surface* when the torch is pointed in the direction of the cut.

Light sheet metal, up to ¼″ thick, is cut with a torch angle of about 45° from the horizontal, in the direction of the cut. With the torch held in this position, thin gauge metal is cut faster than would be possible if the torch was held straight above the work surface.

When cutting metal up to ½" thick, the torch angle should be about 65°, as shown in Fig. 12-60. There is little benefit from cutting a plate over ½" thick with an *acute* (under 90°) lead angle. Therefore, a heavy plate should be cut with the torch held in the vertical position. This gives a 90° lead angle, as in Fig. 12-60.

An angled torch cuts faster on thin gauge metal. The intersection of the kerf and the surface is a knife edge, which is easily ignited. Once the plate is burning, the cut is easily carried through to the other side of the work. When cutting *heavy* plate, however, the welder should keep the torch straight above the work surface and parallel to the starting edge of the work. This helps avoid problems such as incomplete cutting on the opposite side and gouging cuts in the center of the kerf.

Cutting speeds vary for each job and for the thickness of the material being cut. A high quality cut can be made on a plate up to 2" thick if there is a steady purring sound from the torch and the spark stream under the plate is shooting out in the direction of the cut. If the sparks go straight down, or even backwards, it means that the travel speed is too high.

Tip Sizes and Styles

Although any fuel gas will burn in any tip, many welders find that when they use a tip suited for the type of fuel gas being used, they get a better, more economical cut. *Mapp* gas will not operate as efficiently with an acetylene tip because the preheat orifices are not large enough for *Mapp* gas preheating. Natural gas tips, available for cutting with natural gas, would be overheated if they were used with *Mapp* gas. This, in turn, could cause dangerous *flashback*.

Tips are made to match the burning speed, port velocity, and other relationships for each type of fuel gas and orifice size. This helps the welder obtain the correct flame shape and heat transfer for the cutting fuel being used.

Cutting tips for *Mapp* gas come in two basic types—one-piece and two-piece. Each of the two basic tips have two styles each, so there are two kinds of one-piece tips and two kinds of two-piece tips.

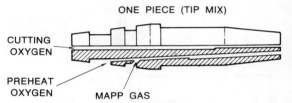

Fig. 12-61. Orifices in a one-piece Mapp® gas cutting tip.

One-Piece Tips—Fig. 12-61 shows the inside of a one-piece *Mapp* gas cutting tip. As usual, the preheat oxygen and *Mapp* (fuel) gas come out of the outer orifices, while the cutting oxygen comes out of the single, large, center orifice. One-piece tips come in two styles—*one-piece/standard pressure* (SP), and *one-piece/high speed* (HS). One-piece tips are also called *tip mix* tips.

The one-piece/standard pressure (SP) tip is used where the fuel gas had previously been acetylene. Here, the welders are accustomed to hand cutting with a one-piece tip, and so they can easily change over to *Mapp* gas cutting. With the SP tip, the cutting oxygen hole (orifice) is straight, as shown in Fig. 12-62A.

The one-piece/high speed (HS) tip is also a one-piece tip, but the front of the cutting oxygen orifice (front center hole) is cone-shaped instead of straight. See Fig. 12-62B. This cone-shaped hole allows a higher gas pressure and speed, while maintaining the required oxygen cutting stinger. HS tips are not recommended for hand (manual) cutting torches, because the welder would not be able to keep up with the higher cutting speed, especially when cutting forms.

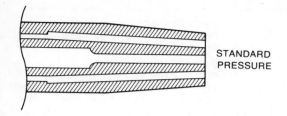

(A) Standard Pressure. SP and FS Tips.

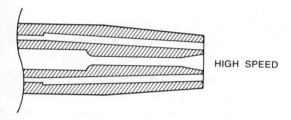

(B) High Speed. HS and FH Tips.

Fig. 12-62. The difference in Mapp® cutting tips is the shape of orifice.

Two-Piece Tips—Fig. 12-63 shows a two-piece *Mapp* gas cutting tip. These tips are also known as *torch mix* cutting tips. They are put together from two parts. Two-piece cutting tips are also available as *standard pressure* (FS) and *high speed* (FH).

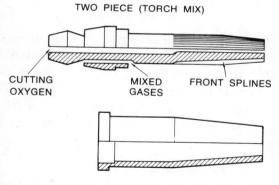

Fig. 12-63. Orifices in a two-piece Mapp® gas cutting tip.

The two-piece/standard pressure (FS) tip requires more skill for hand cutting than does the one-piece SP tip. Since the FS tip has a heavier preheat flame than the one-piece SP tip, it will produce faster starts. The two-piece tip will not stand the abuse that a one-

piece tip will, so the welder must be more careful when cutting with FS tips. The FS tip orifice is shaped like the orifice in Fig. 12-62A.

The two-piece/high speed (FH) tip is generally used only for automatic cutting machines. The oxygen center orifice on the FH tip is the same shape as the orifice in Fig. 12-62B.

Equipment Recommendations—Table 12-2 shows the general recommendations for *Mapp* gas flame cutting with machines. In actual practice, or with manual cutting, the conditions could vary. Therefore, any figures on the table may need to be changed, based on experience in actual cutting operations. The tip sizes shown are the numbered drill sizes of the front center hole. *Mapp* gas cutting tips are compared by the front center hole (cutting oxygen orifice) size, not on the preheat orifice size.

As mentioned previously for welding tips, no industry-wide standards exist for tip sizes. This is also true for cutting tips. For example, #1, #2, etc., tips from different manufacturers usually do not have the same size oxygen orifices. The *Mapp* tips shown in Table 12-2 have a number size that is the same as the number *drill* size of the cutting oxygen orifice. When converting to *Mapp* tips, the welder has to find the number drill size for the tip to be replaced and then use the *Mapp* tip size closest to it.

Oxygen pressures, fuel pressures, and flow rates shown in Table 12-2 are ranges suitable for various tips. If gas pressures are measured at the regulators, rather than at the torch tip, much higher regulator readings than shown in the table may be required to have the correct gas pressure delivered at the tip. This would be especially true if the regulators are attached to long hoses.

Mapp Gas Cutting Techniques

Cutting techniques when using *Mapp* gas are similar to the techniques outlined for oxyacetylene cutting. Cutting can be started on a

Table 12-2. Equipment Recommendations for Using Mapp® Gas as a Cutting Gas Fuel.

Types SP (1 piece) & FS (2 piece) Standard-Speed Cutting Tips

Material Thickness (in.)	Tip Size No.	Hose I.D. Size (in.)	MAPP Pressure (psig)	Oxygen Pressure (psig)
up to 1/8	72	3/16	2—4	20—40
up to 1/4	65	3/16	2—6	30—70
1/4—1	56	1/4	2—8	40—80
1—2	52	1/4	2—10	50—90
2—4	44	1/4	4—12	60—100
4—8	31	5/16	6—15	70—110
8—12	19*	5/16	10—25	80—120

*Type SP not available in size 19.

Type FS (2 piece) Standard-Speed Cutting Tips

Material Thickness (in.)	Tip Size No.	Cutting Oxygen (psig at torch)	Cutting Speed (ipm)
3/16	72	30—40	24—30
1/4	68	35—45	22—28
3/8	65	35—45	21—27
1/2	60	40—50	20—26
3/4	56	40—50	16—21
1	56	40—50	14—19
1 1/4	54	40—60	13—18
1 1/2	54	40—60	12—16
2	52	40—60	10—14
2 1/2	52	40—60	9—13
3	49	40—70	8—11
4	44	40—70	7—10
6	44	40—70	5—8
8	38	40—80	4—6
10	31	40—90	3—5
12	28	40—100	3—5
14	19	40—120	2—4

Type FH (2 piece) High-Speed Cutting Tips

Material Thickness (in.)	Tip Size No.	Cutting Oxygen (psig at torch)	Cutting Speed (ipm)
1/4	65	65—95	25—32
3/8	65	70—100	23—30
1/2	60	75—105	22—29
3/4	56	75—105	20—26
1	56	75—105	18—24
1 1/4	54	75—105	16—22
1 1/2	54	75—105	15—20
2	52	75—105	14—19
2 1/2	52	75—105	12—17
3	49	75—105	10—14
4	44	75—105	9—13
6	44	75—105	7—11
8	38	75—105	6—9

Caution: 1. Guidelines are based on average normal operating conditions.
2. Data are for straight-line cutting using a three-hose torch, perpendicular to the plate in all axes.
3. Preheat pressures given at regulator and based on 25 ft. or less of 3/16-in. I.D. hose supplying a single-torch operation.
4. MAPP Gas: 2-15 psig. Oxygen 10-30 psig (equal pressure torch), or 30-90 psig (injector-type torch).

(Courtesy of Airco Welding Products Division)

plate's edge, or any other point on its surface, by heating the spot to be cut to a cherry red and then opening the cutting oxygen valve.

Piercing—For *piercing* plates, the preheat flames should heat the plate surface while the torch is moved around in circles over the area to be pierced. Once a hot spot is formed, the cutting oxygen is turned on while the torch is slowly raised, as with oxy-acetylene piercing. However, a hole can be cut in thin plates without moving the torch.

When cutting, the welder should protect the oxygen and *Mapp* hoses from fallen metal or hot slag. At the same time, he must point the torch away from his body or flammable material.

Oxygen Lance Cutting

One of the earliest forms of the cutting torch was the *oxygen lance*. By today's standards, this is a crude tool, but it is still often used in demolition work. It is also used in foundry work for cutting off risers or breaking up *spills* (excess material on rough castings).

Oxygen Lance Parts

Basically, the lance consists of a length (or several lengths) of clean steel pipe, 1/8" to 1" in diameter. A T-shaped handle is fastened across the top for guiding and controlling the lance. Near this T-shaped handle, a valve is attached to the end of the pipe. From the valve, a hose leads to a large supply of oxygen for cutting. The setup is shown in Fig. 12-64.

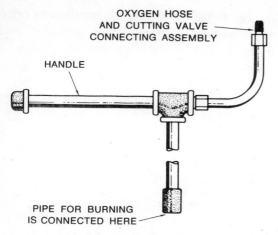

Fig. 12-64. Oxygen lance.

How It Works

This form of flame cutting, like all oxygen-fuel gas flame cutting, is based on the

Usually, a second welder preheats the spot where the cut is to begin, using a welding or cutting torch. Then, when combustion (burning) has started, the open end of the lance is held up to the hot spot and the oxygen is turned on. This provides the large amount of pure oxygen needed to keep the iron burning.

During the lancing process, the lance end also keeps burning, and this helps provide enough heat to keep the cut burning. Although the oxidizing process is *normally* very violent, it can be stepped up even more by using a pipe with heavier wall thickness or by putting iron rods in the lance pipe. These iron rods will provide a greater fuel source for cutting metals that do not oxidize (burn) as easily as does mild steel. When using extra

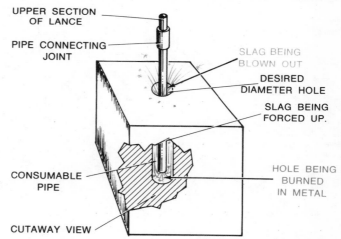

Fig. 12-65. How the oxygen lance produces a larger hole than its own pipe diameter.

rapid burning (oxidation) of iron. When it gets hot enough, iron will burn in oxygen just as other material (such as paper or wood) burns in the *air's* oxygen. Iron, however, needs large quantities of oxygen added to support its combustion. The oxygen lance adds this pure oxygen once the metal is burning.

When the spot has been heated and begins to burn, iron oxides and oxides of other alloying elements mix with the molten iron from the cut to form slag. The oxygen then blows the slag from the kerf while it supports combustion.

rods for fuel, the welder must be sure that he does not restrict the needed flow of oxygen!

The pipe is usually rotated in the burning area to produce a hole that is larger than the original pipe diameter. See Fig. 12-65. This allows slag to be blown out of the hole as the lance burns down into the metal.

Restarting a lance cut in a deep hole can create a problem, since it will be hard to reheat the bottom of the hole. However, the hole *can* be packed with steel wool padding. Then, the top of the padding can be re-lit and burned with the lance, as usual, down to the bottom of the hole.

Using this method, the bottom of the hole can usually be restarted. Whenever possible, the welder should use a lance pipe *long* *enough* to do the job in one continuous operation without stopping. This is easier than trying to restart the hole with steel wool.

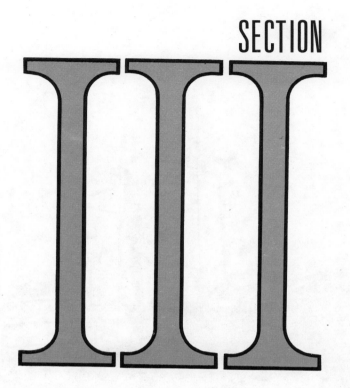

arc welding

One of the most important and widely-used applications of arc welding is in modern ship construction. Many technological advances in ship building would be impossible without arc welding.

13

Principles of Arc Welding

Arc welding is one method of melting two pieces of metal into one solid piece. The edges of the pieces to be joined are heated to a temperature high enough to melt the faces together, known as a *fusion* joining process. On cooling, the melted edges form a single piece of metal. Fig. 13-1 shows two prepared metal coupons before and after welding.

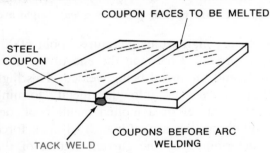

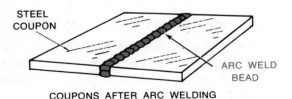

Fig. 13-1. Arc welded coupons, before and after.

The melted faces in Fig. 13-1 could also be mixed with melted filler metal from a welding rod, melted at the same time as were

the faces of the parts. In arc welding, the welding rod is called an *electrode*. Heat from an arc (Fig. 13-2) created by an electric current causes the rod and pieces to melt.

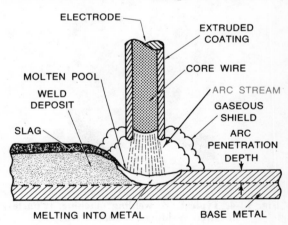

Fig. 13-2. Close-up of the arc welding process at the arc.

The Welding Arc

An electric *arc* is the heat source for many welding processes. The arc is made between the electrode tip and the metal being welded. The arc takes place in an ionized gas, produced when the current is shorted. A *short* is a low-resistance connection between two

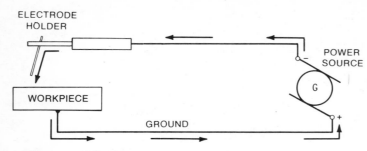

ELECTRODE
HOLDER

POWER
SOURCE

WORKPIECE

GROUND

Fig. 13-3. Basic arc welding circuit.

points in an electric circuit. This short, called the arc, creates the necessary electrical discharge required for fusion by arc welding. The basic arc welding circuit is shown in Fig. 13-3. The very high electrical current needed for arc welding is usually generated in a machine, such as in Fig. 13-4.

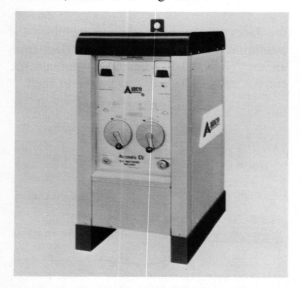

Fig. 13-4. Typical arc welding machine. (Courtesy of Airco Welding Products Division.)

The ionized gas, which contains equal numbers of positive ions and electrons, is known as *plasma*. Positive ions are groups of atoms carrying either positive or negative charges. Electrons are electrical particles with a negative charge. A whole range of complex gas and metal reactions, as well as other metallurgical changes, take place within the arc's envelope of shielding gas. These reactions produce a violent bright arc, as shown in Fig. 13-5.

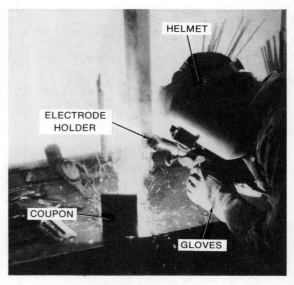

Fig. 13-5. Arc welding produces a very bright arc.

A very high arc temperature, about 6000° F, is produced between the end of the electrode and the metal being welded. This high temperature cuts into the metal, forming either a crater or a molten metal pool, depending on the arc's force. The arc's force is determined by the current setting of the power source (welding machine).

Consumable Electrodes

The core wire of the electrode, as in Fig. 13-2, may also melt. If the electrode melts, it is used up, or *consumed*. It is then called a consumable electrode.

The molten drops leave the electrode due to both electromagnetic (magnetism from electric charges in motion) forces within the molten tip of the rod and the suction effect of the plasma stream. By properly manipulating the electrode, the molten metal from the electrode and parent metal can be moved to where it is wanted on the pieces being welded.

The molten electrode metal, known as *filler metal,* joins the molten metal in the weld crater and solidifies immediately as the arc is drawn away, as in Fig. 13-2. Different oxides and other impurities in the molten metal float to the top and harden after the arc has passed. These are *slag,* which can be chipped off when the weld is cool.

Non-Consumable Electrodes

Electrodes made of tungsten or carbon do not melt away with the arc, and they are said to be *non-consumable.* Tungsten, which melts at almost 6100° F, has the highest melting point of all metals, while carbon melts at about 6512° F. The rods made from these materials are known as *refractory,* because of their heat-resisting quality.

Electrodes with lower melting temperatures than refractory rods melt when an arc is struck. On melting, these rods form the molten drops which are detached by the arc forces. The drops are carried across the arc gap by the plasma jet to the work pieces being welded. These types of electrodes, then, are said to be consumable. Because of the many kinds of arc welding, both types of rods are needed on the market.

Electric Welding Circuits

Arc welding is made possible by either an *alternating current* (AC), or a *direct current* (DC) power source. The current delivered to the welding arc must be controlled, regardless of the type of power source being used. Today, due to the various types of covered electrodes available, alternating current arc stability is comparable to that of direct current.

Alternating Current

In arc welding, the power supply to the arc may be *alternating current.* The direction of

alternating current reverses at regular intervals according to the cycles of current being used. The wave form of the common 60-cycle alternating current is shown in Fig. 13-6.

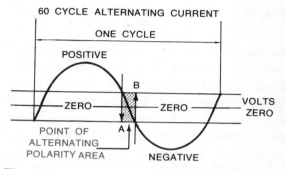

Fig. 13-6. 60 cycle alternating current wave form.

Alternating current passes through 0 twice in every cycle. Each cycle is made up of one positive wave, with the electrons flowing in one direction, and one negative wave, where the electrons flow in the opposite direction. When the frequency of alternating current is increased for Tungsten Inert Gas (TIG) welding, the number of complete cycles in each second are increased. One half-wave of Fig. 13-6 is sometimes considered by welders as DC, straight polarity; the other half is considered as DC, reverse polarity. Polarity is discussed later in the chapter.

In Fig. 13-6, it can be seen that alternating current passes through 0 twice in each cycle. The arc would normally go out at the points of 0 value, except that the *speed* of the change keeps the arc from going out.

Direct Current

In this type of current, the electrons move along the conductor in one direction only. With alternating current, the polarity changes 120 times per second (60 cycle). Direct current, however, flows from one pole continuously. See Fig. 13-7. Many times, alternating current must be changed to direct current. This is done by a *rectifier.*

Rectifier—A rectifier, a device used to convert alternating current to direct (unidi-

Fig. 13-7. Direct current (DC) waveform, fully rectified.

rectional) current, is usually part of today's welding machines. Vacuum- and gas-filled *diodes* are a form of rectifier. The diode is a vacuum tube which permits the stream of electrons to pass in one direction only, from the cathode to the anode. (Cathodes and anodes are the technical names for common electrical poles.)

Semiconductor rectifiers, also known as *solid-state* rectifiers, are used in welding machines rather than tubes. Rectifiers used in welding machines are either *silicon diode* (Fig. 13-8) or *selenium* (Fig. 13-9). The silicon rectifier resembles a spark plug, while the selenium rectifier has a thin layer of the crystalline, non-metallic element selenium. The selenium is deposited on a steel or aluminum plate to dissipate (give off) the heat generated in the rectifier by the current flow.

Electrode Polarity

Polarity is the quality of being attracted to one pole and repelled by the other. Using direct current, polarity can be changed by moving the welding cables on the terminals of the welding machine. See Figs. 13-10 and 13-11. On some machines, a polarity switch on the front of the machine (Fig. 13-12) makes moving the welding cables unnecessary.

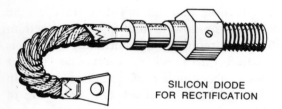

SILICON DIODE
FOR RECTIFICATION

Fig. 13-8. Silicon diode rectifier.

Fig. 13-9. Selenium rectifier. (Courtesy of Miller Electric Manufacturing Co.)

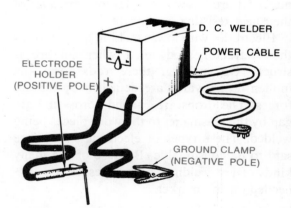

Fig. 13-10. DC welding machine, cables attached for reverse polarity.

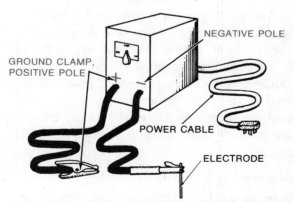

Fig. 13-11. DC welding machine, cables attached for straight polarity.

Since alternating current changes direction (polarity) several times every second, polarity cannot be controlled. As an example, during one cycle in Fig. 13-6, half of the time the current is positive and the other half of the time the current is negative.

Reverse Polarity

The electrode cable for *reverse* polarity is attached to the positive terminal (+), and the ground clamp cable is attached to the negative terminal (−). This will happen, of

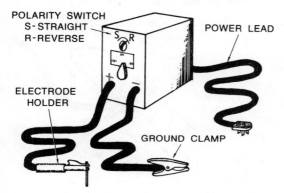

Fig. 13-12. DC welding machine with a switch for changing polarity.

course, only on a direct current welding machine, such as the one in Fig. 13-10. When the electrode holder is used with reverse polarity, more uniform metal transfer is possible with most common metals. In Fig. 13-10, the cables are connected for electrode positive, or reverse polarity.

Straight Polarity

Using *straight* polarity, the cables (leads) carrying the current from the machine to the electrode are reversed from their position shown in Fig. 13-10. Then, they appear on the terminals of the machine as shown in Fig. 13-11. Straight polarity is often referred to as electrode negative. This refers to the cable from the welding machine that is connected to the electrode. If the electrode is thought of as a straight line (—) representing the negative sign (— on the welding machine pole), and if they are connected to each

other, the machine is on straight polarity, as in Fig. 13-11.

Electrode positive would, then, be named reverse polarity. Here, the electrode cable connects to the welding machine's positive (+) poles.

Using Different Polarities

By changing polarity, heat can be transferred to the area where it is most needed. During welding, the work is usually larger than the consumable rod, so the work requires more heat for melting. With coated electrodes, more parent metal penetration is usually possible when the electrode is positive, or on reverse polarity.

On light work pieces that melt easily, straight polarity is desirable. Using straight polarity, burn-through (too deep a penetration on thin metal) is not a problem. About ¾ of the welding heat is liberated (given off) at the positive side of the circuit. Because of this, straight polarity for good penetration should be used with bare or unfluxed rods for any special welding purpose, as in Fig. 13-11. In Fig. 13-13, straight polarity is being used to avoid burn-through on the thin metal. Bare electrodes cannot be used on alternating current machines because the current changes its polarity many times a second.

Arc Blow

A condition known as *arc blow* may be found only when using a direct current machine. This is caused when the current flow in the work sets up magnetic fields, making arc stability and metal transfer difficult. Even a change to another polarity would not remedy the problem. Arc blow may happen to a new welder, so it is a good idea to point out what steps may be taken to control arc blow and what causes it to begin with.

Arc blow is a welding problem created by magnetic fields being set up in the work from the constant flow of current in one direction. These magnetic fields are set up as invisible circles *around* the work, as in Fig. 13-14.

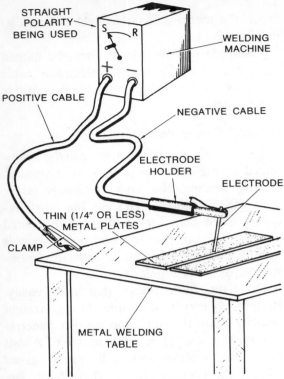

Fig. 13-13. **Using straight polarity to weld thin metal.**

Long periods of continuous welding at high heat will usually create this problem. Molten metal from the rod reacts to these magnetic fields so that the arc can be controlled only after the problem is taken care of. Once the fields are set up, they may repel (push away) the molten arc metal. This causes the metal to be blown to one side, making it very difficult (or impossible) to arc weld under these conditions.

To cure or prevent the problem of arc blow, several steps may be taken, together or individually. Several of the solutions available include the following:

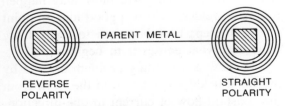

Fig. 13-14. **Magnetic fields are set up around the work when welding with DC current over a long period of time.**

A. Use an AC (alternating current) machine where the current flow reverses evenly. In this manner, the current will be unable to set up magnetic fields.

B. Reduce the DC current. Less current creates less heat.

C. Change the position of the ground clamp on the work. This may help prevent the magnetic fields from being built up due to a certain current pattern. See Fig. 13-15.

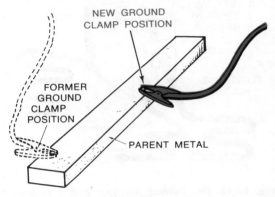

Fig. 13-15. **Changing the position of the ground clamp may help prevent arc blow.**

D. Weld the corners first and then finish the weld using back-step welding. See Fig. 13-16.

E. Wrap the ground cable around the work. This will tend to counteract (work against) the magnetic field. See Fig. 13-17.

F. Use a very short arc and weld toward heavy tack welds. This is being done in Fig. 13-18.

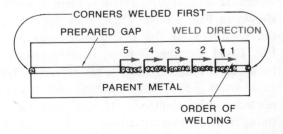

Fig. 13-16. **Back-step welding may help control arc blow.**

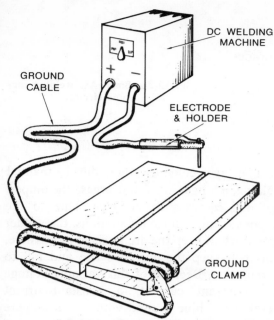

Fig. 13-17. **Wrapping the ground cable around the work may help prevent arc blow.**

Welding Machine Voltage

Alternating and direct current schematic (diagram) circuits are shown in Figs. 13-19

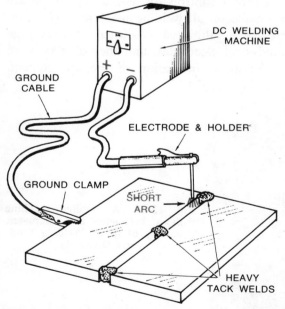

Fig. 13-18. **Arc blow may be cured by using a short arc and welding toward heavy tack welds.**

and 13-20. Fig. 13-19 shows reverse polarity on a direct current (DC) welding machine,

while Fig. 13-20 shows an alternating current (AC) welding machine without any polarity, as usual.

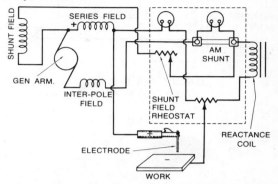

Fig. 13-19. **Basic DC welding machine circuit.**

These circuits represent manually-operated welding machines. These require the welder using the electrode holder to move the electrode for striking and holding a stable arc. During manual arc welding, there is fluctuation (constant small changes) of the amperage and voltage because the length of the arc gap changes as the welder moves his hand.

Voltage Characteristics

The voltage between the electrode and the parent metal forms the *arc stream voltage,* which drives the electrons across the arc through pressure. The arc stream voltage, then, is responsible for good metal flow in the weld. The *volt* is a unit of electrical pressure and can be measured with an instrument known as a voltmeter, Fig. 13-21.

While voltage measures electrical pressure, amperage measures the amount of electrical current actually flowing. An *ampere,* or "amp" for short, is, then, a unit of electrical current.

Amperes are measured with an ammeter. Some welding machines have both a voltmeter to measure volts and an ammeter to measure amps, such as the machine in Fig. 13-22.

The voltage drop across the arc will vary with the length of the arc. A new welder

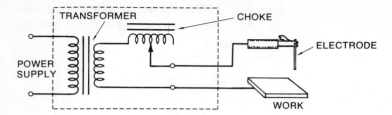

TRANSFORMER
CHOKE
POWER SUPPLY
ELECTRODE
WORK

CIRCUIT FOR THE A.C.
TRANSFORMER WELDER

Fig. 13-20. Basic AC welding machine circuit.

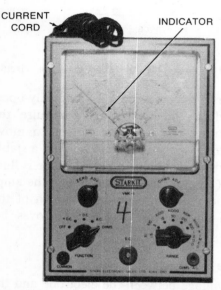

CURRENT CORD
INDICATOR

Fig. 13-21. AC/DC Volt-Ohmmeter. Used to measure voltage in small quantities.

must always remember that, during manual welding, a longer arc will decrease the amperage and increase the voltage. Due to the amperage and voltage fluctuations, the welder should try to keep the arc as short as possible. This will keep the current's rise and fall to a minimum. Because of this variable voltage and constant current, a constant-current welding machine is used with a volt-ampere relationship as shown in Fig. 13-23.

It can also be seen in Fig. 13-23 that the volt-ampere curve drops from the open circuit voltage, representing slight changes in arc current for any variation in what would normally be the open circuit voltage.

Current controls for manual operation are available on all welding machines. The controls are calibrated (adjusted) for average mild steel electrodes, and they give only an approximate indication of the arc current. See Fig. 13-24.

VOLTMETER
AMMETER
CURRENT ADJUSTING CONTROL

Fig. 13-22. A welding machine with a voltmeter and ammeter. (Courtesy of Airco Welding Products Division.)

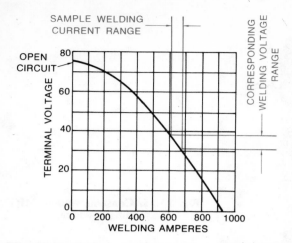

SAMPLE WELDING CURRENT RANGE

CORRESPONDING WELDING VOLTAGE RANGE

OPEN CIRCUIT

Fig. 13-23. With a constant current welding machine, the welding current changes little over the welding voltage range.

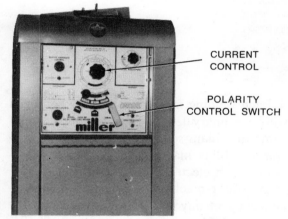

CURRENT CONTROL

POLARITY CONTROL SWITCH

Fig. 13-24. Current controls on an AC/DC welding machine.

Constant Voltage Machines

Automatic welding machines, such as Metal Inert Gas (MIG), are usually constant voltage machines. While welding with these machines, the arc length is kept constant by a motor that feeds in metal filler wire from a spool. See Fig. 13-25. The increase in voltage caused by filler metal burn-off lengthens the arc. This activates the motor speed, increasing the wire feed for a shorter arc. The motor will also slow down the wire filler metal feed when the arc becomes too short.

Electrode Burn-Off

The linear rate of rod consumption is usually expressed in inches per minute and is commonly called the *burn-off* rate. The burn-off rate is affected by electrode polarity, type of flux, and resistance of the wire core, as well as by the current.

Electrode efficiency is measured by the weight of metal deposited in a certain time. The deposit rate is expressed in pounds of metal per hour.

Types of Voltage

Reviewing, *arc voltage* is the voltage between the welding rod (electrode) and work when the arc is struck. Arc voltage does not remain constant, since it rises and falls during welding. The drop in voltage is due to the lead's resistance and the arc fluctuation.

Open Circuit Voltage—When the welding machine is running but no welding is being done, the voltage between the terminals of the welding machine is known as *open circuit voltage*. Since the power of an arc's variations depend on voltage and current changes, adjusting the open circuit voltage can make arc differences possible for various welding jobs.

Arc Voltage—Open circuit voltage is converted into *arc voltage* when the arc is struck. Arc voltage is determined by the type of rod being used and the length of the arc being held. The current will decrease and the voltage will increase when the arc length is increased. If the arc is shortened, the current will increase and the arc voltage will decrease. The open circuit voltage setting will regulate the amount of current change due to changes in the arc length. In Fig. 13-26, the electrode position changes the voltage from open circuit voltage to arc voltage. At position A, the electrode is away from the work, so open circuit voltage can be measured although the machine is on. At position B, an arc has been struck, so arc voltage can be measured.

Welding Accessories

Information about the different welding accessories is presented before the various

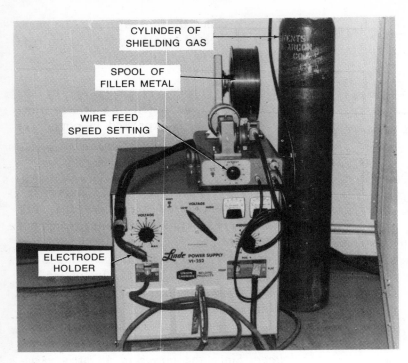

CYLINDER OF
SHIELDING GAS

SPOOL OF
FILLER METAL

WIRE FEED
SPEED SETTING

ELECTRODE
HOLDER

**Fig. 13-25. Constant voltage MIG
welding machine.**

types of welding machines, since the welding machine would be of little value without the accessories. Knowing about the accessories will make it easier to understand welding machines.

Welding accessories include equipment to connect the machine to the work, to hold the electrode, and to make the welder's job safer and easier.

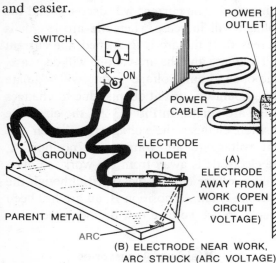

SWITCH

OFF ON

POWER
OUTLET

POWER
CABLE

GROUND

ELECTRODE
HOLDER

PARENT METAL

ARC

(A)
ELECTRODE
AWAY FROM
WORK (OPEN
CIRCUIT
VOLTAGE)

(B) ELECTRODE NEAR WORK,
ARC STRUCK (ARC VOLTAGE)

Fig. 13-26. The electrode position changes the voltage from open circuit voltage to arc voltage when the electrode is moved over to strike the arc.

Each welding machine is usually equipped with a primary lead for connecting the machine to a main source of electricity. A *generator* welding machine has an armature rotating in an electrical field, as in Fig. 13-27. A generator machine can be operated independently of any AC current going into the machine from the primary power lead. See Fig. 13-27. The *transformer* welding machine, however, requires AC current going in to produce power at any time.

Fig. 13-28. Smaller, gasoline-powered welding generator. (Courtesy of Airco Welding Products Division.)

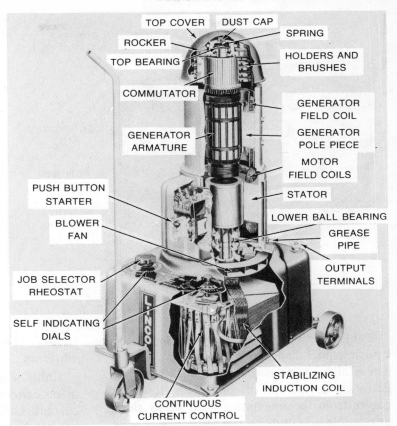

TOP COVER   DUST CAP

ROCKER   SPRING

TOP BEARING   HOLDERS AND BRUSHES

COMMUTATOR

GENERATOR FIELD COIL

GENERATOR ARMATURE   GENERATOR POLE PIECE

MOTOR FIELD COILS

PUSH BUTTON STARTER   STATOR

LOWER BALL BEARING

BLOWER FAN   GREASE PIPE

OUTPUT TERMINALS

JOB SELECTOR RHEOSTAT

SELF INDICATING DIALS

STABILIZING INDUCTION COIL

CONTINUOUS CURRENT CONTROL

Fig. 13-27. Generator welding machine. (Courtesy of Lincoln Electric Co.)

Gasoline, diesel, propane, and even natural gas internal combustion engines can be used to drive welding generators. In Fig. 13-28, a gasoline engine drives the smaller 225/200-amp generator; the large 350-amp generator in Fig. 13-29 is powered by a diesel engine. Larger welding generators are usually powered with diesel engines.

Cables

Large, heavy *cables* carry the current needed for good working heat. These are connected to the positive and negative terminals on the welding machine, as in Fig. 13-30.

Most cables are made of woven, hairlike strands of copper wire covered with insulated rubber. The rubber allows the woven wire to flex, protects it during use under rough conditions, and insulates it from shorts. See Fig. 13-31. Copper connectors fasten the free end of each cable to different kinds of holders, such as in Fig. 13-32.

One of these holders is known as the *electrode holder,* Fig. 13-32A. The other holder clamps or fastens to the workpiece and is known as the *ground clamp,* Fig. 13-32C.

Fig. 13-29. Larger, diesel-powered welding generator. (Courtesy of Airco Welding Products Division.)

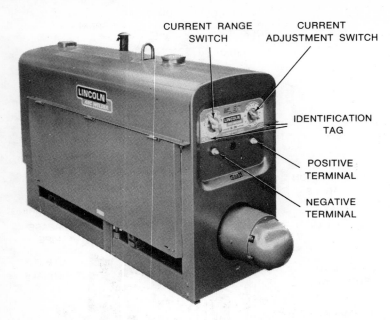

CURRENT RANGE
SWITCH

CURRENT
ADJUSTMENT SWITCH

IDENTIFICATION
TAG

POSITIVE
TERMINAL

NEGATIVE
TERMINAL

Fig. 13-30. The terminals on the welding machine are usually clearly marked, as on this gasoline-powered machine. (Courtesy of Lincoln Electric Co.)

Cable may be spliced or connected easily by using quick-cable connectors, Fig. 13-32B. These allow various lengths of cables to be easily connected and disconnected. However, cables from the machine should be as short as possible, to cut down the resistance of long

Fig. 13-31. Welding cable.

(A) Electrode holder

(B) Quick-cable connector

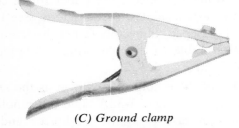

(C) Ground clamp

Fig. 13-32. Welding cable tools.

cables. These would reduce the machine's efficiency.

Cable size must also be selected carefully in order to control the desired current. The current is affected not only by the capacity of the machine, but also by the distance of the work from the machine and the cable size. Table 13-1 gives a recommended cable size for combined lengths of electrode and ground cables for welding machine sizes from 100 amperes to 400 amperes. Cable sizes are given in numbers.

Table 13-1. Welding Cable Number Sizes.

Amps.	Total Cable Length. (Electrode cable length plus ground cable length, in feet.)				
	0-50	50-100	100-150	150-200	200-250
100	8	4	3	2	1
200	2	2	2	1	0
250	3	3	2	1	0
300	0	0	0	00	000
400	00	00	00	000	0000

Ground Clamps

The work on which the welding is being done, usually called the parent metal, forms part of the electrical circuit. Therefore, it is necessary to connect a clamp to the loose end of the cable, so the cable can be attached to the work.

This clamp, known as the *ground clamp,* is available in various shapes or sizes. Two of the most commonly used clamps are shown in Fig. 13-33. Fig. 13-34 shows the details of a simple ground clamp. A connecting wire connects the two halves of the clamp, so that the hinge and strong spring do not have to conduct electricity.

Where both welding cables are the same size and flexibility, either cable can be used

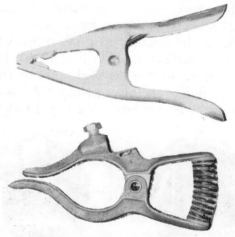

Fig. 13-33. Common ground clamps.

for clamping onto the work. If one cable is less flexible than the other, it is a good idea to attach the ground clamp to the cable with less flexibility. With the electrode holder attached to the more flexible cable, the welder

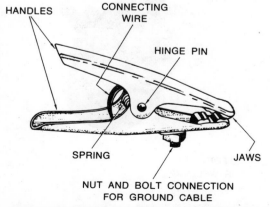

HANDLES

CONNECTING WIRE

HINGE PIN

SPRING

JAWS

NUT AND BOLT CONNECTION FOR GROUND CABLE

Fig. 13-34. Details of a simple ground clamp.

will be able to easily move the electrode and holder.

Electrode Holder

A tool used to hold welding rods, the *electrode holder,* is fastened to the other welding cable, opposite the ground clamp cable. Sometimes, welders refer to the electrode holder as the "stinger end," because of the arc when the rod and work contact.

The holder is built to hold the electrode securely and, at the same time, to offer a quick and simple method of putting in a new rod and/or removing the old rod stub. The uncoated (unfluxed) stub of the consumable rod (electrode) is put in the electrode holder. See Fig. 13-35.

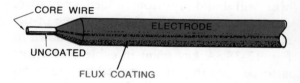

CORE WIRE

ELECTRODE

UNCOATED

FLUX COATING

Fig. 13-35. The bare (uncoated) end of the electrode is put in the electrode holder.

The holder is usually connected to the cable by a copper sleeve tightener inside the body of the holder. The connectors are built to eliminate as much resistance as possible so the holder will stay cool during heavy welding with high amperage. Handles are also insulated and ventilated to reduce heat buildup and any possibility of electrical shock. See Fig. 13-36.

There are several types of holders, many of which are built for a specific type of job. Each common holder used for various weld-

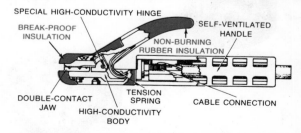

SPECIAL HIGH-CONDUCTIVITY HINGE

BREAK-PROOF INSULATION

SELF-VENTILATED HANDLE

NON-BURNING RUBBER INSULATION

DOUBLE-CONTACT JAW

TENSION SPRING

HIGH-CONDUCTIVITY BODY

CABLE CONNECTION

Fig. 13-36. Cutaway view of a common electrode holder.

ing jobs will be described under special welding procedures as they appear in later chapters. One of the most common, general-purpose holders is shown in Fig. 13-37.

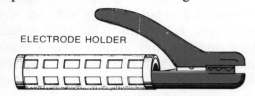

Fig. 13-37. Common electrode holder.

Cable Connectors

Cable connectors are designed so that the cable can be tightly clamped to the welding machine, the ground clamps, or other connectors necessary for extending the cables. If connections are allowed to work loose, heat will build up from added resistance, causing an eventual breakdown. Some of the most common cable connectors are shown in Fig. 13-38. A key point in cable connectors is to be sure that the connections are tight.

Welding Helmets

Fig. 13-39 shows the front and rear of a common *welding helmet*. When arc welding, a welding helmet must be worn to protect the face and head from flying particles and shield the eyes from ultraviolet and infrared

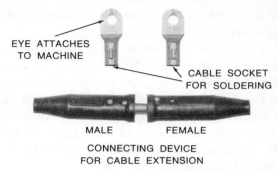

Fig. 13-38. Cable connecting parts.

rays. Helmets are constructed of a light-weight, heat-resistant insulated material. Good helmets have an inner head band which can be fastened to a safety hard hat for added protection, as in Fig. 13-40. Helmets can usually be swiveled upwards from the face due to the pivot point on the head-band of the helmet body. See Fig. 13-41A.

Fig. 13-39. Common welding helmet, front and rear views. (Courtesy of Safety Supply Co.)

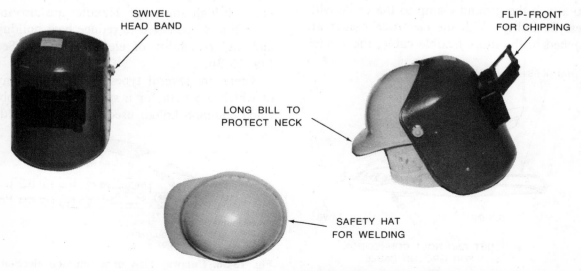

Fig. 13-40. A good quality helmet with an attachment for a hard hat.

Glass plates for eye protection are fastened to the front of the helmet body. These usually consist of an inner rectangular lens of plain glass and a very dark lens toward the outside of the helmet. This dark outer lens is protected from outside weld spatter by two additional plain glass lenses. Gaskets are placed between the glass plates to keep any light rays or dust from leaking into the helmet. By a hinge on the front piece, all but the clear inner glass protecting the eyes can be flipped up. This will allow for quick work inspection and minor chipping. See Fig. 13-41A.

The color and shade number of the filter glass varies for different kinds of welding. Filter glass and cover glass should be cleaned and inspected for cracks every so often. Pitted and dirty cover glasses make it hard to see the weld, which could cause poor workmanship.

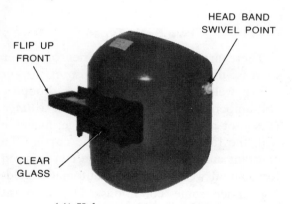

(A) Helmet position for chipping.

(B) Helmet position for welding.

Fig. 13-41. Features of a good arc welding helmet.

Welding Handshields — Welding handshields (Fig. 13-42) are similar in construction to a welding helmet. They are especially suitable for teaching, because the hand grip on the helmet makes protection quick and easy for short bead runs. When using a handshield, however, only one hand is holding the electrode holder; in some welding situations, this would be undesirable and unsafe. For this reason, handshields must only be used for temporary protection.

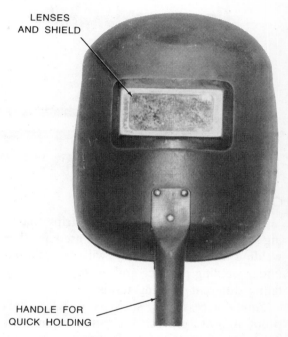

Fig. 13-42. Welding handshield for quick use.

Chipping Hammer

The *chipping hammer* (Fig. 13-43A) is used immediately after the first weld bead is completed. The hammer is used to chip the flux off the filler bead, so that each successive bead will be free of porosity, which could be created by impurities and slag during welding. Carefully cleaning and chipping before laying weld beads will assure the welder of a sound weld. For safety, chipping should always be done with goggles or a welding helmet in place. If the welding helmet is used, it must have a flip-up front for removing only the dark glass.

Wire Brush

A cleaning brush with *wire* bristles, Fig. 13-43B, or a motor-driven wire wheel are important cleaning tools. They are used to brush away the flux and slag particles and any other foreign matter remaining after the beads have been chipped.

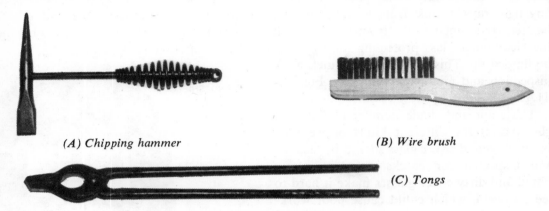

(A) Chipping hammer

(B) Wire brush

(C) Tongs

Fig. 13-43. Welding accessories.

Other Welding Accessories

C clamps, tongs, drills, chill strips, backing strips, and other tools can be used as other welding accessories. They will be covered under welding practice and in the section defining different welding terms.

Safety accessories, such as goggles, fireproof aprons, gloves, and other equipment necessary for the welder's safety, are described in the chapter on welding safety.

SELECTING A WELDING POWER SOURCE[1]

Many different types and sizes of arc welding machines are available today. It is important that welding machine users have enough technical information so that they can wisely select the best machine, one most suited for their particular work.

Arc welding machines can be classified in many different ways—rotating machines, static machines, electric motor-driven ma-

chines, internal combustion engine-driven machines, transformer/rectifier machines, limited input welding machines, conventional, constant voltage welding machines, single operator or multiple-operator machines, and many others. All of these, however, can be generally classified into two basic types.

Two Basic Types

There are two basic categories of power sources: the conventional or *constant current* (CC) welding machine with the drooping volt-ampere curve and the *constant voltage* (CV), or modified constant voltage, machine with the fairly flat characteristic curve. The constant current (CC) machine can be used for manual welding and for automatic welding under some conditions. The constant voltage (CV) machine is used only for continuous electrode wire arc welding, whether operated automatically or semi-automatically.

These machines can be better understood by comparing their characteristic volt-ampere output curves. The curves are made by loading the welding machine with variable resistance and then plotting the voltage at the electrode and work terminals for each ampere output. The graph is plotted with the voltage on the side of the graph and the corresponding amperage on the bottom of the graph.

[1]Courtesy of Hobart Brothers Company

Constant Current (CC) Welding Machines

The conventional or *constant current* (CC) welding machine is used for many kinds of arc welding: *S*hielded (electrode) *M*anual *A*rc *W*elding (SMAW); *T*ungsten *I*nert *G*as (TIG) arc welding; *C*arbon *A*rc *W*elding (CAW); arc gouging; and *S*tud *W*elding (SW). It can also be used for automatic welding with larger-sized electrode wire, but only with a voltage-sensing wire feeder.

The constant current (CC) welding machine produces a volt-ampere output curve as shown in Fig. 13-44. Studying the curve shows that a constant current machine produces maximum output voltage with no load (0 current) and that the output voltage *decreases* as the load *increases*. Under normal welding conditions, the output voltage is from 20 to 40 volts, while the open circuit voltage is from 60 to 80 volts. Constant current machines are available that produce either AC or DC welding power, and some machines can produce both.

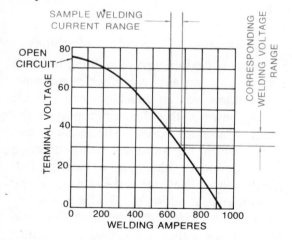

Fig. 13-44. The drooping volt-ampere curve of a constant current welding machine.

When welding with covered electrodes on the constant current (CC) welding machine, the actual arc voltage is controlled largely by the welder himself. This is because the arc voltage is partly determined by the arc length. As the arc *length* is increased, the arc *volt-age* increases. If the arc length is decreased, the arc voltage decreases. By moving the electrode, the welder is able to control the arc length.

Looking at the output curve shows that when the arc *voltage* increases, the welding *current* decreases, or when the arc voltage decreases, the welding current increases. So, without changing the machine setting, the operator can vary the current (welding heat) in the arc by lengthening or shortening the arc.

Constant current (CC) machines can be rotating (generators) or static (transformers or transformer/rectifier) machines. The generator can be powered by a motor for shop use or by an internal combustion engine (gasoline or diesel) for field use. Engine-driven machines can have either water- or air-cooled engines, and many of them provide auxiliary power for emergency lighting, power tools, or other electrical needs.

Constant Voltage (CV) Welding Machines

A *constant voltage* (CV) welding machine is a machine that provides a fairly even voltage to the arc regardless of the arc current. The characteristic (normal) curve of this machine is shown by the volt-ampere curve

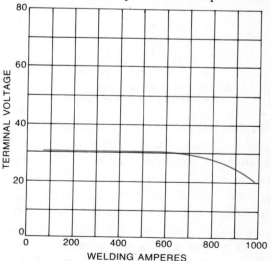

Fig. 13-45. Volt-ampere curve of a constant voltage (CV) welding machine.

in Fig. 13-45. A constant voltage welding machine can *only* be used for *semi-automatic* or *automatic* arc welding using an automatic feed electrode wire. Also, constant voltage machines are made to produce only direct current (DC).

In continuous wire welding, the burn-off rate of a certain size and type of electrode wire is determined by the welding current. In other words, as the welding current increases, the amount of wire burned off also increases. This can be shown on a graph, as in Fig. 13-46. Therefore, if a wire is fed into an arc at a certain speed, it automatically draws from a constant voltage power source a *proportionate* amount of current. The constant voltage welding machine provides the amount of current required by automatically judging the load put on it.

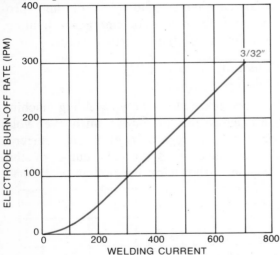

Fig. 13-46. The electrode burn-off rate increases as the welding current increases.

In this way, a very basic automatic welding control can be used. The wire is fed into the arc by a constant feed motor. This feed motor can be adjusted to increase or decrease the wire's rate of feed. Complicated electrical circuits can thus be eliminated. The system, by its design, regulates itself during welding. If the electrode wire is fed in *faster,* the current automatically *increases*. If it is fed in *slower,* the current automatically *decreases*.

The current output of the welding machine is, therefore, set by the speed of the wire feed motor. The voltage of the machine is regulated by an output control on the welding machine itself. In this way, only two controls maintain the correct welding current and voltage when the constant voltage system is used. With this system of a constant speed wire feeder and a constant voltage power source, it is practically impossible to have stubbing (electrode hits the work) or burn-back (electrode is burned too short and the arc is not able to bridge the gap).

The common curves of constant voltage machines have a slight, normal droop. The droop can be increased or the slope made steeper by different methods. Many machines have different tapes or controls for varying the slope of the machine's curve. It is important to choose the slope that is best for the process and type of work being done.

Constant voltage machines can be either generator or transformer/rectifier welding machines. Generator machines may be either motor or engine driven.

Combination CV-CC Welding Machines

The most flexible type of welding machine is what is known as a *combination* machine. Combination welding machines can provide welding power with either a drooping or flat output volt-amp curve. The welder can choose which he wants by using different terminals and/or changing a switch. This type of welding machine is the most universal one available. It allows the welder to use any arc welding process. The combination machine can use either a generator or a transformer/rectifier power source.

Sources of Power

Whether the welding machine is constant current (CC) or constant voltage (CV), the electrical power for the welding arc must come from the welding machine. The welding machine must convert another form of power into the electrical power that the arc needs. This other form of power may be an engine or an AC electrical power line to the work place where the welding machine is.

These "power converters" are generally classified as generator, transformer, or transformer/rectifier types. Any of them may be used to power the multiple-operator welding system discussed later.

Generator Welding Machines

Most dual-control machines are normally *generators,* as shown in Fig. 13-47. On these machines, the slope of the output curve can

Fig. 13-47. Generator welding machine. (Courtesy of Hobart Brothers Company.)

be varied. These are called dual-control machines because they have two controls for the voltage.

The basic control is the range switch. The range switch provides a *coarse* (basic) adjustment of the welding current. After the coarse adjustment is made, a second adjustment for the open circuit, or "no load," voltage is by the *fine* adjustment control knob. This control is also used as a fine current adjustment during welding. Dual-control panels may look like the controls in Fig. 13-48.

By adjusting the fine current while welding, a soft or harsh arc can be made. With the *flatter* curve and its *low* open circuit voltage, a change in arc voltage will produce a greater change in output current. This produces the digging arc that is preferred for pipe welding.

With the *steeper* curve and its *high* open circuit voltage, the same change in arc voltage would produce less of a change in output current. This is a soft, or quiet, arc, useful for welding sheet metal.

In other words, the dual control, constant current, welding machine allows the most flexibility for the welder. These machines can be driven by an electric motor or an internal combustion engine.

The constant current (CC) machine can also be used for automatic welding processes. However, to use a constant current welding machine automatically, the automatic wire feeding tool must allow for changes in arc

Fig. 13-48. Dual welding machine controls for coarse and fine adjustment.

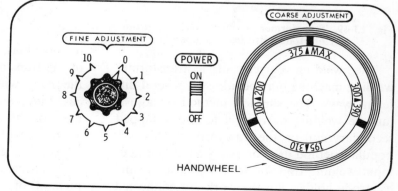

length. This requires complex control circuits that feed back from the arc voltage, called *voltage sensing*. This system, however, is not used for small diameter electrode wire welding.

Transformer Welding Machines

The *transformer* welding machine, Fig. 13-49, is the least expensive, lightest, and smallest of the different types of welding machines. It produces only alternating current (AC), so there is no polarity switch. The transformer welding machine takes power directly from the AC power line and transforms it to the power required for welding. By using various magnetic circuits and inductors, transformer welding machines can provide the volt-ampere characteristics needed for good welding.

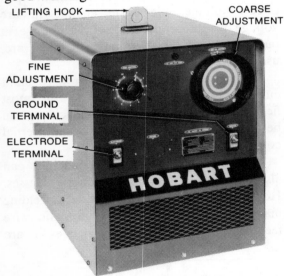

LIFTING HOOK

COARSE ADJUSTMENT

FINE ADJUSTMENT

GROUND TERMINAL

ELECTRODE TERMINAL

Fig. 13-49. Transformer welding machine. (Courtesy of Hobart Brothers Company.)

The welding current of a transformer welding machine may be adjusted several different ways. The simplest method of adjusting the welding current is to use a tapped secondary coil on the transformer. This is a popular method used on many small, limited-input welding transformers. The leads to the electrode holder and the work are connected to plugs that may be pushed into sockets on

the front of the machine to provide the needed welding current. See Fig. 13-50. On some machines, a tap switch is used instead of the plug-in. With either method, it is not entirely possible to adjust the current.

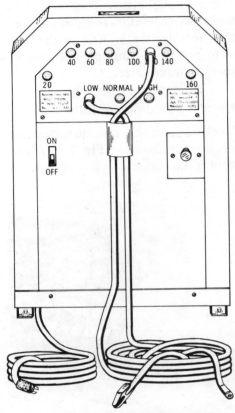

Fig. 13-50. An AC welding machine with taps for current control.

On industrial transformer welding machines, a continuous output current control is usually used. The control can be either mechanical or electrical.

Mechanical Control — The mechanical method may involve either moving the *core* of the transformer or moving the position of the *coils* within the transformer. In either case, any method using mechanical movement of the transformer parts will require a good deal of movement over the full range adjustment. See Fig. 13-51.

Electrical Control—The more advanced method of adjusting current output is by *electrical* circuits. In this method, the core of

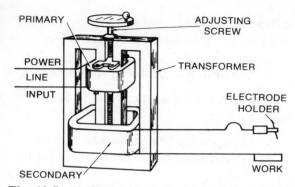

Fig. 13-51. Adjusting the output current mechanically.

the transformer is saturated by another electric circuit from a reactor. This circuit controls the amount of current delivered to the output terminals. By adjusting a small knob, it is possible to provide continuous current adjustment from the minimum output to the maximum. See Fig. 13-52.

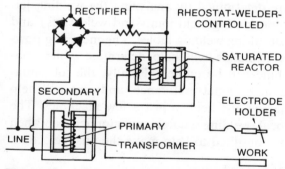

Fig. 13-52. Adjusting the output current electrically.

Although the transformer welding machine has many good points, it also has some limitations. The power required for a transformer welder must be supplied by a single-phase system. This may create an imbalance of the power supply lines, which is not good for most power companies. In addition, transformer machines have a rather low power factor demand unless they are equipped with power factor correcting capacitors. Adding capacitors corrects the power factor under load and produces a reasonable power factor favorable to most electric power companies.

Transformer welders have the lowest initial cost and they are the least expensive to oper-

ate. They require less space and are normally quiet while running. Also, alternating current welding power supplied by transformers reduces arc blow, which can be troublesome on many welding jobs.

Transformer welding machines do not, however, have as much flexibility for the welder as do dual-control generator machines.

Transformer/Rectifier Welding Machines

Transformer welding machines, just described, can provide only alternating current (AC) at the arc. Some types of electrodes, however, can only be operated successfully with direct current (DC) power. One method of supplying direct current power to the arc, other than using a rotating generator machine, is by adding a *rectifier* to a transformer welding machine. (A rectifier is an electrical part that changes alternating current into direct current.) Also, *transformer/rectifier* welding machines can be made to use three-phase input as well as single phase. The three-phase input machine overcomes the power line imbalance problem of a straight transformer machine.

Inside a transformer/rectifier welding machine, the transformers feed into a rectifier bridge; the bridge then produces direct current for the arc. In other cases, where both AC and DC may be needed, a single-phase AC machine is connected to the rectifier. By using a switch, the welder can change the output terminals so they go to either the transformer or the rectifier. He can also select AC, DC straight, or DC reverse polarity current for his welding job.

In some types of AC-DC welding machines, a high-frequency oscillator, as well as water and gas control valves, is installed. This, then, makes the machine ideally suited for *T*ungsten *I*nert *G*as (TIG) welding, along with manual coated electrode welding.

Transformer/rectifier welding machines, Fig. 13-53, are available in different sizes and, as mentioned before, may be set up to use either single phase or three-phase power. They may also be designed for different voltages from different power supplies. The transformer/rectifier machine is more electrically efficient than the generator, and it is usually quieter in operation.

Fig. 13-53. Transformer/rectifier welding machine. (Courtesy of Hobart Brothers Company.)

Multiple Operator Welding System

This system uses a heavy duty, high current, relatively high voltage power source that feeds several individual welding stations. At each welding station, a variable resistance can be adjusted to drop the current to the needed welding range. Based on the duty cycle of the machine (discussed later), one welding machine can supply welding power to a number of welders at the same time. The current supplied at each station has a drooping characteristic curve similar to the single-operator welding machines described earlier. The power source, however, has a constant *voltage* output, not constant current.

The welding machine size must be carefully matched to the number and size of the individual welding stations for an efficient multiple-operator welding system.

Specifying a Welding Machine

Selecting a welding machine is based on several things:

A. The process or processes to be used.
B. The amount of current required for the job.
C. The power available at the job site.
D. Convenience and economy.

Information about what each machine can do and information about each arc welding process will help determine the type of machine needed. The size of a welding machine is based on its *current capacity* and *duty cycle*.

Welding current, duty cycle, and voltage are determined by analyzing the welding job (considering weld joints and weld sizes) and consulting welding tables. The incoming power available will help determine which machine can be chosen. Finally, the job situation, personal preference, and cost will help narrow the field to just a few machines. The local welding equipment supplier can then be consulted to help make the final selection.

Duty Cycle

Duty cycle is defined as the ratio of *arc time* to *total time*. A 10-minute time period is used for rating a welding machine. So, if a welding machine had a 60% duty cycle, the welding load would be on for 6 minutes and then off for 4 minutes. Most industrial, constant current machines are rated at 60% duty cycle. Most constant voltage machines used for automatic welding are rated at 100% duty cycle.

A figure known as *percentage duty cycle* is the ratio of the *square* of the rated current to the *square* of the load current, *multiplied* by the rated duty cycle. This is also known as percent of work time vs. current load, as in Fig. 13-54.

Using the chart in Fig. 13-54, a line can be drawn parallel to the sloping lines, through the intersection of the machine's rated current output and rated duty cycle.

For example, the question might be asked: can a 400-amp, 60% duty cycle machine be used on a fully automatic job of 300 amps for 10 minutes? Since line A crosses the 100% duty cycle (10 minutes) at 310 amps, this *would* be possible.

On the other hand, there may be a need to draw more than the rated current from a welding machine, but for a shorter period of time. Line B, for example, shows that a 200-amp, 60% rated machine can be used at 250 amperes as long as the duty cycle does not exceed 40% (4 minutes out of each 10).

The chart can be used to compare various machines. All machines should be related to the same duty cycle (60%) for a true comparison. See Fig. 13-54.

Ordering a Welding Machine

To order a welding machine, the following data should be given:

A. Manufacturer's type designation or catalog number.

B. Manufacturer's identification or model number.

C. Rated load voltage.

D. Rated load amperes (current).

E. Duty cycle.

F. Incoming power supply voltage.

G. Frequency of the incoming power supply.

H. Number of phases of the incoming power supply.

Fig. 13-54. The duty cycle is used to determine a machine's capabilities.

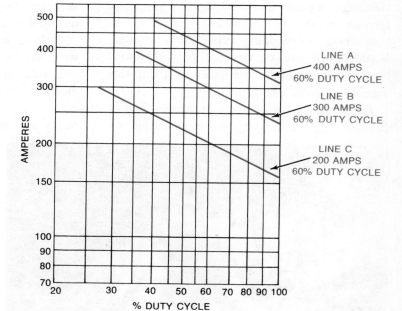

14

Basic Arc Welding Methods

Most welding done today, especially in production, is done with shielded arc welding, either manual or automatic. This chapter, then, will present the various welding procedures needed for most of the welding positions using this method. Shielded arc welding can be used on both ferrous and non-ferrous metals.

Methods of welding other than manual shielded arc will be covered only to show how those methods are different from the manual shielded arc process.

Each step of the welding procedure is discussed using the language covered earlier, so

that a new welder will be able to easily understand the new methods and terms.

Safety Practices

Before actually practicing shielded arc welding, it would be a good idea to review the protective equipment needed during welding, outlined in the chapter on safety.

Helmet and gloves, of course, must be worn all the time when striking the arc and arc welding. New welders, using the safety clothing and equipment for the first time, may notice that the equipment can seem bulky and stiff, especially if it is new.

Fig. 14-1. A well dressed arc welder. (Courtesy of American Optical Corp.)

Even so, once a new welder gets used to working while wearing the safety equipment, it gradually becomes natural and doesn't feel so awkward. It is well worth the little time and inconvenience it causes. In Fig. 14-1, the welder will not be burned or blinded because he is well dressed for the job.

Striking the Arc

The first step to be learned is to strike a smooth arc, without allowing the rod to stick to the metal. When the rod sticks to the metal while striking an arc, it is known as *freezing*. The metal where the bead is to be welded is usually called the *parent* metal. Samples of parent metal for practicing are, of course, called *coupons*.

Preparing for Practice

The parent metal where the bead will be fused (melted in) should be cleaned. Good cleaning allows the current to pass easily through the metal when the ground clamp is

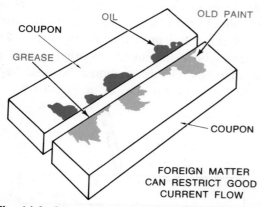

Fig. 14-2. Oil or paint can cause poor arc welding because the electrical contact is poor.

attached and the electrode in the electrode holder comes in contact with the metal. Paint, tar or other materials remaining on the parent metal, especially at the ground clamp or electrode contact points, can restrict the current flow needed for good welding. Dirt or paint put a lot of *resistance* in the welding circuit, as in Fig. 14-2.

When practicing arc striking, a piece of material at least ¼" thick should be placed in the flat position with the ground cable attached as shown in Fig. 14-3. A ⅛" or 5⁄32" electrode is then chosen from the listings in a table, such as Table 14-1. Then, the electrode may be carefully placed in the electrode holder to insure a good electrical contact for the welding circuit.

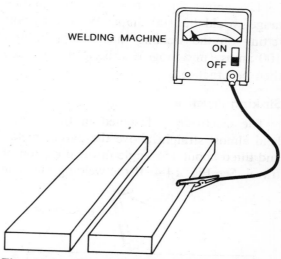

Fig. 14-3. Cleaned practice coupons with the clamp attached.

Choosing an Electrode—A chart similar to Table 14-1 gives the approximate amperage for each size and type of rod and is usually available from all electrode manufacturers. A new welder should become familiar with the charts and should refer to them for the correct information when using unfamiliar rods.

Table 14-1. Suggested Current Ranges for Mild Steel Iron Powder Electrodes (Amperes).

Electrode	Electrode Types					
Diameter	E 6010	E 6014	E 7018	E 7024	E 7027	E 7028
3⁄32	....	80-125	70-100	100-145	....	....
⅛	80-120	110-160	115-165	140-190	125-185	140-190
5⁄32	120-160	150-210	150-220	180-250	160-240	180-250
3⁄16	150-200	200-275	200-275	230-305	210-300	230-305
7⁄32	....	260-340	260-340	275-365	250-350	275-365
¼	....	330-415	315-400	335-430	300-420	335-430
5⁄16	....	390-500	375-470	....	....	....

Adjusting the Current—Depending on the size, each electrode will have a minimum and maximum amperage rating listed on the manufacturer's chart. For practice, it would be a good idea to use mid-range current.

A ⅛″ diameter E 6010 electrode (rod) will usually be rated for flat position welding at about 75-amps minimum to 125-amps maximum. This is listed on all charts as 75-125 amps. So, in this case, the middle range would be 100 amps. With a current setting on the welding machine of around 100 amps (mid-range), striking the arc can then be practiced.

Striking Practice

The electrode is fastened in the holder, held almost straight above the parent metal, and tilted about 15° angle in the direction of travel. See Fig. 14-4. If the welding machine

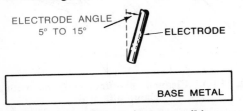

Fig. 14-4. Correct angle to practice striking an arc.

being used has alternating current (AC), polarity will not be a factor, so striking the electrode on the plate will be similar to *striking* a match, as in Fig. 14-5. If, however, the welding machine is direct current (DC), the polarity setting should be reverse polarity for

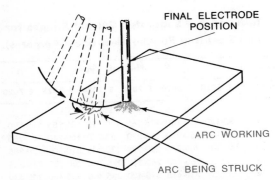

Fig. 14-5. Scratching method used to strike an arc with an AC welding machine.

an E 6010 electrode. On the direct current machine, the rod should be *tapped* to ignite the arc, rather than scratched. See Fig. 14-6.

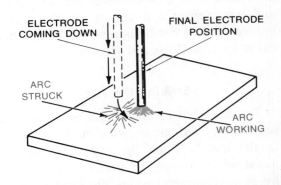

Fig. 14-6. Tapping method used to strike an arc with a DC welding machine.

Adjusting the Arc—As the electrode is brought into contact with the work piece, the arc will try to start. When the arc starts, the rod is raised slightly from the work until it is about ¼″ from the work piece but still carries a long arc. If the arc is held steady in this position, a molten puddle will immediately begin to form below the arc. The electrode should then be lowered toward the surface of the metal until the rod tip is its own diameter above the metal. In other words, a ⁵⁄₃₂″ diameter electrode should be held ⁵⁄₃₂″ above the work surface, as in Fig. 14-7.

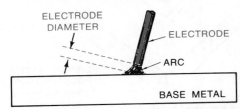

Fig. 14-7. After the arc is struck, the distance from the work to the electrode end should be the same as the electrode's diameter.

Electrode Freezing—Even after a welder has practiced striking the arc many times, he may sometimes have a problem known as electrode *freezing*.

An electrode freezing to the metal plate happens because the electrode touches the plate and sticks to it. The sticking is caused by a *short circuit*. A short circuit is a path of low resistance between any two points in an electrical circuit. It shortens the distance traveled by the current. This happens when the electrode touches the plate and welds itself to the plate. If the electrode is left in this frozen (welded-on) position, it will become red hot. Then, the flux on the rod will become very hot and may even burn, while the frozen rod is overloading the welding machine. For these reasons, rod freezing can be both dangerous and damaging.

Correcting Freezing — Usually a frozen rod can be released immediately by a firm twist on the electrode holder. The electrode holder should not be forced. If the electrode cannot be immediately freed by twisting, the electrode release on the holder should then be used to simply remove the holder from the frozen electrode.

After releasing the holder, the welding machine should be shut off and the electrode allowed to cool in the frozen position. Then, it can be easily removed from the plate by twisting it slightly with a pair of pliers. See Fig. 14-8.

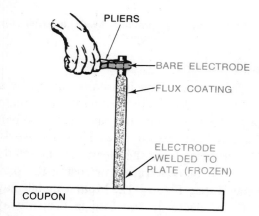

Fig. 14-8. Using pliers to remove a frozen electrode.

Restriking the Arc—After the electrode is freed from the plate, it can be clamped in the electrode holder again and used to strike another arc.

Striking the arc with both the scratching and tapping methods should be practiced un-

til the arc can be struck, raised, and lowered to the welding position without creating a short circuit and freezing the rod to the plate.

The proper (short) welding arc will give a sharp cracking sound similar to eggs frying, especially if the current range on the welding machine has been properly set for the electrode being used.

Depositing a Metal Bead

When an arc is struck, the plate below the arc starts to melt just below the end of the electrode. The metal droplets from the melting electrode mix with the molten puddle in the plate that was created by the arc's heat. Mixing together the melted electrode and the melted base metal is known as *fusion*.

After the arc is struck and the electrode moves across the metal plate, fusion of both the melted rod and melted base metal takes place constantly. This constant fusion causes a metal buildup on the base metal known as a *bead*.

As the electrode melts, of course, it gets shorter. It must be evenly *lowered* all the time as it is moved across the plate, because it is always getting shorter while it produces the bead. See Fig. 14-9.

Bead Practice

A new welder can practice striking the arc and then moving it slowly across the plate. The plate is placed in the flat position, and practice welding is then done from left to right. With practice, a smooth weld bead and good penetration for maximum weld strength can be easily made. As shown in Fig. 14-9,

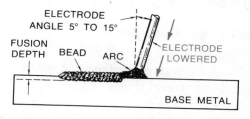

Fig. 14-9. As the electrode melts to form the bead, it must be lowered constantly.

the electrode should be held almost perpendicular (straight up) to the work, and it should be tilted slightly ahead in the direction of travel.

Changing Short Electrodes

The consumed (short, melted-down) rod should be removed from the electrode holder when about 1½″ of the rod remains. Removing the rod before it is completely consumed assures the welder that the holder will not short circuit from touching the work.

However, it may become necessary to break the arc (stop welding) before the end of the weld if the electrode gets too short. When this happens, a new arc should be struck about one inch ahead of where the weld bead stopped. The new arc is then carried *back* to begin welding where the original arc stopped. As the bead is carried along, it will weld over any marks on the metal caused by the scratching or tapping used to start the arc.

If the electrode in Fig. 14-10 were lifted from its present point, the arc and weld would

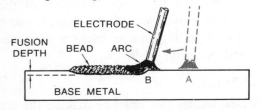

Fig. 14-10. If the arc was stopped at B, it would be restarted at A and moved back to B to continue the weld.

stop at point B. Then, when a new arc was struck, it would be struck at point A and carried back to point B to continue the original bead. See Fig. 14-10.

Producing a Good Bead

A good bead is strong, with deep penetration and an even appearance. A good bead depends on several things:

A. The correct electrode.
B. The correct current for the electrode.
C. Correct polarity in a DC machine.

D. Eliminating any side-to-side motion while welding.
E. Establishing a short arc.
F. Keeping a steady, constant travel speed.
G. Correctly feeding in (lowering) the electrode as it melts.

A welding bead should be slightly higher in the center and taper smoothly toward the edges in a rounded shape. For practice, a series of beads can be deposited against each other. Then, comparing the beads will show that practice will help the welder lower the electrode more steadily as it is melted. This creates the desired bead while still using a short arc. See Fig. 14-11.

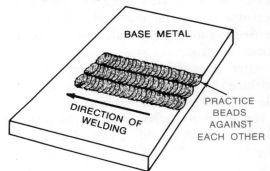

Fig. 14-11. Welding practice beads together for comparison.

Bead Problems

A flat, uneven bead is usually caused by a long arc. A long arc is unable to create enough heat to melt the base metal. When the arc is held too far away from the work, the fluid metal from the electrode cannot do a good job of penetrating or fusing the parent metal. See Fig. 14-12. Molten metal may fly off the electrode and fuse in drops away from the weld. These drops are called *spatter*.

Penetration can be checked by suddenly removing the electrode from the parent metal. Where the rod was removed, a crater is formed in the parent metal. This crater shows the depth of penetration while the bead was being welded. See Fig. 14-13.

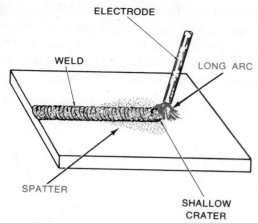

Fig. 14-12. A long arc (electrode too far from the work) creates a shallow crater, poor fusion, and spatter.

Machine Amperage — Poor penetration, poor fusion and humpy beads also show that the machine amperage setting may be too low. On the other hand, high amperage creates troughs at each side of the bead known as *undercutting*. In addition, high amperage creates excessive spatter, as in Fig. 14-14.

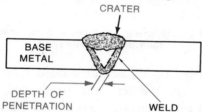

Fig. 14-13. The depth of penetration is shown by the size of the crater when welding is stopped quickly.

Electrode Speed—If the electrode is moved too fast in the direction of travel, the bead will be weak and porous because of poor penetration and fusion. This could also result, of course, from a current setting that is too low.

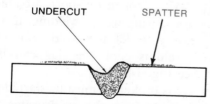

Fig. 14-14. If the welding current (amperage) is too high, the welder will notice undercutting.

If the electrode is moved too slow, it creates high, wide beads. These could overlap and possibly cause *burn-through*, especially if the base metal is thin.

Flux and Slag

The electrode coating, or *flux*, forms the protective shield of gas around the weld. The shield stops the metal from oxidizing. Because of the flux shield's effect, the metal will not become brittle and weak. Chemicals within the flux coating mix with the metal and force impurities to float to the top. Once on top, these impurities solidify into a *slag* as the weld cools. See Fig. 14-15.

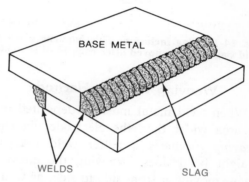

Fig. 14-15. A finished weld with a coated rod will have a thin layer of slag covering the weld.

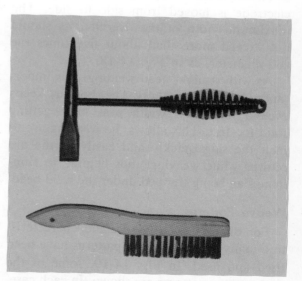

Fig. 14-16. Chipping hammer and wire brush used to remove slag.

The slag, in turn, covers the finished weld and keeps it cool, increasing the metal's ductility. Slag formed on the finished weld or between weld beads should be removed by using a chipping hammer and/or a wire brush, Fig. 14-16. This is especially important when beads will be deposited on top of beads. Leaving slag on the lower beads will produce a poor quality weld. See Fig. 14-17.

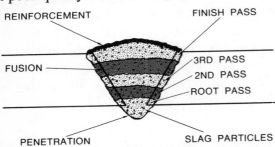

Fig. 14-17. Slag inclusion. This happens when the weld slag is not removed after each bead is welded.

Weaving With the Electrode

When weld metal must be deposited over an area wider than the electrode diameter, *weaving* produces a better weld than the straight bead does. Weaving the electrode means moving it from side to side, as well as along, the weld line.

The weld width depends on how much the electrode is moved from side to side. The maximum width of the overall pass should not exceed more than about five times the rod diameter, as in Fig. 14-18.

As with straight bead welding, it is important to keep the arc short. The short arc keeps even a maximum width pass from getting solid too fast. This allows the weaving arc to melt the slag quickly and easily as the arc returns while weaving, thus helping to eliminate slag being trapped under the weld bead.

Weave Patterns

For different jobs, loads, and characteristics, many different weave patterns have been tried and used. In Fig. 14-19, some of the many weave patterns are shown. In each case, a large dot shows where the welder needs to

hesitate (stop for a moment) while he welds. A new welder may want to practice several weave patterns for different effects. By hesitating for a moment, good penetration and undercutting at the sides is assured.

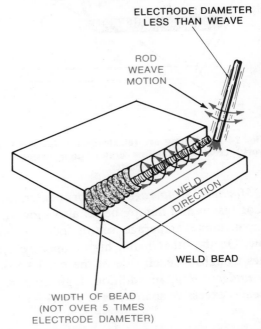

Fig. 14-18. Weaving produces a weld bead wider than the electrode itself.

In Fig. 14-20, actual welded weaves are shown. On the left, the beads are poorly spaced with holes between, indicating too fast a travel speed. On the right, a good weave was made with the proper current and speed.

Weave Welding Practice

When weaving the electrode in the flat position, the rod angle should be the same as it was for the bead practice in Fig. 14-4. Parts of the weave will be poorly formed if the weave motion is uneven.

If the arc is broken or short-circuited for some reason during weave welding, it is usually necessary to let the weld cool. Then, after the weld is cool, the slag can be removed. See Fig. 14-21.

If the arc is broken, it must be started again a short distance from the weld crater, as in bead welding. The arc is first struck a short

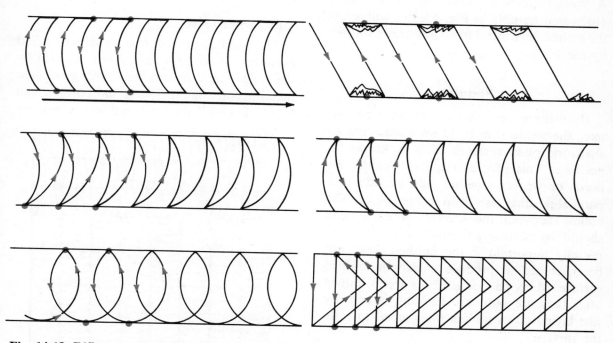

Fig. 14-19. Different weave patterns. In each case, large dots show hesitation points and arrows show the weld direction.

distance from the weld and then returned to the crater where welding in the weave begins again. Restarting is similar to restarting a bead, as was shown in Fig. 14-10.

During weaving, it may be noticed that the penetration is not as good as is the penetration when using the straight bead. The amount of time necessary to make a weave weld, however, will create more heat in the parent metal than if the metal was bead welded. This excessive heat causes the parent metal to try to expand and distort. Like other distortion problems, this condition can be overcome with more practice.

When running weaves, the short arc will try to penetrate the parent metal, and if the welder doesn't pause for a moment at the side

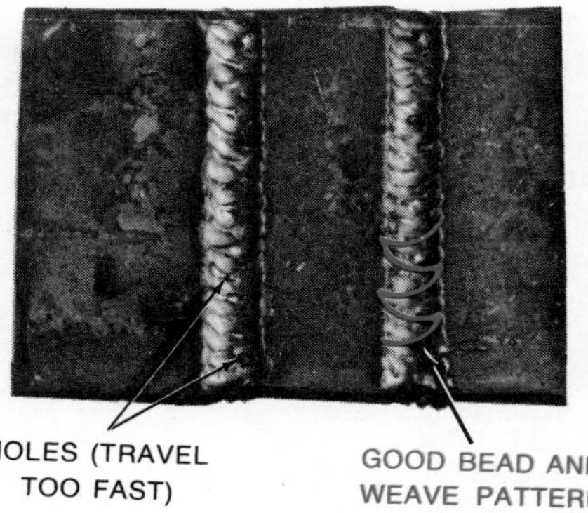

HOLES (TRAVEL TOO FAST)

GOOD BEAD AND WEAVE PATTERN

Fig. 14-20. Weave welding practice plates. Right—good weave weld. Left—poor weld due to travel speed being too fast.

POINTER SHOWING SLAG

Fig. 14-21. Slag particles before being chipped from the weave weld bead.

of the weave, a crater or undercut will result. A short hesitation period during the weave (indicated by dots in Fig. 14-19) will allow the melted filler metal to fill any crater formed by the hot arc.

Padding

By using either the bead or weaving pattern, the welder can build up worn areas on shafts or plate surfaces. This is known as *padding*. This process requires welding the beads or weaves together side by side in a parallel position, as shown in Fig. 14-22. The hollow between the crowns of each bead should be narrow and shallow. This is especially important if the finished weld will be machined to produce a smooth finished surface. Many padded surfaces are machined after they are built up. Padding is actually just laying down molten filler metal to raise the surface.

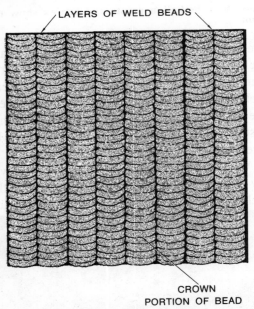

Fig. 14-22. Padding. Building up a surface by laying beads or weaves next to each other.

If a greater amount of buildup is required for increased thickness, alternate layers can be placed across the first layer, as in Fig. 14-23. Of course, each bead should be de-

slagged with the chipping hammer and wire brush before the next bead is applied.

Padding Practice

The padding procedure is sometimes called *buttering*. It requires continuous practice by the welder until he can weld with good base metal fusion. Slag inclusion and porosity can be eliminated if the bead and weave welding

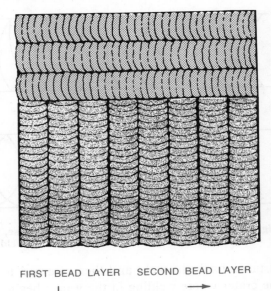

FIRST BEAD LAYER   SECOND BEAD LAYER

Fig. 14-23. Heavy padding. Adding more metal to the original beads, but running across the first beads in the opposite direction.

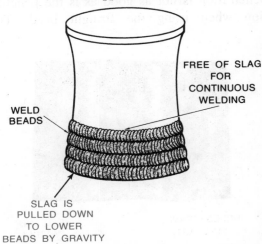

Fig. 14-24. Vertical padding may not need frequent slag cleaning because gravity tends to pull the slag down to the bottom beads.

practice is carefully done and the slag is removed after each pass.

The size of the padding bead or weave is governed by factors such as welding speed, electrode size, polarity, and current settings. For general practice, it is a good idea to use a typical electrode, such as an E 6010, from ⅛″ to ⁵⁄₁₆″ diameter. Various amperage settings can then be used and the deposited bead results compared. After some practice, the welder will automatically be able to set the machine amperage to give the desired bead or weave results for the diameter of the rod being used. Since the E 6010 is a typical electrode used for padding, the current values in Table 14-1 can be used for different electrode diameters and experiments with various settings.

Padding in the Vertical Position

If the welder needs to pad (build up) parts in the vertical position, slag removing can be eliminated by using the proper procedure.

The effect of gravity on the hot fluid metal and slag will cause the slag to spill over on the finished weld *if* the weld beads are welded from the bottom of the piece to the top. This slag spill-over leaves the parent metal free of slag where the next weld bead is to be welded, as in Fig. 14-24.

When padding on flat material, it is a good idea to place the parent metal on an angle of about 45°, and then lay the beads from the bottom edge of the material to the top in order, as in Fig. 14-25. This procedure may eliminate the need for deslagging. By using this padding method whenever possible, the experienced welder will save valuable welding time by not having to stop welding for deslagging.

Practicing Fusion Welding

When a welder has practiced running beads and weaves (and possibly done some padding), he will be ready to actually weld separate pieces of metal together. This can be done by starting with simple joints in the flat position and working up to complex joints in difficult positions.

Practice Welding Equipment

Most welders find that *butt joints* are the most common joints used for fabricating and repair work. The thickness of the metal being used for fabricating or being repaired determines the type of joint to be used and the welding technique necessary for full strength. Fig. 14-26 shows two simple butt joints.

The butt joint is widely used in the four basic welding positions: Flat, Vertical, Overhead, and Horizontal. For demonstrating butt welding in all the positions, two differ-

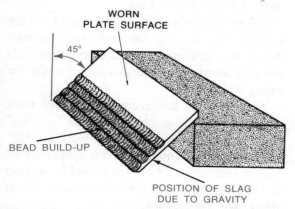

Fig. 14-25. To save deslagging time, a flat plate may be leaned on a vertical surface so the slag will fall down from gravity.

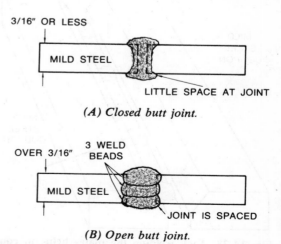

(A) Closed butt joint.

(B) Open butt joint.

Fig. 14-26. Simple butt joints and their positions.

ent thicknesses of metal should be used for welding practice.

Welding Coupons—Pieces of plate commonly known as welding coupons are most often used for welding demonstration and practice (Fig. 14-27). They should each be about 2″ × 4″. Each set of coupons consists of two pieces of metal, preferably mild steel of the same thickness. They should be either ¼″ or ⅜″ thick, depending on whatever welding process is to be demonstrated or practiced.

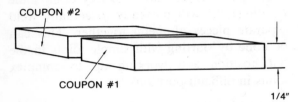

Fig. 14-27. Sample welding coupons, ¼″ thick.

Distortion—To prevent distortion, especially on plates requiring long welds, the plates will have to be tacked together as shown in Fig. 14-28. Distortion known as *expansion* is created, of course, by the heat of the welding process.

Due to the fact that expansion might cause warping or misalignment, spot or tack weld

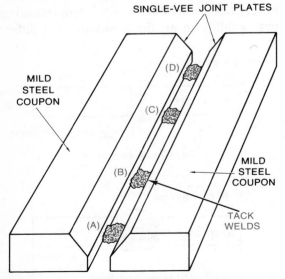

Fig. 14-28. Tack welding the plates helps to control distortion.

both the coupons before welding. A slightly larger gap near the end of the coupons where the weld will end should prevent a closed gap from forming in that area due to the plate expansion. See Fig. 14-29.

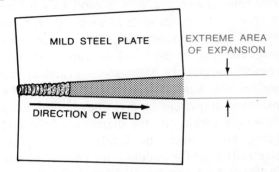

Fig. 14-29. Spacing the plate with a large gap where the weld will end. After tacking, this will help guard against expansion that could close the gap as the weld is made.

As welding is done from the starting point, the heat going into the metal from welding will increase. The increase in heat causes each plate to expand to the point where the end gap will be closed unless the coupons are tacked.

Electrode and Current—The amperage and voltage adjustment on the machine depends on the size and type of electrode being used, as well as the thickness of the material being welded. An E 6010 electrode can be used for practice, and the current setting for the diameter of electrode being used can be learned from Table 14-1.

Before rods other than E 6010 are used for welding practice, the electrode requirements should be considered so that changes in voltage and amperage settings can be made for strong welds.

The electrode size recommended for most practicing on these coupons is ⅛″ or ⁵⁄₃₂″ diameter. For the butt joint weld, a ⁵⁄₃₂″ electrode is first used with an amperage of 120 to 160 amps; 140 amps would be a good first trial, since it is half way between the two limits. See Table 14-1.

If the machine has a manual voltage control, a setting of 24 to 26 volts is recommended for practice. The voltage range for a 1/8″ electrode is the same as required for a 5/32″ electrode, but the amperage range is only 80 to 120 amps, due to the smaller rod diameter.

Sample Beads—With the machine on open circuit voltage, the ground clamp attached to the work, and the electrode fastened securely in the electrode holder, it is a good idea to first run some sample beads on scrap stock to test the 140-amp setting. Then, if the current is not correct, an adjustment can be made before actually making the practice butt joint.

Practice Welding Procedure

Although the basic technique is the same in each kind of practice welding, there may be differences in the preparation of the cou-

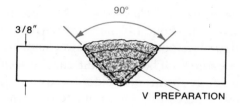

Fig. 14-30. A single-Vee preparation. Used on metal up to 3/8″ thick.

pons, the electrode current setting, or the electrode angle. A major difference is how the coupons are prepared. They may be prepared with either a single-Vee, Fig. 14-30, or a double-Vee, Fig. 14-31. Or, they may not be prepared at all, which is the simple butt joint shown in Fig. 14-32. The simplest

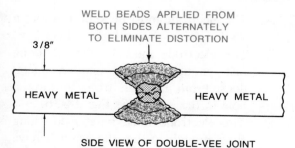

Fig. 14-31. A double-Vee preparation. Used on metal 3/8″ thick or over.

practice welding is done on unprepared butt joints. These may be either closed or open butt joints, as was shown in Fig. 14-26.

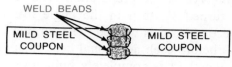

AREA OF FUSION

PREPARATION WELDED FROM THE TOP IN THREE PASSES

Fig. 14-32. Unprepared butt joint.

Closed Butt Joint—Fig. 14-26A shows a closed butt joint. This is the usual method of welding metal 3/16″ thick or less. In the closed butt joint, the edges of the two pieces of metal are positioned together to eliminate the space or gap at the fusion point.

Open Butt Joint—When the pieces to be welded together are over 3/16″ thick, a gap should be left at the joint. This *open* butt joint position allows the arc to penetrate the joint, forming a strong bond on the underside of the two parts joined, Fig. 14-26B.

The gap between the parts to be welded with an open butt joint should be from 1/16″ to 1/8″, depending on the thickness of the pieces. Heavier parts to be welded need a larger gap between the parts for deep penetration.

Welding Round Shafts

When welding a broken round shaft, the double-Vee preparation is recommended with welding in the proper bead sequence to eliminate distortion. In Fig. 14-33, the proper weld sequence and preparation are shown for successfully welding round shafts.

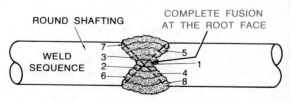

DOUBLE-VEE PREPARATION

Fig. 14-33. Weld preparation joint and welding order for a round shaft.

Round shafting can be easily aligned in a piece of angle iron, as in Fig. 14-34. Then, the first weld bead is laid in the bottom of the top Vee of the preparation. The shaft is then turned ½ turn in the angle iron and the second bead is welded. The turning continues before each additional bead is applied until the shaft is completely built up to the required size, ready for use or machining.

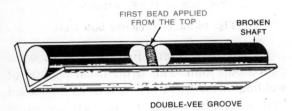

FIRST BEAD APPLIED FROM THE TOP   BROKEN SHAFT

DOUBLE-VEE GROOVE

Fig. 14-34. Aligning and welding a broken round shaft in a piece of angle iron.

Before each bead or weave is applied, the preceding bead or weave must be cleaned by chipping and brushing. This will ensure complete fusion without trapping any slag in the weld that could cause porosity. Fig. 14-35

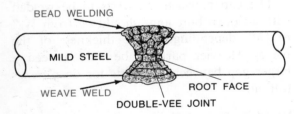

BEAD WELDING

MILD STEEL

WEAVE WELD   ROOT FACE

DOUBLE-VEE JOINT

Fig. 14-35. A broken shaft repaired with bead welding on one side and weave welding on the other.

shows one side of a shaft welded with beads and the other side welded with weaves. Properly applied, either method will give a strong weld.

Flat Position Welding

Most arc welding practice is done in the flat position. The flat position is the easiest and cheapest position to use, making it very popular in industry also. Special fixtures are often built to revolve a piece so that it can be welded in the flat position.

The following section discusses the different flat position joints and their practice. In each case, the recommended electrode, current, coupon preparation, and special features about each joint will be noted.

Butt Joint

This is the easiest and simplest arc welding joint to practice welding. Since it may be a new welder's first try at actually arc welding two coupons together, he may want to spend a lot of time experimenting with several coupons, electrodes, angles, and welding currents. These recommendations will give him a good, average place to start.

Coupon Preparation — Using ¼″ thick coupons with no preparation, the welder is ready to start practicing when the coupons are placed in an open butt joint, about ⅛″ apart, as was shown in Fig. 14-26B.

Electrode and Current—A ⁵⁄₃₂″ electrode is used with about 140 amps of current. The E 6010 electrode, as usual, is recommended.

Procedure—The butt joint can be welded successfully from one side by using three passes, if necessary, for complete penetration. See Fig. 14-32.

Since there are so many points to remember when first practicing arc welding, it would be a good idea to review the safety practices before practice starts. That way, a new welder will not accidentally burn himself or damage anyone's eyes, including his own. Then, when the arc is struck, a new welder will be able to concentrate on several things that are needed for a good arc weld:

A. The electrode must be kept at about a 15° angle, as shown in Fig. 14-4.

B. The electrode must be moved along fast enough so the molten metal doesn't fall through from burning a hole in the work. See Fig. 14-36.

C. The electrode must be moved along slow enough so that the arc will have time to melt the sides of the coupon for good penetration. See Fig. 14-37.

D. The electrode must be lowered as it melts away, so that the arc length stays the same.

E. The electrode must not lean toward one plate or the other, or penetration will be unequal. See Fig. 14-38.

Since this may be a new welder's first practice at welding two plates together, he may want to check carefully for good penetration

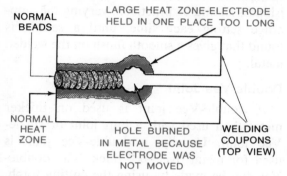

Fig. 14-36. If the electrode is moved too slow, a hole may be burned in the plate.

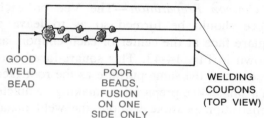

Fig. 14-37. If the electrode is moved too fast, there will be poor fusion between the plates.

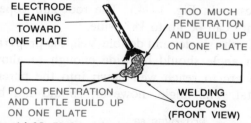

Fig. 14-38. If the electrode leans toward one plate, penetration and buildup will be uneven.

and weld strength. Several of these methods are mentioned in the chapter on weld testing.

Welding from both sides may be necessary for the best penetration. When welding from both sides, of course, each bead must be chipped and wire-brushed before the next bead is welded on top.

Single-Vee Joint

While a butt joint is suitable for thinner metal, a single-Vee joint is usually used on thicker metal. The single Vee is also used for added strength where a butt joint would not be strong enough.

Coupon Preparation—Beveled coupons at least ¼″ thick are used for the single-Vee joint. Most frequently, they are beveled with a root face (Fig. 14-39), but they may be

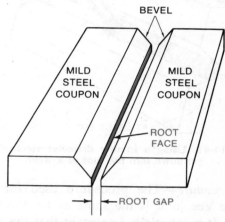

Fig. 14-39. Mild steel coupons prepared with a root face for a single-Vee joint.

beveled without one. A root face is used to keep the metal from falling through when high welding amperage is used. Without the root face, the hot weld metal and coupon sides might fall through, as in Fig. 14-40. After the ¼″ coupons are prepared, they can be placed up to ⅛″ apart for practice welding.

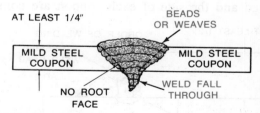

Fig. 14-40. Without a root face, molten coupon metal and electrode metal may fall through.

Electrode and Current—The electrode and current used for single-Vee practicing may be the same as for butt joint practicing. Sometimes, it may be a good idea to use a smaller ⅛″ electrode (with less current) than the ⁵⁄₃₂″ electrode. The smaller ⅛″ electrode may be able to get farther down into the root than the larger ⁵⁄₃₂″ electrode. See Fig. 14-41.

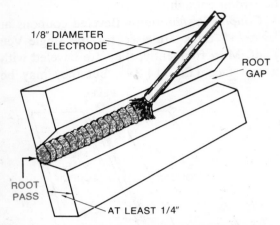

Fig. 14-41. Using a smaller diameter electrode to get down into the root of a weld.

Procedure—The procedure used for the single-Vee joint is the same as for the butt joint. It is especially important that the electrode's side-to-side angle be kept straight, to avoid the problem shown in Fig. 14-38.

Before applying each bead, of course, the existing beads should be chipped and wire brushed.

After the first bead is applied, it should be chipped, cleaned, and brushed. Then, a weave bead can be applied with a larger diameter rod if desired.

The welder should be sure that the first bead and the side of each coupon are com-

pletely fused, as shown in Fig. 14-42 for the lap weld to be studied later.

Either beads or weaves can both be used successfully, although weaving is usually used on heavier, thicker metal.

A *cover coat* (sometimes referred to as a wash coat) is then weaved at a lower amperage range, with the electrode still in the vertical position, for the finished pass. Such a finish pass should be practiced, varying the amperage setting each time, until a setting is found that gives a smooth finish on the welded metal.

Double-Vee Joint

A double-Vee joint is used on thicker metal than used with a butt joint or single-Vee joint. Usually, a double-Vee joint is used for metal over ⅜″ thick. The double-Vee can be made by using the cutting torch, chipping, or grinding.

Coupon Preparation—The Vee on each piece should be formed so as to leave a square face at the center of each coupon, as shown in Fig. 14-43. This square-faced surface serves the same purpose as the root face in a single-Vee preparation, making complete penetration possible without the weld metal falling through. The coupons should be at least ⅜″ thick, positioned with open root faces as in Fig. 14-43. The root gap (open joint) can be up to ⅛″ wide.

In the single- and double-Vee, the preparation angle should be wide enough to allow the arc to reach well down into the parent metal. The angle, however, should not be

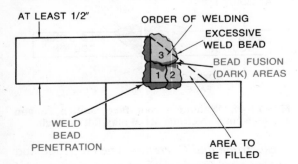

Fig. 14-42. Good penetration from each bead shown on a lap weld.

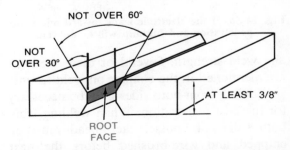

Fig. 14-43. The root face on a double-Vee preparation.

more than 30° on each piece of metal (Fig. 14-43). When the two coupons are butted together, the included angle will be approximately 60°, as also shown in Fig. 14-43.

Electrode and Current—The electrode and current used for double-Vee joint practice can be the same as for single-Vee practice. Again, a smaller ⅛″ electrode may be used for good root face penetration on the first pass.

Procedure—In welding the double-Vee groove, the first bead is welded the same as for a single-Vee preparation. The second bead is then welded to the opposite side. Each bead is then made on the opposite side to balance the pull from warpage. As usual, the slag must be chipped off before each new bead is applied. Either weaves or beads may be used with the double-Vee joint, as in Fig. 14-44, although weaving is usually used toward the top of the joint because the top is fairly wide.

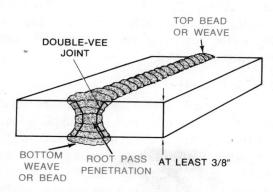

Fig. 14-44. Double-Vee joint weld on both sides.

Single-Bevel Joint

A single-bevel joint is used on special applications where one piece is stressed (loaded) differently than the other, or where the metals to be welded are not of the same thickness. In this case, the *thicker* plate would usually be bevelled, as in Fig. 14-45.

Coupon Preparation — As mentioned above, only the thicker coupon is prepared, using a 30° bevel to produce a root face on one side. For practice, ⅜″ and ¼″ coupons

may be used. They are positioned as shown in Fig. 14-45, with an open joint about ⅛″ wide.

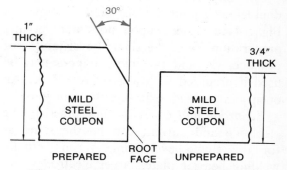

Fig. 14-45. Single-bevel joint preparation.

Electrode and Current — The recommended electrode and current for the single bevel joint are the same as for the single-Vee joint mentioned earlier.

Procedure—The welding method used with the single-bevel joint is similar to that used with the single-Vee joint. Welding the single bevel should be completed from one side. Either the bead or weave method can be used successfully, depending on the thickness of the metal. However, regardless of whether the welder prefers bead passes or weave passes, the root pass must be welded as a *bead* pass for complete penetration. As usual, the slag must be chipped off each pass before welding on the next layer of metal.

Double-Bevel Joint

The double-bevel joint may be used on special applications where the metal is too thick to use a single bevel or is stressed differently.

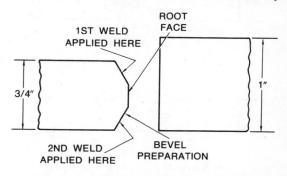

Fig. 14-46. Double-bevel joint preparation.

A common double-bevel joint is shown in Fig. 14-46.

Coupon Preparation — Coupons for the double-bevel joint are prepared as shown in Fig. 14-46. One coupon does not get any preparation, as in the single bevel joint. For practice, ½″ and ⅜″ thick coupons may be used if prepared with a good root face. An open joint is used, exposing the root joint.

Electrode and Current—For practice, the electrode and current can be the same as for the single-bevel joint. When actually welding heavier plate, larger electrodes and higher currents are used, as recommended in Table 14-1.

Procedure—The double-bevel joint can be welded with the methods used for the double-Vee joint described earlier. Distortion can be partly controlled by welding the beads opposite each other on the double bevel, as indicated in Fig. 14-46. As usual, the slag must be chipped off each pass before the next one is welded.

Single-U Joint

The single-U joint is used for careful filling and reinforcement for maximum strength on heavy stock.

Coupon Preparation—Coupons for a single-U groove are prepared as shown in Fig. 14-47. This must be done with a grinding wheel using a round stone, to produce a smooth U shape between the coupons. Heavier coupons, possibly ⅜″ thick, would be used for practicing. They are then positioned with an open joint exposing the root faces, about ⅛″ apart.

Electrode and Current—Again, a 5⁄32″ E 6010 electrode can be used with 140 amps of current for beginning practice. The electrode should be held above the plate and angled to the rear, as in Fig. 14-4. The amperage may be varied for good results as the welder practices.

A thinner ⅛″ electrode, with less current, may be used for the root pass to get down into the root faces for good penetration.

Procedure—This prepared joint can be welded with either the bead or weave methods as shown in Fig. 14-47. Penetration must be completed in the same manner as for welding done on single-Vee joints. Fusion on the pieces to be joined must be completed from the bottom of the root face to the face of the weld. See Fig. 14-47.

Before welding each additional bead or weave, the slag must be removed from the bead just completed.

Double-U Joint

The double-U joint is used on thicker metal than is the single-U joint. The applications are similar; only the metal thicknesses vary.

Coupon Preparation—Coupons for a double-U joint are prepared with the same machine used for single-U joint coupons, only those for a double-U are prepared on both sides of the coupon, as in Fig. 14-48. Although fairly thick coupons could be used,

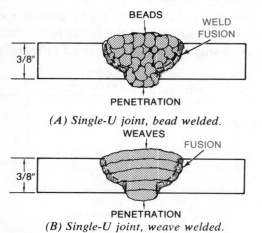

(A) Single-U joint, bead welded.

(B) Single-U joint, weave welded.

Fig. 14-47. Single-U joint, prepared and welded.

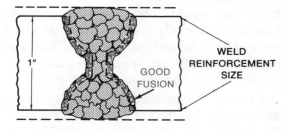

Fig. 14-48. Double-U joint, prepared and weave welded.

good practice can be done with thinner coupons, carefully prepared.

The coupons are then placed about ⅛" apart for an open-root joint, exposing the root faces for good penetration.

Electrode and Current—The electrode and current recommendations for the double-U joint are the same as for the single-U joint just discussed.

Procedure—The welding methods used for double-Vee and double-bevel joints can be used for the double-U joint. To control distortion, the beads or weaves must be welded on alternate (opposite) sides, with the slag being chipped off each pass before welding the next bead. Again, a finish or wash coat at low amperage can be used for final buildup and a smooth appearance. This also gives a good weld reinforcement, as in Fig. 14-48.

Single-J Joint

A single-J joint might be used for more strength than a single-bevel joint, where the metals are also of different thicknesses. In some places, the single J might give a better weld shape than the single bevel, or other desirable features.

Coupon Preparation—In a single-J joint, only one coupon is prepared, as in Fig. 14-49. Usually, it would be the thicker coupon. For practicing, coupons ⅜" and ¼" thick could be used. They should then be placed about ⅛" apart, to allow the electrode to melt down into the root face on the prepared coupon.

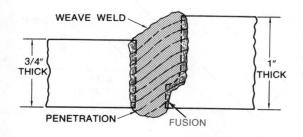

Fig. 14-49. Single-J joint, prepared and weave welded.

Electrode and Current—The recommended electrodes and currents for the single U joint can also be used for the single-J joint.

Procedure—When welding the single-J joint, the welder must make sure that his first pass thoroughly penetrates the root face of the prepared coupon. Then, he can continue to build up the weld with weaves until he is up to the height of the higher plate, as in Fig. 14-49.

In this type of joint, he may want to angle the electrode slightly toward the prepared coupon for good penetration, if that coupon is thicker. Of course, slag must be chipped off before the next pass is welded.

Double-J Joint

The double-J joint might be used in a limited application where only one piece of metal could be easily prepared. In this case, the easily moved part would be the one prepared for a double-J joint.

Coupon Preparation—The easily moved coupon is prepared with a J on each side, as in Fig. 14-50. The coupons are then placed about ⅛" apart in an open joint.

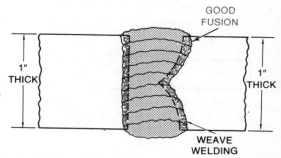

Fig. 14-50. Double-J joint, prepared and weave welded.

Electrode and Current—The recommended electrodes and currents for a single-U joint can also be used for a double-J joint.

Procedure—Again, this weld must be completed from alternate sides to control distortion. The first pass may be made with a smaller electrode for good penetration.

After the first weld is made with a bead down the root joint, weaves can be welded in for good appearance and good penetration, as in Fig. 14-50. As usual, the slag must be chipped off each pass before welding the next bead or weave.

Backup Strips for Flat Welding

Where the root gap of any flat position joint, just discussed, is wider than it should be for strong penetration, a *backup strip* may be used to reinforce the weld.

Backup strips, whether permanent or temporary, allow the welder to put a good hot bead in the groove without losing any filler metal because of the bead falling through. Fall through, shown in Fig. 14-51, usually happens in welds where the fit-up was not close enough, or the root face was not large enough, to control the heat of the close arc. There are two kinds of backers: *permanent* and *temporary*.

Permanent Backers—Permanent backers become part of the welded structure and help increase the strength of the finished weldment. Extra hot beads, created by increasing the amperage, are applied to fuse the backer to the parent metal parts. Using permanent backers, beads can be applied in both the flat and overhead position, Fig. 14-52, especially where the gap between the pieces is very large.

Temporary Backers—Temporary backers are usually made out of carbon or some other material that will not melt into the root joint when the root pass is made.

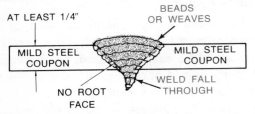

Fig. 14-51. Weld fall through due to too much heat, not enough root face, or too large a root gap.

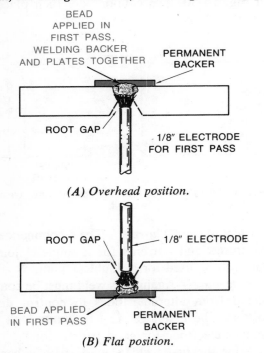

Fig. 14-52. Using a permanent backer to reinforce a root gap.

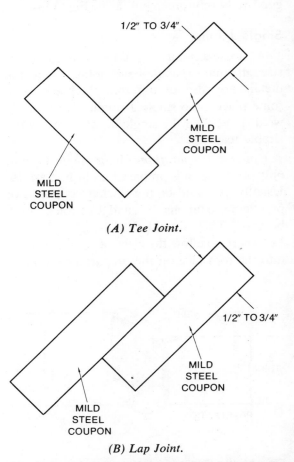

Fig. 14-53. Coupons positioned for fillet welding in the flat position.

When temporary backers are used, it is usually a good idea to weld back over the root face *after* the backer has been removed and the welded structure has been chipped and brushed.

Increasing the amperage before putting on such a pass is usually recommended to insure complete fusion and penetration.

Fillet Welds

Welds made on two pieces of metal as shown in Fig. 14-53 are known as *fillet* welds. These welds can be made in all positions on tee, lap, and corner joints. In Fig. 14-53, the joints have been placed in position for flat fillet welding.

Fillet welds are included in this section about the flat welding positions only, but fillet welds *can* be done in any position. For this reason, fillet welding procedures for the *other* welding positions will be fully explained in the section about each position.

Flat Position Fillets—In the flat position, the V-shaped joint for a fillet weld can be made by any two pieces of metal. In this case, they must be positioned so that the welder can place the first pass in the V-shaped

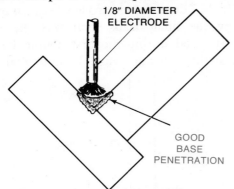

1/8″ DIAMETER ELECTRODE

GOOD BASE PENETRATION

Fig. 14-54. The first pass through the fillet weld is usually done with high current for good base penetration.

corner. In most cases, the first bead pass should be very hot for good penetration. See Fig. 14-54. For this reason most welders increase the amperage for the first pass so that the heat from the arc will be sure to

well penetrate the two pieces of metal being joined.

Coupon Preparation—The Tee and lap joints shown in Fig. 14-53 should be mild steel plate about ½″ to ¾″ thick. Then the coupons can be positioned as shown for practicing.

Electrode and Current—At this point, a new welder should try welding with either a ³⁄₁₆″ or ¼″ diameter electrode. Although the E 6010 or E 6011 position electrodes can be used for this exercise, an AWS class E 6020 electrode will also work. By changing to different electrodes, a new welder can see the working differences between electrodes.

Using the E 6020 electrode with the same amperage used for an E 6010 or E 6011, a new welder will immediately recognize the need for more amperage. For this reason, the manufacturer's amperage recommendations should be followed when changing electrode types or diameters.

An amperage range of 175-250 for a ³⁄₁₆″ E 6020 electrode and a range of 275-375 for a ¼″ E 6020 electrode should be used. The middle of each range should be "sampled" on a piece of scrap before starting the weld. The voltage for these rods will be from 30 to 36 volts.

Direct current welding machines can use either polarity (reverse or straight) with an E 6020 electrode. The E 6020 electrode is unlike the E 6010 or E 6011 electrodes that are used with only reverse polarity. The E 6020 electrode, though, can only be used in both the horizontal and flat positions.

Procedure—With the machine set on reverse polarity, a new welder can practice running flat position beads on scrap pieces of mild steel. While running these beads, the amperage should be changed several times to discover which current range gives the best bead shape and the best penetration.

The machine can then be set on straight polarity and the same practice repeated to find which polarity better suits the welder.

The E 6020 electrode should be tilted slightly ahead of the weld, as in Fig. 14-4, with the electrode perpendicular to (straight above) the metal. Although the E 6020 electrode produces heavy slag, the finished bead will have an excellent contour. When welding with the E 6020 electrode, the slag should not be allowed to run ahead of the arc.

When welding is to be done inside an angle between two pieces of metal, as in Fig. 14-55, the rod should be held so that the arc will strike in the center of the angle.

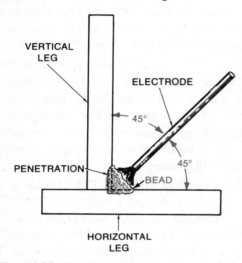

Fig. 14-55. Electrode position when welding inside an angle.

However, it may be necessary to change the electrode angle when one of the pieces being welded is thicker than the other. In those cases, the electrode is pointed more toward the heavier piece, as in Fig. 14-56. This puts more heat toward the heavy piece, so that it melts at a rate more equal to the melting rate of the thin piece. Thus, the two pieces melt at about the same rate, giving better and equal fusion.

The electrode angle is also changed to point directly into any metal having an *undercut,* as shown in Fig. 14-57. By pointing the arc toward the undercut, more weld metal will be deposited in the undercut for a stronger weld. However, the E 6020 electrode has less penetration than the E 6010 or

E 6011 electrode, so it will deposit metal and leave little or no undercutting by comparison.

The fillet weld, as was shown in Fig. 14-54, can be welded the same way as flat welding butt joint preparations.

Some welders use a $\frac{3}{16}''$ diameter electrode for the first two or three layers, followed by a $\frac{1}{4}''$ diameter electrode for the other weld passes.

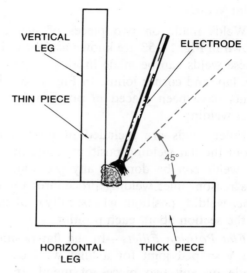

Fig. 14-56. When welding unequal pieces, the electrode is pointed at the heavier piece.

The $\frac{1}{4}''$ diameter electrode is weaved more during the final beads than was the $\frac{3}{16}''$ diameter rod during the first couple beads. The slight weaving, while hesitating at the

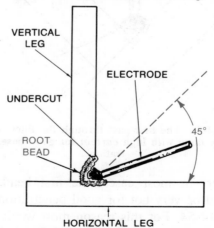

Fig. 14-57. When one piece of metal is undercut, the electrode is pointed more toward the undercut.

point where each plate is touched, gives good fusion.

Vertical Position Welding

Welding in the *vertical* position is done either up or down on an upright surface, as shown in Fig. 14-58.

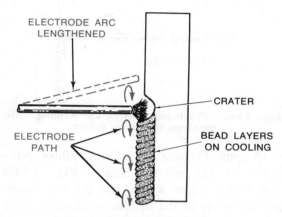

Fig. 14-58. Vertical welding.

Vertical position welding with different amperage settings must be practiced in order to learn about the effect of gravity. Gravity will pull down on the molten metal from both the electrode and the plates being welded. The molten metal, which tends to run down, can be controlled only by adjusting some method of the welding process.

If good welding procedures are used, vertical welding can produce high-strength welds comparable to flat position welding. The welding speed for the vertical position is naturally slower than for welding done in the flat position.

Basic Types of Vertical Position Welding

The most common type of vertical welding is called *vertical-up* welding, as shown in Fig. 14-60. Vertical-up welding is started from the bottom of the pieces to be joined and is better for welding materials ³⁄₁₆″ thick or heavier. Vertical-up welding gives good penetration with less danger of porosity and slag inclusion.

When the weld bead is started from the *top* of the joint and welded down to the bottom, it is called *vertical-down* welding. See Fig. 14-59. Vertical-down welding is better for

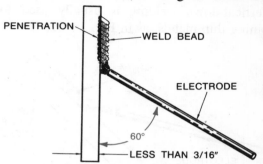

Fig. 14-59. Vertical-down welding is used on metal under ³⁄₁₆″ thick.

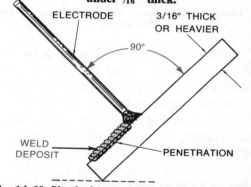

Fig. 14-60. Vertical-up welding is used on metal ³⁄₁₆″ thick or over.

light materials. The welding speed needed to keep ahead of the melted flux and metal in vertical-down welding is fast enough to eliminate burn-through.

Vertical Position Electrodes—Metal less than ⅛″ thick can be welded vertical-down without burn-through if the welder uses an E 6013 electrode. This electrode is used on straight polarity for less penetration and little chance of burn-through on thin metals.

The #3 flux (designated by the fourth digit of the E 6013 rod) makes the molten electrode metal adhere (stick) to the base metal before gravity can pull the molten metal down.

While the E 6013 electrode is good for vertical-down welding, an E 6010 or E 6011 electrode is better for vertical-up welding.

These rods give good penetration at low amperage and have good "fast-freeze" characteristics to counteract gravity.

Practicing Vertical-Down Bead Welding—Vertical-down welding is usually used for joining thin metals, as in Fig. 14-61.

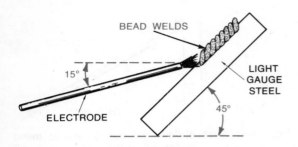

Fig. 14-61. Vertical-down welding, as used on very thin metal.

Coupon Preparation—Thin coupons, less than $\frac{3}{16}''$ thick, are used for vertical-down welding practice. These can be positioned on either a 90° angle, as in Fig. 14-59, or on a 45° angle, as in Fig. 14-61. Because this is a bead practice, no additional preparation is needed.

Electrode and Current—An E 6013 electrode can be used for practice with a current setting near the bottom of the electrode range. With a lower current setting, it will be harder to burn through the thin metal. A $\frac{1}{8}''$ diameter electrode should work well for practicing.

Procedure—In vertical-down welding, the electrode holder should be held so that the rod will make an angle of 60° with the work, as in Fig. 14-59.

The downward feed of the electrode will burn through the joint or pieces being welded if the speed is too slow or the amperage is too high. On the other hand, too fast a downward electrode travel will create a joint with poor or incomplete penetration.

By practicing, a new welder will be able to establish the right amperage and welding speed for good, strong, vertical-down welds.

Practicing Vertical-Up Bead Welding—The welder must use a very short arc when welding a bead in the vertical-up position be-

cause the molten metal will try to run down the plate and onto the arc. The run-down, of course, is caused by gravity. See Fig. 14-62.

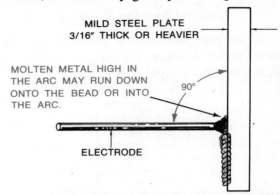

Fig. 14-62. Practicing vertical-up welding with problems caused by gravity.

Coupon Preparation—Using plates positioned at 45° and 90° angles, Figs. 14-60 and 14-62, a new welder can practice establishing an arc at the bottom of the plate. The coupons are otherwise the same as for vertical-down welding practice.

Electrode and Current—A new welder, for practice, could use a $\frac{1}{8}''$ all-position electrode, such as the E 6010 or E 6011, set at about 100 amperes. The 100 amp setting is somewhat less than the setting for vertical-down welding.

Procedure—With the arc held close and steady, the molten bead can be welded up the plate with the rod held as close to 90° as possible, as in Fig. 14-60. Then, the bead is welded steadily upward until the finished bead resembles the bead formed by welding in the flat position.

When vertical-up welding is practiced, the welding heat must not be hot enough to cause overheating and slag run-down, but it must be hot enough for good penetration. It may be necessary to adjust the welding machine amperage for the desired effect.

Vertical Position Weave Welding

When a new welder is able to run beads up and down the plate successfully, he is ready to practice *weave welding* in the vertical position. Actually, weaving in the vertical position is little more than making very wide beads without having the metal fall down on the vertical plate.

Coupon Preparation—Coupons are positioned for weave welding practice as they were for vertical bead welding. The coupons may be ⅛″ or ³⁄₁₆″ thick, positioned at a 90° angle as in Fig. 14-62.

Electrode and Current—The electrode and current recommended for vertical-down welding may also be used for vertical position weave welding.

Procedure—The electrode must have the 90° angle to the plate and a short arc must be held against the plate.

The vertical weave bead generally used when welding up is shown in Fig. 14-63. The semicircular-shaped downward weave shown in Fig. 14-64 is also fairly common.

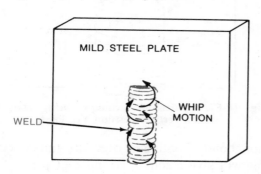

Fig. 14-63. Common vertical weave pattern used in vertical-up welding.

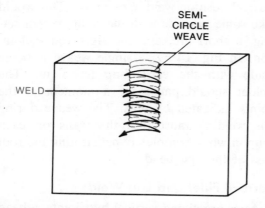

Fig. 14-64. Semicircle weave often used in vertical-down weave welding.

Weave designs, as shown in Fig. 14-65, should also be practiced on 45° and 90° positioned scrap plates. These weld designs will be useful for vertical-up welding on different joints explained in the chapter.

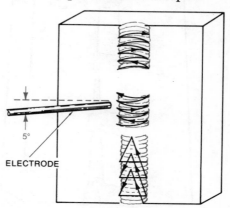

Fig. 14-65. Common weave designs used in vertical position weave welding.

During weave welding, there should be a slight pause at the end of each pass before returning to the starting point, as was shown in Fig. 14-19. This pause, while holding a short arc, will assure a good weld with no undercut. The pause at each end of each pass allows the melting rod to fuse and fill the crater melted in the metal.

Vertical Butt Joints

All the different butt joints welded in the flat position can also be welded in the vertical position by using the same basic techniques with extra care.

Coupon Preparation—Coupons about ¼″ thick can be used with or without joint preparation. Practice coupons for vertical butt joints should be tack welded and then placed in the vertical position with the wide gap at the top for vertical-up welding. The wide part of the gap is to the bottom in the vertical-down welding position, as in Fig. 14-66.

Thicker parent metal surfaces to be welded can be prepared in any desired shape. The value of one preparation over another, however, requires special consideration when the weld is to be stressed in an unusual way. In

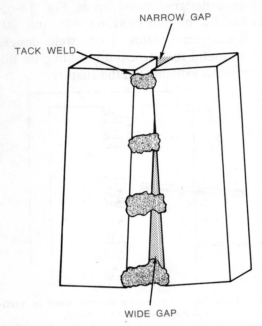

Fig. 14-66. Coupons positioned for a vertical-down butt weld.

most cases, it is possible to eliminate distortion and residual (internal) stress by using a different joint preparation. The U preparation is better than the Vee preparation on especially thick metal. This well help overcome the above welding problems.

Electrode and Current — The amperage used with ⅛″ diameter electrodes on ⁵⁄₁₆″ thick coupons should be from 90 to 100 amperes. If a ⁵⁄₃₂″ diameter electrode is used on ½″ thick metal, the current can vary from 150 to 160 amperes. After some practice, the machine should be set to the amperage that worked best during practice for this type of weld.

Procedure—Again, before beginning the weld, the vertical bead technique should be practiced with the machine on different amperage settings, until the best setting is found.

While vertical-down welding can be used on most metal thicknesses, vertical-up welding is always preferred for thicker metal. The first pass (root pass) is usually a stringer bead.

When the arc is struck and shortened, it is moved into the prepared groove. The metal deposited by the rod is allowed to cool by moving the rod up from the work *twice* the length of the arc and slightly up the groove. Lengthening the arc and moving the rod slows down the cooling of the melted electrode deposit.

Returning the arc close to and above the newly hardened deposit metal will deposit more metal. This deposit fuses the original electrode metal and parent metal in a com-

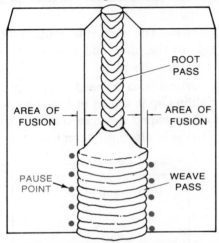

Fig. 14-67. Vertical-up technique being used to weld a vertical position Vee joint.

mon bond. Continued over the full length, this will give a sound, strong root pass.

Using up to ¼″ thick coupons, it is possible, with practice, to complete a single-Vee, vertical position weld in one pass. This would take some practice with the right current setting, a short arc, and the right rod inclination. In Fig. 14-67, a single weave is being made with the vertical-up technique. The welder should pause for a moment at the points indicated by dots. The weave during this welding must fuse the walls on each coupon while completely penetrating the root pass of the first bead.

Vertical Fillet and Lap Welds

After practicing vertical butt joints, a bead or weave weld can be easily applied with little practice on vertical fillet or lap joints. See Fig. 14-68.

Preparation—For these welds, the coupon, electrode, and current recommendations are the same as for the earlier vertical joints. Of course, some changes may need to be made as the welder becomes more experienced and can make fine adjustments on the machine current and electrode selection.

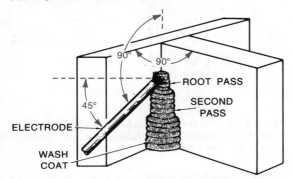

(A) Tee joint, vertical-up welding.

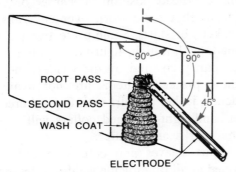

(B) Lap joint, vertical-up welding.

Fig. 14-68. Electrode angle for vertical welds.

Procedure—As with the vertical-up butt joint weld, the electrode contacts the bottom of the Vee formed by the plates to strike the arc. The angle formed by the plates is usually 90° If the pieces of metal are the same thickness, the electrode divides (splits) the 90° angle of the Vee, as shown in Fig. 14-69.

If one of the plates is *thicker* than the other, the electrode should be pointed more directly toward the thicker material, as in Fig. 14-70. By pointing it at the thicker

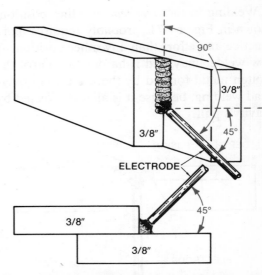

Fig. 14-69. If the steel plates are the same thickness, the electrode is positioned evenly between the plates.

metal, the thicker metal receives more heat, so it melts as fast as the thinner metal.

By moving the electrode, it is also possible to deposit (weld) an equal amount of filler metal on each plate for maximum strength.

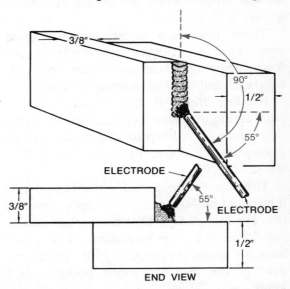

Fig. 14-70. If one plate is thicker than the other, the electrode is pointed more toward the thicker plate.

Overhead Position Welding

Welding in the *overhead* welding position, shown in Fig. 14-71, probably requires more practice and effort than any other position. A new welder may find it harder to control the molten metal formed by the arc during overhead welding, because it is always affected by gravity's pull.

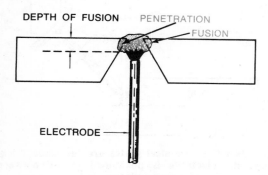

Fig. 14-71. Simple overhead weld.

After practicing vertical-up welding with the E 6010 and E 6011 electrodes, a new welder may want to practice overhead welding using these same electrodes. That way, he will already know about the fast-freeze quality of these rods.

These fast-freeze electrodes give good penetration for full-strength welds at the lower amperage setting necessary for both overhead and vertical welding positions.

Welding Heat—If the amperage is too high during overhead welding, "icicle" beads may result. This happens due to the pull of gravity on the hot molten metal, as in Fig. 14-72.

The heat (amperage) for overhead welding should be *high* enough for good penetration but *low* enough to prevent overheating, run-down, and icicle formation. These problems are often experienced while practicing most of the vertical welding positions and joints.

Personal Preparation—Proper and protective clothing is absolutely necessary when doing overhead welding. Falling sparks and spatter, due to gravity, are very common in overhead welding. See Fig. 14-73. The welder's arms, face, and head must be very well protected, as in Fig. 14-74, to prevent serious burns from falling slag and sparks.

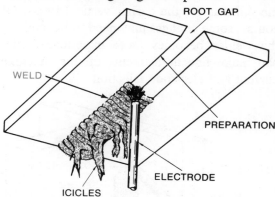

Fig. 14-72. Icicles in overhead welding happen when the hot molten metal falls down and "freezes" like an icicle, due to gravity.

For protection, especially when welding in the overhead position, a leather or asbestos jacket should be worn over coveralls, cloth jackets, or shirts. Some protective jackets are made with a high, tight-fitting collar that can be partly covered by the welder's helmet. If these protective items are not available, a denim jacket or shirt with flap-covered pockets and tight-buttoned collars will help shed the sparks and spatter.

Shirt sleeves and collars should be buttoned at all times while welding. Pants having cuffs should not be worn during welding because the cuffs may trap hot particles and burn the welder's ankles and feet. For that reason, gloves, boots, and other protective clothes should be designed to shed sparks and resist heated particles, as in Fig. 14-75.

The welder must position himself so that he can see the weld area and at the same time avoid being in a direct line with falling slag or molten metal from the overhead weld.

Constant welding creates fatigue, so a new welder should practice welding in different positions, changing his position every so often to avoid fatigue. Many times, welds have to be made on scaffolds or lying in tunnels, so many positions will be necessary.

At times, a welder may need to use both hands on the electrode holder for correct electrode movement, especially when he has been forced into many awkward overhead welding positions. Even so, once the overhead welding position has been practiced extensively while standing up, a new welder will have little trouble in overcoming new positions, such as kneeling or lying down while welding.

Fig. 14-73. Overhead welding produces more problems than usual because gravity can pull sparks, slag, and molten metal down on the welder.

Fig. 14-75. Asbestos leggings and heavy duty shoes protect the welder's legs and feet.

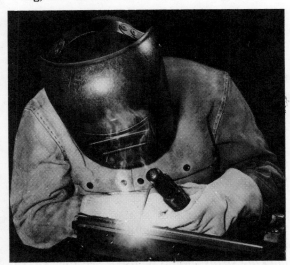

Fig. 14-74. The welder's arms, hands, face, and eyes are well protected from the welding heat, brightness, and sparks.

Coupon Preparation—Steel plate the same thickness and size as that used for the flat position is also good for overhead welding practice. The practice coupons are usually prepared with a single-Vee preparation and may be ¼″ thick.

Once the coupons are prepared, they are clamped in a fixture so that the welder can stand directly under and slightly to one side of them. See Fig. 14-76.

Electrode and Current — For overhead welding practice, an E 6010, ⅛″ diameter electrode should be used on current between 80 and 120 amps, 24 to 26 volts.

When using the same E 6010 electrode as above, but in a ⁵⁄₃₂″ diameter, the current setting should be from 120 to 160 amperes, but the voltage setting is basically the same.

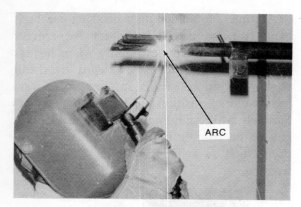

Fig. 14-76. Fixture used to hold the weld in an overhead position.

Procedure—The arc is struck in a method similar to that used for flat and vertical welding. Once the arc is working, however, it must be kept very short all the time. Sometimes, the electrode will have to be pulled back slightly to help transfer the filler metal to the desired position.

The electrode must be placed well into the weld crater, at the bottom of the Vee preparation, using a short, steady arc. As metal is melted off the electrode, the rod is then moved about ⅛" to ⁵⁄₃₂" away from the crater. Moving the electrode this far away is just enough to allow the molten metal to cool. This helps eliminate metal being lost through gravity, forming icicles.

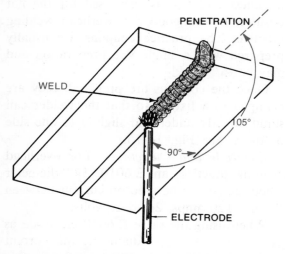

Fig. 14-77. Electrode position for overhead welding.

The angle of the electrode during overhead welding is basically the same as the angle used for flat position welding. See Fig. 14-77.

After one small weld is made, the electrode is returned to a position slightly ahead of the last weld. There, it is allowed to melt and fuse into the last crater before again being moved away. This continuous action is done rapidly without losing arc stability and is known as a *whip* action. It should be continued until the bead is completely welded.

Weaving skill on filler beads and cover passes can also be developed after bead depositing has been learned. Excessive weaving, however, forms large molten pools that can be difficult to control. Fig. 14-78 shows the weave bead and pattern for overhead weaving.

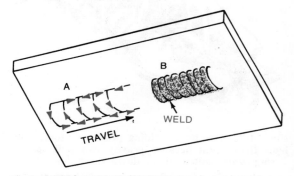

Fig. 14-78. Weave welding in the overhead position. a) Weave pattern used. b) Weld bead formed.

Bead depositing can be practiced from left to right as well as from right to left. When overhead beads with good penetration and contours have been completed, a new welder is ready to try running beads away from him and toward him. Then he will realize what some of the problems are with overhead welding.

Backing Strips

Since it is often difficult to weld with good root penetration on an overhead butt joint, *backing strips* should be used. See Fig. 14-79. For overhead welding, the Vee joint preparation and root spacing should be large

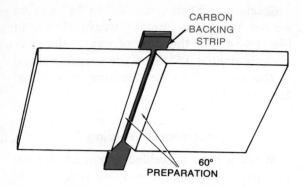

Fig. 14-79. Single-Vee joint with a carbon backing strip for welding overhead.

enough to allow the electrode enough space for a close arc. This will ensure complete penetration. Poor spacing allows a long arc, creating little penetration and rough bead deposits.

Practicing—By welding ¼″ plate in the overhead position while using backing strips, beads are formed as shown in Fig. 14-80. More bead passes are used with heavier plates, forming the passes needed for full strength in the overhead position. See Fig. 14-81.

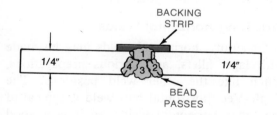

Fig. 14-80. Bead passes using a backup strip when welding in the overhead position.

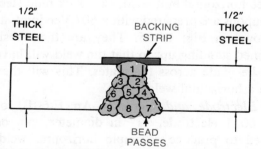

Fig. 14-81. Overhead welding thick plates, using many passes.

As usual, the angle of the electrode is important while welding in the overhead position, as was shown in Fig. 14-77.

Overhead Fillet Welding

The technique for welding Tee and lap joints in the overhead position usually depends on the metal's thickness. These joints can be made with or without weaving.

Coupon Preparation—Coupons used with fillet overhead welding practice should be ¼″ thick. Tee and lap joints, as were shown in Fig. 14-53, don't usually need any edge preparation unless the metal is very thick. Practice coupons ¼″ thick do not need any preparation. The coupons can then be positioned for whatever joint is going to be practiced.

Electrode and Current—The electrode and current used for overhead position butt welds is also good for practicing overhead position fillet and lap welds.

Procedure—The electrode is held at a 60° angle from the horizontal leg for the first pass, as in Fig. 14-82. Then, on the second pass, the electrode is held at a 60° angle from the vertical leg. The second pass fuses the first bead and the horizontal leg. The size of the finished fillet should not be over ¼″. See Fig. 14-82.

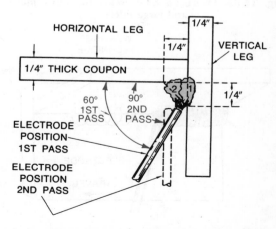

Fig. 14-82. Overhead welding a simple fillet Tee joint.

Generally, the deposited metal should be kept at a minimum thickness and width. Deposited beads should be parallel to each other while using a short arc for overhead fillet welds.

Where overheating and undercutting can be avoided, weaving may be practical if the molten metal can be controlled.

Other Fillet Reminders—On metal ½″ to ¾″ thick, ⅜″ fillet welds should be used. These should be made in three passes, as in Fig. 14-83. If the metals are about the same thickness, fillets up to ½″ can be formed by using up to six passes. See Fig. 14-83.

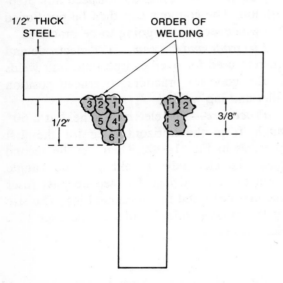

Fig. 14-83. Overhead fillet welds for thick metal and large fillets.

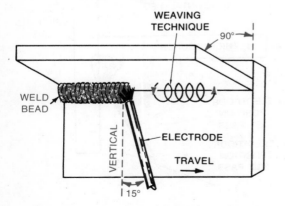

Fig. 14-84. Electrode angle and technique for a single pass, overhead fillet weld.

Single pass fillets from ¼″ to ⅜″ can be welded using the circular weave shown in Fig. 14-84. For this type of weave, the electrode is slanted 15° from the vertical, ahead of the weld and in the direction of welding. See Fig. 14-84.

Horizontal Welding

When a bead is welded across an upright (vertical) surface, the welding is known as *horizontal welding.*

If not properly controlled, molten metal will not fuse properly to upright surfaces during horizontal welding. This results in incomplete fusion and poor looking welds.

Many experienced welders use the force of the arc to control gravity's downward pull on the molten metal. By directing the arc slightly upwards when welding on an upright surface, the molten metal can be controlled and welded where it is wanted.

A fast whipping motion, similar to that used in vertical welding, stops the bead puddle from getting too large. A large puddle may run down the work and leave the weld unfilled, weak, and porous.

Practicing Horizontal Welds

Generally, horizontal welds may be made with laps, fillets, and various preparations, much like the other weld positions. The single-Vee, horizontal butt weld is typical of the other horizontal welds, so it is a good practice weld to begin with.

Coupon Preparation—To make a single-Vee horizontal butt weld, ¼″ thick mild steel coupons are prepared with a 60° Vee joint, as shown in Fig. 14-85. They are then positioned standing up, so that the weld will have to be made across the plates. This will then be a horizontal weld.

Electrode and Current—An E 6010 or E 6011 electrode, ⅛″ in diameter, can be used to practice a simple horizontal weld. The same current (amperage) may be used as was used for vertical weld practice. In this

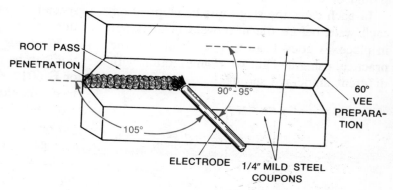

Fig. 14-85. Welding the practice horizontal weld. A single-Vee preparation with the correct electrode angle.

case, a good starting amperage would be 100 amps. As usual, the current may be changed slightly as the welder compares the results of several different settings.

Procedure—The first pass (root pass) is welded with the electrode right between the plates' 60° angle, as shown in Fig. 14-85. (Some welders may want to tack weld the plates first.)

The electrode should be inclined 15° ahead of the weld in the weld direction. (This rod angle is similar to that used for flat position welding.) However, the electrode should not slant down more than 5° from a 90° angle with the upright plate. See Fig. 14-85.

A prepared joint (such as the Vee joint) can be welded in the horizontal position if the first bead is placed at the bottom of the Vee.

The small diameter electrode and a close arc will help weld the first bead with good penetration and fusion.

A second bead is then welded below the first bead. It must completely fuse the lower plate and the first bead, and is welded by holding the electrode at the angle shown in Fig. 14-86.

The third bead is welded by holding the electrode as shown in Fig. 14-87. This rod angle assures complete fusion between the top plate, the first bead, and the second bead.

For welding thicker materials in the horizontal position, more passes (forming additional weld layers) can be welded above the bottom welds.

Backing strips, as shown in Fig. 14-88, depend on the joint fit-up. Where joints are

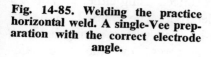

Fig. 14-86. Rod angle for welding the second bead after the root bead has been welded.

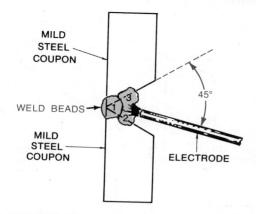

Fig. 14-87. Rod angle for welding the third bead after the bottom two beads are welded.

poorly aligned or gapped, backing strips should be considered for horizontal welds.

In each case, the slag must be chipped off each weld before the next weld bead is put in place. A good bead will result after some practice, using a slight whipping motion with different amperage settings.

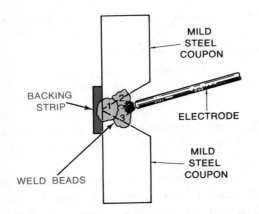

Fig. 14-88. Using a backing strip in the horizontal position.

Horizontal Position Fillet Welds

Many times, horizontal position fillet welds may look like flat position Tee welds. For this reason, the weld position is known as horizontal when a Tee or lap joint has its lower leg in a horizontal (flat) position, as shown in Fig. 14-89. This is a fairly easy joint to practice, although it can often give normal horizontal position problems due to gravity.

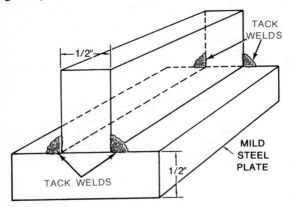

Fig. 14-89. Positioning and tack welding for a horizontal tee joint.

Coupon Preparation—Steel plate ½″ thick is good to use for practicing this weld. No special preparation is needed, except to position the coupons to form the Tee shape.

Electrode and Current—This welding position can be practiced using either an E 6010 or E 6012 electrode. The E 6012, ⁵⁄₃₂″ electrode should be used with a setting between 110 and 190 amperes. Once the correct amperage has been found, from 18 to 24 volts is a good setting for depositing metal.

As with the other weld positions, practicing with different amperage settings within the suggested range will help the welder choose the best setting for good welds under his conditions.

The ⁵⁄₃₂″ diameter E 6010 electrode, using an amperage between 120 and 160, is easier to handle in the horizontal position once the E 6012 electrode has been used successfully.

The E 6012 electrode is used on *straight* polarity because of the makeup of the flux covering, a high-titania sodium coating. This electrode gives medium penetration. The weld bead made by this electrode has a convex, rippled surface and heavy slag.

The E 6010 electrode, while having the same core wire as the E 6012, is used on *reverse* polarity. Due to the E 6010's high-cellulose sodium flux coating, it gives deep penetration with a flat, wavy surface and thin slag.

Procedure—As in other positions, tack welds will hold the plates in position for good welding. Tack welds are first applied to the joint as in Fig. 14-89.

A short arc is started at the beginning of the weld, in a technique similar to that used in the flat welding process. After the short arc is started, a single pass is made at the correct angle, as shown in Fig. 14-90.

The speed used to move the electrode should be such that the bead contour (shape) is similar to the shape explained earlier in the chapter.

When welding a horizontal Tee joint fillet weld, the electrode is usually pointed down the center of the angle. This way, the electrode stays the same distance from each plate, so each plate receives the same amount of filler metal and the same amount of the arc's heat.

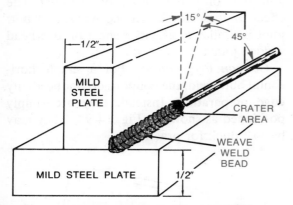

Fig. 14-90. Welding a horizontal fillet weld.

However, if one leg is thicker than the other, as in Fig. 14-91, the electrode angle must be changed so that the electrode points more toward the thicker leg. When the electrode points more toward the thicker piece, the thicker piece receives more heat. Thus, it will melt at about the same time as the thinner piece, so they will be equally melted, and at the same time, for good fusion. Naturally, the finished weld is built up more on the thicker piece, because it received more weld metal when the welding was being done. See Fig. 14-91. The length of electrode travel time, then, will determine the height of the weld bead.

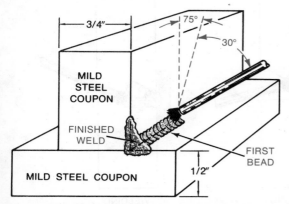

Fig. 14-91. Welding a horizontal fillet weld when the legs are unequal in thickness.

Weave-Welded Horizontal Fillet—The circular weave for a horizontal fillet is produced by an electrode tilted as much as 15° in the direction of travel, splitting the angle of the plates.

The welder should practice these weaves on horizontal fillet fit-ups until the weld compares in shape and size to the weave welds produced on flat plates. Two or more layers of circular weaving, known as multiple-pass horizontal fillet welds, can be welded over the root pass.

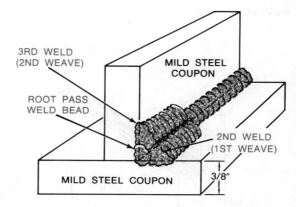

Fig. 14-92. Horizontal fillet made with weave welding.

Where a wider layer is needed than one pass would make, multiple passes can be welded with a circular weave pattern as shown in Fig. 14-92.

Horizontal Position Lap Welds

Lap welds in the horizontal position are relatively easy. Fig. 14-93 shows the features and technique for a horizontal position fillet weld.

Coupon Preparation—Mild steel coupons ¼″ thick can be used to practice the horizontal lap weld. No edge preparation is needed, but the coupons must be positioned and clamped so that the weld can be made as in Fig. 14-93.

Electrode and Current—A ⅛″ diameter E 6010 or E 6012 electrode can be used to practice this weld. Amperage for the E 6010 is 80 to 120, depending on which setting produces the best welds. For the E 6012, the current should be set as much as 20 amps higher for the same job conditions.

Procedure—Tack welds on each end, and in the center, will help control distortion. After the tack welds are made, the arc is re-struck and held close for good penetration in the root pass.

When the root pass is done, the electrode is held at the same angle for the second and third passes. As usual, each pass must be chipped and brushed before the next is welded. See Fig. 14-93.

Horizontal Position Butt Welds

The forces of surface tension and gravity have a greater effect on the horizontal butt joint than on either the lap or Tee fillet. The effect of these forces on the horizontal butt joint is similar to gravity's effect on overhead welded joints.

Coupon Preparation—For a simple horizontal butt weld, the coupons do not need any edge preparation. Instead, they are simply positioned as shown in Fig. 14-94. They may be ¼″ thick for practice.

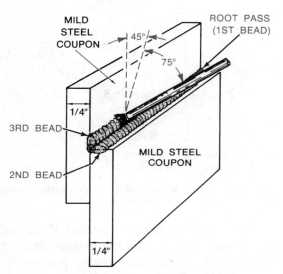

Fig. 14-93. Horizontal position lap weld.

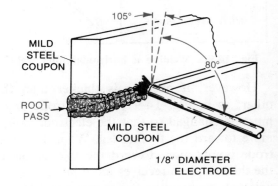

Fig. 14-94. Simple horizontal position butt weld with no joint preparation.

Electrode and Current—The electrode and current for this weld may be the same as for the single-Vee, horizontal butt weld mentioned earlier.

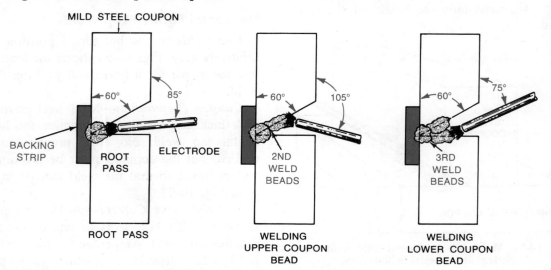

Fig. 14-95. Single-bevel, horizontal butt weld.

Procedure—The welds are first tacked in the center and on each end. Then, the arc is struck at one end and carried along as in Fig. 14-94. The arc must be kept close and the electrode pointed up slightly for good root pass penetration. If needed, then, other passes may be made after the root pass is cleaned and checked.

For a simple butt joint, however, the welder should know that weaving will not give a good butt weld in the horizontal position. The weave creates large puddles that are affected by gravity more than with the bead weld.

Single-Bevel Horizontal Butt Weld

As shown in Fig. 14-95, this joint often uses a backing strip for good root joint penetration. It may be easier to practice than other horizontal welds because of the "ledge" formed in the weld by not preparing the lower coupon.

Coupon Preparation—The upper coupon is prepared by grinding a bevel on it as shown in Fig. 14-95. The coupons should be ⅜" thick for practicing this weld.

Electrode and Current—A 5⁄32" diameter E 6010 electrode should be used to practice this joint. Although the amperage range for this electrode is 150 to 210 amperes, an amperage of 140 to 160 should be used for practice. This is lower than would be used for fillet and lap welds in the horizontal position.

Procedure—The electrode angle needed for each bead is shown in Fig. 14-95. After the joint is tacked, each bead can be welded in order, as in Fig. 14-95. Here, it is especially important to remove all the slag before making each new pass.

Basic Arc Welding Review

This chapter has given a new welder his first experiences in arc welding. Explanations have been given concerning arc welding for different joints. Any problems or mistakes encountered by the new welder in these prac-

tice welds may be eliminated through more practice.

Overall Reminders

Although there is nothing quite like practicing arc welding for experience, several important reminders are necessary.

A. Arc welding rays are very dangerous. An arc welding hood must always be worn to protect the welder's eyes.

B. Arc welding produces sparks and hot metal. Clothes must be chosen with that fact in mind. They must be worn at any time when arc welding is going to be done.

C. Chipping an arc weld will produce hot, sharp slag. Clear goggles, of course, must be worn when chipping slag.

D. Beginning welders should follow the electrode manufacturer's specification chart when welding with an unfamiliar electrode.

E. All adjustments to the arc welding machine (voltage, amperage, polarity) must be made with the machine turned off.

F. Generally speaking, a shorter arc is used for root passes; also in any welding position where gravity could pull the molten metal down, away from the weld.

G. The force of the arc may be used to move the molten metal where it is needed within the weld area.

H. When arc welding several beads in one joint, each bead must be carefully chipped free of slag before the next bead is welded. Otherwise, the finished joint will have slag and air pockets trapped in it.

I. As the electrode melts, it gets shorter. For this reason, the electrode must be evenly and constantly (all the time) lowered during arc welding.

J. While the simple butt joint may serve many welding jobs, edge preparation may be needed for thick metals or for joints that will be highly stressed.

K.   Good arc welding practice depends on trying slightly different current set-tings, electrodes, and electrode positions. In this way, any welder, new or experienced, develops his own personal technique for good arc welding.

15

Electrodes

Both welding rods and consumable welding electrodes add filler metal to the weld. Usually, the filler metal in oxy-acetylene welding is called a *filler rod*. In arc welding, the filler metal is called an *electrode*.

As with filler rods, there are many different types of electrodes on the market. There are even more types of electrodes, though, because the electrode must conduct current as well as melt into the weld. For this reason, an electrode needs more properties than does a simple filler rod.

Classifying Electrodes

AWS Number Codes—Electrodes used in the United States and Canada are approved by the American Welding Society. The electrode number on the box means that the electrodes have met the standards set up by the AWS. A code number is used to identify each box of electrodes and certain colors are used to identify each property.

The American Welding Society establishes the classifications for various types of electrodes. Usually, an electrode is placed in only one classification, even though it may meet the requirements of several other classifications. The AWS allows certain exceptions for some rods, in which case both classifications are given.

NEMA Color Codes—The National Electrical Manufacturers Association (NEMA) establishes various color codes that are painted on the electrodes to distinguish the various types. These types are painted with small dots of the color, as shown by the letters E, S, and G in Fig. 15-1.

Not all electrodes have classifications or color codes. Some types have color codes, but not classifications. In some cases, there are two different electrodes with the *same* classification, color code, and coating color. Only their diameters are different.

The prefix "E" (electric) says a rod is used for arc welding. In mild and low alloy steel electrodes, the first two digits (such as 60) indicate the minimum tensile strength (TS) in thousands of pounds per square inch (psi). The first two digits of four-digit numbers and the first three digits of five-digit numbers both indicate tensile strength. See Table 15-1.

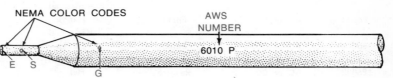

Fig. 15-1. Markings on a typical electrode.

Table 15-1. Tensile Strength as Indicated by Different Electrode Prefixes.

Electrode Prefix	Tensile Strength (psi)
E60	60,000
E70	70,000
E80	80,000
E90	90,000
E100	100,000
E110	110,000

The third digit indicates in what position the rod can be used. For instance, if the third digit is 1, the rod can be used in all positions (flat, vertical, horizontal, and overhead). If the third digit is 2 (E6020) the rod will operate in only two positions: horizontal and flat. If the third digit is 3 (E6030) the rod will work in only the flat position.

The fourth digit is used for some special characteristic of the electrode. What the fourth digit refers to depends entirely on what digit it is. The fourth digit may be 0, 1, 2, 3, 4, 5, or 6. Each of these is defined below:

0—This electrode will leave flat or concave beads with deep penetration. It produces a high-quality weld and

should be used with DC straight polarity. If the last *two* digits are 20 or 30, either polarity DC, or AC, may be used.

1—This electrode will produce flat to slightly concave beads with deep penetration. It will produce high-quality welds when used on AC or reverse polarity DC.

2—Medium-quality deposits, a medium-sized arc, and fair penetration are produced with this electrode. It has convex bead heads and may be used on DC straight polarity or AC.

3—This electrode has a soft arc, producing only shallow penetration with slightly convex beads. AC current or either polarity DC current may be used with this electrode to produce medium-to-high quality deposits.

4—This electrode is used for deep groove butt, fillet, and lap welds with medium penetration and easy slag removal. It may be used on either DC positive or AC current.

5—This electrode should be used on DC reverse polarity. It will produce flat to

Table 15-2. Welding Requirements of Common Electrodes.

Rod	Type of Current	Polarity	Position	Flux Coating	% Iron Powder
E6010	DC	Reverse	All	Cellulose-sodium	0-10
E6011	AC or DC	Reverse	All	Cellulose-potassium	0
E6012	AC or DC	Straight	All	Rutile-sodium	0-10
E6013	AC or DC	Either	All	Rutile-potassium	0-10
E6014	AC or DC	Either	All	Rutile-iron powder	25-40
*E6015	DC	Reverse	All	Low-hydrogen-sodium	0
*E6016	AC or DC	Straight	All	Low-hydrogen-potassium	0
*E6018	AC or DC	Reverse	All	Low-hydrogen-iron powder	25-40
E6020	AC or DC	Straight	Flat or Horizontal	Iron oxide-sodium	0
E6024	AC or DC	Either	Flat or Horizontal	Rutile-iron powder	50
E6027	AC or DC	Either	Flat or Horizontal	Iron oxide-iron powder	50
*E6028	AC or DC	Reverse	Flat or Horizontal	Low-hydrogen-iron powder	50
E6030	AC or DC	Straight	Flat	Iron oxide	

*Low-Hydrogen Rods

slightly convex beads with high-quality deposits and a soft arc. Also known as *lime ferritic,* it has only moderate penetration.

6—This electrode is similar to an electrode with a #5 last digit, except that the #6 should be used with an alternating current.

In Table 15-2, common electrodes are listed with their welding requirements.

When comparing electrodes, keep in mind that the wire in all of the above electrodes is the same, including the wire in the E6015 and E6016 electrodes. It is the *coating* (flux) that causes the different handling and deposited weld metal qualities. It should also be noted that the elongation (stretch) ability goes down as the tensile strength goes up. Therefore, a welder should not expect an E80XX or E90XX electrode to have the elongation (ability to stretch) of an E60XX electrode.

Coated Electrodes

Electrodes with a flux around the wire are called *coated* electrodes. The coating may be applied by spraying, painting, or dipping. The purpose of the flux is to unite with the impurities in the weld metal and float them away as slag or gas. Oxides and nitrites that would normally reduce the weld strength are partly excluded by a short arc, but the flux added to the rod further protects the metal with a gaseous shield. Each manufacturer develops his own special electrode coating.

Flux not only takes out impurities but also adds new elements necessary for strong welds. The weld speed is increased by the flux, and the flux's cleaning action helps produce good finished products.

Manufacturing Coated Electrodes

Each type of electrode is manufactured differently because of the difference in the coating materials. A coating may represent as little as 1% of the electrode's weight. The coating may be applied by dipping the wire into a quick-drying liquid flux. This type of coating is then known as the medium, or semi-coated.

Heavily coated electrodes are made by a type of extrusion where the coating is forced onto the wire as it passes through a press. Another method is by dipping the wire into a liquid flux several times, with a drying time between each dipping.

Binders—A solution of sodium silicate, gums, lacquers, and various glues are the most common materials used for *binding* the flux to the wire.

Coating—The *coating* on the electrodes is either cellulosic (made of cellulose) or mineral. The flux on some electrodes contains a mixture of both types.

Cellulosic coatings usually burn away to form a gas in the arc. These coatings are being replaced by coatings made from sodium silicate or titanium dioxide that do not burn away so quickly.

Mineral coatings, on the other hand, are made of silicates and metallic oxides. Rods covered with mineral coatings produce a great deal of slag over the weld area, protecting it from the harmful effects of the atmosphere.

Types of Flux Covering (Electrode Coating)

High Cellulose Content—Electrodes have rapid burn-off and deep penetration with this flux. Welds with a high cellulose flux electrode can be made in all positions, but they are usually used for vertical-down pipe welding and structural work. Usually, rods with this type of flux are used with DC reverse polarity.

Titania Content—Flux containing titania (as a rutile) makes welding easier. Using this flux, the molten metal is supported by a thick, heavy slag, making the electrode suitable for vertical and horizontal fillet welds. Rutile (titanium dioxide) is fast becoming the most important substance for mineral-coated electrodes.

Titania and Basic Compounds—Flux containing titania and added basic compounds (ferro-manganese, sodium silicates, feldspar) are similar to the one above. However, they have a more fluid slag that produces a smooth arc with medium penetration, suitable for all-position welding.

Electrodes with the above two types of flux can be used on DC machines with either polarity, or AC.

Manganese, Iron, and Silicates—This flux consists of *oxides* or *carbonates* of manganese, iron, and silicates, making slag removal easy. Electrodes containing this flux are used in the flat position, usually with DC reverse polarity.

Iron Oxide Content—Iron oxide flux gives low penetration resulting in lower tensile strength. However, the welds made with iron oxide flux are smooth in appearance.

Calcium Carbonate Fluxes—This is a low-moisture flux made of calcium carbonate with fluoride as limestone and fluospar. These are low-hydrogen electrodes, popular when welding high-carbon or high-sulfur steels. Calcium carbonate fluxes must be kept dry. When iron powder is added, higher heats are possible with smaller electrode cores.

The main stabilizers in these coatings are calcium carbonate, feldspar, and titanium dioxide. To control porosity, a deoxidizer of ferro-manganese is added. Molybdenum or other alloys may also be added to the coatings.

Choosing An Electrode

Electrodes are usually chosen on the basis of job requirements. A welder must look the job over carefully to figure out just what the electrode must be able to do.

Within the selected group, an electrode should be chosen that has the physical properties and operating characteristics that the job needs. Special characteristics of other electrodes in different groups should be checked to be sure that all possible choices have been considered. If more than one electrode appears suited for the job, it should be tried on the job. One electrode may prove to be better for that particular work.

Basically, an electrode should be chosen by looking at these qualities of the welding job:

A. Properties of the base metal.
B. Position of the joint.
C. Type of joint.
D. Amount of welding required.
E. Tightness of the joint's fit-up.
F. Type of welding current available.

E6010 (Fast-Freeze) Electrode

E6010 electrodes have a high cellulose content in the coating, producing large volumes of carbon dioxide and water vapor. This protects the deposited metal from the atmosphere even though the slag deposit is very low. In most cases, the slag can be removed easily with a wire brush. Since the slag deposit is low, the weld cools quickly, making the electrode easy to use in vertical and overhead positions.

The E6010 and E6011 electrodes are known as *fast-freeze* electrodes. They have the ability to deposit a weld that solidifies (freezes) rapidly. This is important where there is some chance of slag or weld metal spilling out of the joint, as in the vertical or overhead positions.

Joints requiring deep penetration, such as square edge butt joints, are welded in the flat position with the larger sizes of these electrodes. Galvanized steel is best welded with these electrodes because the forceful arc bites through the galvanizing and the light slag reduces bubbling while helping prevent porosity.

Sheet metal edge and butt welds on 10-to-18-gauge steel are welded with these electrodes while using straight polarity. This produces a fine "spray" arc with little penetration and excellent *fast-follow* ability. (Fast-follow is the ability to deposit a small bead on 10-to-18-gauge sheet steel with high speeds and few skips.) This rod can also be used on straight polarity for light sheet metal while using very low heat.

Since the quality of the deposited metal is good, the electrode is used for structural steel erection and pipe welding. General-purpose welding is usually done with these electrodes, especially when most of the work is out of position, dirty, or greasy. Deep penetration produces the best possible results under these adverse conditions, with the added benefit of only light slag.

Since this rod is listed as an E6010 electrode, it should have 60,000 psi tensile strength. However, the deposited weld metal, unless stress-relieved or annealed, will probably have a higher tensile strength. It may be from 65,000 to 75,000 psi, or even higher if the weld was cooled quickly.

The elongation will probably be 22% to 28% in two inches. If stress-relieved, the tensile strength will be 60,000 to 72,000 psi, and the elongation will be about 29% to 37% in two inches.

Stress-relieving a weld may help prevent cracking while the weld is in service. The tensile strength will still be above the classification range of 60,000 psi. In Fig. 15-2, the welder is using the ball end of a peening hammer to stress-relieve the weld.

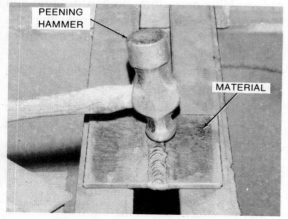
PEENING HAMMER

MATERIAL

Fig. 15-2. Peening a weld to relieve stress.

This applies to all welding, but the changes in the tensile strength and elongation vary with each different rod classification.

E6011 (Fast-Freeze) Electrode

E6011 electrodes are made to be used with an AC welding machine. They produce results equal to the E6010 rod, which will not work with an AC machine. The E6011 rod will also work with a DC machine and can usually be used on any work where an E6010 rod is specified.

Fill-Freeze Electrode

Fill-freeze is the name given to electrodes that combine both fast-freeze and fast-fill characteristics to some degree. There is considerable difference among all the fill-freeze electrodes. Some electrodes are mostly fast-fill with some fast-freeze, while others have less fast-fill and much more fast-freeze. Several electrodes in the fill-freeze group are also called fast-follow, as discussed earlier.

Fill-freeze electrodes are usually classed into four groups: E6012, E6013, E6014, and E7014.

E6012 Electrode—The E6012 electrode can be used in all positions. It is a reliable general-purpose and production electrode with a forceful arc. The E6012 electrode should be used with straight polarity DC. Iron provides very good deposit rates for the E6012, and the rod also operates well on AC. Excellent for low-current applications, it resists sticking with a smooth and steady arc. The low-current E6012 gives minimum spatter and allows easy slag removal.

Although this electrode does not have the stretching qualities of an E6010, it has a very useful place in welding because its heavy slag covers the weld, protecting it from the atmosphere. If the E6012 is properly applied, slag removal is easy, leaving the weld metal bright and clean. Penetration is somewhat lower than the E6010, but the burn-off is faster; this makes it economical for production work.

Although the E6012 is an easy electrode to handle, slag can be trapped in the weld if the welder isn't careful. Then, the root of the weld may not be welded. On large weldments and in cold weather, the first and second beads may crack if they are not preheated. The E6012 is used extensively for building up mild steel shafts, sheet metal, fabricating structural steel, and general welding.

Although the composition of the core wire is the same as the E6010, the *as-welded* results are different. As welded, the tensile strength is about 70,000 to 82,000 psi. Elongation in 2 inches is about 18% to 23%. Stress-relieving the weld, the approximate results would be: tensile strength, 65,-000 to 80,000 psi; elongation in 2 inches, 24% to 27%.

E6013 Electrode—The E6013 electrode is very popular because of its flexibility in operation. It provides excellent AC operation and is softer and steadier with less sticking than the E6012 electrodes; however, it is somewhat slower. It is widely used on sheet metal when appearance and ease of operation are more important than speed. It is also used for general-purpose welding with smaller, limited-input, low open-circuit voltage welders. Spatter loss is low, and the beads are

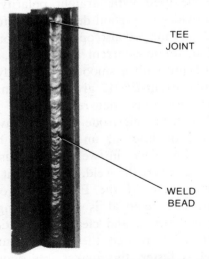

TEE JOINT

WELD BEAD

Fig. 15-3. Tee joint fillet weld, using an E6013 electrode.

bright, smooth, and flat, with easily removed slag. See Fig. 15-3.

E6014 and E7014 Electrodes—E6014 and E7014 electrodes contain iron powder, providing the best fast-fill ability of all the electrodes in this group. Many welders like these electrodes, and their main use is on production welding of irregular-shaped parts where some vertical-down welding must be done. This electrode is actually a slightly changed E6012 and E6013 electrode, but it has better physical properties with lower spatter loss. Also, the E6014 will stand more heat and has a better rate of deposition.

Joints with poor fit-up can be welded with the E6014 and E7014 fill-freeze electrodes when fast-fill electrodes (which would normally be used with good fit-up) may burn or spill through. These electrodes are often used in general-purpose welding, particularly when only one or two electrodes will be used for all welding jobs. High-speed lap and fillet welds on 10-to-20-gauge metals are welded with these electrodes. On those jobs, fast-follow ability becomes important.

Low-Hydrogen Electrode

Low-hydrogen is a term used to describe electrodes with coatings containing practically no hydrogen. These electrodes produce welds that resist underbead and micro-cracking and have exceptional ductility. They simplify welding procedures on hard-to-weld and high tensile alloy steels by reducing the amount of preheat needed. Another feature of low-hydrogen electrodes is that porosity is eliminated when welding steels with high sulfur content.

E6015 Electrode—This type of electrode should be used on direct current, reverse polarity. It is very similar to the E6016, with the exception of the current used. The qualities of both the E6015 and E6016 will be described together under the E6016.

E6016 Electrode—This electrode is similar to the E6015, mentioned above. The

E6016, however, can also be used with DC straight polarity or AC.

The E6015 and E6016 electrodes were first developed for welding steels that cracked in the bead area when welded with ordinary electrodes. Steels that were then classified as hard-to-weld contained small percentages of chromium, nickel, molybdenum and other alloying elements that produce medium- and high-carbon steels. It was found that hydrogen was responsible for the bead cracking in these welds. Therefore, the flux on the E6015 and E6016 electrodes consists of materials that produce practically no hydrogen when the flux is burned.

With E6015 and E6016, high heats can be used with little or no undercutting and little spatter loss. Both preheating and post-heating are recommended when materials likely to crack are being welded. Although these electrodes are all-position types, it is recommended that electrodes over $5/32''$ in diameter not be used for vertical and over-head welding.

E6018 Electrode—This electrode, like the other low-hydrogen electrodes, depends on the flux for its capacity to weld hard-to-weld materials. The core wire, as usual, is the same as the core wire used in mild steel electrodes. The E6018 is used with either direct current, reverse polarity or alternating current. Penetration is shallow, even when using a short arc with little spatter loss.

Other Low-Hydrogen Electrodes—E7018 and E6018 electrodes also meet the requirements of E7016 and E6016 electrodes and are for out-of-position (other than flat) work. Iron powder in the coating gives high deposit rates, considering that the welding is being done out of position. The E7028 and E6028 electrodes have a high iron powder content that is responsible for the deposit rates and welder popularity of fast-fill electrodes; yet these electrodes retain the quality of low hydrogen deposits. The slag cleans easily, and they are used mainly on flat and horizontal welds.

Welds can be made successfully on all types of joints on hard-to-weld and high-tensile steels, including out-of-position work, when low-hydrogen electrodes are used. Using a low-hydrogen electrode reduces or eliminates the need for preheating carbon or alloy steels and produces porous-free welds on sulfur steels. A low-hydrogen electrode will also help eliminate hot-shortness in steels containing phosphorus.

E6020 Electrode

This rod should be used only in the horizontal welding positions. Due to its extremely good finish and ease of handling, it is often used for fillet welding. The penetration is excellent, so the fit-up must be good for best results. Due to the excellent quality of the weld metal, it is highly recommended for X-ray work on vessels. Larger rods can be used due to the high heat input, making flat welding faster and more economical. The weld qualities with this electrode are excellent, with little spatter loss.

Iron Powder Electrode

Using *iron powder* in electrode coatings has opened up a new field for the welding industry, and some experiments have been very successful.

The iron powder in the coating speeds up the weld deposit. Although some electrodes have as much as 60% of the coating in iron powder, others have as little as 3%. When large amounts of iron powder were added to the E6012-E6013 group, it became E6024. The E6024 electrode, however, cannot be used in all positions. When iron powder was added in large quantities to the E6020, it became E6027. When a small amount was added to the E6015 and E6016 electrodes, the new electrode became known as the E6018. When large amounts were added, it became the E6028 and could not be used for all-position work.

E6024 Fast-Fill Electrode—This is an iron powder, flux-coated electrode that welds equally well on reverse or straight polarity direct current or on alternating current. It has been found that the E6024, when used on mild steel for horizontal fillets and flat welds, gives a good convex bead appearance. Penetration is shallow and slag removal is easy.

E6027 Fast-Fill Electrode—This rod has a high percentage of iron powder in the flux. It operates well on alternating current or either polarity direct current for flat fillet welding. Either straight polarity direct current or alternating current is more satisfactory when welding in the horizontal position.

The E6027, E6024, and E7024 electrodes will all deposit metal rapidly. Their fast-fill might be considered the opposite of fast-freeze and is the most outstanding characteristic of this group of electrodes. E6024 and E7024, the heavily coated, iron powder electrodes, have high deposit rates, produce exceptionally smooth beads, and have a thick, dense slag that tends to peel off the weld. Operating qualities are the biggest difference between the two electrodes.

Small sizes ($\frac{1}{8}$″-$\frac{3}{16}$″) of E6024 and large sizes ($\frac{7}{32}$″-$\frac{5}{16}$″) of E7024 have faster speeds, higher deposit rates, and smoother arc action than their counterparts. By contrast, small sizes of E7024 and large sizes of E6024 offer greater arc force, better control of the molten pool during arc blow or when the work is in the vertical position, and a flatter bead shape.

The E6027 also has a heavy iron powder coating and high deposit rates. Although the E6027 has an excellent bead appearance, the bead is not quite as smooth as the bead of the electrodes above. Its slag is crumbly and easily removed from any joint. The E6024 electrodes penetrate lightly, so there is little pickup of alloy from the base metal. They have high-strength deposits and the deposits, in crack resistance, are almost as good as those of low-hydrogen electrodes.

E6030 Electrode—This electrode is recommended for heavy flat welding only. The E6030 is being replaced, to a large extent,

by the E6027 because the speed of the E6027 is faster. The E6030 will work on either AC or DC and provides excellent results when welding under X-ray conditions. The slag comes off easily, and the finish is smooth, clean, and bright.

70XX, 80XX, 90XX, 100XX Series Electrodes

Rods with 7010 and 70XX ratings are usually alloyed. The alloys are often mixed in the flux coating where they will mix with the molten metal in the weld puddle. For this reason, a welder must carefully try for good puddle control on out-of-position welds for good results. It should also be noted that, even with the increased tensile strength, the ductility *as welded* is approximately the same as the E6012 electrode. The tensile strength range for the 7010 is 77,000 to 82,000 psi, while the E6012 is from 71,000 to 82,000 psi. An even tensile strength is assured with the E6012 electrode if it is properly welded.

80XX, 90XX, and 100XX electrodes are alloyed, special-purpose electrodes. It is always safe to use the current and positions as recommended by the manufacturer.

Practice Review Using Different Electrodes

In the earlier chapter on basic arc welding, practice welds were generally made with the same electrode. This electrode would be chosen as an "all-purpose" electrode, for common practice experiences in arc welding.

During arc welding, there are many ways to change a welder's results. Earlier, different currents and polarities were discussed, to show how they affected the weld joint. Also arc length, electrode diameter, and welding positions were discussed. All these factors are very important for good welding and for correcting any mistakes made while welding. They might be called *variables*.

A variable is something that can be changed, because to *vary* something means to *change* it. A welder can vary the electrode diameter by changing electrodes, he can vary the current by adjusting the welding machine current control, and so on. All these factors that he can change are the variables in arc welding.

An important variable is the electrode itself. Because different electrode types (not diameters) have different qualities, a welder can choose an electrode to produce different results.

This part of the chapter will deal with practicing the same welds using different electrodes. Many of the arc welding procedures will be a review from the earlier chapter on basic arc welding. These will provide a good basis for better understanding of the newest variable: the *electrode*.

Mild Steel Electrodes

Most electrodes used are made of mild steel and vary only in the flux covering. Since almost all electrodes are mild steel, welding with them varies only because the flux has different characteristics. Each electrode, or groups of two or three, can be looked at individually.

E6010 and E6011

These are the most common electrodes used for welding practice. For practicing, the E6010 electrode should be used on DC reverse polarity. Although the E6011 electrode is highly rated on AC machines, DC reverse polarity also works well. The current for DC welding should be 10% less than the current used on the AC machine.

Flat Position—In the *flat* position, a close arc with fast enough travel speed to stay ahead of the molten pool is required. In Fig. 15-4, the welder is set up to practice a flat weld with an E6010 electrode. Note his electrode position and safety equipment.

Vertical Position—The *vertical* position gives better penetration when a vertical-up

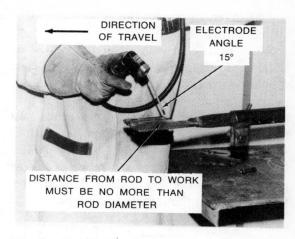

Fig. 15-4. Set up to practice with an E6010 electrode.

weld is made. Vertical-down, however, allows faster welding. For that reason, pipeline welders often use the vertical-down method. See Fig. 15-5.

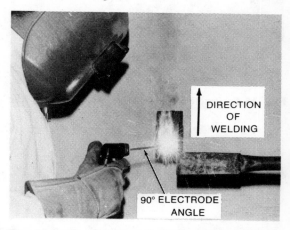

Fig. 15-5. Practicing vertical-up welding with the E6010 electrode.

Overhead and Horizontal Positions—Practicing *overhead* and *horizontal* butt welds with the E6010 electrode is shown in Figs. 15-6 and 15-7. These welds can best be made with stringer beads.

E6012 and E6013

Both of these electrodes use DC straight polarity for good performance on all applications except where arc blow is a problem. Where arc blow is a problem, several steps can be taken to correct the situation, as discussed earlier.

When practicing with either the E6012 or E6013, forehand welded stringer beads should be used for the first pass, although on joints with poor fit-up, a slight weave is preferred. Either stringer or weave beads can then be used for succeeding passes with a 1/8″ or shorter arc. As usual, tip the holder in the direction of travel and move as fast as possible while keeping a desired bead size. Use currents in the middle to high portion of

electrode current range, as is being done in Fig. 15-8.

E7014

This is an all-position, high-speed iron powder electrode. It can be used with normal arc techniques, or it can be used with the *drag* technique shown in Fig. 15-9. With the drag technique, the electrode is held at more of an angle and pulled along inside the groove. This electrode has very easy slag removal with very low spatter loss. It is used

Fig. 15-6. Practicing overhead welding with the E6010 electrode.

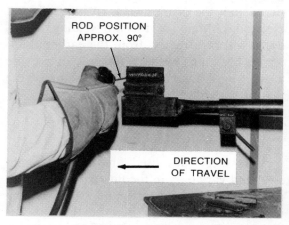

Fig. 15-7. Practicing horizontal welding with the E6010 electrode.

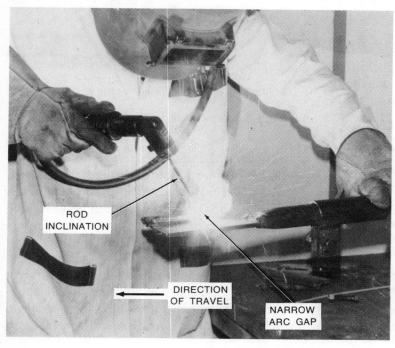

Fig. 15-8. Using the E6012 electrode to practice forehand arc welding in the flat position.

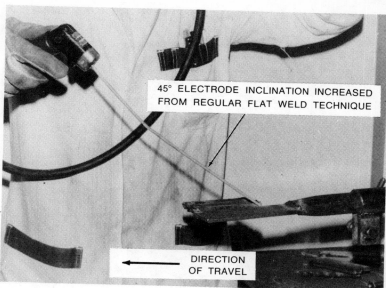

45° ELECTRODE INCLINATION INCREASED FROM REGULAR FLAT WELD TECHNIQUE

DIRECTION OF TRAVEL

Fig. 15-9. Using an E7014 electrode with the unusual drag arc welding technique.

for many out-of-position welding jobs, especially vertical down.

Because of low spatter loss and the adding of iron powder through the coating, higher current rates and welding speed are possible on production work.

The E7014 electrode may be used on AC or either polarity DC. On thin materials, DC straight polarity is preferred, especially for fillet or lap welds. A higher inclination angle can be used on fillet welds to help prevent the problem of double beading.

E6024, E6027, and E6074

With these rods, AC current produces the best speeds and operating characteristics. Reverse polarity DC may also be used, although it may cause arc blow and complicate control of the molten puddle.

When welding in the flat position with these electrodes, use the drag technique and deposit stringer beads. The electrode should be held perpendicular (straight up) with the holder tipped forward about 30° in the direction of travel.

E6018, E6028, E7018, and E7028

These low-hydrogen electrodes must be kept dry. They should either be used as soon as they are taken from freshly-opened con-tainers, or they should be stored in a warm, dry place. If these steps are not taken, they will pick up moisture from the air and lose some of their low-hydrogen properties. An old upright refrigerator with a small heat lamp inside will serve the purpose. See Fig. 15-10.

These rods may be used with alternating current or reverse polarity direct current. They should usually be welded using the maximum amperage that the job will permit, within the recommended range. Because they are low hydrogen, they should be welded with a short arc to keep atmospheric gases out of the molten metal.

In practicing with these electrodes, drag the electrode lightly or hold a shorter arc, not over ⅛". A long arc should not be used at any time, since these electrodes rely on molten slag for shielding. Stringer beads or small weave passes are better than wide weave passes.

When these electrodes are used in the vertical position, they should be welded vertical up with ⅛" of 5/32" diameter electrodes and a triangular weave. A shelf of weld metal is built and, with a weave technique, layer upon layer of metal is deposited as the weld progresses up the joint, as shown in Fig. 15-11.

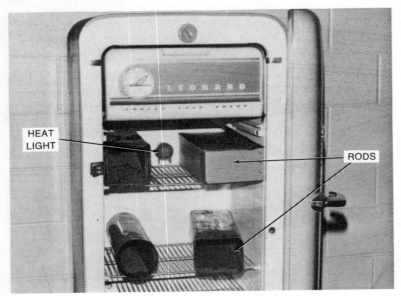

HEAT LIGHT

RODS

Fig. 15-10. An old refrigerator converted to dry storage for welding rods, with a heat lamp.

In overhead practice with these electrodes, ⅛" or ⁵⁄₃₂" diameter electrodes should be used with a slight circular motion in the crater. A short arc is maintained. Motions should be slow and deliberate. The electrode is pointed directly into the joint and the tip holder tilted slightly forward in the direction of travel. Practice moving fast enough to avoid spilling the weld metal, although the slag will spill to some extent. Use welding currents in the lower portion of the recommended range.

Stainless Steel Electrodes

Excellent welds can be made on most grades of stainless steel while using only a few standard electrode types. Generally, higher alloy content means higher corrosion-resistance. Therefore, the properties of the base metal are often *exceeded* by welding with a higher alloy content electrode. The high welding heat sometimes causes changes in the base metal that can seriously reduce the steel's ductility or corrosion-resistance.

Using an electrode with a different type of stainless steel than the parent metal will sometimes eliminate the need for heat treating before or after welding. Special rods are made that match the characteristics of most different types of stainless steels. However, the majority of welding jobs for the general welder can be easily welded if he uses general-purpose stainless steel electrodes.

Before presenting the usual physical properties and practice methods used with these electrodes, it would be a good idea to review the information on the two general categories of stainless steels. The two basic groups of

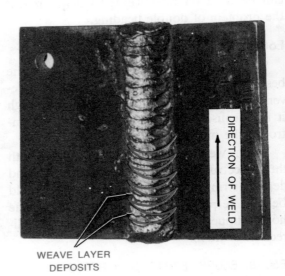

DIRECTION OF WELD

WEAVE LAYER DEPOSITS

Fig. 15-11. Vertical weld made with weaving technique.

stainless steels are *straight chromium* and *chromium-nickel*.

Straight Chromium Stainless Steels

There are two straight chromium types of stainless steel: ferritic chromium and martensitic chromium.

Ferritic Chromium—These stainless steels have low carbon content compared with their chromium content, so they have a ferritic structure at room temperature. After these alloys are heated, they usually cannot be hardened by heat treatment.

When heated to high welding temperatures, all ferritic chromium stainless steels may become brittle and lose their corrosion-resistance. When this happens, the result is a large-grained weld zone that is relatively soft *and* brittle. The grain size remains coarse even after annealing, although the brittleness may be removed.

Martensitic Chromium — These stainless steel alloys have properties like an alloy steel with high hardenability. Austenite is formed when they are heated. During rapid cooling, the austenite is transformed into hard and brittle martensite. These alloys can be heat-treated to any specific hardness required for a certain application.

The hardness of martensitic chromium stainless steel increases as the carbon content increases. As hardness increases, the tendency to crack as the weld cools also increases. To control the tendency to crack, preheating and postheating are often required.

Chromium-Nickel Stainless Steels

The *chromium-nickel* steels are alloyed so that the austenite is stable at room temperature. Since the austenite does not change when the steel is cooled, these alloys cannot be hardened by heat treating. Because the austenite is tough and ductile *as welded*, chromium-nickel grades are the best stainless steels for welding. No annealing is necessary.

Basically, there are four grades of chromium-nickel stainless steels:

1. The *extra low-carbon* grades. They have a maximum carbon content of 0.03%.
2. The *stabilized grades*. They contain either titanium or columbium, or a combination of both. They are more corrosion-resistant than low-carbon grades, especially for continuous service.
3. The *high-manganese* grades. They are basically 18-8 type austenitic stainless steel with a higher percentage of manganese than nickel.
4. *Molybdenum grades*. They are used to help resist corrosion from chemicals.

Coatings for Stainless Steel Electrodes

Stainless steel electrodes are available with either *lime* or *titania* coatings. The lime-coated electrode is used with DC reverse polarity, while the titania-coated electrode can be used with DC reverse polarity or with AC.

Lime Coated—These electrodes give the best penetration and weld easily in all positions. When welding vertical down, smaller diameter electrodes generally work better.

Titania Coated—These AC-DC electrodes weld with a smooth, stable arc that has very little weld spatter. The slag from this electrode is easily removed, making slagging and cleaning much easier than with the lime-coated electrode. The titania-coated electrode is not recommended for vertical-down welding.

To minimize (control) chromium loss when welding stainless steel, welding should be done with a short arc and high travel speeds. When welding stainless steel, the loss of alloy is greater when using titania-coated electrodes.

Stainless Steel Practice

Before practicing with stainless steel electrodes, the work must be *thoroughly* clean, especially on high-quality jobs where maximum corrosion-resistance is needed. As usual, each bead must be thoroughly cleaned before welding over it.

Flat Position—In the flat welding position, stringer beads are used for the first pass, except on joints with poor fit-up. A slight weave pattern would be better on joints with poor fit-up, as in Fig. 15-12. Then, a stringer or weave bead is used for the other passes to finish the weld.

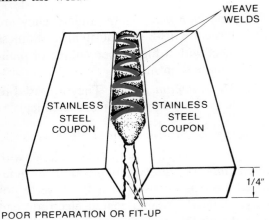

Fig. 15-12. Practicing stainless steel welding on a joint with poor fit-up.

Vertical Position—In vertical-down welding, stringer beads or a slight weave are used. The electrode holder is tipped down so that the arc force will push molten metal back up into the joint.

For welding vertical up, a wide triangular weave is used with a $5/32''$ or smaller electrode. A weld shelf is made at the bottom of the joint, followed by layers on top of it.

Overhead Position—In the overhead position, a whipping technique is used with a slight circular motion in the crater, making stringer beads instead of weaves. The holder is tipped slightly forward in the direction of travel with a short arc.

In all the above positions, currents as low as possible should be used, as long as there is good arc action and proper fusion.

General Notes for Stainless Steel Welding

Generally speaking, arc and oxy-acetylene welding are the more primitive (early) meth-

ods of welding stainless steel. These methods are more difficult than the advanced methods of welding stainless steel: Tungsten Inert Gas (TIG) and Submerged Arc Welding (SAW).

TIG and submerged arc produce easier and better-looking stainless steel welds. These methods are thoroughly discussed in other chapters.

Table 15-3 gives the physical properties, as welded, and the weld deposit analysis of typical austenitic AWS class stainless steel. Table 15-4 shows the same properties for martensitic stainless steel.

Cast Iron Electrodes

Electrodes used to arc weld cast iron are not made out of cast iron itself. Many properties of cast iron would make it a poor choice for a current-carrying electrode.

Instead, cast iron is arc welded with electrodes made of several different types of metal.

Nickel Electrodes (For Cast Iron)

A coated *nickel* electrode is used to repair cast iron when the deposit must be machinable. Nickel electrodes are very ductile and they stretch as the cast iron repair cools. The stretching, therefore, relieves the stress that was set up in the welded cast iron. The stretch is opposite the separation that would happen if most filler rods were used on cast iron.

The arc should be held the same as if welding on mild steel and can be used in all positions for both beading or weaving. DC reverse polarity gives a flat, smooth bead with the nickel electrode on cast iron.

Table 15-5 has the classifications, chemical analysis, and recommended current ranges of nickel rods for arc welding cast iron.

Steel Electrodes (For Cast Iron)

Steel electrodes with special coatings can be used for welding cast iron, but the deposit cannot be machined. The deposits are dense, strong, and tough. Since mild steel has a higher tensile strength than cast iron, the deposits are actually stronger than the cast iron itself.

For arc welding cast iron with a steel electrode, DC reverse polarity or AC may be used. When welding, the arc should be held as short as possible without allowing the coating to touch the work.

Bronze Electrodes (For Cast Iron)

When *bronze* electrodes are used to weld cast iron, the welds resist cracking fairly well,

Table 15-3. Austenitic Stainless Steel Characteristics.

AWS Class	Typical Physical Properties, As Welded	Typical Weld Deposit Analysis
E-308-15, 16	Tensile strength 85,000 to 95,000 p.s.i. Elongation (2") 40% to 50%	Carbon 0.07% Max. Chromium 19.0 % Nickel 9.5 %
E-308Le-15, 16	Tensile strength 80,000 to 90,000 p.s.i. Elongation (2") 40% to 50%	Carbon 0.04% Max. Chromium 19.0 % Nickel 9.5 %
E-309-15, 16	Tensile strength 85,000 to 95,000 p.s.i. Elongation (2") 35% to 45%	Carbon 0.10% Max. Chromium 23.0 % Nickel 13.0 %
E-309Cb-15, 16	Tensile strength 85,000 to 95,000 p.s.i. Elongation (2") 30% to 40%	Carbon 0.10% Max. Chromium 23.0 % Nickel 13.0 % Columbium .80%
E-309Mo-15	Tensile strength 85,000 to 95,000 p.s.i. Elongation (2") 35% to 45%	Carbon 0.10% Max. Chromium 23.0 % Nickel 13.0 % Molybdenum 2.2 %
E-310-15, 16	Tensile strength 85,000 to 95,000 p.s.i. Elongation (2") 35% to 45%	Carbon 0.20% Max. Chromium 26.0 % Nickel 21.0 %
E-310Cb-15, 16	Tensile strength 85,000 to 95,000 p.s.i. Elongation (2") 30% to 40%	Carbon 0.12% Max. Chromium 26.0 % Nickel 21.0 % Columbium .80%
E310Mo-15, 16	Tensile strength 85,000 to 95,000 p.s.i. Elongation (2") 35% to 45%	Carbon 0.12% Max. Chromium 26.0 % Nickel 21.0 % Molybdenum 2.0 %
E-312-15, 16	Tensile strength 110,000 to 120,000 p.s.i. Yield strength 80,000 to 90,000 p.s.i. Elongation (2") 22% to 25%	Carbon 0.15% Max. Chromium 29.0 % Nickel 9.5 %
E-316-15, 16	Tensile strength 85,000 to 95,000 p.s.i. Elongation (2") 35% to 45%	Carbon 0.07% Max. Chromium 18.0 % Nickel 13.0 % Molybdenum 2.25%
E-316ELC-15, 16	Tensile strength 80,000 to 90,000 p.s.i. Elongation (2") 35% to 45%	Carbon 0.04% Max. Chromium 18.0 % Nickel 13.0 % Molybdenum 2.25%
E-317-15, 16	Tensile strength 85,000 to 95,000 p.s.i. Elongation (2") 35% to 45%	Carbon 0.07% Max. Chromium 19.0 % Nickel 13.0 % Molybdenum 3.25%
E-317ELC-16	Tensile strength 85,000 to 95,000 p.s.i. Elongation (2") 40% to 50%	Carbon 0.03% Max. Chromium 18.0 % Nickel 13.0 % Molybdenum 3.25%
E-318-15, 16	Tensile stength 85,000 to 95,000 p.s.i. Elongation (2") 30% to 40%	Carbon 0.07% Max. Chromium 18.0 % Nickel 12.0 % Molybdenum 2.25% Columbium .80%
E-330-15, 16	Tensile strength 75,000 to 85,000 p.s.i. Elongation (2") 25% to 35%	Carbon 0.25% Max. Nickel 35.0% Chromium 15.0%
E-347-15, 16	Tensile strength 85,000 to 95,000 p.s.i. Elongation (2") 35% to 45%	Carbon 0.07% Max. Chromium 19.0 % Nickel 9.5 % Columbium .80%

so the weld length can be increased. The wire core is phosphor bronze, covered with a flux that gives a fairly stable arc. However, striking the arc may take some practice.

With a bronze electrode, it is important that each bead be cleaned before fusing in the next bead. Peening each bead and allowing the bead to cool before welding on the next pass will help eliminate pinholes and other problems in the finished cast iron weld.

Low-Hydrogen Electrodes (For Cast Iron)

Although *low-hydrogen* electrodes are basically used on steel with less than 1.5% carbon content, they have been used on many cast iron welding jobs with good results. As mentioned earlier, the heat input of low hy-

drogen electrodes is so great that residual (leftover) stress will be in the weld joint after welding. Then, the weld will crack unless the casting is heavy enough to resist the stress. The finished product will not be machinable.

Aluminum and Bronze Electrodes

Both aluminum and bronze metal can be arc welded with electrodes made from the same metal. In both cases, it is very important that the base metal be cleaned thoroughly.

Aluminum Electrodes

These dip-coated electrodes can be used on aluminum with either metallic or carbon arc welding. The flux coating holds down the oxidation and dissolves the aluminum oxides that easily form. During welding with a short arc on DC reverse polarity, just touching the molten pool gives very high-quality aluminum arc welds. The melting rate of aluminum electrodes is high. For this reason, it may be necessary under certain conditions to preheat aluminum before arc welding because the welding puts so little heat into the parent metal.

The electrode in aluminum arc welding should be held straight above the work, so that the flux melt-off will be effective. Slag

Table 15-4. Martensitic Stainless Steel Characteristics.

E-410-15, 16	Tensile strength 85,000 to 90,000 p.s.i.	Carbon 0.10% Max.
	Yield strength 55,000 to 60,000 p.s.i.	Chromium 12.5 %
	Elongation (2") 30% to 35%	
E-430-15, 16	Tensile strength 75,000 to 80,000 p.s.i.	Carbon 0.10% Max.
	Yield strength 40,000 to 45,000 p.s.i.	Chromium 16.0 %
	Elongation (2") 30% to 35%	
E-502-15	Stress relieved at 1325° F. Tensile strength 95,000 p.s.i.	Carbon 0.05% Max.
		Chromium 5.10%
	Annealed from 1550° F. Tensile strength 79,000 p.s.i.	Molybdenum 0.56%
	Elongation (2") 22% to 35%	Manganese 0.55%
E-502-16	Welds on aircraft-type steels may be heat treated to 180,000 p.s.i. tensile strength	Carbon 0.10% Max.
		Chromium 5.0 %
		Molybdenum 0.50%

Table 15-5. Nickel Electrode Characteristics for Arc Welding Cast Iron.

| Electrode Class | Chemical Analysis—%* | | | Recommended Current Range | | |
	Nickel	Iron	Carbon	Electrode Diameter (in.)	Amperes AC	Amperes DC
E Nickel	92.9	3.16	1.24	3/32	50-70	40-70
				1/8	90-110	80-100
				5/32	120-140	100-130
				3/16	130-160	120-150
E Nickel-Iron	53.4	42.6	1.83	3/32	50-90	40-80
				1/8	90-120	80-110
				5/32	120-150	100-140
				3/16	130-170	120-160

*Balance of Content to 100%: Silicon, Manganese, Titanium, and Magnesium.

should be removed completely, using hot water and wire brushing if necessary. Out-of-position welding with aluminum electrodes is not recommended.

Bronze Electrodes

Bronze electrodes are also coated, and they should be used on DC reverse polarity for welding ferrous metals, bronze, or copper. Where the parent bronze or copper metal is a heavy material, it is usually necessary to preheat the metal, because of its high heat conduction away from the weld.

High heats are recommended on these metals because they tend to produce gases that could cause porosity. The heat will help the gases to be released quickly. This can also be helped by holding the electrode so that the arc tends to flare back over the molten metal. This will keep the metal molten long enough so that the trapped gases can escape. If the metal is to be machined after welding, it should be undercut before welding so that the machining work will be done on deposited metal free of porosity.

Hardsurfacing Electrodes

The process of applying special wear-resistant alloys to any part that may be subjected to heat, impact, corrosion, and abrasion is known as *hardsurfacing*. These alloys are applied by welding alloy electrodes to the base metal in the form of beads or weaves. Care must be taken in analyzing each job, so that the alloyed rod best suited for the job will be used.

Hardsurfacing *powders* are available that are spread over the area to be surfaced and are then fused to the surface by heat from a carbon arc or carbon torch. These powders

have minimum admixture and high abrasion-resistance, and they can be deposited in very thin layers or on thin edges.

Maximum Abrasion Resisting Electrodes

These electrodes are made of a hollow chrome carbide tube, filled with carbide particles as shown in Fig. 15-13. (Uncoated rods are also available for oxy-acetylene hard surfacing, in sizes from ⅛″ to ¼″.) Coated electrodes from ⅛″ to ¼″ diameter are available and should be used on DC reverse polarity or on AC welding machines. Deposits of these electrodes cross-check as they cool, relieving shrinkage stresses. The deposits have good resistance to corrosion and high temperature oxidation. These electrodes are designed for hardsurfacing on thin cutting edges, reamers, plow shares, and various types of knives. Fig. 15-14 shows a plow bottom built up with hardsurfacing. The cross-hatch pattern on the tool shows the pattern of how the weld weaves and beads were laid down.

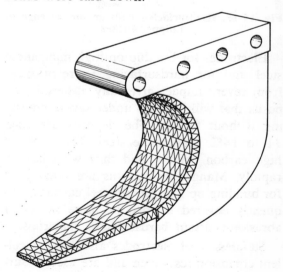

Fig. 15-14. Hardsurfacing built up on the edge of a plow bottom.

Fig. 15-13. Chrome carbide welding rod, used for hardsurfacing.

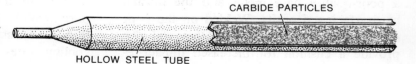

CARBIDE PARTICLES

HOLLOW STEEL TUBE

Severe Impact Resisting Electrodes

These electrodes can be used on DC reverse polarity or on AC welding machines. The deposited metal work-hardens under impact and can be used for building up high manganese steels. Severe impact resisting electrodes are available in ³⁄₁₆″ and ¼″ sizes and are used for building up crusher jaws, hammer mill parts, and dipper parts, as shown in Fig. 15-15.

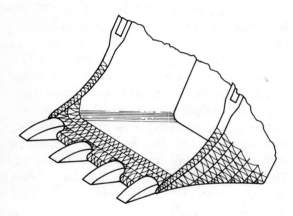

Fig. 15-15. Hardsurfacing built up on the edge of a dipper bucket.

Electrodes for buildup on high manganese steel and for hardsurfacing to resist wear from severe impact give fully austenitic deposits that will hold up under severe pounding without failure. The deposits are then 11 to 14% manganese steel. They have a high carbon content and they work-harden rapidly. Manganese deposits are often used for building up manganese steel castings, frequently covered by two or more layers of abrasion-resistant hardsurfacing materials.

Stainless steel electrodes also have excellent corrosion-resistance and are used as surfacing deposits to resist corrosion. Also, they are frequently used for a good bond between a poor-quality base metal and the normal hardsurfacing deposits. Welded on the poor metal under the hardsurfacing, they will also provide a cushion for increased impact-resistance.

High-Temperature Corrosion and Abrasion-Resisting Electrodes

These rods are available for both oxy-acetylene and arc welding. They are made from a nonferrous cast alloy containing cobalt, tungsten, and chromium. The finished weld surface resists both corrosion and abrasion at high temperatures and will withstand medium impact. These electrodes are available in ⅛″, ⁵⁄₃₂″, ³⁄₁₆″, and ¼″ diameters. They are commonly used for facing exhaust valves and valve seats where the temperature and abrasion may be very high.

Severe Abrasion and Moderate Impact-Resisting Electrodes

This electrode is made of a high-speed steel core wire with additional alloys in the flux coating. The type of core and high iron powder content make the electrode good for hardsurfacing oil well drill collars, feed mill hammers, and other parts subject to severe high- or low-stress abrasion, moderate impact, or a combination of both.

Abrasion with Moderate to Heavy Impact-Resisting Electrodes

These shielded arc electrodes are manufactured in four sizes, from ⅛″ to ¼″ in diameter. The self-hardening property of the weld metal makes it an excellent surfacing material for scarfer teeth, dipper teeth, and rock crushers.

Abrasion and Impact-Resisting Electrodes

These electrodes, resisting wear from both abrasion and impact, produce a semi-austenitic deposit with good resistance to both abrasion and impact. Its *abrasion* resistance is usually unaffected by the welder's technique, but its *hardness* may vary considerably, depending on the cooling rate. Preheating the part to 600° F will help produce a deposit that is almost entirely martensitic. Faster cooling rates produce deposits that have more austenite.

Welding Wires

Welding wires are used in many kinds of automatic arc welding machines. Since these wires conduct electricity and melt into the weld, they are very much a kind of electrode. The main difference is that single electrodes are stiff and coated, while welding wires are usually sold in long rolls of wire without a heavy flux coating. The wire and arc are usually shielded from the atmosphere by the shielding gas coming out of the welding gun.

How Welding Wires Are Made

Welding wires are made by drawing hot rolled steel, alloy rods, and flat strips (with their fluxing materials) into various wire sizes. Drawing wire means pulling the metal through a die to give it a wire shape.

The wire must meet rigid specifications for size and physical properties. Careful attention is given to tempering the wire, so that the wire will give trouble-free performance in all types of automatic and semi-automatic wire feeding mechanisms.

Wires that are lightly coated for good electrical contact will resist atmospheric conditions and give consistent welds even after the machine has been shut down for some time. Regardless of the type of wire coating used, it must adhere (stick) well enough so that it will not chip or flake. This could cause problems in spooling or coiling, and it

WELDING
WIRE SPOOL

Fig. 15-16. Welding wire spool for MIG welding.

would affect the wire feed mechanism during welding.

Welding wires are packaged in spool sizes from a four-inch diameter spool for hand welding guns up to spools as large as 750 pounds. Fig. 15-16 shows a smaller-sized wire spool.

Types of Welding Wires

As with electrodes, there are also many kinds of welding wires. Welding wires are available with different metals and different alloys for many different kinds of welding jobs.

Mild- and Low-Alloy Steel Welding Wires —welding wires for use on mild- and low-allow steels with the gas-shielded metal arc are available in sizes from 0.025" to 0.1875" in diameter. Sizes for submerged arc welding range from 0.045" to 0.3125" in diameter.

The mechanical properties of the weld depend on the wire analysis, as shown in Table 15-6. Wires developed for welding medium

Table 15-6. Mild- and Low-Alloy Steel Welding Wire Characteristics.

	Molybdenum Content Welding Wire	Standard Welding Wire
Mechanical Properties:		
Yield Strength (psi)	93,000	56,000
Tensile Strength (psi)	103,000	74,200
Elongation in 2"	23%	26%
Wire Analysis (%):		
Carbon	0.11	0.095
Manganese	1.92	1.60
Phosphorus	0.012	0.018
Sulfur	0.017	0.022
Silicon	0.80	0.97
Molybdenum	0.50	—

carbon-sulfur, free-machining steels are shown with molybdenum added. These steels are usually welded with carbon dioxide as the shielding gas.

Stainless Steel Welding Wires—Stainless steel welding wires, like mild- and low-alloy welding wires, must have a uniform finish with no foreign matter. Foreign matter on any wire could adversely affect the wire feed mechanism and the welding arc, possibly creating a poor weld.

Gas shielded metal-arc stainless steel welding wires are available in 0.30″, 0.035″, 0.045″ and 0.625″ diameters. The wires used in submerged arc welding can be purchased in several sizes from 0.045″ to 0.15625″ in diameter.

Stainless steel welding wire contains elements such as carbon, manganese, chromium, nickel and silicon. The percentage of each of the elements is shown below:

Carbon	0.05%
Manganese	1.53%
Chromium	24.57%
Nickel	13.28%
Silicon	0.35%

These stainless steel wires are used for welding clad material, for joining stainless steel to mild steel, and for stainless steel overlay work. Sound ductile welds can also be made in chromium grades of stainless steel with this welding wire.

Copper and Copper Base Wires—These wires are produced for the gas metal arc process and are used with either manual or automatic equipment. The wire can also be used as a filler metal during the gas tungsten-arc welding process.

All of the usual copper base alloys, except beryllium copper, can be welded with copper base wire. Beryllium copper cannot be welded with this wire bcause the beryllium is toxic.

Aluminum bronze, phosphor bronze, and dissimilar metals can be welded with copper base wire rods while using helium, argon, or a combination of both as a shielding gas. The usual wire sizes for these methods with copper welding wires are 0.030″, 0.035″, 0.045″, 0.0625″, and 0.09375″.

Aluminum Wires—These high-quality wires are made with an extremely clean surface by using a shaving process that removes the surface oxides. The wires must then be foil covered and individually packed to prevent exposure while they are shipped and stored.

Aluminum wires are available with different properties for welding plates, sheets, and

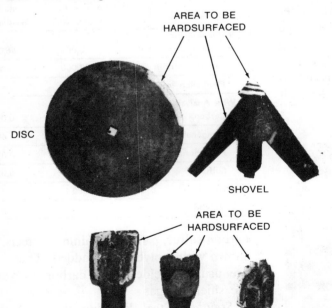

AREA TO BE HARDSURFACED

DISC

SHOVEL

AREA TO BE HARDSURFACED

OIL INDUSTRY STARTER BIT

OIL INDUSTRY ROCK BIT

OIL INDUSTRY FINGER BIT

Fig. 15-17. Common industry tools are hardsurfaced to resist wear.

many aluminum alloys. The wire is available in assorted sizes, ranging from 0.030″ to 0.125″ in diameter.

Buildup and Hardsurfacing Wires—Tubular alloy wires for open arc welding and automatic submerged-arc welding are available for metal buildup and hardsurfacing. These have increased efficiency up to 30% over manual buildup and hardsurfacing. Fig. 15-17 shows several parts that may be built up for hardsurfacing.

The wires are available in $\frac{7}{64}$″, $\frac{1}{8}$″ and $\frac{5}{32}$″ sizes. They are alloyed with elements such as chromium, manganese, molybdenum, silicon, and carbon. The specific welding job helps decide how much of each element will be added.

Flux-cored Wires—These are wire rods with a flux *core* containing deoxidizing ingredients. The ingredients will help remove impurities in the weld metal like the flux coating does for a standard electrode.

These wires are used for both automatic and semiautomatic welding equipment with a shielding gas. The flux-core materials are nonhydroscopic (they will not absorb moisture) so they may be stored for some time.

Flux-cored wires are used for welding on all grades of mild steel, chrome moly steel castings, pressure vessels, and pipe.

16

Arc Welding Common Metals

After a new welder has mastered several techniques for welding mild steel, he should not have much trouble arc welding metals other than mild steel. Stainless steel, quenched and tempered steel, manganese steel, copper, aluminum, and alloys of these metals are all common to the welding trade. With practice, they can be easily welded after learning new procedures.

It would be impractical to try to discuss, in one text book, all the various arc welding techniques and the metals and alloys that can be welded with those techniques. With that in mind, this chapter will try to cover those metals and alloys most commonly welded with the manual shielded arc welding method.

Preparations and welding procedures used on these metals and their alloys will usually be similar to those used for mild steel. Methods of welding each metal other than manual shielded arc will be mentioned briefly throughout the chapter, when the other method would specifically apply to welding that particular metal.

Manual Arc Welding Stainless Steel

Stainless steels are divided into two basic groups. Because the welding method is different for each group, it would be a good idea to review the two groups and their makeup. The two groups of stainless steels, as mentioned earlier, are known as the *straight chromium* and the *chromium-nickel* types. They may be reviewed here to discuss their different properties as they apply to modern arc welding.

Straight Chromium Stainless Steel

Straight chromium stainless steel is made in two different grades, ferric chromium and martensitic chromium.

Ferric Chromium Grades — The ferric chromium type of stainless steel, such as type 430, cannot be hardened by heat treating. This type of stainless steel has a higher percentage of chromium than carbon. Due to the contrast in the amounts of these elements (14% to 18% chromium and a maximum of 0.12% carbon), the steel has a ferritic structure at room temperature. While heating straight chromium stainless steel, there is no recrystallization or austenite being formed. The steel, therefore, cannot be hardened by heat treating.

During welding, all ferritic chromium stainless steels are likely to lose their corrosion-resistance, and the weld zone will then become brittle and soft. Some improvement can be made in the area where the steel is brittle if annealing is done at a temperature of 1300° to 1450° F. The exact heating time will depend on the thickness of the material being annealed. A handy estimate for the heating time would be one hour of time for each inch of metal thickness. After the correct heating period, the metal can be quickly cooled by cold air.

Martensitic Chromium Grades—The chromium martensitic grade may be hardened by heat treating in a range of 1700° to 1850° F. During welding, austenite is formed in the martensitic chromium grade. On cooling, it transforms into hard, brittle martensite. Martensitic chromium stainless steel increases in hardness as the carbon content is increased, similar to carbon steels. Increasing the carbon, however, increases the susceptibility (likelihood) of cracking as the weld cools.

Because the alloy may crack, preheating and postheating are often considered. Using these processes, however, will depend on the size of the metal being welded. Light, thin sections would not require preheating, especially where the <u>metal's</u> carbon content is low.

Stainless steel weld cracking may also be eliminated if the welder uses a small diameter electrode while making multi-pass welds.

Chromium-Nickel Stainless Steels

Chromium-nickel stainless steels are alloyed so that the *austenite,* Fig. 16-1, is stable

Fig. 16-1. Austenitic stainless steel, magnified 100 times. (Courtesy of Bethlehem Steel.)

at room temperature. Since the austenite does not transform when the steel is cooled, these alloys cannot be hardened by heat treating. Because the austenite is tough and ductile as welded, chromium-nickel stainless steels are the most suitable for welding. They do not require annealing after welding if they are to be used in normal atmospheric or mildly corrosive conditions.

The most common problem when welding on these alloys is *intergranular corrosion*. This is a corrosion between the grains of the stainless steel, and it results from the carbide precipitating (changing position) during welding. This is called carbide precipitation, explained as follows:

Heating unstabilized chromium-nickel stainless steel to its sensitizing temperature (800° to 1500° F) causes the chromium carbides to precipitate (change position) near the grain boundaries. This carbide precipitation removes chromium from the area near the grain boundaries. With no chromium for protection, the steel is then subject to intergranular (between the grains) corrosion.

The effects of carbide precipitation can be removed by annealing after welding. To avoid the need for annealing, special alloys may be used that are designed to prevent intergranular corrosion. They can also be used if the weldment will be subject to sensitizing temperatures in service.

Extra Low Carbon (ELC) grades and columbium or titanium stabilized grades are the two stainless steel alloys designed to prevent intergranular corrosion, even in severe service.

ELC Grades—These have a maximum carbon content of 0.03%. They are usually welded with an ELC electrode to produce welds that are not as crack sensitive on heavy sections and that have better impact-resistance at extremely low temperatures. Stabilized electrodes, as in Table 16-1, can also be used on ELC grades and will produce welds with physical properties similar to ELC rods.

Table 16-1. Stainless Steel Electrode Recommendations.

To select the proper electrode for a particular application, you must first know the following:
1. The type of stainless steel base metal to be used.
2. Whether the weldment is to be used in the as-welded condition or after heat treatment.
3. The physical and corrosive conditions of its service.

Martensitic-Chromium Grades

Base Metal AISI Designation	Popular Designation	Conditions where weldment will be placed in service.	Recommended Electrode AWS①-ASTM Designation
403	12Cr		E308 / E309 / E310
410	12Cr	As Welded	
414	12Cr-2Ni	Heat Treated	E410
416	12Cr FM②	Not Recommended for production welding.	
420	13Cr	As Welded	E308 / E309 / E310
		Heat Treated	E410 / E430
431	16Cr-2Ni	As Welded	E309 / E310
		Heat Treated	E430
440A,B,C	17Cr and High Carbon	Not Recommended for production welding.	

Ferritic-Chromium Grades

Base Metal AISI Designation	Popular Designation	Conditions where weldment will be placed in service.	Recommended Electrode AWS①-ASTM Designation
405	12Cr-A1	As Welded	E308 / E309 / E310
		Annealed	E430
406	12Cr-4A1	Not recommended for production welding. Use E310 for repair welding.	
430	16Cr	As Welded	E308 / E309 / E310
		Annealed	E430
430F	16 CR FM②	Not recommended for production welding.	
430Ti	16CrTi	As Welded	E308 / E309 / E310
442	18Cr	As Welded	E308 / E309 / E310
		Annealed	E442 / E446
446	27Cr	As Welded	E308 / E309 / E310
		Annealed	E446

Table 16-1. Stainless Steel Electrode Recommendations, continued.

Chromium-Nickel Grades

Base Metal AISI Designation	Popular Designation	Conditions where weldment will be placed in service.	Recommended Electrode AWS①-ASTM Designation
201	17-4 Mn		
202	18-5 Mn		
301	17-7		
302	18-8		E308
304	19-9		
305	18-10		
308	20-10		
302B	18-8 FM⑤		E309
304L	19-9 ELC		E308L
			E347
303	18-8 FM②	Not generally recommended for production welding. If welding is necessary, use E309 or E308 electrode.	
309	24-12		
309S	24-12		E309
309Cb	24-12 Cb		E309Cb
310	25-20		
310S	25-20		E310
316	18-12 Mo		E316③
316L	18-12 Mo ELC		E316L③
317	19-13 Mo		E317 / E316③⑥
317L	19-13 Mo ELC		E317L / E316L③⑥
318	18-12 Mo Cb		E318
D319	19-13 Mo		E316
321	18-8 Ti④		E347
347	18-8 Cb		
348	18-8 Cb		E347

Courtesy of Lincoln Electric Co.

① Where more than one AWS-ASTM electrode type is listed as suitable for the same base metal, the first one is better for most applications.
② FM indicates a free machining steel.
③ In certain conditions E309Cb should be used.
④ E321 electrode is not regularly manufactured because the titanium is lost in the arc.
⑤ Silicon is used in E302B Steel to improve the scaling resistance. To match the scaling resistance, the higher alloy content of E309 electrode is necessary.
⑥ E316L can be used but the weld will not match the molybdenum content of 317 base metal.

Stabilized Grades—These contain certain amounts of either titanium or columbium, or a combination of both. These elements have a stronger affinity (attraction) for carbon than does chromium. The elements in the stabilized grades combine with the carbon and prevent it from changing position near the chromium grain boundaries while the stainless steel is in the sensitizing temperature range.

Columbium or titanium stabilized grades of stainless steel are used for continuous service under severe corrosive conditions. Under those conditions, the ELC grades would be unsuitable. The stabilized grades must be welded with stabilized E 347 electrodes.

Stainless Steel Electrodes

Only a few of the more widely used types of stainless steel electrodes are made as standard production items. Because of this, excellent welds with high corrosion-resistance can be made on most grades of stainless steel while using only a few standard electrode types. The properties of the metal being welded may be exceeded by using an electrode with a higher alloy content. This generally increases the weld metal's resistance to corrosion. Table 16-1 shows the stainless steel electrodes and recommendations from one welding equipment manufacturer.

The high welding heat may also cause metallurgical changes that seriously reduce the metal's ductility or corrosion-resistance. To minimize the effects of the heat without expensive preheat or postheat treatments, it may be better to use a stainless steel electrode made of a different type of stainless steel than is the base metal.

Standard production electrodes with the same chemistry as all the different kinds of stainless steel are not always available. To weld any special kinds of stainless steel, then, it is sometimes necessary to use special high-priced electrodes matching the corrosion-resistance of the base metal. However, the electrodes in Table 16-1 provide excellent and economical welds for most stainless steel applications.

Fluxes on stainless steel electrodes are *lime* or *titania* coatings. These will stabilize the arc and help provide maximum mechanical strength and corrosion-resistance.

Lime-Coated Electrodes—The *lime* flux coating is used with reverse polarity DC and is indicated by the suffix (—15) on the AWS-ASTM electrode designation. These electrodes are designed for the best penetration and operating characteristics in all welding positions, including vertical-up and vertical-down. The 1/8″ diameter and smaller electrodes are especially good for welding vertical-down. The slag from the lime coating provides a rapid wetting action, assuring good wash-in and preventing undercutting in all positions. The coating also fluxes impurities from the weld metal, pro-

ducing a deposit with little or no porosity and with maximum mechanical strength and corrosion-resistance. Lime-coated electrodes also produce slightly convex beads, desirable for large root passes to help prevent weld cracking.

Titania-Coated Electrodes—The *titania* flux coating is used for welding with either DC reverse polarity or AC and is indicated by the suffix (—16) on the AWS-ASTM electrode designation. Titania-coated AC-DC electrodes are designed to be popular with most welders. The arc is smooth and stable, and the electrode produces uniform beads with minimum spatter. The slag is easily removed with ordinary hand tools and does not leave a secondary film. This, together with the smooth, slightly concave bead, means less cleaning, grinding and/or polishing than with lime-coated electrodes. These electrodes produce welds with excellent mechanical strength and corrosion-resistance in all positions except vertical-down.

Chemical Composition of Stainless Steel Electrodes

When welding with a stainless steel electrode, a certain amount of alloy is lost when the metal is transferred from the electrode to the base metal through the arc. This is especially true for the alloy chromium. This loss can be held down by welding with a short arc at high travel speeds. Also, to eliminate the danger of excessive alloy loss, many electrodes are made with a higher alloy content than the specifications require for the finished deposit.

Titania-coated electrodes have a greater tendency to lose alloy than do lime-coated electrodes. For this reason, many titania-coated electrodes have ferro-chromium added to the coating to supplement (add to) the chromium in the core wire. For instance, the chromium content of the E 308 and E 347 electrode is at least 1.9 times the nickel content. See Table 16-1. This ratio will always make the weld deposit contain some free ferrite (Fig. 16-2) with the austenite, thus reducing the weld's cracking sensitivity. The properties of these structures are fully covered in the chapter on metallurgy.

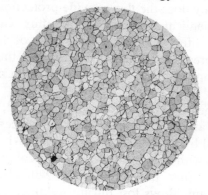

Fig. 16-2. Ferrite grains in stainless steel, magnified 100 times. (Courtesy of Bethlehem Steel).

Welding Straight Chromium Stainless Steel

If a straight chromium weldment is to be heat-treated (or if service at high temperatures will produce a similar effect), the weld must be made with a straight chromium electrode. Table 16-1 lists several straight chromium electrodes. Then, postheating must be done to remove the brittle effects of this welding process. For example, a postheat treatment as high as 1600° F can be used when welding type 446 stainless steel, so it is preferable to use a chromium-nickel E 309 or E 310 electrode when welding type 446. If high-chromium stainless steel is welded with straight chromium electrodes, the weld will have poor mechanical properties.

If a straight chromium steel weldment is to be used without postheating, it should be welded with a chromium-nickel electrode. These weldments should be used either at or very near room temperature.

Using a chromium-nickel electrode produces a strong, ductile weld bead. Even though the heat-affected metal next to the weld bead may become brittle, the ductile bead holds up better against impact and deformation than if the entire weld area were

brittle. The parent metal will become less brittle if welding is done in several passes using small diameter electrodes with low current and stringer beads.

Welding Chromium-Nickel Stainless Steel

Austenitic (chromium-nickel) stainless steels are very low-carbon steels, so they are easily carburized when they contact any foreign material. Materials such as protective paper, marking paint, pencil marks, soap film, grease, oil, or dirt left on the surface of these steels will cause the steels to be carburized while they are being welded. This, in turn, will cause a loss of corrosion-resistance and carbide precipitation. The effect can also be produced by cleaning these stainless steels with *carbon steel* brushes, since small pieces of the carbon steel brush wire may remain on the surface. Only stainless steel brushes should be used for cleaning chromium-nickel stainless steels.

After complete cleaning in the weld area, this type of steel may be readily welded by the shielded metal-arc method. The molten weld metal must be protected from the atmosphere during welding, so that the important alloying ingredients will not be oxidized. Should the alloying ingredients oxidize, the corrosion-resistant properties of the weld would be limited. The electrode fluxes must be free of carbonaceous materials (materials that contain carbon) that could increase the carbon content of the weld. The fluxes should also have the fluidity and dissolving power needed to melt undesirable impurities from the weld area.

The welder should use a chipping hammer and wire brush to deslag any tack welds on the parts before proceeding with the main welding. To help reduce distortion, the backstep welding sequence may be used effectively. Using stringer beads instead of weaving will tend to eliminate cracking which, in turn, assures the welder of more corrosion-resistant welds.

The joint preparations and precautions before welding, for the most part, are the same as those used to weld mild steel. Joint preparations, again, are similar to those explained in other chapters and will depend on the size and thickness of the metal being welded. The electrode manufacturer's chart should be used to locate the correct electrode type and size and the recommended polarity.

Other Methods of Welding Stainless Steel

Of the many welding processes mentioned earlier, most can be used successfully to weld stainless steel. The more important TIG welding and oxygen/gas welding chapters fully describe how those procedures are used for welding stainless steel and other metals.

Other welding processes, however, are also used to weld stainless steel. For this reason, the more important information on *atomic-hydrogen, submerged arc,* and *resistance welding* is presented here as it relates to welding stainless steel.

Atomic-Hydrogen Welding — Atomic-hydrogen welding is unique because the heat generated by the arc does not perform the actual fusion of the stainless steel. Instead, a reaction is created that brings about heat absorption by hydrogen atoms. The intense heat from *that* process not only welds the stainless steel but also distorts the components if they are not clamped to control the warpage. This method, however, may be used successfully on light-gauge austenitic chromium-nickel steels to produce smooth, well-formed beads.

Stainless steels containing selenium and sulfur cannot be welded with atomic-hydrogen welding because of severe porosity that will form in the weld metal. The porosity will form even though the hydrogen atmosphere has the additional advantage of holding down any normal oxidation.

Submerged Arc Welding—This process is either automatic or semiautomatic. Unlike shielded arc welding, submerged arc welding introduces the flux and filler metal into the weld area separately. Austenitic chromium-nickel steels can be successfully welded by this method partly because the flux allows complete protection from the atmosphere. Not only does the flux provide protection, but the chemical makeup of the flux can be changed for welding different types of metals. Powdered metals may be added to the flux to compensate for any alloys lost from oxidation.

Types 302 and 304 stainless steel, both heavy and light gauge, have been successfully welded by this method. See Table 16-1. Electrodes with higher chromium and columbium content than the parent metal will often replace any alloys lost during welding.

The joint designs used for submerged arc welding on mild steel can also be used with stainless steel. When welding stainless steel, though, the current settings should be about 20% less than for the same thickness of mild steel.

Resistance Welding—Here, the stainless steel parts are fused by using heat and pressure *combined*. Stainless steel parts joined by resistance spot or seam welding compare in strength to low-alloy steel joints made by the same process.

Closely controlling the amperage setting and contact pressure time is more critical with stainless steel than with low-alloy steels. The amperage setting and contact pressure must be controlled to have very corrosion-resistant joints. Welding current for stainless steel seam welding is less than for mild steel seam welding because the high electrical resistance of the chromium-nickel alloys *adds* to the current heat. By practicing on scrap stainless, a new welder can get the experience needed to make strong, corrosion-resistant stainless steel joints. This is especially true for spot or seam welding metal over 1/8″ thick, because it is very susceptible to carbide precipitation. With thinner gauges of stainless steel, however, the welder will have little trouble keeping the weld time below the danger point of harmful carbide precipitation.

Spot-welded austenitic chromium-nickel steels will give a good appearance at the joint. Also, little discoloration will happen if the welding pressure is delayed until the metal cools below the oxidizing temperature. After the steel has been welded, the seams or spots should be peened if they were warped or distorted during welding.

Manual Arc Welding Quenched and Tempered Steels

An important part of today's technology is the constant striving to develop steels with greater strength, better ductility, increased toughness, more resistance to corrosion, and greater fatigue strength. As these steels are developed, new designs and fabricating procedures are needed along with new welding techniques and processes. The new welding processes usually require special electrodes to develop enough heat for welding and preheating.

Quenched and tempered construction steels have a yield strength as high as 135,000 psi. They have created refinements in both material design and workmanship, and they have brought together new welding techniques and the equipment needed to weld these new metals.

These steels can be cut with conventional oxygen-gas equipment without preheating or postheating. Machining these steels is no problem if the cutting speed is reduced to about ⅓ the speed required for carbon steels. Punching and shearing is possible on these steels as long as they are not over 1″ thick.

Sample Steel—Bethlehem RQ-100A is a typical roller-quenched, tempered alloy steel with a minimum yield strength of 100,000 psi. This discussion will be based on that particular Bethlehem steel as an example. The Bethlehem steel and other types of quenched and tempered steels have been used for bridges, buildings, pressure vessels, earth-moving equipment, and other high-strength applications.

Welding Procedures

The first requirement of any construction steel is that it must be able to be readily and reliably joined. The main method of joining is welding. When two pieces of this steel are to be joined by welding, filler metal must be

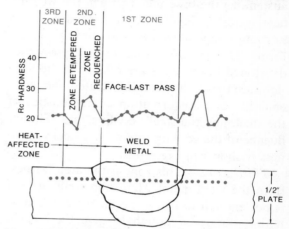

Fig. 16-3. The three zones of ½″ Bethlehem RQ-100A alloy steel, as welded.

added to the joint along with the heat used to provide fusion. This results in three major areas (zones) in the weld. These are shown in Fig. 16-3 and may be individually discussed as they have differing properties.

First Zone—This zone is the weld metal that solidifies from the molten pool of filler metal. The properties of this zone are made up from the composition of the filler metal, the melted base metal from nearby areas, and the conditions while the metal is becoming

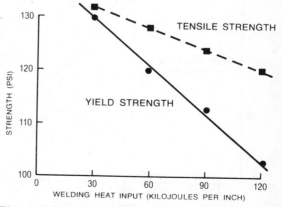

Fig. 16-4. Welding heat effect on the properties of the weld (base) metal.

solid. The effect of the welding heat on the properties of the weld metal is shown in Fig. 16-4.

Second Zone—This zone is the area next to the weld metal where the base metal has not melted, but there *has been* enough heat to austenitize the base metal. When the welding heat is removed, the austenitized area cools to room temperature and the austenite transforms. The microstructure and the properties developed in this zone are related to the temperature to which the steel is heated, the hardenability or transformation characteristics of the steel, and the steel's cooling rate. The influence of the cooling rate on the steel's hardness is shown in Fig. 16-5.

Third Zone—This zone is the area farthest from the weld metal. Here, the welding heat is not enough to cause the base plate to austenize, but it is enough to give the steel a second, short-term tempering. The effects of this mild tempering are shown in Fig. 16-6. Remember that this information is for Bethlehem RQ-100A alloy steel—other steels would likely have other curves.

The properties in each zone may vary somewhat, depending on the conditions involved. In any case, however, the welding conditions must be established for the steel being welded, so that the overall properties of the welded joint will be satisfactory. This means carefully controlling the properties in the weld metal and heat-affected zones *and* closely controlling the hydrogen content of the weld and the residual (leftover) stresses in the assembly. Failure to control the last two items can result in cracking.

Electrode—For manual arc welding practice on these steels, a new welder should use an E 110 or 120 low-hydrogen electrode. Using low-hydrogen electrodes helps eliminate underbead cracking if the electrodes have been properly stored and dried to keep the moisture content to a minimum. Welding on these steels with the right electrodes eliminates the need for stress relieving. In some cases, stress relieving could cause stress-relief cracking.

Technique—*Stringer bead welding* is always preferred over wide band weaving. Using the narrow stringer beads will help prevent the welder from accidentally doubling or tripling the weld heat input. This heat may cause problems in the heat-affected zones.

Generally, quenched and tempered steels require greater care than lower strength steels for fabricating and set-ups. Even so, the welding requirements, preparations, and procedures for welding quenched and tempered steels are similar to those for mild steel.

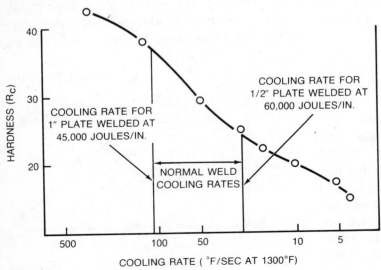

Fig. 16-5. How the cooling rate affects the hardness and strength of the austenized part of the heat-affected zone.

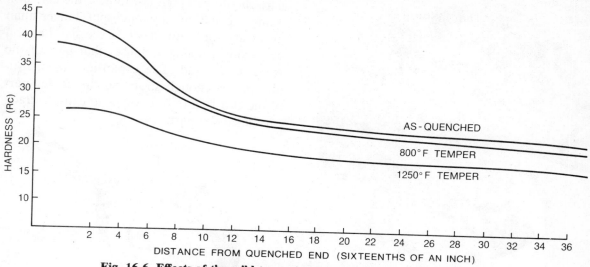

Fig. 16-6. Effects of the mild tempering heat farthest from the weld area.

Other Welding Methods

Quenched and tempered steels can also be successfully joined by submerged arc and gas metal-arc welding methods, as well as by manual arc welding.

In Table 16-2 are the suggested minimum preheats for the above-mentioned arc welding processes. Heavy sections may require additional preheat, but the welder should keep the preheat temperature low, allowing the widest range of weld heat input.

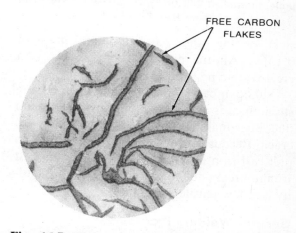

Fig. 16-7. Plain cast iron before heat treating, multiplied 500 times. (Courtesy of American Cast Iron Pipe Co.)

Table 16-2. Bethlehem RQ-100A Minimum Preheat Temperatures.

| | Minimum Preheat (°F) | | | |
| | | | Submerged Arc | |
Plate Thickness	Shielded Metal Arc	Metal Arc Gas	Alloy Wire Neutral Flux	Carbon Steel Wire Alloy Flux
1″ or less	50	50	50	150
1 1/16″ to 1 1/4″	100	100	100	200

Manual Arc Welding Cast Iron

Cast iron is a term used for a wide variety of alloys containing iron, silicon, and carbon. Fig. 16-7 shows plain cast iron multiplied 500 times.

Carbon is a main part of cast iron. There are two forms of carbon in cast iron: *combined* and *free*. Combined carbon is the carbon that has combined chemically with the iron. It tends to make the metal brittle (unable to bend) and very hard.

Free carbon is present in the form of flakes and is also known as *graphitic* carbon. The free carbon is distributed among the grains of iron and creates the weakness in gray cast iron. See Fig. 16-8.

The carbon content of cast iron is a *minimum* of 1.5%. The carbon content of mild

Fig. 16-8. Cast iron containing pearlite and graphite, multiplied 750 times. (Courtesy of American Cast Iron Pipe Co.)

steel, on the other hand, is a *maximum* of 0.15%. All the extra carbon in cast iron creates welding problems.

Although cast iron has a high compression strength, it is very low in tensile strength and ductility, making the metal very likely to crack. Because of this, any stress that tends to pull cast iron apart may easily cause the cast iron to break, due to cast iron's lack of ductility and low tensile strength.

General Welding Procedures

Cast iron is usually difficult to arc weld. This is due to the violent thermal process of the rapid heating by arc welding, followed by rapid cooling when the heat of the molten pool is absorbed by the cold cast iron. Carbon pickup from the parent metal is then absorbed into the weld deposit and, through dilution, may also make the weld metal very sensitive to cracking. Since most of the absorbed carbon is deposited at the end of the bead, the fracture usually begins at that point. See Fig. 16-9.

To help overcome cast iron's crack sensitivity, *preheating* the entire part is desirable. Welding should then be followed by a slow, uniform cooling. If preheating is not done,

care should be taken not to heat the cast iron being welded in any one spot longer than necessary for good fusion. Excessive heat on cast iron in one spot causes that part of the metal to expand. The expansion, then, usually places enough strain on the cooler part of the cast iron to fracture or to cause a crack on the cooler part.

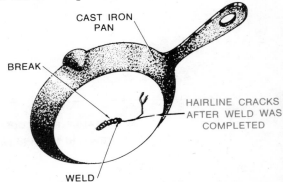

Fig. 16-9. If hairline cracks form at the end of a cast iron weld, the piece was probably not preheated correctly.

Restrained expansion and contraction may create major problems for a new welder. When the expansion is restrained, especially in all directions, residual (internal) stress is a factor to overcome, especially in the more ductile metals. In brittle materials such as cast iron, the internal stresses may cause cracks to take place in either the weld or adjacent areas. Restricted contraction can also cause enough stress to distort the areas surrounding the weld.

Example—Fig. 16-10 illustrates a cast iron wheel broken at point *X* and prepared for welding. Without some knowledge of expansion and contraction, the welder would find it hard to weld the wheel without causing *more* cracks.

In Fig. 16-10 the preheat areas on the cast iron wheel *must* be preheated before welding the prepared break at point *X*. The preheating will compensate (allow for) the expansion and contraction creating stress. Preheating the areas before welding *creates* expansion in the prepared crack. Then, while the prepared crack is expanded, the welder

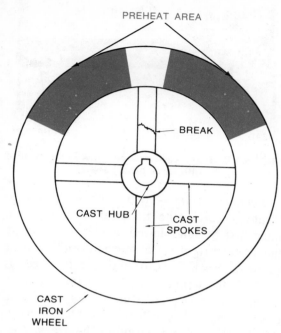

Fig. 16-10. The broken spoke can be repaired successfully only if the areas above the spoke on the wheel are preheated.

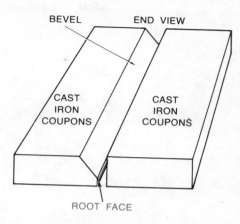

Fig. 16-11. Cast iron should be prepared with a root face to eliminate fall through.

can apply the filler metal and heat with manual shielded arc welding.

When it cools, then, the material will contract. In this way, any stress set up will be a *compression* stress as the contraction tries to press in on the wheel. Rigid cast iron can best withstand compression. This common welding procedure for the "broken wheel" should be clearly understood before other cast iron welding jobs are attempted.

Rules for Welding Cast Iron

Clean It—Cast iron pieces to be welded must be clean and free of rust. Whenever possible, cast iron that has been cleaned and prepared by grinding should be filed or wire brushed over the entire prepared surface before the welding is started. Although this cleaning procedure is not important before arc welding cast iron, it *is* required before brazing cast iron. In either case, the filing and brushing helps to free some of the graphite that adheres to the ground surfaces of the parts being prepared.

Prepare It—A root face, as in Fig. 16-11, should be left on prepared cast iron to create extra body material. This will help eliminate fall-through when the cast iron is being welded.

Preheat It—Before starting a weld on light cast iron, the welder must preheat not only the area to be welded but, if possible, the entire casting. If the casting is too large to fit in a preheating furnace, an oxy-acetylene torch or other heating device may be used to preheat the entire casting. The preheat temperature should be about 500° overall.

Complete the Weld Once It Is Preheated—Welding should be completed once it has been started, since cooling and reheating will create uneven shrinkage stresses. These, in turn, could cause cracking in the finished weld. The welded part should be allowed to cool slowly when the job is completed.

On either small or large repairs, the welder may need to weld without preheating. This may be done on typical cast iron jobs such as automobile cylinder blocks or heads.

To successfully weld with little or no preheating, the welder must keep both the weld and the work cool at all times. The beads should be kept very short, not more than 2" long. Cracking and shrinking may be prevented during this type of welding if the 2" bead is peened immediately after welding, as shown in Fig. 16-12.

Fig. 16-12. Peening the cast iron weld immediately
may help prevent cracking.

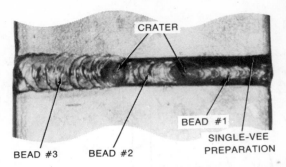

Fig. 16-13. Heavy cast iron coupons being welded
with three bead passes.

Use the Right Technique—The crater of
each bead should lie on top of the previous
bead. This may be done by the backstep
welding method to produce results as shown
in Fig. 16-13. The next 2″ bead should not
be started until the previous bead is completely cooled.

Hairline Cracks

Hairline cracks in cast iron are very fine
cracks, so small that many times they cannot
be seen with the eye. Because they *are* cracks,
though, they are defects in the cast iron and
may be repaired.

Quite often, hairline cracks are caused by
poor welding repair, so the earlier methods of
welding cast iron must be reviewed to guard
against forming hairline cracks. To find hairline cracks when they can't be seen, several
methods and materials may be used.

Ferromagnetic Powder—This powder is
shaken onto the magnetized cast iron part.
Then, the particles in the powder are attracted to tiny magnetic poles on each side
of the crack, as in Fig. 16-14. In turn, these
particles form a pattern around the crack
because the magnetic field is broken or distorted by the crack.

Temper Colors—To use temper colors to
locate a crack, the material to be welded
must first be sandblasted or cleaned to leave
the metal bright. Heat is then applied near
where a crack is suspected. As the heat is
applied, the temper colors will fan out in a
normal color sequence until they reach the
crack. There, they will be disrupted, separa-
ting the colors and outlining the crack.

Fluorescent Inspection—This is used on
nonferrous as well as ferrous metals. Here,
fluorescent materials are rubbed over the
suspected crack areas and then rubbed off.
Ultraviolet light is then used to expose the
fluorescent materials that will be flowing inside any cracks. If needed, fluorescent material can be combined with ferro-magnetic
powder to better outline the cast iron cracks.

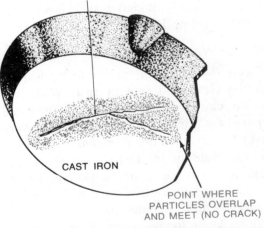

Fig. 16-14. Ferromagnetic powder is used to locate
hairline cracks in cast iron by watching the pattern that the particles make on the piece.

Patches for Broken Cast Iron Containers

The welder must always keep in mind the effects of expansion and distortion when using a *steel* patch to repair cast iron. The steel patch used in the repair must be much lighter than the cracked cast iron. This will eliminate as much pull from contraction stress as possible.

The design of the steel patch should allow for dimensional changes due to heat. This way, the steel patch will be able to compensate (allow) for any weld pull that might crack the less ductile cast iron. See Fig. 16-15. In Fig. 16-16, the patch has been partly welded on the repair and the welding heat has caused it to bend up. Simply pounding the soft, warm, sheet metal back down allows the repair to continue.

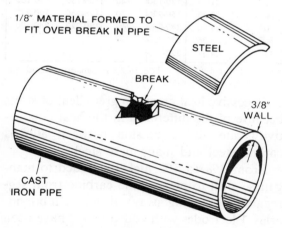

Fig. 16-15. Patch prepared to repair a break in a cast iron pipe.

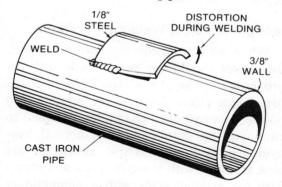

Fig. 16-16. Distortion may happen part way through a repair using light sheet metal. The condition is not serious and is easily repaired.

The dished patch will be flexible under pull if it is made of light material, and it will be strong enough for most repair jobs. Lap welds should only be used on curved surfaces and all flat surface butt welds. Flanged patches are best prepared for use in edge welds.

Testing and Sealing Cast Iron Containers

After welding a cast iron water jacket or other container, the repaired container can be safely tested with cold water. Cast iron, unlike ductile materials, cannot be peened to stop a small seepage leak in the weld area.

It is a good idea to use external or internal sealants instead of more welding if small leaks appear. This is easier than reheating and rewelding the whole piece. Many *internal* sealants on the market do an excellent job of sealing seepage in cast iron containers without any additional welding.

When the inside cannot be reached for repair, an *external* sealant may be used. External sealants include products known as smoothon cements, monoclinic sulfur, epoxies, and fast-drying enamels. Any of these will usually penetrate and seal minor weld seepage.

Manual Arc Welding Austenitic Manganese Steel

Manganese steel, a nonmagnetic material, is in great demand because of its work-hardening characteristic. The work-hardening quality of manganese makes it good for manufacturing pieces such as scarifier, bucket, or dipper teeth. See Fig. 16-17.

When a manganese product is cast and slowly cooled in the mold without heat treating, it becomes hard and brittle. The casting is composed of austenite, as in Fig. 16-18, and free cementite or carbide. The hard, brittle cementite surrounds the austenite grains distributed throughout the metal. To overcome the weakness in the casting due to the cementite, the casting is placed in a furnace and the entire piece heated to about 1950°F.

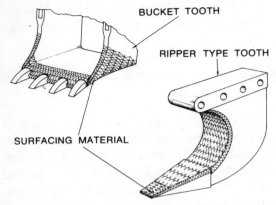

Fig. 16-17. The edge of a dipper bucket may be built up with manganese steel.

When the temperature has been reached, the material is then quenched in cold water. During the heating, the austenite absorbed the freed carbide (cementite) and formed a solid solution. The solid solution is retained by quenching. After quenching, the metal has high ductility and toughness.

The soft and tough characteristics of the metal are further changed by impact. Repeated blows, such as received by manganese bucket teeth, make the material harder and tougher. Because of this toughness, manganese castings are almost impossible to machine. Grinding, oxy-acetylene cutting, or arc-air cutting will then be the only effective methods of preparing the metal for welding.

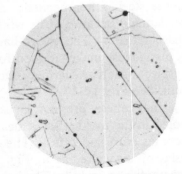

Fig. 16-18. Austenite, as present in manganese steel, multiplied 500 times.

How Welding Heat Affects Manganese Steel

Reheating toughened manganese steel causes the steel to lose some of its tensile strength and ductility. For this reason, arc welding is recommended over other welding methods. Manganese steel heated over 750° causes the carbides to be liberated (freed) from the solid solution. When this happens, the toughened steel within the heat-affected zone changes back to the *as cast* condition.

Table 16-3. Recommended Welding Currents for Atom Arc 7018 and 7018 Mo Electrodes.

Diameter	Volts	Amperes Flat		Amperes Vertical	Amperes Overhead
		Range	Preferred		
3/32"	21-24	65-110	80	65- 80	65-110
1/8"	21-24	120-160	140	100-130	120-160
5/32"	22-25	150-220	170	120-160	150-220
3/16"	23-26	240-300	260	180-220	240-300
7/32"	23-26	280-350	300		
1/4"	24-27	320-400	350		

Courtesy of Canadian Liquid Air.

Excessive heat creates a great deal of stress because of the difference in the heat conductivity rate and expansion between alloyed carbon steel and manganese steel.

From 3% to 5% nickel is added to manganese steel to prevent the carbide from segregating (breaking away) at the grain boundaries. Electrodes with added nickel have been developed with the nickel in either the flux or core wire, or both.

Manganese Steel Welding Procedures

When welding medium manganese steel, the welder should use an electrode with the properties of the *Atom Arc 7018* or *7018 Mo*. These electrodes were the original iron powder low-hydrogen electrodes, designed to eliminate the shortcomings of normal low-hydrogen electrodes. These electrodes are now available with strength levels and chemistry for making welds with a minimum of 120,000 psi tensile strength. Although these electrodes do not usually require preheating to prevent underbead cracking, the welder should preheat highly hardenable steels before they are welded. Preheating highly hardenable steels and other heavy sections will prevent hard heat-affected zones from forming and will prevent the tendency toward quench cracking as the weld cools.

Table 16-3 shows the recommended welding currents for the Atom Arc 7018 and 7018 Mo electrodes. The recommended current is for AC or reverse polarity DC. Table 16-4 shows the physical properties of manganese steel after being welded with the Atom Arc 7018 electrode. Table 16-5 shows the information if the metal were welded with a 7018 Mo electrode.

Preparation—The manganese steel casting to be welded is usually prepared by oxyacetylene cutting, as discussed in the oxyacetylene cutting chapter. An automatic cutting machine (Fig. 16-19) is usually used in industry. After preparation, the cut should be thoroughly ground and wire-brushed to remove all the dirt and accumulated oxides.

Then, the manufacturer's specification chart can be referred to for the size of electrode to use, current range, and suggested polarity for the best results. Fig. 16-20 is an example of one manufacturer's electrode specification chart.

Procedure — While welding manganese steel, the welder will get the best results by using a short arc and a small diameter electrode. After the root pass has been applied, an experienced welder may use a larger diameter electrode and apply short beads to finish the weld. In many cases, stress may be

Table 16-5. Typical Physical Properties of Manganese Steel Welded With an Atom Arc 7018 Mo Electrode.

Property	As Welded	Stress Relieved 1 hr @ 1150° F
Yield Point (psi)	68,000	68,500
Tensile Strength (psi)	79,000	79,500
% Elongation in 2″	31	32
% Area Reduction	72	74
Chemical Analysis of Weld Metal			
C	Mn	Si	Mo
0.05%	0.75%	0.56%	0.53%

Courtesy of Canadian Liquid Air.

partly eliminated by avoiding long bead welds. Peening, shown earlier in Fig. 16-12, flows the outer surface and makes it smooth, relieving the tension that could cause cracks. For this reason, all beads should be peened while they are hot.

If the sections are thick enough to require a weave bead, the weaves should be made with the crescent weave technique. See Fig. 16-21. Using this technique, the electrode motion should not exceed 2½ times the electrode diameter. While welding, it is a good idea to allow the manganese steel to cool somewhat after each bead or weave is welded. This will cut down on the total heat input.

Manganese steel can also be successfully welded by using the *cascade* technique. Here, shorter successive beads are fused to each other much like the method successfully used to weld cast iron.

Manual Arc Welding Aluminum

Aluminum is one of the newest and most widely used basic construction materials. Aluminum's advantages over steel in corrosion-resistance, weight, and ease of handling have caused the use of aluminum to increase greatly in recent years.

Table 16-4. Typical Physical Properties of Manganese Steel Welded With an Atom Arc 7018 Electrode.

Property	As Welded	Stress Relieved 1 hr @ 1150° F
Yield Point (psi)	68,500	62,000
Tensile Strength (psi)	75,000	72,000
% Elongation in 2″	34.0	38
% Area Reduction	75.5	77
Chemical Analysis of Weld Metal		
C	Mn	Si
0.06%	0.78%	0.58%

Courtesy of Canadian Liquid Air.

The science of welding aluminum is very challenging, since welding aluminum is a major new construction method. Most alu-minum welds are made with Tungsten Inert Gas (TIG) welding, but arc welding may be used for repair work and small jobs. This

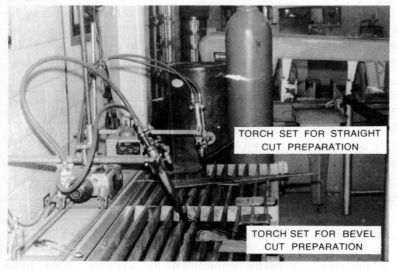

TORCH SET FOR STRAIGHT CUT PREPARATION

TORCH SET FOR BEVEL CUT PREPARATION

Fig. 16-19. An automatic oxy-ace-tylene cutting machine may be used in industry to prepare manganese steel for welding.

ELECTRODE	CONFORMS TO TEST REQUIREMENTS OF AWS CLASS	QUALIFICATIONS	POLARITY DC (+) = "REVERSE" DC (−) = "STRAIGHT"	SIZES AND CURRENT RANGES (Amps.)* (electrodes are manufactured in those sizes for which current ranges are given)							
				5⁄64" Size	3⁄32" Size	1⁄8" Size	5⁄32" Size	3⁄16" Size	7⁄32" Size	1⁄4" Size	5⁄16" Size
MILD STEEL											
Fleetweld® 5	E6010	A—B—C—D	DC (+)			75-130	90-175	140-225	200-275	250-325	280-400
Fleetweld 5P	E6010	A—B—D	DC (+)		40-75	75-130	90-175	140-225			
Fleetweld 7	E6012	A—B—C	DC (−); AC		55-80	80-130	120-180	155-250	225-290	270-340	340-430
Fleetweld 35	E6011	A—B—C—D	AC; DC (+)			70-110	80-145	110-180	135-235	170-270	
Fleetweld 35LS	E6011	D	AC; DC (±)				110-150				
Fleetweld 37	E6013	A—B—C—D	AC; DC (−)		60-90	80-125	125-180				
Fleetweld 57	E6013	B—D	AC; DC (±)	40-75	70-95	90-135	135-180	180-235	225-280	270-330	
Fleetweld 47	E7014	A—B—D	AC; DC (−)			100-145	135-200	180-250	235-305	285-380	
Fleetweld 180	E6011		AC; DC (+)		40-80	55-110	105-135	145-190			
Jetweld® 1	E7024	A—B—D	AC; DC (+)			120-180	180-230	220-280	280-360	330-400	
Jetweld 2	E6027	(1) A—B—C—D	AC; DC (±)				175-215	230-270	270-340	315-405	475-600
Jetweld 3	E7024	A—C—D	AC; DC (±)		60-110	100-160	160-215	270-340	270-340	315-405	450-600
Jetweld LH-70	E7018	(2) A—B—C—D	DC (+); AC		70-100	90-150	120-190	160-280	190-310	230-360	375-475
Jetweld LH-3800	E7028	A—D	AC; DC (+)			170-240	210-300	260-380	360-520		
LOW ALLOY, HIGH TENSILE STEEL											
Shield-Arc 85	E7010-A1	(3) A—C—D	DC (+)		50-90	75-130	90-175	140-225			
Shield-Arc 85P	E7010-G		DC (+)					140-225			
Shield-Arc HYP	—		DC (+)			75-130	90-185	140-225			
Jetweld 2-HT	E7020-A1	(4)	AC; DC (−)				175-215	230-270	250-320	330-370	
Jetweld LH-90	E8018-B2	(5)	DC (+); AC			90-150	110-200	160-280			
Jet-LH 8018-C1	E8018-C1	D	DC+; AC			90-150	110-200	160-280		230-360	
Jet-LH 8018-C3	E8018-C3	C—D	DC+; AC			90-150	110-200	160-280	210-330	230-360	
Jetweld LH-110	E11018-M	A—C—D	DC (+); AC			95-155	120-200	160-280	190-310	230-360	
BRONZE & ALUMINUM											
Aerisweld®	E-CuSn-C	C	DC (+)			50-125	70-170	90-220	For Copper Alloys		
Aluminweld®	Al-43	C	DC (+)		20-55	45-125	60-170	85-235	For Aluminum Alloys		
FOR REPAIRING CAST IRON											
Ferroweld®	ESt		DC (+); AC			80-100			Non-Machinable Deposit		
Softweld®	EN-Ci		DC (±) AC			60-110 65-120	100-135 110-150		Machinable Deposit		

NOTES:

A — Approved by the U. S. Coast Guard and ABS‡
B — Approved by Lloyds‡
C — Conforms to Military Specification‡
   ‡See price book for cert. numbers, limitations, etc
D — Certified by Bureau of Public Roads

(1) Also meets requirements of E6020
(2) Also meets requirements of E7016
(3) Also meets requirements of E7010-G, and E6010 in 3⁄32" size
(4) Also meets requirements of E7027-Al
(5) Also meets requirements of E9018-G
   * Current ranges listed for DC only; AC current ranges run approximately 10% higher than DC

Fig. 16-20. Typical manufacturer's electrode specification chart for welding manganese steel. (Courtesy of Lincoln Electric Co.)

section will deal on the use of arc welding for joining aluminum.

Aluminum Arc Welding Procedures

Factors to be considered before welding aluminum and its alloys are distortion, fluxes, and filler metal. To some degree, arc welding

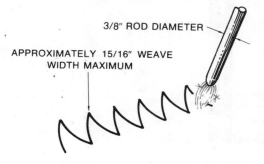

APPROXIMATELY 15/16" WEAVE WIDTH MAXIMUM

3/8" ROD DIAMETER

Fig. 16-21. The weave width must be over 2½ times the electrode diameter when weave welding manganese steel.

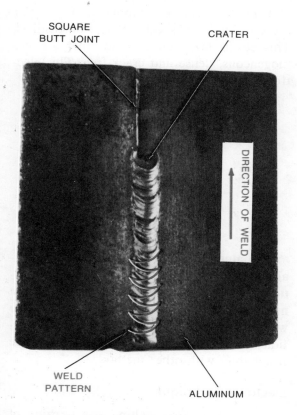

SQUARE BUTT JOINT

CRATER

DIRECTION OF WELD

WELD PATTERN

ALUMINUM

Fig. 16-22. Vertical weld on ⅛" aluminum coupons with no joint preparation.

prevents excessive expansion and distortion that aluminum usually has when it is heated.

Joint designs depend on the gauge of the material to be welded. Aluminum and the various alloys can be satisfactorily welded with a square butt joint if the metal is not over ¼" thick. See Fig. 16-22.

Metal over ¼" thick can be prepared with the single-Vee or double-Vee butt joint, as shown in Fig. 16-23. To control the weld penetration, back-up strips are recommended.

Cleaning—After preparing the material for good fit-up, the metal must be thoroughly cleaned of oxide and other films.

These surface contaminations (dirt) reduce the weld's strength. They can be removed and the aluminum cleaned by using alkaline or commercial degreasing solutions. The solutions should not emit (give off) toxic fumes during the welding process. Drying the metal completely after solution cleaning is important, to prevent porosity from forming in the weld metal.

Preheating—On large aluminum surfaces, it may be necessary to preheat the weld area if the area refuses to melt properly. This may be due to the welding heat being conducted into the air faster than the arc can supply heat to the weld metal.

The amperage can be increased to make up for this heat loss, but, in most cases, preheating with an oxy-acetyelene welding torch will produce a better bead. See Fig. 16-24.

Fluxed Electrode—At welding temperatures, aluminum and its alloys have a strong tendency to form harmful refractory oxides. For strong aluminum welds, these must be removed by using a flux, either from a can or on the rod. The flux during arc welding is often in the form of an electrode coating. For this reason, the manufacturer's recommendation chart should be followed for choosing the right aluminum electrode. Fig. 16-25 shows several aluminum electrodes, with and without a flux coating.

Here is an example of manufacturer's specifications for an AWS-ASTM E A1-43

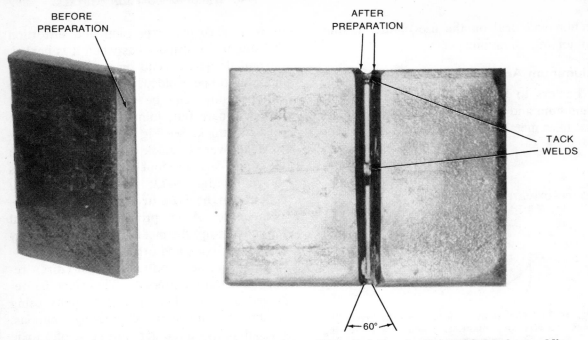

Fig. 16-23. Prepared aluminum coupons forming a 60° included angle, tack welded before welding.

aluminum electrode for arc welding with reverse polarity direct current.

The manufacturer's description is: A dip-coated aluminum shielded arc electrode with a coating that will dissolve any aluminum oxides that may be formed. The coating also assists in giving a smooth operating arc to provide good quality weld metal.

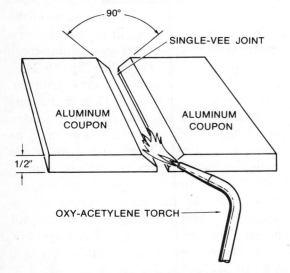

Fig. 16-24. Preheating thick aluminum coupons with an oxy-acetylene torch.

The manufacturer's application is: For use on aluminum sheets, plates, or castings. This electrode produces welds that are homogeneous, dense, and crack-free. This electrode will not discolor the base metal.

The current recommendations are:

Diameter	Amperage
$\frac{1}{8}''$	20- 40
$\frac{3}{32}''$	30- 60
$\frac{1}{8}''$	75-125
$\frac{5}{32}''$	125-150
$\frac{3}{16}''$	140-175
$\frac{1}{4}''$	170-225

The aluminum arc welding process is usually done only in the horizontal and flat positions. Distortion and excessive expansion are more limited when aluminum welds are done with the electric arc method. Joint preparation is also simpler and welding can be done faster than with other methods.

Electrode Technique

Due to the rapid melting and freezing of aluminum electrodes, the arc should be started only by the scratching method commonly used when welding with alternating current machines. Since the arc can be broken only by excessive movement, it should be struck on the seam of the weld, near the area where the weld is to start.

Using a short arc during welding will increase the arc stability. Should the arc short out during welding because it contacted the metal being welded, accumulated flux will form on the end of the rod; this should be removed before restriking the arc. The flux can be removed by tapping the end of the rod with a chipping hammer.

As when welding on mild steel, the welder must adjust the amperage and develop the proper electrode travel speed for the best results. This way, he will weld with a sound bead that is free of porosity and excessive penetration. The electrode angles while welding in different positions should be the same as the electrode angles previously mentioned for welding mild steel. See Fig. 16-26.

Manual Arc Welding Magnesium and Magnesium Alloys

The skill of welding commercial magnesium is as important to the welder as is the welding of aluminum and aluminum alloys.

Magnesium is a light, high-strength material that is becoming increasingly popular in manufacturing. When alloyed with zinc, manganese, and aluminum, magnesium is especially corrosion-resistant, making it useful for various types of arc welded structures. Because the weight of magnesium is only 2/3 the weight of aluminum, magnesium is valuable for strong, lightweight castings.

Common Magnesium Alloys

There are two common magnesium alloys used for making magnesium plate and sheet that are commonly welded in industry.

AZ31X—This magnesium alloy is made of 3% aluminum, 1% zinc, and 0.3% manganese, plus magnesium. AZ31X has a tensile strength of 38,000 psi.

M1—The other common magnesium alloy, M1, is made of 1.5% manganese and has a tensile strength of 33,000 to 37,000 psi.

The AZ31X alloy, due to its strength, ductility, and toughness, is used for high-strength arc welded structures. The M1 alloy is preferred when stress relieving after welding is not necessary.

Welding Precautions

Due to the fact that magnesium oxidizes (burns) rapidly when heated to its melting

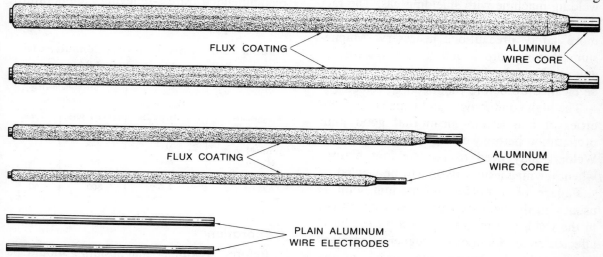

Fig. 16-25. Aluminum electrodes.

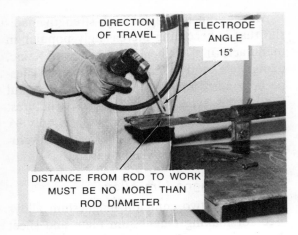

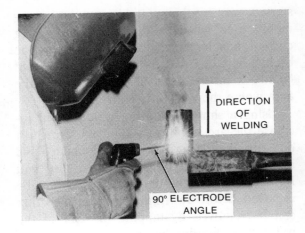

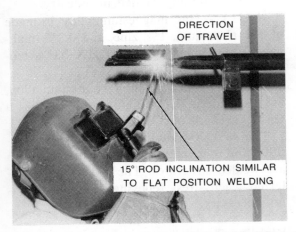

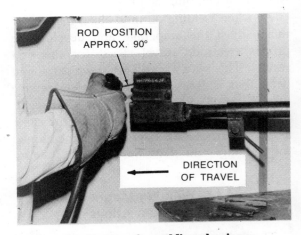

Fig. 16-26. These electrode angles for welding mild steel are also the same for welding aluminum.

point in air, it is necessary to use an arc welding machine with shielding gas. The fact that magnesium burns rapidly also makes more safety precautions necessary. Magnesium shavings are a definite fire hazard, so any machine shavings from magnesium must be cleaned up before the metal is welded.

The high conductivity and expansion properties of magnesium mean that good joint preparation before welding is very important. Welders may need to use welding fixtures whenever magnesium parts will be welded.

Failure of a welded magnesium joint is usually in the heat-affected zone, rather than in the weld, due to grain growth in the heat-affected zone. Some of the higher alloy magnesium alloys, however, may crack in the weld area. For that reason, stress-relieving

after welding is required. Table 16-6 shows recommended stress-relieving temperatures and time for the common magnesium alloys.

Table 16-6.   Stress-Relieving Conditions for Common Magnesium Alloys.

Magnesium Alloy	Magnesium Electrodes Used	Stress-Relieving Conditions	
		Temperature (°F)	Time (Min.)
M1 (A)	M1	500	15
M1 (H)	M1	400	60
AZ31A	AZ61	500	15
AZ31H	AZ92	300	60

Preparation

Before welding, magnesium should be cleaned by using steel wool, abrasive cloth, or chemical cleaners; this is essentially the same method recommended for aluminum and aluminum alloys. Emery cloth should not be used to prepare magnesium, because it may leave foreign particles in the metal.

After cleaning, the magnesium may be positioned as needed for welding. Tack welding, if done, must use the same welding method as for the final weld.

Procedure

Magnesium cannot be welded with the manual shielded arc method that uses the electrode flux as the shielding gas. Magnesium must be welded with the Tungsten Inert Gas (TIG) welding method, with a steady flow of shielding gas to keep the air's oxygen from contacting the molten magnesium.

When welding with the Tungsten Inert Gas method, a flux is not required. The shielded gas eliminates any impurities from the atmosphere, as well as the oxygen, that could affect the molten metal. The Tungsten Inert Gas method increases the welding speed, creates less distortion, and can be used for welding thicker materials. This is true not only on magnesium and magnesium alloys, but on aluminum alloys as well.

The exact procedure for TIG welding is covered in another chapter.

Other Methods of Welding Magnesium

Magnesium may also be welded through resistance welding, seam welding, and pressure gas welding. Brazing and soldering can also be used to join magnesium.

In both brazing and soldering, of course, the base metal is not melted. In this case, the base metal is magnesium. Since molten magnesium burns easily, it is important that the metal not get hot enough to melt while brazing or soldering. If the metal became molten, it would easily burn because soldering and brazing are done without a shielding gas.

Manual Arc Welding Copper and Copper Alloys

Many manufacturers consider copper and copper alloys second only to iron and steel in commercial importance. There are two basic classifications of copper: oxygen free and oxygen bearing. The differences between the two classes of copper are not so great that the welding procedures vary widely. Generally speaking, the welding procedures are similar for both basic classes of copper and may be discussed together.

Welding Methods

Welding copper and copper alloys may be done by TIG, carbon arc, resistance, submerged arc, and brazing processes. With practice, oxy-acetylene and shielded metal-arc welding processes also give excellent welding results on copper parts.

Due to copper's high heat conductivity, only light gauge copper can be satisfactorily welded with the shielded arc method, unless the copper is preheated.

Copper Electrodes

The electrodes used in copper metal-arc welding are somewhat limited. Covered electrodes are available that give good weld strength but poor ductility, due to the porosity set up in the grain boundary areas. Manufacturers have, however, developed for light gauge copper alloys a bronze-aluminum alloy electrode that requires little or no preheating. Electrodes alloyed with phosphor bronze

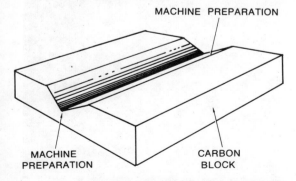

MACHINE PREPARATION

MACHINE PREPARATION

CARBON BLOCK

Fig. 16-27. A machined groove in a carbon block, to be used as a back-up block.

are also suitable for welding with the copper metal-arc method.

Technique

Copper and copper alloys can be welded with either the bead or weave techniques, using joint preparations similar to those for iron and steel. When welding copper or copper alloys ¼" thick, a 60° angle in a single-Vee joint preparation will give good weld results. The angle of the Vee joint should be increased up to a maximum of 85° when welding metals over ¼" thick. A lesser angle may be used for thinner copper Vees.

Copper should then be welded only in the flat position, using the shielded arc welding process. Using fixtures will help eliminate metal distortion *during* welding, but they will create extensive internal stress in the weld area. Metals free from fixture-like restraints are relieved of stress through distortion. Metals fastened in fixtures may crack in the weld if the copper is held so as to restrict all movement.

Backing Strips—Temporary *backing strips* made of carbon or copper may be used. These may be employed on ferrous and nonferrous metals for eliminating fall through, a common problem when welding copper or copper alloys. The strips may be machined or ground to form grooves of various shapes. See Fig. 16-27. When the metal is welded with a back-up bar, the bead on the root pass looks better than a bead welded without a root pass, when the back-up bar is removed. See Fig. 16-28.

Heavy Sections—During welding of heavy sections, heat should constantly be applied at temperatures between 800° F and 1000° F. Large diameter electrodes in the high current range should be used to melt the parent metal and the electrode together carefully. This will hold down any slag inclusions in the weld. As usual, electrode manufacturers recommend current ranges for each size electrode. Table 16-7 shows the chemical composition and current recommendation for a common copper electrode.

Table 16-7. Recommendations for AWS-ASTM Class ECuSn-A Copper Welding Electrodes.

Chemical Composition		Electrode Diameter	Amperage
Element	Percentage	(Inches)	
Copper	94.3 %		
Tin	5.0 %	⅛	80-120
Lead	0.20%	⁵⁄₃₂	130-180
All Others	0.50% Max.		

UNEVEN
PENETRATION

EVEN
PENETRATION

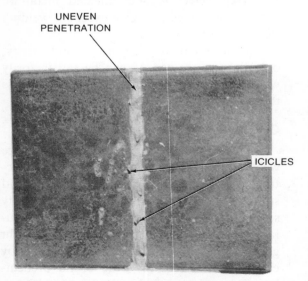

ICICLES

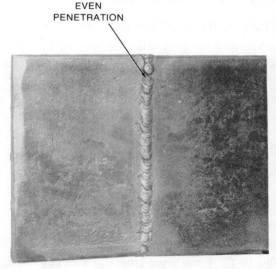

Fig. 16-28. Sample coupons welded without a back-up bar (left) and with a back-up bar (right).

Manual Arc Welding Clad Steel

A need for more corrosion- and/or heat-resistant metals, other than regular galvanized steel, led to developing a process known as *cladding*. Clad steel is made by cladding.

How Clad Steel Is Made

Clad steel is made of a steel *core* (usually mild steel) covered on one or both sides by other metals or *alloys*. The type of covering (coating) depends on where the corrosion-resistant or heat-resistant material is required on the finished product. The coating thickness also depends on the use of the finished product, but it is usually not over 0.030″. The core, sometimes referred to as back-up metal, gives the required strength to any weaker cladding matrial. However, in many cases, clad materials have a tensile strength as high as, or higher than, the core metal.

The cladding material may be as little as 5% and up to 20% of the total plate thickness. Cladding material is bonded to the steel core (base metal) by a process of first melting the two metals, with their sides together, followed by heating and rolling to the required thickness. Cladding applied in this way will not separate from the core as a result of pressure of temperature changes or bending while being formed.

Characteristics of Clad Steels

Clad steels give welded results that are as good as stainless steels *if* the welder knows and follows the special welding procedures for clad steel. This way, the welded clad steel will keep the corrosion- and heat-resistant properties of the base clad steel.

Galvanized sheet metal is one of the most common clad metals. Galvanizing is a method of protecting steel from corrosion and produces a clad steel that can be readily spot or seam welded. The zinc cladding on galvanized sheet metal, applied by hot dipping, will burn off during arc and oxy-acetylene welding. When the zinc coating burns away, the

weldment can again corrode easily. If it is possible to dip the weldment in the hot zinc bath *after* fabrication (assembly), the welds will be coated and will have good corrosion-resistance properties. If a total dip is not possible, the parts may be metal sprayed.

A coating of 80% lead and 20% tin (a soldering alloy) has been used to cover one or both sides of iron or steel for manufacturing corrosion-resistant tanks. The coated steel, known as *terne steel,* is best joined by seam welding or soldering.

Aluminized steel, covered by dipping in an aluminum coating, is used to manufacture such items as automobile mufflers and roofing sheets. This form of clad steel must be welded by resistance welding if the corrosion-resistance coating is to be retained. The low melting temperatures of the coating, compared with the high melting temperature of the steel core, makes welding difficult with the arc or oxy-acetylene methods. If either of these welding methods were the only method available to repair aluminum-clad steel, the welder would need inert gas to keep the aluminum cladding from oxidizing quickly.

Stainless-clad steels are made when some type of low-stainless steel is rolled on one or both sides of mild steel that is thicker than the aluminum or zinc coatings just mentioned.

Monel-clad steel has the high-strength, corrosion-resistant, nickel-copper alloy *monel* clad to the mild steel. This is done in the same manner as stainless steel cladding and is welded in the same manner as stainless-clad steels, with the exception of the filler metal.

Welding Precautions

When welding clad steels, it is important to fuse the metal to be welded so as to provide both good strength *and* corrosion-resistant qualities within the weld itself.

Earlier, because the melted filler metal was diluted with molten high-carbon cast iron, the weld bead could not be machined when mild steel was used to weld cast iron. This bead hardening happened because excessive carbon was picked up through the mixing caused by dilution. A similar dilution problem can happen when clad steel is welded, if good welding procedures are not understood or followed.

Submerged arc welding with very high currents will quickly deposit filler metal. If the amount of base melting becomes *greater* than the amount of filler metal melting, the filler metal will be greatly diluted. If submerged arc welding is used on clad steel, the welder must realize that the bead analysis would depend more on the base metal than on the filler metal, since more of the base metal is melted.

Common Joint Preparations for Clad Steels

On heavy clad steel, a double-Vee or U-groove butt joint should be used. The preparation on each side of the metal should be *un*equal, as shown in Fig. 16-29. If the shallow preparation is placed on the clad side, less *alloy* filler metal will be needed for the weld.

NOTE: When welding any clad steel, the welder must use an *alloy filler metal rod* against the cladding whenever a bead is needed against the cladding. Mild steel filler metal rods must *only* be used where the joint does not have any clad material, as shown in Fig. 16-30.

If a single-Vee butt joint can be prepared and welded only from the steel side, either copper or alloy backing strips must be used. See Fig. 16-31. When welding this preparation where the clad side is down, the welder should use an alloy electrode to weld the entire cross-section of the joint for maximum strength and good weld appearance.

Typical lap joints for clad steel (Fig. 16-32) must be welded with alloy rods wherever the bead will come in contact with the clad side.

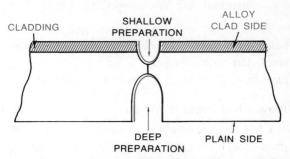

Fig. 16-29. A shallow preparation is used on the clad side with a double-U preparation on clad steel.

Corner joints with clad steel require special welding for each form of joint.

In any form of clad welding, accessible (open, not clad) steel can be welded with a steel rod up to $\frac{1}{16}''$ from the cladding material. Keeping $\frac{1}{16}''$ away from the cladding will assure that dilution does not take place. See Fig. 16-33.

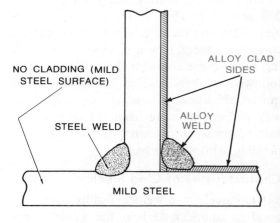

Fig. 16-30. Alloy weld material must be used when the weld contacts any alloy clad metal.

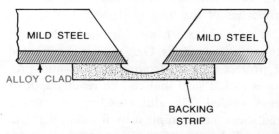

Fig. 16-31. Alloy or copper back-up bars must be used if clad steel is to be welded with the clad side down.

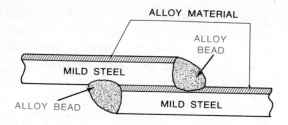

Fig. 16-32. An alloy electrode must be used whenever the bead will contact the clad side.

Corner joints, shown in Figs. 16-34 and 16-35, are always welded throughout with alloy electrodes. After welding the clad side, the steel side is usually ground and welded with alloy electrodes. The corner joint shown in Fig. 16-36 is usually welded completely with an alloy electrode. The welder may prefer to weld the outside of the corner joint in Fig. 16-37 to within 1/16″ of the cladding with a large diameter steel electrode. This will dilute and dissipate (spread out) the pickup of the clad material. The inside of the joint in Fig. 16-37 would then be welded with an alloy electrode.

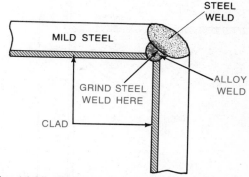

Fig. 16-35. Corner joint originally welded entirely with a steel electrode. Then, the corner inside the clad area was ground out and welded with an alloy electrode.

Clad Steel Welding Methods

Light-gauge clad metals are usually welded with the TIG or atomic-hydrogen methods. In some cases, copper-clad metals may be brazed. Gas welding and resistance welding are seldom used, especially on clad metals with extreme variations in heat conductivity between the two metals.

Submerged arc welding can be used for welding heavier gauge clad steel metals, especially for the weld on the steel side of a double-U preparation, as in Fig. 16-33. Shielded metal arc welding is preferred for welding clad joints with 18 gauge or thicker material.

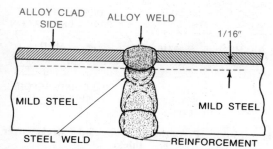

Fig. 16-33. Mild steel electrode may be used up to 1/16″ from the clad material.

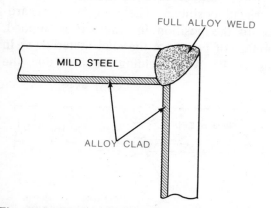

Fig. 16-34. Corner joint welded with a full alloy weld.

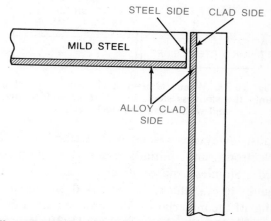

Fig. 16-36. This type of corner joint is usually welded throughout with an alloy electrode.

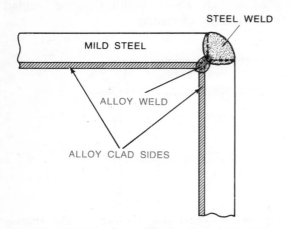

Fig. 16-37. Prepared clad corner joint, welded on the back with a mild steel electrode and on the front with an alloy electrode.

Shielded Metal Arc Welding

To successfully weld clad steel, the welder must take extra care in fitting up the joints. Where fit-up is poor, the clad material could penetrate the weld, creating a hard, brittle weld with low ductility. If good fit-up is not possible, the welder would use an E 6013 electrode on the steel side. The E 6013 electrode has low penetration, reducing any chance of dilution.

Light-gauge, stainless-clad steel, from the lighter 18 gauge to the heavier 11 gauge, is prepared like any other light-gauge clad steel, as in Fig. 16-38. However, stainless steel

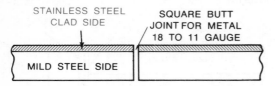

Fig. 16-38. When stainless steel is used for cladding, the stainless steel side must be welded first, and with a stainless steel electrode.

must *always* be welded with a stainless steel electrode and initially from the stainless side. Stainless and other clad steels from 10 gauge to $\frac{3}{16}''$ thick, are prepared on the steel side of the material. A Vee groove with a 60° included angle, shown in Fig. 16-39, is used for preparing the joint. Using the thickness

of cladding material as a guide, the *first* weld is welded from the steel side with a stainless steel electrode. Then, the stainless steel side can be welded with the same electrode.

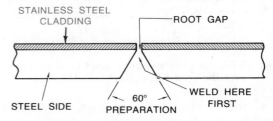

Fig. 16-39. Preparing and welding mild steel clad with stainless steel.

Clad steels heavier than $\frac{3}{16}''$ are prepared as shown in Fig. 16-40. The weld can then be completed entirely on the steel side, using an electrode with the same characteristics as the backing metal (metal on which the clad is placed). For flat position welding, an E 6020 or E 6030 electrode is usually recommended. The E 6010 or E 6011 electrodes,

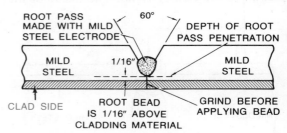

Fig. 16-40. Preparation for clad steels over $\frac{3}{16}''$ thick.

however, should be used in the vertical or overhead positions. The root bead, regardless of the position in which it is applied, should not penetrate the clad material, as in Fig. 16-40. After welding the steel side, the

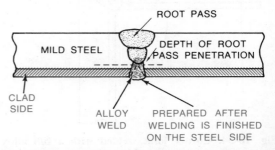

Fig. 16-41. Welding the clad side of thick steel.

clad side can be prepared by grinding or chipping, as shown in Fig. 16-41. The preparation should be deep enough so that any unfused section of the root bead is ground free

of slag or other foreign bodies. The clad side is then welded with the correct alloy electrode for the cladding alloy being welded. See Fig. 16-41.

17

Arc Cutting

Arc welding equipment may also be used to cut metal. This may be needed when metal is being prepared for welding or for demolition work. In arc cutting, as in arc welding, the violent force of the electric arc is still used to melt the metal. Then the molten metal is either oxidized or blown away, depending on the method being used.

Arc-Cutting Electrodes

Arc-cutting material such as copper, cast iron, low-or high-carbon steel, and stainless steel may be done with either a general-purpose electrode or a metal-cored electrode made for this specific purpose. Generally speaking, the electrode is usually moved up and down, allowing the arc to cut.

Special coated electrodes are able to cut, pierce, gouge, and chamfer when using high amperage. The welder has to practice with various heat settings to find the best setting for whatever type of cutting is being done on his particular job. It is important that the electrode coating serve as an *insulator* (nonconductor) to allow the electrode to go into the gap of the cut without shorting out due to arcing on the sides of the electrode. For this reason, an iron-powdered electrode is not recommended. The coatings on the special electrodes act as an arc stabilizer, concentrating the arc while intensifying the arc's action.

Special cutting electrodes are usually used on straight-polarity DC. Standard E 6010 or E 6011 electrodes may be used for arc cutting, although they should be used on reverse-polarity DC because they penetrate (cut) better on DC reverse.

Cutting with these electrodes should only be used for preparing metal if arc-air equipment is not available.

Arc-Air Cutting

The arc-air torch is a combination arc and air cutting torch. It may be used to cut, bevel, gouge, groove, pierce, or flush away any kind of metal. See Fig. 17-1.

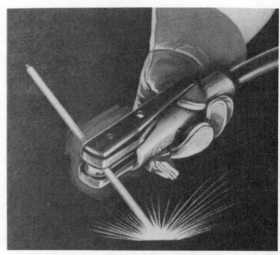

Fig. 17-1. An arc-air cutting torch.

This torch uses current from a welder, a special electrode, and ordinary compressed air. It is versatile because it can be used in all work positions. On most models, the angle of the electrode can be changed without destroying the air jet flow to the molten metal. Both current and air are fed through the same line to the torch electrode and to the metal being cut. This is shown below in Fig. 17-2.

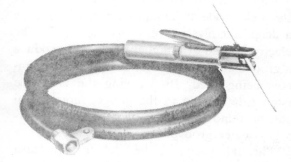

Fig. 17-2. Arc-air cutting torch with hollow welding cable for carrying compressed air through the torch tip.

Power Source

An ordinary welding machine is usually the power source for arc-air cutting. It must be large enough to produce the current needed for the penetrating done while arc cutting, such as the machine in Fig. 17-3. Heavy-duty applications usually require a constant voltage welding machine. In this type of cutting, reverse polarity is required. Recommended current ranges are given in Table 17-1. The welder should experiment with the amperage settings shown in Table 17-1 while varying the air pressure from 80 to 100 psi.

Electrode

A carbon-graphite electrode will work well for almost all arc-air cutting jobs. These electrodes are available in both plain and copper clad forms. The electrodes covered with copper are able to carry higher currents than the plain uncovered type, and they will

keep their original diameters for improved grooving and piercing.

Fig. 17-3. A typical welding machine with enough power for arc cutting.

Table 17-1. Recommended Current Ranges for Arc and Air Cutting.

Electrode Diameter (In.)	Current Range (Amperes)
5⁄32	80- 150
3⁄16	100- 200
1⁄4	150- 350
5⁄16	200- 450
3⁄8	300- 550
1⁄2	400- 800
5⁄8	600-1000
3⁄4	800-1600

Procedure

Air flow in the arc-air torch is controlled by an on-off valve in the torch handle, as in Fig. 17-4. The torch head can be rotated to position the electrode at any desired angle.

An arc is struck between the electrode and the metal to be cut or gouged. When the arc is struck, the parent metal melts and is re-

Fig. 17-4. A simple on-off valve allows the welder to control the compressed air through the torch tip.

moved by the compressed air jet pointed at the arc.

On critical work, where no porosity or cracking can be allowed, the arc and air torch grooves out cracks and removes porous areas quickly. It can also cut, groove, or flush off any surfacing material because it is *not* an oxidizing process, as is oxy-acetylene cutting. The low heat input does not usually affect the properties of the base metal. Where it is necessary to undercut the parent metal before welding an initial deposit, the electrode should be held at a fairly flat angle to the work and then moved from side to side for the desired width. This procedure gives a clean, ductile base metal for the weld deposit.

Gouging

In gouging, the metal is not cut into two pieces. Instead, a groove is cut in the metal to a certain depth, depending on the job.

Procedure

The electrode position is sloped back from the direction of travel when gouging is done, as in Fig. 17-5. The air jet must be positioned so that the compressed air will blast from *behind* the electrode.

The width and depth of the groove depends on the angle of the electrode, the electrode diameter, and the welder's travel speed. A small angle and fast travel speed produce a wide, shallow groove, as in Fig. 17-6. A steep electrode angle and slow speed produce a deep, narrow groove, as in Fig. 17-7. The electrode used at a higher angle will produce a groove about 1/8″ wider than the rod diameter being used. By weaving the electrode from side to side in the gouge, more metal can be removed by the compressed air, giving a wider groove.

Through practice, the welder will learn the importance of varying the travel speed according to the diameter of the electrode, the metal being cut, and the amperage setting.

Flat Gouging—When the arc is struck, it must be kept short and steady without contacting the work, and it must be moved in the direction of the cut for *flat* gouging. The compressed air should be turned on as soon as the arc is struck and not turned off until the arc is broken.

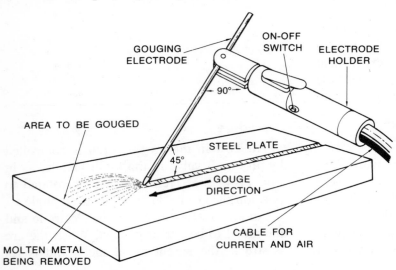

Fig. 17-5. Common gouging procedure.

Vertical Gouging—When gouging must be done in the *vertical* position, the direction of cut should be downward, as in Fig. 17-8. This will allow the pull of gravity to help the compressed air remove the molten metal.

Horizontal Gouging—This can be done by gouging out the molten metal from either direction across the plate, as in Fig. 17-9.

Overhead Gouging—This is the most difficult gouging position, especially for a new welder. Although gravity helps remove the metal, a new welder may have some difficulty at first in establishing and holding a short, steady arc.

For most flat, horizontal, and vertical gouging, the electrode should be positioned in the holder about half-way up the electrode, at an angle of about 90°, as was shown in Fig. 17-5. Varying the angle for certain preparations in the above positions can be done by turning the electrode in the swivel holder as shown in Fig. 17-10.

For the overhead position, however, the electrode is placed as straight forward in the holder as the electrode swivel will allow. See Fig. 17-11. In this position, the welder will avoid being hit with the dripping molten metal and he will have a better view.

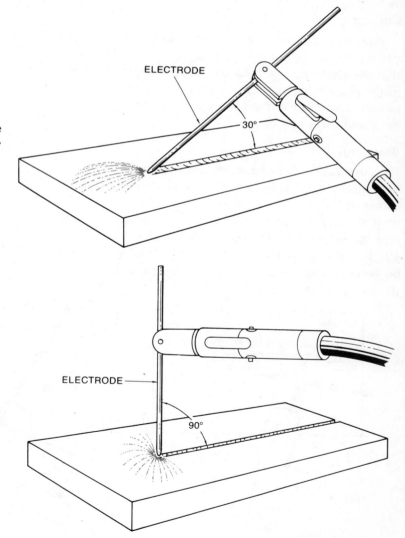

ELECTRODE

30°

Fig. 17-6. A lower electrode angle produces a wide, shallow groove.

ELECTRODE

90°

Fig. 17-7. A higher electrode angle produces a deep, narrow groove.

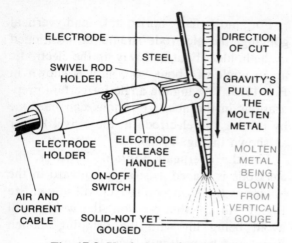

Fig. 17-8. Vertical position gouging.

Cutting

Basically, the same technique is used for cutting metals as is used for gouging, except that the electrode is tipped at a steeper angle to the plate being cut. This allows the welder to melt and remove the metal by sawing in a back and forth motion.

Beveling

Coupons or other pieces for weld preparation can be beveled by drawing the electrode across the edge of the work at the bevel angle desired. The electrode should be positioned at a 90° angle from the holder. See Fig. 17-12. For this preparation, the air jet should be between the electrode and the surface of the metal being beveled.

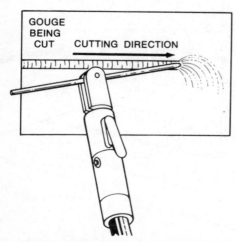

Fig. 17-9. Horizontal position gouging.

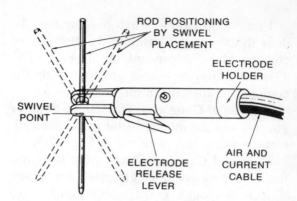

Fig. 17-10. For certain jobs, the electrode can be swivelled in the electrode holder.

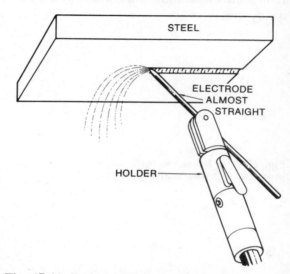

Fig. 17-11. Position of the electrode for overhead gouging.

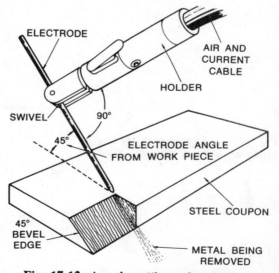

Fig. 17-12. Arc-air cutting a bevel edge.

Plasma-Arc Cutting

Plasma-arc cutting is a high-speed cutting technique that can cut very difficult metals, such as stainless and hard-to-cut alloy steels. Plasma-arc cutting is also fairly accurate, capable of cutting to tolerances within $\frac{1}{16}''$. Also, due to the fast cutting speed, the heat-affected zone formed along the edge of the kerf is usually less than 0.05".

How It Works

Plasma-arc cutting is done by super-heating an air or gas to convert it totally or partially into ions, through ionization. When an air or gas is heated high enough to convert it to ions, the air or gas changes to a form of matter called *plasma*.

Most people know the three common states of matter: solids, liquids, and gases. Plasma is considered to be a fourth state of matter, higher than a common gas. Super-heating an air or gas changes the air, for instance, from the gas state to the plasma state. This is the same way that heating changes water from a solid (ice) to a liquid. In general, heat is needed to change anything from one state of matter *up* to another. In this way, heat is used up.

In turn, when something goes from one state of matter *down* to a lower state, heat is given off. For instance: when water freezes, the warm liquid water has to give off heat to become ice. So, if a gas was in a *plasma* state and it changed back to the *gas* state, heat would be given off. In fact, a great deal of heat would be given off, since it takes so much heat to convert the gas to plasma to begin with. In this way, heat is given off during the plasma conversion.

In plasma arc welding or cutting, the plasma welding machine supplies the energy needed to change the gas being used into plasma. Then, when the gas comes out of the nozzle in the form of plasma, it changes back to a gas and gives off the heat it took to become plasma. See Fig. 17-13. Also, an arc is established between the cathode and the work, so that the normal arc heat *and* the heat of the plasma gas work to cut the metal. Together, they produce more than enough heat to cut even very hard metal.

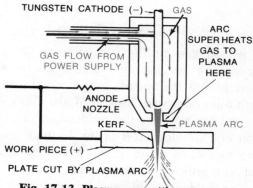

Fig. 17-13. **Plasma-arc cutting process.**

The heat from the total plasma arc is about ten times as hot as the oxy-acetylene cutting flame. The oxy-acetylene cutting flame produces temperatures of about 6300° F, so the plasma-arc temperature may reach over 60,000° F.

Types of Plasma Torches

Plasma cutting torches come in two basic types: transferred and nontransferred. The *transferred* type holds the arc between the torch and the work; that is, the arc is transferred to the work piece.

The *nontransferred* type, however, uses the DC current to hold an arc *within* the torch itself, between the negative torch electrode and the positive torch nozzle. In this way, the arc force is used to cut the metal but is not established with the metal. With either type of torch, heat formed by the arc and the plasma melts and a jet stream forces the molten metal from the cut.

Gases Used with Plasma-Arc

The gases commonly used for plasma-arc cutting include hydrogen, argon, nitrogen, and helium. Compressed air (thermal air), being cheaper than the above-mentioned gases, is also frequently used.

The compressed (thermal) air cutting process is cheaper than the oxy-acetylene process for cutting metals up to 1″ thick. For metal thicker than 1″, nitrogen must be used in place of compressed air.

Arc-Oxygen Cutting

Arc-oxygen cutting is similar to arc and air cutting because the electric arc is still used to cut the metal, but in combination with a jet of *oxygen* instead of air. Using the arc-*oxygen* process, the welder can cut cast iron and stainless steel up to 2″ thick. Nonferrous metals up to 1″ thick can be easily cut with arc-oxygen equipment.

The welder should consider using arc-oxygen cutting when the oxy-acetylene process will not work. This may be when cutting metals such as stainless steel, bronze, brass, and cast iron.

Torch Construction

The electrodes used in this cutting process consist first of a metal or ceramic *core* that is heat- and corrosion-resistant. The core is then covered with an inert (nonconducting) *matrix* so that it can be inserted into deep grooves. The electrode core conducts the current and holds an arc that brings the work piece up to a white heat.

Torch Operation

While holding the arc, oxygen is discharged from a hole in the center of the electrode. In turn, this ignites the white hot metallic core of the electrode and the molten metal being cut, similar to the action of an oxygen lance. The kinetic energy (energy associated with motion) in the regulated oxygen stream then removes the products of combustion. With careful control, the arc-oxygen process leaves the required cut.

Ceramic core electrodes are usually used for underwater cutting, and, like metal core electrodes, they need only minor control from the welder. Mild steel over 4″ thick, and most

nonferrous metal up to ¾″ thick, can be cut by using metal core electrodes, dragging the electrode across the line to be cut with a slight pressure.

Piercing—Plate *piercing* is easily done by using the arc-oxygen cutting process. Once the arc contacts the material to be pierced, the oxygen can then be released and the electrode pushed through the metal. The outer cover of the electrode prevents the core from shorting out on the side of the hole being pierced.

Like other arc cutting processes, arc-oxygen cutting will leave rough, uneven edges on the kerf. In many cases, these edges will be so rough that they must be ground smoother before welding.

Carbon-Arc Cutting

Carbon-arc cutting is similar to the cutting discussed earlier under Arc Cutting Electrodes. With carbon-arc cutting, however, the carbon rods being used are consumed differently and not as fast as are the coated electrodes for arc cutting.

How It Works

In the carbon-arc cutting process, an arc is struck between a carbon (or graphite) electrode and the metal to be cut. The high heat required for cutting usually causes the carbon electrodes to be quickly consumed, especially if continuous cutting is being done. Because carbon electrodes are used up so quickly, graphite electrodes are usually used where extremely high current is needed for cutting. This is because graphite electrodes set up lower resistance to the current flow, creating less heat in the electrode itself. Although alternating current can be used, the preferred current is straight-polarity DC.

Torch Operation

Cutting with a carbon arc can be done in any position. With the electrode held either vertically or slightly forward in the horizontal position, the welder receives help from gravity in removing the metal without being hit by falling metal. This is a problem that the operator would face with carbon-arc cutting in the overhead position.

The width of the cut depends on the electrode's size and the welder's skill. Since metal must be removed to form a kerf, a new welder will soon discover that the thicker the metal, the wider will be the groove or cut. With some practice, the welder can satisfactorily cut most metals if the current recommendations shown in Table 17-2 are used.

Carbon-arc cuts, as should be expected with arc cutting, will be rough. They will usually require more preparation, such as grinding, before welding. In carbon-arc and coated electrode arc cutting, the cutting must be done where there is good ventilation for the welder's safety.

Table 17-2. Current Recommendations for Carbon-Arc Cutting.

Graphite Electrode Diameter (In.)	Current Range Recommendations (Amperes)
¼- ⅜	Up to 300
½- ⅝	300- 600
¾- ⅞	600- 900
1-1½	900-1200

Tungsten-Arc Cutting

Tungsten-arc cutting is commonly used on metals such as stainless steel, copper, magnesium, and aluminum, although most other metals can also be cut by this process.

How It Works

A compressed arc is first formed between the work piece to be cut and a nonconsumable tungsten electrode. While the compressed arc is melting the metal, a high-velocity gas stream is used to blow the molten metal away, producing the cut.

The arc, at a temperature of about 28,-000° F, melts the metal when concentrated on a small area. The gas, expanded by the heat of the arc and forced through a shielded constricting reducing cup, then removes the molten metal to form a kerf. See Fig. 17-14.

Results

The sides of the kerf are very smooth due to the force of the arc stream. The arc stream's force also protects the heated metal from the oxidizing elements in the air.

Other Cutting Procedures

Either automatic or semiautomatic equipment has been used satisfactorily in all of the arc cutting processes just discussed. Generally speaking, however, the coated electrodes

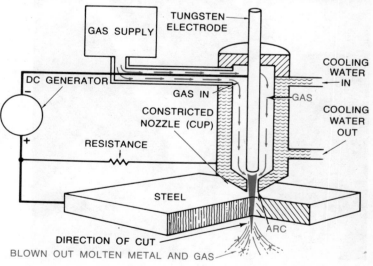

Fig. 17-14. Tungsten-arc cutting process.

Fig. 17-15. One manufacturer's assortment of grinding products that may be used for weld metal preparation. (Courtesy of The Norton Company.)

Fig. 17-16. A motor–driven grinding wheel. Although this wheel is being used for weld clean-up, it could also be used for metal preparation. (Courtesy of The Norton Company).

Fig. 17-17. Using an emery stone being driven by an electric motor on a flexible shaft. (Courtesy of The Norton Company.)

for arc cutting are seldom used, due to their fast rate of electrode consumption. Coated electrodes are usually only used for piercing. Joint preparations for welding, however, can often be made by methods other than the arc or flame cutting processes.

Grinding

Many different *grinding* stones and emery wheels can be used for preparing metal for welding and cutting out various shaped joints. See Fig. 17-15. Many of these abra-

Fig. 17-18. Mechanical tools that may be used for weld metal preparation.

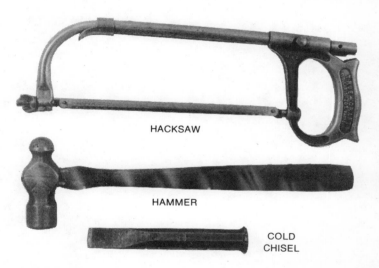

HACKSAW

HAMMER

COLD CHISEL

sive stones are motor driven, as shown in Fig. 17-16.

Other small wheels or stones available are mounted on flexible driven shafts. As in Fig. 17-17, the stones on drive shafts allow the welder to grind in small areas that are hard to reach.

Machining

Weld joint preparations can also be *machined*. Here, the needed preparation of cutting, finishing, or shaping is done by machines. The lathe, milling machine, shaper, and even a shop drill press may be suitable, especially on heavy but portable materials.

Mechanical Means

The oldest and simplest method of preparing a weld joint is by simple mechanical means. A hacksaw may be used to cut the metal before welding, and a hammer and cold chisel may be used to bend, cut, or chip parts for better weld results.

Any form of weld preparation will involve bending or removing some metal. With this in mind, a welder must always wear safety glasses and take any other necessary safety precautions while setting up to weld. See Fig. 17-18.

SECTION

IV

tungsten inert gas (tig) welding

The TIG welding process being used to weld aluminum coupons. (Courtesy of Aircom Metal Products, Inc.)

18

Introduction to TIG Welding

Tungsten Inert Gas (TIG) welding is an arc welding process where the fusion is shielded from the air by an inert gas shield. Metal fusion is created by the heat of an arc formed between a nonconsumable (does not melt) electrode and the work being welded.

Commercially, the TIG welding process is often referred to as *Heliarc*® welding. Heliarc® is a trademark of the Linde Welding Products Division of Union Carbide Corporation. Other manufacturers also make excellent TIG welding equipment.

As with MIG welding and other shielded gas welding processes, TIG welding makes use of inert gas or inert gas mixtures. An inert gas is neutral, that is, it will not react with the welding process. This gas or gas mixture usually consists of helium and/or argon acting as a protective shield around the molten metal, protecting the metal from coming into contact with the atmosphere.

Filler metal may or may not be used. When using a filler metal with the TIG process, the welder feeds in the filler rod with one hand while he holds the torch with the other hand, somewhat like oxy-acetylene welding. The torch is held steady, however, and must not be moved back and forth, as in oxy-acetylene welding. Movement takes place only with the

molten puddle as it moves forward. Fig. 18-1 shows the basic parts of the TIG welding process.

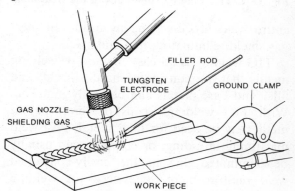

Fig. 18-1. The basic TIG welding process. (Courtesy of Hobart Brothers Company.)

Advantages of TIG Welding

The Tungsten Inert Gas process can be used for welding aluminum, magnesium, stainless steel, copper and copper alloys, nickel and nickel alloys, and a wide range of different metal thicknesses in mild steels. The top quality welds made in the above metals will need little, if any, cleaning after welding.

Welding with the TIG process can be done more easily in all positions, and free of spatter, since there is *no* metal passing through the arc. Welds made by TIG welding are also free of slag inclusions, since slag is never produced. TIG welding conditions are shown in Fig. 18-2. Here, it can be seen that the

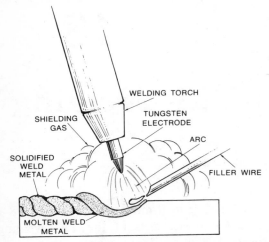

Fig. 18-2. TIG welding conditions in the weld area.

entire weld process happens inside an inert gas shield, eliminating flux and slag.

TIG welding is an easy method of welding metals that are considered hard-to-weld, and filler and base metals can be easily matched. With TIG welding, strips of scrap parent metal may be used for filler metal. Post-weld machining, grinding, or chipping can usually be eliminated due to the easily-controlled weld reinforcement. The need for welding flux is eliminated, even on hard-to-weld metals such as aluminum.

A dense weld deposit, free of porosity and slag inclusions, can be made while using high welding speeds. This lowers the overall welding costs. Although recent developments (plasma arc, electron beam, and special applications of gas metal arc welding) are being used to some extent, conventional Tungsten Inert Gas welding remains the most widely used process where excellent weld quality is most important.

TIG Welding Machines and Current

The TIG welding machine is sometimes called a power source. It may be either AC

or DC. The type of machine for a particular TIG job will depend on both the materials being welded and the arc characteristics desired. A typical TIG welding machine is shown in Fig. 18-3.

Fig. 18-3. A typical TIG welding machine. (Courtesy of Hobart Brothers Company.)

DC Polarity

With the direct current machine, the welder has a choice of either reverse or straight polarity, as discussed with conventional arc welding. Reverse polarity, however, is seldom used for TIG welding, due to its poor electrode current-carrying capacity, arc instability, and shallow penetration. Reverse polarity current, however, does have an oxide cleaning action that may be an advantage when welding metals that are likely to oxidize, such as aluminum and magnesium.

Stainless steel, copper, nickel, carbon steels, low-alloy steels, and any other metals not requiring the oxide cleaning needed for aluminum and magnesium, are better welded with straight polarity DC. Most welders find that the straight polarity DC arc is more stable and has deeper penetration than the reverse polarity arc. Less distortion and stress are set up in the weld metal, since narrow beads can be best welded with straight polarity DC.

AC Current

Alternating current, of course, is a combination of both straight *and* reverse polarity direct current, as in Fig. 18-4.

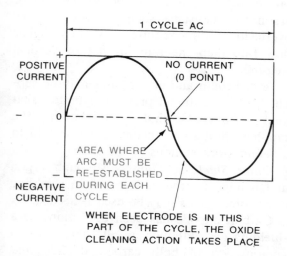

Fig. 18-4. Characteristics of the AC current in TIG welding.

Like reverse polarity DC, AC has the oxide cleaning action during the half cycle when the AC is on reverse polarity. See Fig. 18-4. For this reason, AC is preferred over reverse polarity DC for welding aluminum, magnesium, and beryllium copper. With these metals, the oxide cleaning action is very im-

portant for assuring the welder of sound, clean welds.

AC machine problems—The conventional AC welding machine does not produce a voltage that is high enough to reestablish the arc in the *inert* atmosphere when the voltage

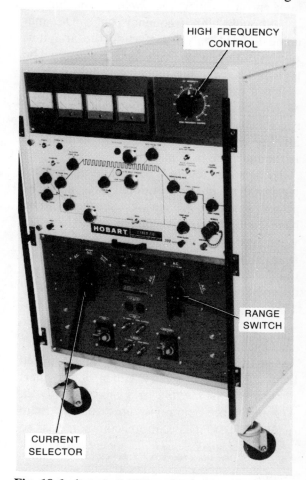

Fig. 18-6. A typical TIG welding machine. (Courtesy of Hobart Brothers Company.)

Fig. 18-5. A higher-frequency current is superimposed on the original frequency to eliminate the zero value of each cycle in the original frequency.

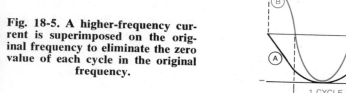

goes through the zero point of the AC cycle. See Fig. 18-4. In order to produce a good, stable arc, a high frequency current is *super-imposed* across the arc, as shown in Fig. 18-5. This eliminates the troublesome zero value. In Fig. 18-6, the high frequency control can be seen on another TIG welding machine.

Although high frequency on a DC machine is used only for starting the arc, it is used continuously with AC current. The typical TIG welding machine should operate within a range of 3 to 350 amperes, with 10 to 35 volts at a 60% duty cycle. Machines specifically designed for TIG welding can also be used for stick electrode welding, as is the machine in Fig. 18-6.

High-frequency units are available that can be installed on conventional AC and DC power sources. AC power sources should have an open circuit voltage of at least 75 volts, and they must be insulated for high-frequency current before a high-frequency unit is installed. The characteristics of the different welding currents are shown in Fig. 18-7.

TIG Shielding Gas

The arc and molten metal are shielded from the atmosphere by either of the inert gases *argon* or *helium*, or a mixture of both. Argon is more commonly used for two main reasons. First, it is a more common gas, so it can be easily found. Secondly, argon is heavier than helium, so it can provide better shielding at lower flow rates.

When helium is added to argon, the mixture increases the heat generated by the arc, although the current and arc length remain the same. For this reason, using the helium-argon mixture reduces the welding time needed for thicker pieces, due to the extra heat input. Where a higher heat is needed, pure helium can be used. However, the flow rate of helium gas should be double that of argon for the same shielding effect. Also, overhead welding requires a slightly higher flow rate than the rate for flat or vertical welding.

Both helium and argon gases may be supplied in conventional pressure cylinders that look like oxygen cylinders. Or, they may be supplied through a pipeline and manifold to different work stations in the shop.

Other Shielding Gases—Although argon and helium are the most common shielding gases, other gases may be used if needed.

Carbon dioxide has been used alone as a shielding gas, although an alternate gas is a combination of both carbon dioxide *and* argon. Altogether, some five gases and gas mixtures may be used to shield the arc for TIG welding:

1. Pure Argon
2. Pure Helium

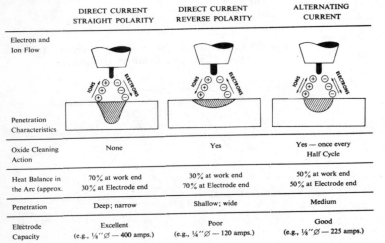

	DIRECT CURRENT STRAIGHT POLARITY	DIRECT CURRENT REVERSE POLARITY	ALTERNATING CURRENT
Electron and Ion Flow			
Penetration Characteristics			
Oxide Cleaning Action	None	Yes	Yes — once every Half Cycle
Heat Balance in the Arc (approx.	70% at work end 30% at Electrode end	30% at work end 70% at Electrode end	50% at work end 50% at Electrode end
Penetration	Deep; narrow	Shallow; wide	Medium
Electrode Capacity	Excellent (e.g., 1/8" ∅ — 400 amps.)	Poor (e.g., 1/4" ∅ — 120 amps.)	Good (e.g., 1/8" ∅ — 225 amps.)

Fig. 18-7. Characteristics of welding current sources for TIG welding. (Courtesy of Canadian Liquid Air.)

3. Pure Carbon Dioxide
4. Argon/Helium Mixture
5. Argon/Carbon Dioxide Mixture

Fig. 18-8. Successful TIG welding depends on good equipment and a modern TIG torch. (Courtesy of Airco Welding Products Division.)

TIG Torches

In Fig. 18-8, the welder is easily holding a modern, lightweight TIG torch. The TIG torch holds the tungsten electrode and funnels the shielding gas and electric welding power to the arc. Fig. 18-9 shows a common, medium-size TIG welding torch.

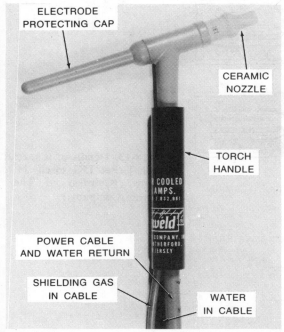

ELECTRODE PROTECTING CAP

CERAMIC NOZZLE

TORCH HANDLE

POWER CABLE AND WATER RETURN

SHIELDING GAS IN CABLE

WATER IN CABLE

Fig. 18-9. A common TIG welding torch. (Courtesy of Hobart Brothers Company.)

Cooling

Because the TIG welding process produces so much heat, the TIG welding torch must be cooled. This makes it safer and more comfortable for the welder and helps the torch parts last a long time.

Low-current TIG torches are usually air-cooled, since they produce only a small amount of heat. Fig. 18-10 shows a low-current, air-cooled TIG torch.

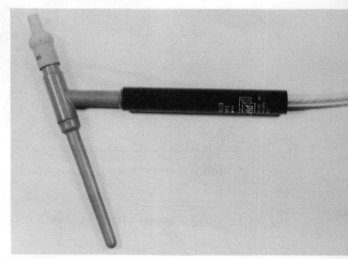

Fig. 18-10. A small, air-cooled TIG torch. (Courtesy of Hobart Brothers Company.)

Large, high-current TIG torches, however, must be water-cooled. Water-cooled TIG torches may have a safe amperage capacity up to 500 amperes, such as the water-cooled torch shown in Fig. 18-9. Fig. 18-11 shows the paths followed through the torch by the cooling water, welding current, and shielding gas.

Flexible Jacket—A special type of TIG torch is available with a flexible jacket. With this feature, the jacket and torch tip may be bent to any special angle needed for a certain job. In Fig. 18-12, the welder is making a TIG weld with such a torch. In Fig. 18-13, the special torch is shown in cutaway. Here, the water and gas lines form the conductors for the welding current. See Figs. 18-12 and 18-13.

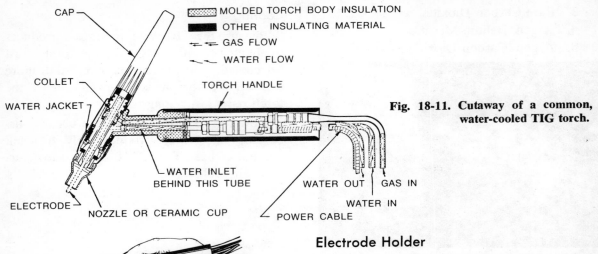

CAP

COLLET

WATER JACKET

MOLDED TORCH BODY INSULATION

OTHER INSULATING MATERIAL

GAS FLOW

WATER FLOW

TORCH HANDLE

Fig. 18-11. Cutaway of a common, water-cooled TIG torch.

WATER INLET BEHIND THIS TUBE

ELECTRODE

NOZZLE OR CERAMIC CUP

WATER OUT

GAS IN

WATER IN

POWER CABLE

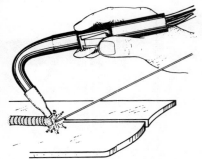

Fig. 18-12. Using a special TIG torch with a flexible jacket.

Electrode Holder

The electrode is held in place by *collets* screwed into the inside of the torch. See Fig. 18-11. A *collet* is a kind of nut with a hole in the center. When the nut is tightened, whatever is in the hole of the nut is squeezed by the nut and held in place. In the TIG torch, the electrode is the piece slid inside the nut.

The size of the collet depends on the diameter of the electrode being used. Elec-

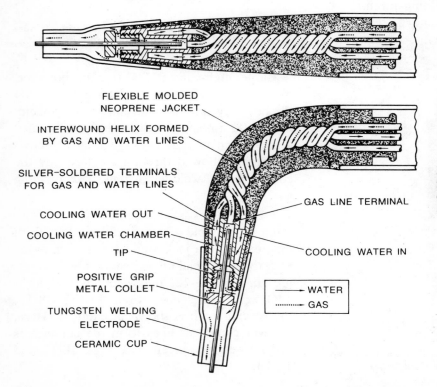

FLEXIBLE MOLDED NEOPRENE JACKET

INTERWOUND HELIX FORMED BY GAS AND WATER LINES

SILVER-SOLDERED TERMINALS FOR GAS AND WATER LINES

COOLING WATER OUT

COOLING WATER CHAMBER

TIP

POSITIVE GRIP METAL COLLET

TUNGSTEN WELDING ELECTRODE

CERAMIC CUP

GAS LINE TERMINAL

COOLING WATER IN

WATER

GAS

Fig. 18-13. Details of a special flexible-jacket TIG torch in cutaway. (Courtesy of The Falstrom Company.)

Fig. 18-14. A gas lens and other equipment to control the gas delivery. (Courtesy of Union Carbide Canada Ltd.)

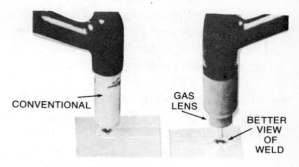

Fig. 18-15. With a gas lens, the welder can see the weld more easily than with a conventional TIG torch nozzle. (Courtesy Union Carbide of Canada Ltd.)

trodes may sometimes be easily removed from the collet with a quick-release mechanism on the torch.

The direction and amount of inert gas blown around the weld is controlled by gas cups or nozzles threaded into the torch head.

Gas Lens—Sometimes, when necessary, a gas lens is used to control gas turbulence (uneven delivery) on the end of the torch. See Fig. 18-14. With the gas lens, welding can be done with the nozzle as high as 1″ above the work. This improves the welder's ability to see the weld puddle and allows him to reach difficult places, such as inside corners. See Fig. 18-15.

Including the DC welding supply, inert gas, cooling water, filler rod, and torch assembly, the TIG welding process appears as in Fig. 18-16. Obviously, the cables and conductors to the torch must be very carefully designed to be sturdy, yet lightweight and easy for the welder to handle. The cables to the torch must deliver welding current and shielding gas, and they must deliver and pick up any cooling water used in the torch.

TIG Welding Electrodes

With the TIG welding process, the electrode must not melt as it does in shielded arc welding. Therefore, the electrode metal must have a very high melting point to conduct the welding current without melting itself.

Pure Tungsten Electrodes

The original welding electrode for the TIG process was the pure metal *tungsten,* with a high melting point of 6170° F. Tungsten's high melting temperature made these electrodes practically nonconsumable. Since then, newer alloyed tungsten electrodes have been developed. They are being used because they are more suitable than tungsten for various welding applications.

Fig. 18-16. The complete TIG welding setup.

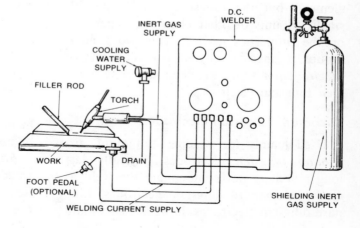

However, the pure tungsten electrode can still be used with any TIG welding machine, and is mainly used for welding aluminum. When using the lower-priced pure tungsten electrode, the welder should use an AC current for the best TIG welding results. The current ranges for the different diameter pure tungsten electrodes are shown in Table 18-1.

Like all TIG electrodes, pure tungsten electrodes are color coded with a color on the tip. Pure tungsten electrodes are color coded with a green tip. The letters indicating a pure tungsten electrode are *EWP*.

Under most welding conditions, either of the two newer tungsten alloy electrodes will produce better TIG welds. Which of the two

Table 18-1. Recommended Welding Current for Pure Tungsten Electrodes.

Pure Tungsten Electrode Diameter (Inches)	Recommended AC Welding Current Range (Amperes) (For welding any base metal with high frequency stabilization)
0.020	0 - 7
0.040	10 - 40
1/16	30 - 70
3/32	70 - 100
1/8	100 - 150
5/32	150 - 225
3/16	200 - 300
1/4	275 - 400

alloyed electrodes to use will generally depend on how much current is required for the welding job at hand. The two alloyed electrodes still contain tungsten as their basic element, but the tungsten is alloyed with small amounts of other elements, depending on the properties needed. The two general groups of tungsten alloy electrodes are known as *zirconium* tungsten electrodes and *thoriated* tungsten electrodes.

Zirconium Tungsten Electrodes

Used mainly for welding aluminum, *zirconium tungsten* electrodes generally last longer and do not contaminate the weld as easily as do pure tungsten electrodes. Like pure tungsten electrodes, however, zirconium

tungsten electrodes are better used with AC current. Zirconium tungsten electrodes produce very high quality welds. The welds will test out well under X-ray testing.

The color code on the tip of a zirconium tungsten electrode is brown, and the letter designation is *EWZr*. Table 18-2 shows the recommended AC welding currents for different diameters of zirconium tungsten TIG electrodes.

Thoriated Tungsten Electrodes

These electrodes are used for good welds on all types of low-alloy steels. Generally, *thoriated tungsten* electrodes are used at a lower amperage than are pure tungsten electrodes, and they should be used with straight polarity DC current.

Because they are used with a lower amperage than pure tungsten electrodes, thoriated tungsten electrodes do not deform due to heat as quickly as do pure tungsten electrodes. Also, thoriated tungsten electrodes will not contaminate the work as much if they are accidentally shorted out against the

Table 18-2. Recommended Welding Current for Zirconium Tungsten Electrodes.

Zirconium Tungsten Electrode Diameter (Inches)	Recommended AC Welding Current Range with High Frequency Stabilization (Amperes)
0.020	0 - 7
0.040	0 - 30
1/16	20 - 115
3/32	100 - 185
1/8	150 - 225
5/32	190 - 300
3/16	200 - 340
1/4	300 - 445

work while welding. Thoriated tungsten electrodes are available in two types, 1% thoriated and 2% thoriated, and the recommended currents shown in Table 18-3 may be used for either type.

First Type—1% thoriated tungsten electrodes have the color code *yellow* on their tip for identification. Their letter code is *EWTh1*.

Second Type — 2% thoriated tungsten electrodes have the color code *red* on their tip for identification. Their letter code is *EWTh2*.

Selecting Tungsten Electrodes

Whether pure, zirconium, or thoriated, tungsten electrodes are available in two finishes: *standard* and *ground*. Standard finished electrodes are either chemically cleaned or etched; then they are sized to an approximate diameter for good TIG welding.

Ground electrodes, however, are used for more precision work. The diameter of a ground electrode is more carefully controlled; that is, it is held to a closer tolerance. The ground finish costs about half *again* as much as a standard-finish electrode, thus making the ground electrode more expensive.

All types of tungsten electrodes are available with either the standard or ground finish in diameters ranging from 0.020″ to 0.250″. They are available in lengths of 3″, 6″, 7″, and 18″. In some instances, 24″-long electrodes are available.

As mentioned earlier, argon is by far the most common shielding gas used with TIG welding. For this reason, most charts and tables recommending welding currents for the TIG process are made considering argon as the accepted shielding gas. Tables 18-1,

18-2, and 18-3, accordingly, give recommended currents with argon shielding gas.

Table 18-3. Recommended Welding Current for Thoriated Tungsten Electrodes.

Thoriated Tungsten Electrode Diameter (Inches)	Recommended Straight-Polarity DC Current Range (Amperes)
0.020	0 - 25
0.040	12 - 100
1⁄16	20 - 190
3⁄32	35 - 325
1⁄8	50 - 475
5⁄32	65 - 600
3⁄16	(not given)
1⁄4	(not given)

Helium may also be used as a shielding gas, but the electrodes will not be able to use as much current. As a general rule, the current range on any recommendation chart must be reduced 10% if *helium* gas is being used for shielding.

TIG Welding Regulators and Flowmeters

Accurately regulating the shielding gas pressure and flow is very important for good TIG welding. For this reason, the welder should use only regulators and equipment designed for the gases being used.

As with the compressed gases used for oxygen–gas welding, the shielding gas must be controlled and regulated as it leaves the supply cylinder or source. With TIG welding, the single shielding gas cylinders, piping systems, and manifolds will need a regulator that is able to supply an even, steady flow of shielding gas.

Regulators commonly used for TIG welding are slightly different from those used for oxygen–gas welding. With oxygen–gas welding, the pressure going to the torch (the delivery pressure) is indicated in psi on a second round gauge. With TIG welding, however, the shielding gas going to the torch is not measured in psi, but in *cfh*. Cfh stands for *cubic feet per hour* and is another standard measurement for any gas delivery.

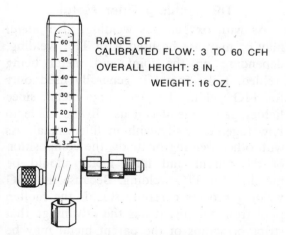

RANGE OF
CALIBRATED FLOW: 3 TO 60 CFH
OVERALL HEIGHT: 8 IN.
WEIGHT: 16 OZ.

Fig. 18-17. A common TIG welding flowmeter for shielding gas.

Cfh is measured by a gauge called a *flowmeter*. The flowmeter in Fig. 18-17 can measure from 3 to 60 cfh and looks like a round gauge redesigned to read in a straight line. As the gas flow increases, an indicator ball is pushed farther up the scale to read how many cubic feet of gas per hour are going past the flowmeter.

Like the regulators used with oxygen–gas welding, a second gauge is used to indicate how much shielding gas is still in the tank or cylinder. Fig. 18-18 shows a combination regulator and flowmeter that is made in one unit, ready for mounting on the argon cylinder. Designed specifically for TIG welding, it is preset to send the gas to the flowmeter at a pressure of exactly 20 psi.

Fig. 18-19. Regulator and flowmeter with a gauge to read shielding gas delivery in psi, and a flowmeter to read shielding gas delivery in cfh. (Courtesy of Chemtron Corporation.)

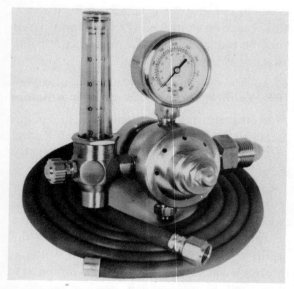

Fig. 18-18. Combination regulator and flowmeter for argon shielding gas in TIG welding, with delivery hose. (Courtesy of Hobart Brothers Company.)

In Fig. 18-19, a flowmeter has been attached to a conventional-looking regulator with two round gauges. Here, both the delivery *pressure* in psi and the delivery *flow* in chf can be seen. This particular regulator and flowmeter is designed to be used only with argon.

Flowmeters for accurately measuring the gases are also available separately to be used where several welding stations are connected to one shielding gas supply. Direct reading in cfh for all the shielding gases can then be easily seen from several sides, as with the flowmeter in Fig. 18-17.

A major difference between flowmeters and working pressure gauges is their operating position. Working pressure gauges will operate in any position but standard flowmeters *must* be standing up straight (vertical) in order to read accurately.

TIG Welding Filler Metal

As with oxygen–gas welding, *filler metal* may or may not be used with TIG welding, depending on the joint and metal being welded. Filler metal is generally used only for TIG welding heavier gauge metal, since lighter gauge metal can usually be made to flow together easily without filler metal. As with other welding methods, the composition of the parent and filler metal should be matched for TIG welding. Because the TIG welding process carefully shields the molten pool from the air, it has the advantage that strips or scraps of the parent metal may be used for filler metal.

Although it is possible to use an automatic electrode feed with TIG welding, most TIG welding is done manually. The size of the filler rod or metal depends on the welding current being used and the thickness of the metal being welded. Table 18-4 shows the recommended TIG welding conditions and filler rod diameter for welding common aluminum plate.

TIG Welding Safety

Generally speaking, even greater care must be taken during TIG welding than during manual shielded arc welding. This is because the radiant energy produced by TIG welding is *twice* as much as is the energy produced by stick electrodes being welded with the same current.

Helmet and Lens

A good welding helmet, such as the helmet in Fig. 18-20, must always be worn while TIG welding. The darkest lens available should be used in the helmet body for TIG welding. The welder in Fig. 18-21 is metal arc welding, not TIG welding, high on a construction job with the correct helmet and lens for TIG welding.

Fig. 18-20. A welding helmet with a flip-up front lens holder. Only the darkest lens should be used for TIG welding.

Clothing

Protective asbestos clothing must be worn while TIG welding. Because the TIG arc has so much energy, extra attention must be paid to covering the feet, legs, and arms of the welder.

Another problem with TIG welding is that the bright arc radiation may be reflected up onto the welder's face *behind* the protective

Table 18-4. TIG Welding Recommendations for Common Aluminum Plate.

Aluminum Plate Thickness (Inches)	Type of Weld—Any Position	Pure Tungsten Electrode Diameter (Inches)	Filler Rod Diameter (Inches)	Nozzle Inside Diameter (Inches)	Argon Shielding Gas Flow (cfh)	AC-HF Welding Current (Amperes)	Number of Passes	Travel Speed (Inches Per Minute)
3/64	Square Groove	1/16	1/16	1/4	19	40 - 60	1	16
1/16	Square Groove	3/32	3/32	5/16	19	70 - 90	1	11
1/16	Fillet	3/32	3/32	5/16	15	70 - 90	1	9
3/32	Square Groove	3/32	3/32	5/16	19	90 - 110	1	11
3/32	Fillet	3/32	3/32	5/16	16	95 - 115	1	9
1/8	Square Groove	1/8	1/8	3/8	20	115 - 135	1	11
1/8	Fillet	1/8	1/8	3/8	19	120 - 140	1	10
3/16	Fillet	5/32	5/32	7/16	25	180 - 200	1	10
3/16	Vee Groove	5/32	5/32	7/16	25	160 - 180	2	11
1/4	Fillet	3/16	3/16	1/2	30	230 - 250	1	10
1/4	Vee Groove	5/32	5/32	7/16	30	200 - 220	2	9
3/8	Fillet	3/16	3/16	1/2	35	250 - 310	2 or 3	(as needed)

Fig. 18-21. A well-dressed welder high on a construction job. (Courtesy of Lincoln Electric Co.)

helmet and lens. This may happen when the arc is reflected up from light-colored clothing. To guard against this reflection, it is a good practice to wear *dark* outer clothing.

Fig. 18-23 shows a welder making a practice TIG weld on aluminum coupons. Fig. 18-24 shows a common TIG practice weld, as would be made by the welder in Fig. 18-23.

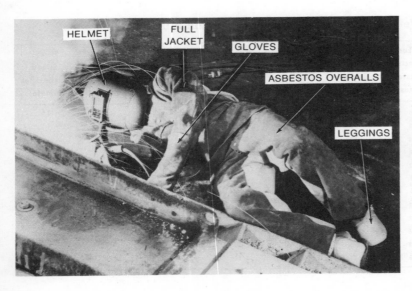

HELMET
FULL JACKET
GLOVES
ASBESTOS OVERALLS
LEGGINGS

Fig. 18-22. A well-dressed welder on production welding. With the correct lens, he would be prepared for TIG welding. (Courtesy of American Optical Corp.)

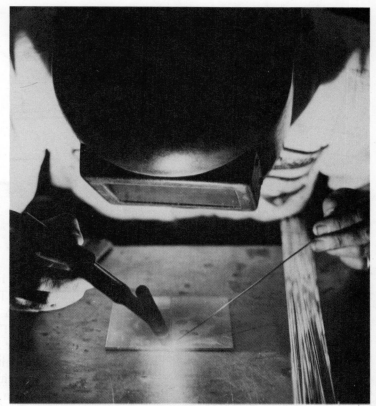

Fig. 18-23. Practicing TIG welding on aluminum coupons with an aluminum filler rod. (Courtesy of Aircom Metal Products, Inc.)

Fig. 18-24. A common TIG practice weld, as being made in Fig. 18-23. (Courtesy of Aircom Metal Products, Inc.)

19

Basic TIG Welding

For practicing TIG welding, a new welder will have little difficulty setting up and adjusting the equipment if he follows the manufacturer's specifications and a few simple setup operations. Many times, it is not possible for the welder to use the recommended type of current for the metal being welded. However, the setting up and shutting down operations are very similar for either AC or DC TIG welding machines.

This chapter will deal basically with TIG welding aluminum. Aluminum is usually TIG welded in both production and repair work. It is the easiest and most common metal to TIG weld, easily lending itself to practice welding.

Setting Up Equipment

After being set up, the equipment necessary for TIG welding will resemble the setup shown in Fig. 19-1 with the available power source. The shielding gas is supplied to the regulator and flowmeter from either cylinders or a manifold system. The cooling water may be supplied from a regular water line in the shop or a circulating pump system controlled by a pressure reducer. The torch is connected to the hoses and cable with marked connections for easy, safe assembly.

Welding Machine, Current, Shielding Gas

The welding machine for TIG welding has a control panel with remote controls for the

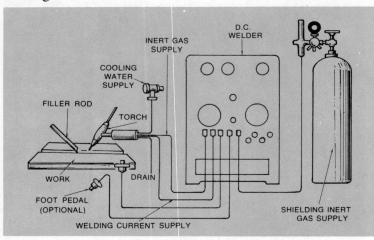

Fig. 19-1. A typical TIG welding setup.

434

shielding gas, cooling water, and welding current. Most controls allow the shielding gas and cooling water to stay on for a short time after the arc is broken, as a safety measure. This control can be made by a hand or foot remote control switch. Lighter, air-cooled torches are protected from excessive heat by a similar control system. These controls can be seen on the welding machine in Fig. 19-2.

The machine (or setting) needed for TIG welding any certain metal depends entirely on what the metal is. Therefore, the first choice to be made before practicing TIG welding is the welding machine type or current setting. A table such as Table 19-1 must be used to find the correct current setting.

Setting the power source (welding machine) for a continuous high-frequency boost

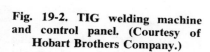

Fig. 19-2. TIG welding machine and control panel. (Courtesy of Hobart Brothers Company.)

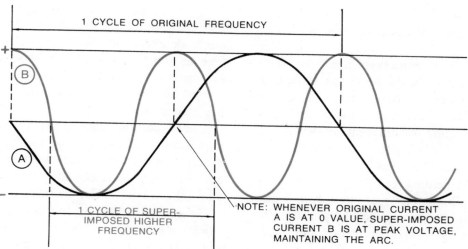

1 CYCLE OF ORIGINAL FREQUENCY

1 CYCLE OF SUPER-IMPOSED HIGHER FREQUENCY

NOTE: WHENEVER ORIGINAL CURRENT A IS AT 0 VALUE, SUPER-IMPOSED CURRENT B IS AT PEAK VOLTAGE, MAINTAINING THE ARC.

Fig. 19-3. A higher-frequency current is superimposed on the original to eliminate the zero value in each cycle of the original frequency.

is necessary when using AC. The superimposed high-frequency current, Fig. 19-3, is required to positively restrike the arc when the voltage goes through the zero point of the AC cycle. In Fig. 19-4, a TIG welding machine is shown set on AC. Here, the high-frequency control is listed as a spark density control. In Fig. 19-2, a DC TIG welding machine is shown with a high-frequency control. The high-frequency control for *DC* welding is used only to help strike the arc.

Electrode

When choosing an electrode, the size and type of electrode will depend on the type of metal being welded and its thickness. The electrode is then pushed into the right size collet. See Fig. 19-5. When inserted in the torch, the electrode should be centered in the orifice (hole) of the gas nozzle. An off-center electrode may damage the gas nozzle through overheating and create a poor shielding gas flow.

Fig. 19-4. TIG welding machine set on AC. (Courtesy of Lincoln Electric Co.)

When the DC arc is working, the high-frequency control and boost are not needed or used.

Current—The amperage used for TIG welding practice should be in the middle of the range. After practicing with the current in the middle of the range, the welder may change his setting for different results.

Shielding Gas—Argon is usually the recommended shielding gas for TIG welding aluminum. Shielding gas flow, of course, is given in cfh—*cubic feet* per *hour*—and should be set in the middle of the range for TIG practicing. For other metals and welding conditions, tables will recommend the gas and gas flow needed for a certain application.

When installing the electrode in the collet and collet body, the welder should keep the electrode extension that is beyond the nozzle face as short as possible. A short electrode *extension* (how much it sticks out) will allow a good flow of shielding gas around the molten metal.

Under certain conditions, the electrode may have to extend beyond the nozzle. One example might be when welding in tight corners. Here, the electrode extension would allow the welder better visibility.

Generally, the electrode should not extend (stick out) more than $\frac{1}{16}''$, especially for butt welds. However, the electrode may stick out up to $\frac{1}{4}''$ for fillet welding. The generally recommended stickout is shown in Fig. 19-6.

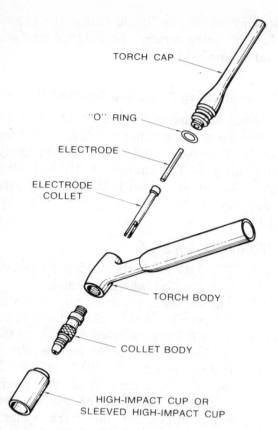

TORCH CAP

"O" RING

ELECTRODE

ELECTRODE
COLLET

TORCH BODY

COLLET BODY

HIGH-IMPACT CUP OR
SLEEVED HIGH-IMPACT CUP

Fig. 19-5. Parts of the TIG torch. The electrode is held tightly by the electrode collet when the collet is screwed into the collet body. (Courtesy of Union Carbide Canada Ltd.)

Connections

With TIG welding, many connections must be made between the equipment sup-

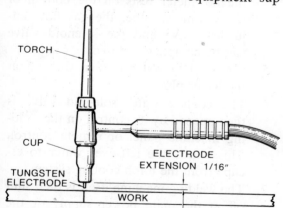

TORCH

CUP

TUNGSTEN
ELECTRODE

ELECTRODE
EXTENSION 1/16"

WORK

Fig. 19-6. The electrode extension should be about ¹⁄₁₆" for butt welding.

plies, welding machine, and TIG torch. While these connections do not have to be made in any specific order, common safety rules should be followed when the equipment is to be connected as in Fig. 19-1. Here are some general rules for connecting TIG equipment:

A. When connecting shielding gas regulators and flowmeters, follow the steps for connecting oxygen regulators.

B. Be sure that all valves are closed and the machine turned off—preferably unplugged—while connections are being made.

C. All electrical connections must be clean and tight everywhere in the TIG welding circuit.

D. Route the water drain to a safe place where the water will not back up onto the floor or welding machine.

E. Secure (fasten) the shielding gas cylinder upright so that it cannot fall over if it is bumped.

F. Route all the many cables where they will be safe from welding sparks and out of the welder's way, so that he cannot trip or step on them.

G. Double check all routing and connections when the setup is finished.

Turning on the Equipment

When the entire TIG setup has been made and all connections double checked, the equipment is ready to be turned on. See Fig. 19-1. While the exact order of turning units on and opening valves is not critical, there is a general order of those operations that must be done before others when turning on the TIG equipment.

Welding Machine Controls

The controls on a TIG welding machine may vary from one machine to another. Basically, on any TIG welding machine, three things need to be controlled: the electric arc current, the shielding gas flow, and the cooling water flow. Each of these may be controlled individually on the welding machine front panel, or they may be controlled by the remote control foot valve.

Usually, at least the electric arc current on-off is controlled by the foot valve. On some remote (foot) controls, the *amount* of electric arc current can be controlled as well as whether the current is on or off. These, of course, are the most convenient—and most expensive—controls available. Since solenoid (electrical) valves control the water and gas flow, foot controls can be easily adapted to work these solenoids in addition to the electric arc current. Sometimes, remote control switches may be mounted on the torch handle itself.

Generally speaking, these steps should be followed, in order, for turning on the TIG equipment. (NOTE: Before running a bead, see *Shutting Down the Equipment*, discussed next.)

Basic Operation Steps

1. Place the tungsten electrode and TIG torch where they cannot strike an arc when the welding machine is turned on. This must be away from the work piece.
2. Open the water control valve from the water supply.
3. Open the shielding gas cylinder valve *slowly*. By opening the valve slowly, the regulator will not be damaged by the rush of gas from the cylinder.
4. Hold the TIG torch in one hand and turn on the welding machine.
5. With the welding machine on, check the water return line to be sure that the cooling water is flowing.
6. With the welding machine on, check the flow of argon by opening and closing the hand or foot valve.
7. With the argon flowing, adjust the flowmeter to supply the cfh of argon required.

When the above steps have been completed, the welder is ready to strike an arc and practice TIG welding. However, *before*

striking an arc, he should review the procedure for safey turning the welding machine and equipment off.

Shutting Down the Equipment

After completing the final weld, or when reviewing before practice welding, the welder should thoroughly understand how to safely shut down the TIG equipment. The following steps should be followed, in order, for safe shutdown of the TIG equipment. See Fig. 19-1.

Shutting Down Steps

1. The TIG torch should first be placed in a position where it cannot arc.
2. The shielding gas valve on the top of the cylinder is then closed hand tight.
3. The shielding gas solenoid valve on the welding machine or the remote control is then opened. This allows the shielding gas to escape from the torch, "bleeding" the shielding gas air passages throughout the torch system. With the solenoid valve still open, step #4 is done.
4. The gas flow adjusting valve on the welding machine is then slowly closed. This bleeds the shielding gas from the hose between the gas cylinder and the welding machine. When the ball in the flowmeter rests on the bottom of the flowmeter tube, the gas flow adjusting valve and the solenoid valve can be completely closed.
5. The cooling water is then shut off at the water pipe.
6. The cooling water solenoid valve is then opened by its control on the welding machine panel or remote control. This allows the left-over water to escape from the torch cooling system.
7. The welding machine itself is then shut off at the main power supply or the control switch on the machine's front panel.

When a new welder fully understands the safe procedure for both setting up and shutting down TIG welding equipment, he is ready to safely practice TIG welding.

TIG Welding Set-up

TIG welding is most often used for joining aluminum from 1/32" to 1/8" thick. Although heavier sections can be joined by TIG welding, other processes are usually more economical.

The TIG welding process can be used to weld joints in aluminum, copper and copper alloys, magnesium, stainless steel, silicon bronze, titanium, and carbon or low alloy steels. It requires a certain amount of skill from the welder, who has to learn to carefully coordinate the movement of both hands. Therefore, a new welder should first understand and learn TIG welding procedures for aluminum. After the new welder has successfully TIG welded aluminum, he should have little difficulty successfully welding other metals.

For this reason, welding practice with TIG equipment is best done on aluminum. This chapter, and the practice outlines, will deal entirely with TIG welding aluminum.

Power Source (Welding Machine)

The welder should use an AC welding machine for the best results when welding aluminum. Reverse polarity DC should not be used for welding aluminum due to the danger of weld contamination (dirt). As mentioned earlier, the alternating current cycle gives an arc action that acts like an oxide cleaner.

Electrodes

A high-quality zirconium tungsten electrode is recommended for practice welding aluminum. With a zirconium tungsten electrode, there is less electrode contamination (dirt) due to shorting, and it will carry higher currents than a lower-priced pure tungsten electrode of the same diameter. Thoriated tungsten electrodes are not recommended for TIG welding aluminum with AC current.

Table 18-4 gives the information needed for choosing an electrode with the correct diameter for different thicknesses of aluminum, along with the other information needed for good welds.

TIG Torches

Where the welder will be required to use only low amperage (under 100 amperes), the torch may be air-cooled. For higher amperages (100 to 250), a water-cooled torch and handle, as in Fig. 19-7, must be used. These torches may be equipped with automatic controls to stop the water flow when the torch is not in use, as in Fig. 19-8. Where portable cooling systems are being used, it

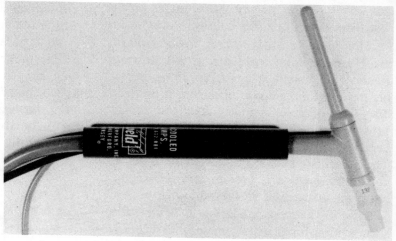

Fig. 19-7. Water-cooled TIG torch. (Courtesy of Hobart Brothers Company.)

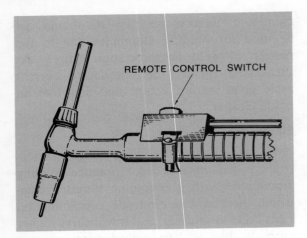

REMOTE CONTROL SWITCH

Fig. 19-8. Water-cooled TIG torch with a remote control switch located on the handle.

may be necessary to use antifreeze in the torch cooling system if welding is being done at temperatures below freezing.

Shielding Gas

Argon is considered to be the best shielding gas for welding aluminum. A black coating will form on the welded section of aluminum when not enough shielding gas is being used to keep out the atmospheric contamination. For this reason, the welder should always refer to a gas flow specification chart, such as Table 18-4, to find the correct shielding gas flow for the job being done. With enough shielding gas flowing, the joint will not have any black coating from the atmosphere. The welder may have to increase the gas flow to allow for any shielding gas being lost from drafts or other causes.

Afterflow—In order to keep the hot end of the electrode from oxidizing, it is necessary to allow the shielding gas to continue escaping for a few moments after the arc is broken. The flow of shielding gas after the arc is broken is called *afterflow,* and should be adjusted ahead of time on the welding machine. The afterflow allows the electrode time to cool after the arc is broken and before the electrode is exposed to the atmosphere.

Most TIG recommendation charts have a specification for afterflow in seconds accord-

ing to the welding current being used and the electrode diameter. In Fig. 19-4, the shielding gas afterflow control can be seen on the control panel of the welding machine.

Cleaning

When aluminum welding is critical, requiring oxide-free welds, the welder may have to clean the aluminum before striking an arc. The cleaning may be easily done with a stainless steel brush that leaves no contaminating particles. Or, the welder may want to use a caustic soda solution to etch (eat away) the top surface of the parts to be welded. This is an excellent cleaning procedure.

Safety TIG Welding Procedures

Before practicing TIG welding, a new welder should equip himself with the proper protective clothing, including good welding gloves, cap, long-sleeve shirt, and helmet. A dark filter glass lens should be placed in the helmet for TIG welding.

Fitup

Where a joint has poor fitup, the welder will not be able to complete a high-quality TIG weld. Before welding, it is usually a good idea to space the aluminum plates and tack weld them in one or more places to eliminate distortion. When tack welding joints, the welder should be sure that the edges being joined remain parallel while tack welding. This will ensure that penetration will be good during the final weld.

To ensure good penetration, fusion, and proper weld metal composition, it may be necessary to prepare the edges of the metal as specified for various joints and welds. Fig. 19-9 shows a typical weld specification chart for manual TIG welding aluminum butt joints in the flat position.

Position and Spacing

When welding all types of materials, the *flat* position is preferred. The flat position will increase the welding speed and quality

Metal Thickness (in.)	Weld Pass Number[1]	Electrode Dia. (in.)	Filler Rod Dia. (in.)	Tack Spacing (in.)	Alternating Current[2] (amp)	Arc Travel Speed (ipm)	Argon (cfh)	Shielding Gas Nozzle Dia. (in.)	Joint Details and Order of Welding	AWS Weld Symbol
3/64	1	1/16	3/32	3	55	11	10	3/8	 0 MAX. ROOT OPENING	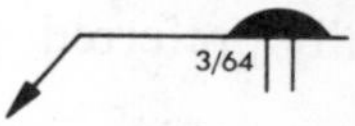 3/64
1/16	1	3/32	3/32	3	90	12.5	10	3/8		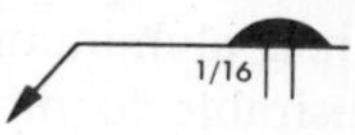 1/16
5/64	1	3/32	3/32	3	105	18	16	3/8		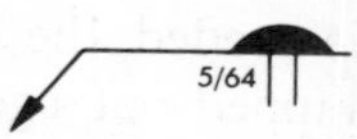 5/64
3/32	1	1/8	1/8	3	120	12	16	3/8	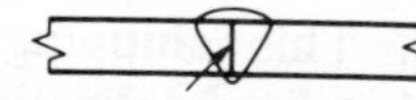 1/32 MAX. ROOT OPENING	 3/32
1/8	1	1/8	1/8	4	155	10	16	3/8		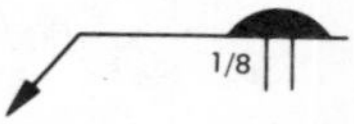 1/8
3/16	1 / B2R	3/16	3/16	4	230 / 230	12 / 12	16	1/2	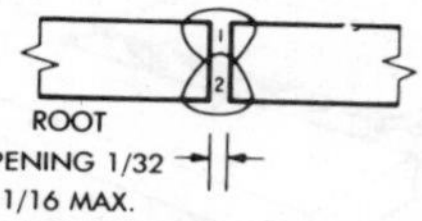 ROOT OPENING 1/32, 1/16 MAX.	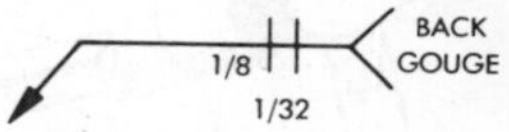 1/8, 1/32, BACK GOUGE
3/16	1 / 2	5/32	5/32	4	190 / 190	8 / 8	16	1/2	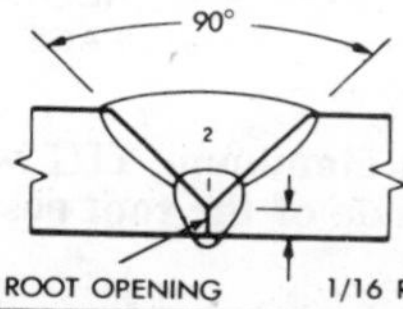 90°, 0 MAX. ROOT OPENING, 1/16 ROOT FACE	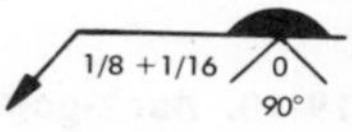 1/8 +1/16 0 90°
1/4	1 / B2R	3/16	3/16	4	270 / 270	11 / 11	20	3/8	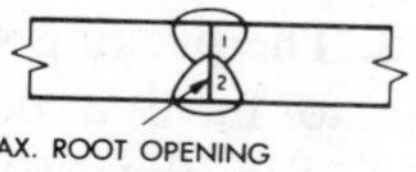 1/16 MAX. ROOT OPENING	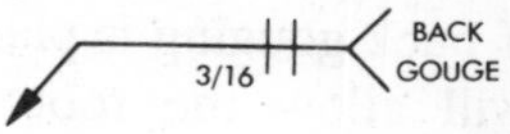 3/16, BACK GOUGE
1/4	1 / B2R	3/16	3/16	4	240 / 240	9 / 9	20	3/8	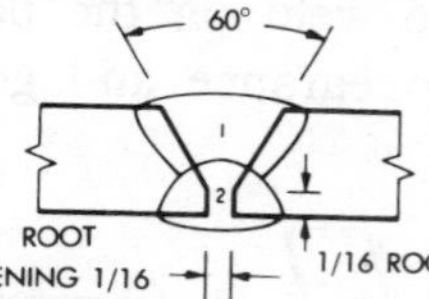 60°, ROOT OPENING 1/16, 3/32 MAX., 1/16 ROOT FACE	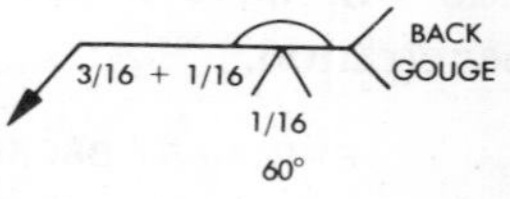 3/16 + 1/16, 1/16, 60°, BACK GOUGE
1/4	1 / 2	3/16	3/16	4	220 / 260	6 / 6	20	1/2	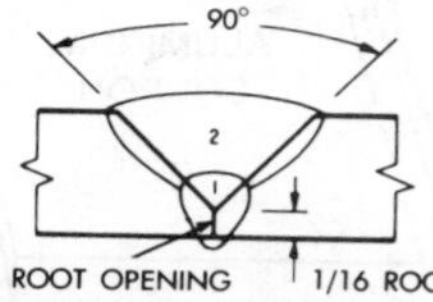 90°, 0 MAX. ROOT OPENING, 1/16 ROOT FACE	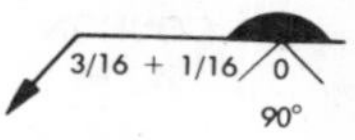 3/16 + 1/16 0 90°

[1]B indicates back gouge before making pass.
R indicates weld pass made on the reverse of the side on which the first pass was made.
[2]Current should not vary more than 3% during welding.

Fig. 19-9. Typical weld specification chart. (Courtesy of Alcan Aluminum Ltd.)

while decreasing the welding cost. When welding aluminum, the welder will find that butt, lap, and Tee joints welded in the horizontal or flat position give high-speed, quality welds. Butt joints in the flat position are especially preferred.

Backing Strips

When aluminum is being welded, it may be desirable to turn the weldment over and reweld the root face from the back side. If this is needed, the root face back side should be chipped out (gouged) before the weld is made on the back side. This chipping is called *back gouging,* and it is usually done with a hammer and chisel, as in Fig. 19-10.

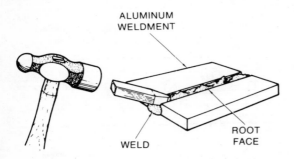

Fig. 19-10. Back-gouging an aluminum TIG weldment to reweld the back side of the root pass.

Temporary *backing strips* should be used if back gouging is planned. The backing strip will allow the root pass to have a better shape when it is welded. See Fig. 19-11. Then, after gouging, the weld on the back side will have a neat appearance and good penetration.

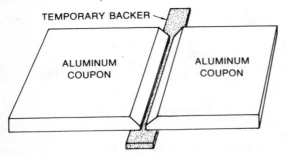

Fig. 19-11. A temporary backer will give the root pass a better shape.

Permanent backing strips are sometimes used with TIG welding when welding carbon or low-alloy steels. They are especially valuable when welding these steels in the overhead position.

Fixtures

Where tack welds alone do not hold or align the parts well enough, the welder may have to use fixtures or clamps to hold the work for welding. Although clamps are usually strong enough for holding small jobs, special fixtures made of scrap material would save welding time on large jobs. Also, on production welding where good joint accuracy is needed, fixtures would save time in the long run.

Running Aluminum Beads

Practicing TIG welding by running beads, a new welder will be able to learn the procedure for starting and breaking the arc as well as the angles for holding the torch and filler metal.

Electrodes

The tungsten electrodes for welding aluminum or aluminum-magnesium alloys are ball-shaped on the end. See Fig. 19-12. For carbon steel or stainless steel welding, the electrodes are sharpened on the end as in Fig. 19-13.

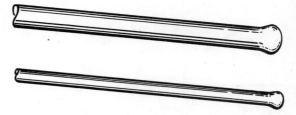

Fig. 19-12. Tungsten electrodes for TIG welding aluminum are shaped with a ball on the end.

When the correct electrode has been chosen, the electrode should be tightened in the collet about 1/8″ below the nozzle for practicing bead welds, as in Fig. 19-14.

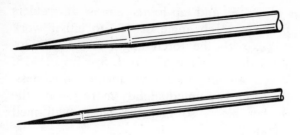

Fig. 19-13. Tungsten electrodes for TIG welding carbon steel or stainless steel are sharpened on the end.

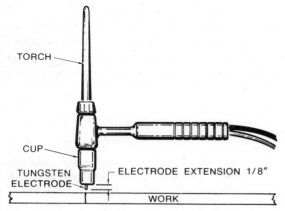

Fig. 19-14. For bead practice on aluminum, the electrode should extend ⅛″.

Striking the TIG Arc

If either AC or DC current with superimposed high frequency is used for striking the arc, the arc can be started *without* having the electrode come in contact with the work. By swinging the torch directly into the joint close to the work, the arc will be created due to the high-frequency voltage. The superimposed high frequency jumps the gap between the electrode and the work, starting the arc.

As with manual shielded arc welding, the arc should be allowed to establish itself before adjusting it to the desired arc length. The arc should be started near the starting point of the weld until a bright, clean pool of molten metal the required size is formed. Then, the torch can be moved in the direction of the weld.

For DC welding without superimposed high frequency, the arc can only be started by having the electrode come into contact

with the metal. Since thoriated electrodes improve arc starting, they are preferred by most welders over pure tungsten for this job. To eliminate possible contamination from starting the arc, especially when welding aluminum, the welder should start the arc on a *starting block* made of copper. See Fig. 19-15.

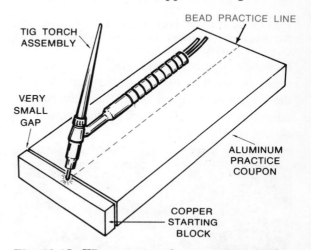

Fig. 19-15. When starting the arc, contamination from the atmosphere can be eliminated by using a starting block to establish the arc before moving over to the practice coupon.

TIG Arc Length

A good guide to arc length, suitable for most metals being TIG welded, is 1½ times the electrode diameter being used. This short arc length produces narrow welds with deep penetration due to the concentrated arc heat. As the arc length is increased, the heat concentration is decreased, creating more shallow weld penetration.

Breaking the Arc

The welder should *in*crease the welding speed before breaking the arc to help eliminate *crater cracking* at the end of the weld. Many welders like to break the arc and then immediately restrike it, melting the filler rod to fill out the formed crater.

Torch Angle

The torch angle above the work is normally 90° for butt joints, as in Fig. 19-16.

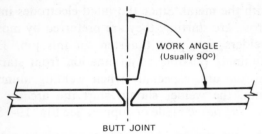

Fig. 19-16. Normal work angle for butt joints and flat welding. (Courtesy of Alcan Aluminum Ltd.)

However, some welders prefer to slant the torch as much as 60° from the horizontal in the direction of travel, with the filler rod inclined as low as 20° from the horizontal, as shown in Fig. 19-17.

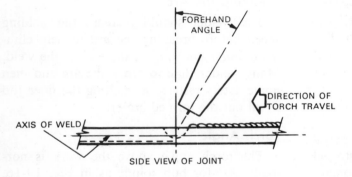

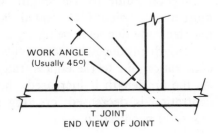

For Tee and lap joints, almost all welders prefer a torch angle of 45° to either work piece, with the torch angle 5° to 15° toward the direction of travel. See Fig. 19-18. When welding pieces of unequal thickness, the torch is pointed slightly more toward the thicker piece, so that the pieces will melt equally.

The number of aluminum alloys used in industry today and differences in filler alloy composition have a definite effect on the burn-off rate. In turn, these factors help determine how the alloys should be welded. For good results, a new welder should refer to the manufacturer's specification chart for

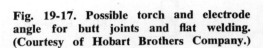

Fig. 19-17. Possible torch and electrode angle for butt joints and flat welding. (Courtesy of Hobart Brothers Company.)

Fig. 19-18. Work angles for Tee and lap joints. (Courtesy of Alcan Aluminum Ltd.)

Procedures for Welding Alloys: 1100, 3003, 5052, 5083, 5454, 6061, 6063, 6351

Metal Thickness (in.)	Electrode Dia. (in.)	Filler Rod Dia. (in.)	Tack Spacing (in.)	Alternating Current (amp)	Arc Travel Speed (ipm)	Argon (cfh)	Shielding Gas Nozzle Dia. (in.)	Joint Details	AWS Weld Symbol
1/16	3/32	3/32	3	90	8	10	3/8	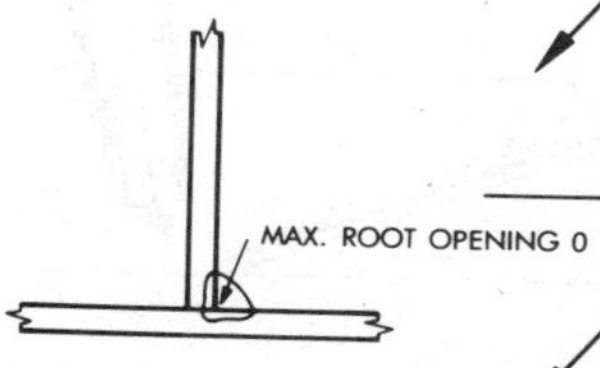	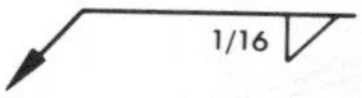 1/16
								MAX. ROOT OPENING 0	
5/64	3/32	3/32	3	110	7.5	16	3/8		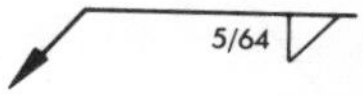 5/64

Fig. 19-19. Sample specification chart for welding the alloys shown. (Courtesy of Alcan Aluminum Ltd.)

a certain alloy whenever possible. Fig. 19-19 shows a sample specification chart for horizontal Tee joints on the aluminum alloys shown.

Fig. 19-20. A well-equipped, portable TIG welding outfit. (Courtesy of Hobart Brothers Company.)

Bead Procedure

Fig. 19-20 shows a typical, complete, portable TIG welding outfit. The unit has been completely connected and is ready to be used to practice TIG welding. On the rear is a tank for shielding gas on the left and another tank for cooling water on the right. The torch handle contains an on-off remote control switch. For practice bead welding, it would be a good idea for a new welder to have an outfit that is this well equipped.

With the equipment set up as just outlined, then, the new welder is ready to practice running beads on aluminum with TIG welding.

Adjusting the Equipment—A new welder can best practice by using a 3/32″ zirconium tungsten electrode, an argon shielding gas flow of 15 cfh, and a welding machine current of about 165 amperes. Then, the welder should switch the power source to open circuit voltage by moving the power switch to the "on" position.

Striking the Arc—The torch should be first held almost straight above the aluminum coupon, with the electrode ready to move into the molten puddle at about a 20° angle to the horizontal plane. See Fig. 19-21. Then the

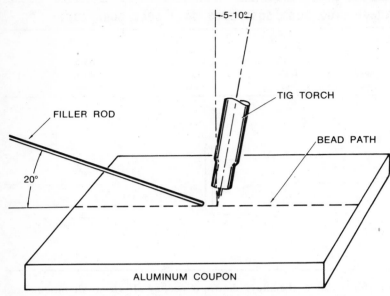

FILLER ROD

20°

5-10°

TIG TORCH

BEAD PATH

ALUMINUM COUPON

Fig. 19-21. Correct position of the filler rod and TIG torch for practice bead welding.

helmet can be lowered and the arc started by turning on the foot or hand remote-control switch. When the arc has been struck and stabilized, and the required size puddle formed, filler metal can then be added.

Adding Filler Metal—Adding filler metal can best be done by first forming a molten puddle at the start of the weld, allowing for proper penetration. See Fig. 19-22.

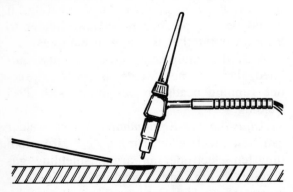

Fig. 19-22. Forming the molten puddle. (Courtesy of Canadian Liquid Air.)

After the molten puddle has been formed to the desired size, the arc is drawn to the *back* of the puddle. As the arc is withdrawn, the end of the filler rod is dipped into the *front* of the puddle at the angle shown in Fig. 19-23.

During TIG welding, the welder should *not* use the arc to melt the filler rod at any time. If the rod is melted with the arc, it may ball up in front of the puddle. By dipping the

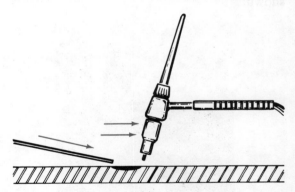

Fig. 19-23. Moving the torch back to allow the filler rod to be moved into the front of the puddle. (Courtesy of Canadian Liquid Air.)

filler rod into the *front* of the puddle, the filler rod melts enough for a good bead contour.

After the bead size has been completed in the melted section, the filler rod is taken out and the torch is moved to the front end of the puddle to melt the next metal along the weld line. See Fig. 19-24.

The procedure is then repeated many times, in order, until the bead has been completed along the weld line. After the bead is completed, the arc is broken and the equipment shut down, as outlined earlier.

Before practicing aluminum joint welds, the welder should continue bead practicing

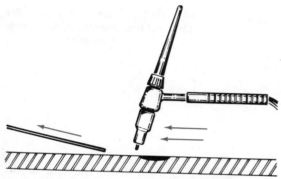

Fig. 19-24. After the filler rod is withdrawn, the torch is moved to the front of the puddle to repeat the process. (Courtesy of Canadian Liquid Air.)

until a clean, bright weld bead with smooth edges and the correct size can be easily welded. To analyze his weld beads, a new welder can compare the beads that he has done with the welding problems shown in Fig. 19-25. When he is able to weld beads easily, a new welder can then try to weld some of the simpler TIG aluminum welds.

Square Groove Butt Joints

Once a welder is able to comfortably TIG weld beads on aluminum plate, he is ready to practice welding together pieces of aluminum plate. As usual, the beginning practice weld should be a simple butt joint with the unprepared square groove, as in Fig. 19-26. These welds may be practiced in the flat, vertical, and horizontal positions. TIG welding aluminum butt joints in the overhead position is not recommended.

WELD QUALITY INSPECTION

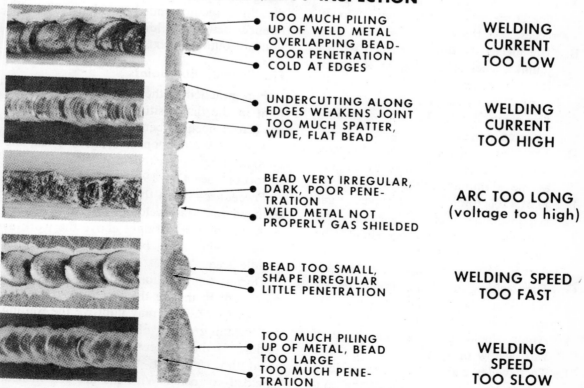

TOO MUCH PILING UP OF WELD METAL
OVERLAPPING BEAD—POOR PENETRATION
COLD AT EDGES

WELDING CURRENT TOO LOW

UNDERCUTTING ALONG EDGES WEAKENS JOINT
TOO MUCH SPATTER, WIDE, FLAT BEAD

WELDING CURRENT TOO HIGH

BEAD VERY IRREGULAR, DARK, POOR PENETRATION
WELD METAL NOT PROPERLY GAS SHIELDED

ARC TOO LONG (voltage too high)

BEAD TOO SMALL, SHAPE IRREGULAR
LITTLE PENETRATION

WELDING SPEED TOO FAST

TOO MUCH PILING UP OF METAL, BEAD TOO LARGE
TOO MUCH PENETRATION

WELDING SPEED TOO SLOW

Fig. 19-25. Common TIG welding problems and their causes. (Courtesy of Hobart Brothers Company.)

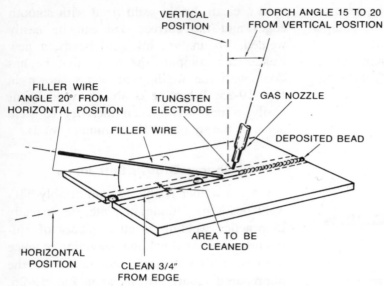

VERTICAL POSITION

TORCH ANGLE 15 TO 20 FROM VERTICAL POSITION

FILLER WIRE ANGLE 20° FROM HORIZONTAL POSITION

TUNGSTEN ELECTRODE

GAS NOZZLE

FILLER WIRE

DEPOSITED BEAD

AREA TO BE CLEANED

HORIZONTAL POSITION

CLEAN 3/4″ FROM EDGE

Fig. 19-26. Important points for TIG welding aluminum in a simple butt joint.

Practicing Factors—For practicing in each position, several things may be used for all three practice positions. Generally speaking, the following factors remain constant for the simpler TIG butt joint practice:

A. The recommended aluminum practice coupons should be about 2″ x 4″ pieces, ³⁄₃₂″ thick.

B. The pieces should be brushed with a stainless steel brush at least ¾″ back from the edges to be welded. This will insure clean weld conditions.

C. The electrode should be a zirconium tungsten type, ⅛″ diameter.

D. The filler rod should be ⅛″ diameter with the same aluminum alloy number as the aluminum coupons.

E. The TIG torch should have a gas nozzle diameter of ⅜″.

F. The electrode extension should be about ⅛″ from the end of the gas nozzle, as was shown in Fig. 19-14.

G. The welding machine should be set for superimposed high-frequency AC current with 120 amperes.

H. Argon shielding gas should be used with a flow of 16 cfh.

I. Tack welds should be made at least ¼″ to ½″ from the end of the coupons to be welded. By placing the tack welds away from the end of the coupons, the welder will not have to remelt the tack welds when he is trying to start the arc.

The above factors may be used for practicing the square groove butt joint in all of the three recommended positions. The differences between welding set-up in each position are outlined next.

Flat Position Butt Joints

For practicing the square groove butt joint in the flat position, the welder should use a root opening between the plates that is as wide as the coupons are thick. Then, the coupons may be tacked in position to offset distortion. See Fig. 19-26.

Procedure—With the equipment set up as outlined for bead welding, an arc is started with the torch as straight above the work as possible while still being able to see the puddle area.

During flat position welding, the welder may want to incline the torch up to 15°, as in Fig. 19-27. This may improve the wetting and cleaning action of the arc ahead of the puddle. If the welder inclines the torch beyond the 15° limit, the weld will not be shielded correctly. With the arc established, the welder may proceed along the weld line by first melting the weld puddle and then inserting the filler rod when the torch is moved back and the puddle is molten. The procedure is repeated many times in sequence, the same as for the bead welding shown in Figs. 19-22, 23, and 24. At the end of the weld, the arc is broken as outlined earlier.

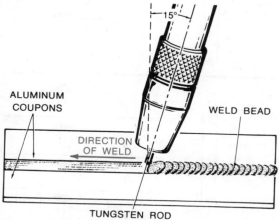

Fig. 19-27. Inclining the TIG torch up to 15° may improve the wetting and cleaning action of the arc puddle.

Butt joints in different thicknesses of aluminum can be welded with or without a temporary grooved backer, as was shown in Fig. 19-11. As usual, a welder should refer to the aluminum manufacturer's specifications when changing from one thickness of aluminum to another, to determine the right current setting, gas flow, electrode diameter, filler metal diameter, and other information necessary for good TIG welds.

Butt joint welds made on aluminum less than 1/8" thick should have good penetration under the bottom of the material being welded. Penetrating the weld beyond the

Fig. 19-28. An example of penetration fall-through. The underside bead should be smooth and free of lumps.

thickness of the aluminum is commonly known as *penetration fall-through* by TIG welders. The penetration should be smooth and free from lumps, as shown in Fig. 19-28. A good butt joint weld in the flat position on aluminum will be twice as thick from top to bottom as is the aluminum metal. See Fig. 19-29.

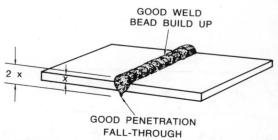

Fig. 19-29. A finished butt weld on aluminum plate should be twice as thick as the plate itself, if done correctly with TIG welding.

Flat position aluminum butt joints up to 1/8" thick are welded from one side only. Root joint openings should be used for material as thin as 5/64" thick. Without backing, aluminum from 3/16" to 1/4" thick can be welded with the square butt joint or single-Vee preparation. Welding different thicknesses of aluminum is usually well illustrated in manufacturer's specifications for different alloys, as in Fig. 19-30. In Fig. 19-30, recommendations are made for welding the alloys shown in the horizontal position. The joint details, root opening sizes, and weld pass sequences (the order of weld passes) are all illustrated in the AWS weld symbol.

Vertical Position Butt Joints

After successfully TIG welding aluminum in the flat position, most welders will have

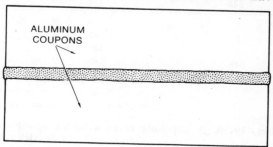

Procedures for Welding Alloys: 1100, 3003, 5052, 5083, 5454, 6061, 6063, 6351

Metal Thickness (in.)	Weld Pass Number[1]	Electrode Dia. (in.)	Filler Rod Dia. (in.)	Tack Spacing (in.)	Alternating Current[2] (amp)	Arc Travel Speed (ipm)	Argon (cfh)	Shielding Gas Nozzle Dia. (in.)	Joint Details and Order of Welding	AWS Weld Symbol
3/64	1	1/16	3/32	3	55	11	10	3/8		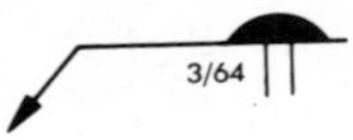
1/16	1	3/32	3/32	3	90	12.5	10	3/8		
5/64	1	3/32	3/32	3	105	18	16	3/8		
3/32	1	1/8	1/8	3	120	12	16	3/8		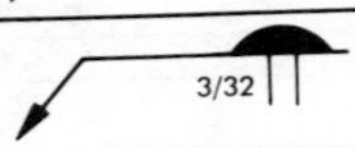
1/8	1	1/8	1/8	4	155	10	16	3/8		
3/16	1 B2R	3/16	3/16	4	230 230	10 10	16	1/2		
1/4	1 B2R	3/16	3/16	4	260 260	9 9	20	1/2		
5/16	1 2 3 B4R	1/4	5/32	6	275 275 275 275	9 9 9 9	20	1/2		
3/8	1 2 3 B4R	1/4	3/16	6	340 340 340 340	9 9 9 9	20	1/2		

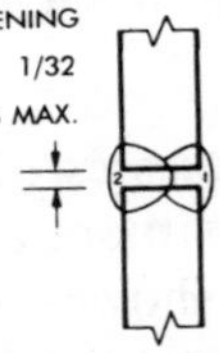

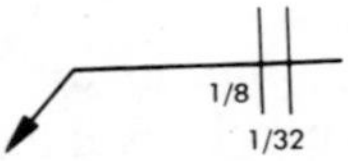

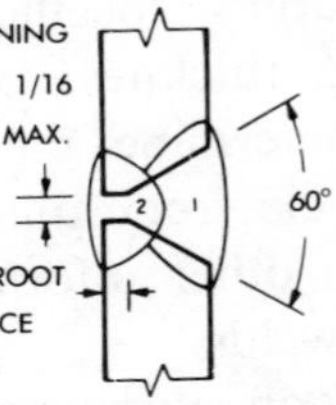

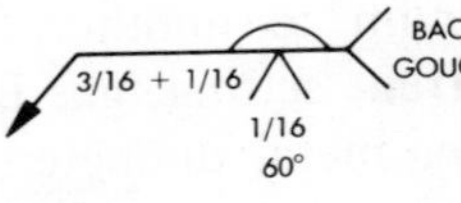

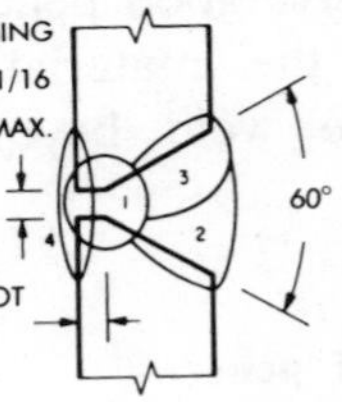

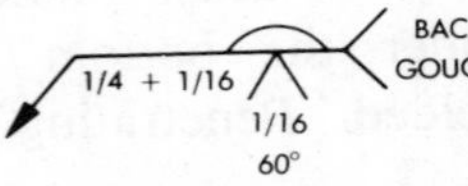

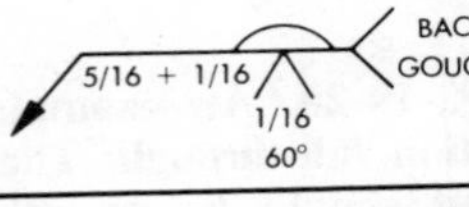

[1]B indicates back gouge before making pass.
R indicates weld pass made on the reverse of the side on which the first pass was made.
[2]Current should not vary more than 3% during welding.

Fig. 19-30. A complete manufacturer's specification chart for the alloys and joints listed. (Courtesy of Alcan Aluminum Ltd.)

little difficulty with vertical position butt joints. The most common thicknesses of aluminum welded in this position are between $3/64''$ and $3/8''$.

The first step for vertical-position butt welding is to clean and tack two pieces of aluminum the same size and thickness as used for practicing flat position welding. After tacking the aluminum coupons, they should be held in the vertical position by a fixture. When welding aluminum in the vertical position, a backing strip is not required. As indicated in the specifications, the welder should use a square butt joint, a root opening up to $1/32''$, and one bead pass for vertical position practice with this thickness of aluminum.

Vertical TIG welding is usually done from the bottom to the top for the best fusion and penetration. Some welders prefer to put a finish coat over the final bead by welding down. Fig. 19-31 shows a vertical position butt weld being welded from the bottom to the top.

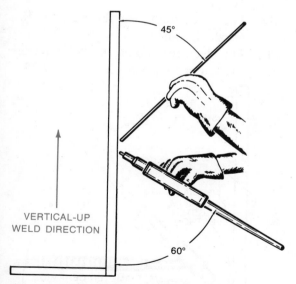

45°

VERTICAL-UP
WELD DIRECTION

60°

Fig. 19-31. Vertical position butt joint being welded from the bottom to the top. (Courtesy of Hobart Brothers Company.)

Procedure—After the equipment is adjusted to the proper settings, the welder can start an arc a short distance up from the bottom of the joint. Then, after starting the arc, the welder should quickly bring the arc back to the bottom of the weld joint to produce a molten puddle and adjust the arc length. When a clear molten puddle is formed, the filler metal can be melted in, as was done for flat position welding. The angle used for the filler metal during vertical welding is shown in Fig. 19-31.

With vertical position welding, it may be necessary for a new welder to remove the torch for just a moment to allow the molten metal to become solid. By doing this, gravity will not be able to cause the molten metal to run off, as may happen to a new welder when practicing vertical position TIG welding.

Vertical position butt welding aluminum less than $3/16''$ thick can be successfully done without machine preparation. Bevel preparations, however, are usually needed for metals over $3/16''$ thick. For vertical joint details and weld pass sequences with different thicknesses of aluminum, the welder should refer to a chart such as Fig. 19-30, only for *vertical* position TIG welding. (Fig. 19-30 shows recommendations for the horizontal position.)

Horizontal Position Butt Joints

After a welder has controlled the effects of gravity on vertical position TIG welding, he will be able to use a similar torch movement for horizontal position TIG welding.

Aluminum butt joints that are TIG welded in the horizontal position usually do not need backing strips. Aluminum from $3/64''$ to $3/8''$ thick is easily TIG welded in the horizontal position.

Horizontal position butt welding practice should be done with the same size and thickness of aluminum that was recommended for flat and vertical position practice. Using the information listed in manufacturer's specification charts for welding $3/32''$ thick aluminum in the horizontal position should help eliminate trouble. See Fig. 19-30.

Procedure—With the welding machine switched on, the welder should start the arc about ½″ from the right end of the horizontal coupon, in line with the joint to be welded. If the coupons have been cleaned and tacked properly, there should be no weld contamination or excessive warpage.

After starting the arc, the welder moves it to the right end of the joint to form the first molten pool, similar to the procedure used with flat welding. The torch and rod should be held at the angles shown in Fig. 19-32. The filler rod should then be dipped into the front of the puddle as the torch is moved to the rear of the puddle. Dipping the rod while withdrawing the arc will allow the molten metal to solidify.

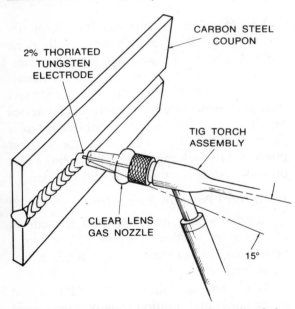

Fig. 19-32. Horizontal position butt weld being welded from the right to the left.

Horizontal position welding, like flat and vertical, may be done from both sides for additional strength. If the back side of the weldment is to be welded, it is a good idea to back gouge a narrow path on the back of the root face before welding on the back side. This will assure that no contamination will be trapped because of poor shielding on the root face during the first welding pass.

Lap and Tee Joints

When a welder feels that he can do a good job TIG welding aluminum plates in a butt joint, the butt joint and bead running practice will be useful for learning how to TIG weld lap and Tee joints. Like butt joints, lap and Tee joints may be welded in the vertical, horizontal, or flat positions. Unlike butt joints, however, lap and Tee joints may be *carefully* welded in the overhead position, too.

Practicing Factors—For practicing lap and Tee welds in each position, several things may be used for all four practice positions. Generally speaking, the following set-up factors should be used for all lap and Tee joint practice:

A. Aluminum coupons used for practice should be ³⁄₃₂″ thick with no preparations.

B. The welding machine should be set on AC with superimposed high frequency, 140 amperes current.

C. The shielding argon gas flowmeter and regulator should be set to deliver 16 cfh.

D. The TIG torch should have a ³⁄₈″ nozzle installed.

E. The electrode should be ⅛″ diameter zirconium tungsten. See Fig. 19-33.

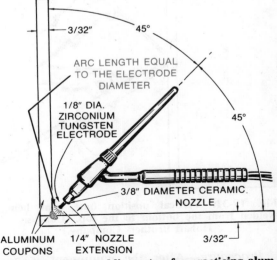

Fig. 19-33. TIG welding setup for practicing aluminum lap and Tee joints.

F. The electrode should have a ¼″ stickout. See Fig. 19-33.

G. The filler rod used should be a larger ³⁄₃₂″ diameter.

H. The torch should always be 45° from either side of the 90° angle being welded. See Fig. 19-33.

Each of the lap and Tee positions may have different procedures to be used. Once these techniques and methods have been learned, a new TIG welder will be able to carefully weld specific TIG jobs, such as the job being done in Fig. 19-34.

Procedure — With the material tacked, cleaned, positioned as shown in Fig. 19-35, and the equipment set as outlined in the manufacturer's specifications, the welder is prepared to start an arc for completing a flat fillet weld. The torch must be angled so that the tungsten electrode makes a 45° angle with either plate, as in Fig. 19-33. If one plate is thicker than the other, the electrode is pointed more toward the thicker plate for more uniform melting and to help eliminate undercutting. This will aid both the strength and appearance of the finished weld.

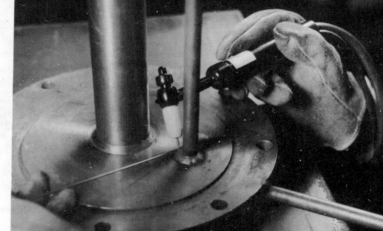

Fig. 19-34. The setup for TIG welding a precision part. (Courtesy of Airco Welding Products Division.)

Flat Position Tee and Lap Joints

When welding Tee and lap joints in the flat position, as in Fig. 19-35, the welder should try to *avoid* penetration fall-through, even though it is important when butt welding thin aluminum in the flat position. The welding technique for lap and Tee, as well as all types of TIG welds, continues to be melting the filler rod in the molten weld puddle as the torch moves along. The torch is usually kept steady during welding to produce a sound, uniform weld.

By using ³⁄₃₂″ thick aluminum coupons for all practice welding, a new welder can become familiar with the amount of heat needed to melt aluminum for fusion.

To make the welding procedure easier, the welder should always try to position the joint so that the welding can be done from right to left.

With the arc started and established at the right end of the joint to be welded, the welder can then use the technique developed for flat position aluminum welding practice. A slight angle on the torch, as in Fig. 19-27, will increase the welder's visibility. The filler rod can be melted into the puddle when the puddle appears to have the right fusion and penetration. As with flat position butt welding, the filler rod should not be melted by the arc itself; if it is, it will ball up in front of the puddle.

By following the specifications outlined for flat position Tee and lap fillet joints and some practice, most welders will have little difficulty welding these joints in the flat position on aluminum from 1/16″ to 3/8″ thick.

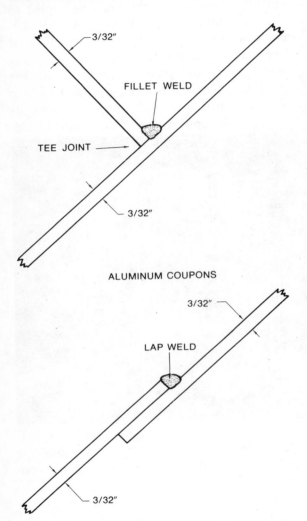

Fig. 19-35. TIG welded aluminum coupons with lap and Tee fillet welds.

Horizontal Position Tee and Lap Joints

Using the same thickness of aluminum plate as for the flat position, the welder should clean and tack the parts to form either a Tee or lap joint for horizontal position practice. Then, they may be positioned for welding in the horizontal position by using a fixture or vise. See Fig. 19-36.

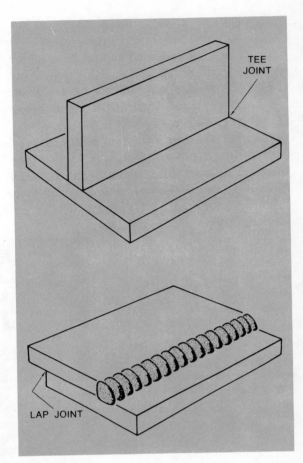

Fig. 19-36. Tee and lap joints positioned for horizontal TIG welding.

Procedure—For the horizontal position fillet weld on a Tee or lap joint, the torch should, as usual, be held at a 45° angle to either plate. The welder should angle the torch slightly in the direction of weld travel. This will allow for good fusion and penetration, as shown in Fig. 19-37. Lap and Tee fillet joints in this position can be welded on aluminum from 1/16″ to 3/8″ thick.

The arc is ready to be started after all the settings have been adjusted for the thickness of metal being welded if aluminum other than 3/32″ is being used. As in Fig. 19-33, the welder should hold the arc length equal to the electrode diameter. By holding this arc length, possible undercutting can be eliminated while still welding with complete penetration.

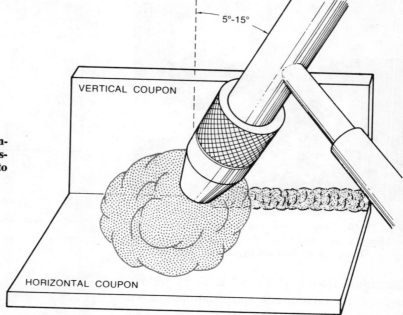

Fig. 19-37. Side view of a horizontal fillet weld. For good welder visibility, the torch is angled 5° to 15° in the weld direction.

The welding speed will still help determine the size of the bead. For this reason, a new welder should try to weld fast enough for complete penetration into the joint *and* a bead face about twice the electrode diameter being used, Fig. 19-38.

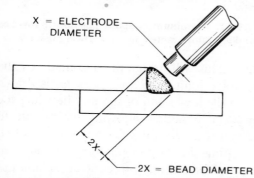

Fig. 19-38. With a TIG welded lap weld, the bead face should be about twice the diameter of the electrode being used.

When welding lap joints in the horizontal position, the welder should usually try to complete both sides of the fillet joint by dipping the filler rod in and out at the *high*

side of the leading edge of the puddle, as in Fig. 19-40. Welding both sides of the joint with the filler rod angled as shown in Fig. 19-40 will ensure the welder of a high-strength joint with no undercutting.

Vertical Position Tee and Lap Joints

Fillet welds in the vertical position on Tee and lap joints can be easily TIG welded on

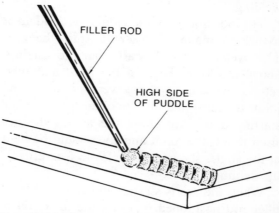

Fig. 19-39. The filler rod should be dipped in and out at the high side of the puddle with horizontal position fillet welding.

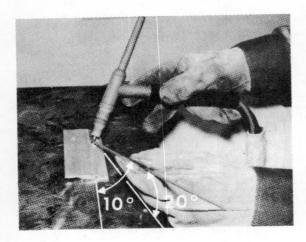

Fig. 19-40. Correct rod and torch angle for TIG welding a lap weld in the horizontal position. (Courtesy of Hobart Brothers Company.)

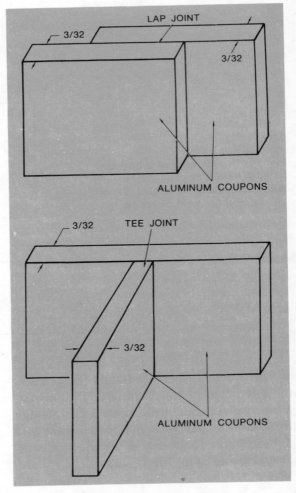

Fig. 19-42. Aluminum coupons positioned for Tee and lap joint practice in the vertical position.

aluminum from 1/16″ to 3/8″ thick by following the manufacturer's specifications shown for each joint. Fig. 19-41 shows the specifications for TIG welding the alloys listed in vertical position Tee joints.

Procedure—After positioning two pieces of clean aluminum to form a Tee or lap joint, the pieces should be tacked together for welding. After tacking the lap or Tee, they are positioned for welding in the vertical position as in Fig. 19-42. When this has been done, the welder is ready to start an arc at the bottom of the joint for vertical-up welding.

The positioned parts should be held in a fixture to make it easier to start the arc. The welder should start the arc slightly above the lower edge of the joint and then return the arc to the lowest point of the joint to form the first molten pool. After starting and establishing the arc at the desired point, the torch and rod should be held at the angles shown in Fig. 19-43 for vertical-up welding.

When the arc is started and the torch and filler rod positioned to weld, the filler rod is added as was described for vertical butt welds *except* that the filler rod is dipped back and forth on *each side* of the front of the

puddle. This will bridge the gap made by the right angle joint. The arc, as usual, is drawn back to the lower end of the puddle while the filler metal is added, allowing the puddle to solidify before welding further up the joint. By keeping the molten puddle small at first, a new welder can learn to control the amount of filler metal added and will learn to control the problem of gravity. With practice, the welder will be able to weld a strong joint on vertical position Tee or lap joints with one pass of the TIG torch.

The welder should allow the torch tip to stay 45° from either plate except when one of the aluminum pieces is heavier than the

Procedures for Welding Alloys: 1100, 3003, 5052, 5083, 5454, 6061, 6063, 6351

Metal Thickness (in.)	Electrode Dia. (in.)	Filler Rod Dia. (in.)	Tack Spacing (in.)	Alternating Current[1] (amp.)	Arc Travel Speed (ipm)	Argon Flow (cfh)	Shielding Gas Nozzle Dia. (in.)	Joint Details	AWS Weld Symbol
1/16	3/32	3/32	3	90	8	10	3/8		1/16
5/64	3/32	3/32	3	110	7.5	16	3/8	MAX. ROOT OPENING 0	5/64
3/32	1/8	3/32	3	140	9	16	3/8		3/32
1/8	3/16	1/8	4	180	9	16	3/8	MAX. ROOT OPENING 1/32	1/8
3/16	3/16	3/16	6	280	9	16	3/8		3/16
1/4	1/4	3/16	6	320	8	20	1/2		1/4
5/16	1/4	3/16	6	350	6	20	1/2	MAX. ROOT OPENING 1/16	5/16
3/8	1/4	3/16	6	380	6	20	1/2		3/8

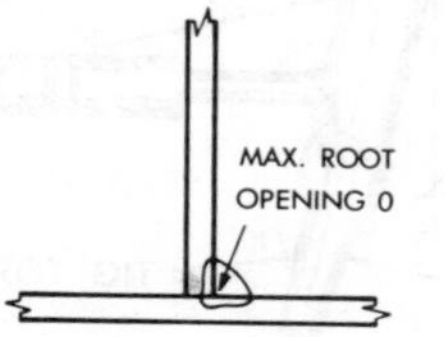

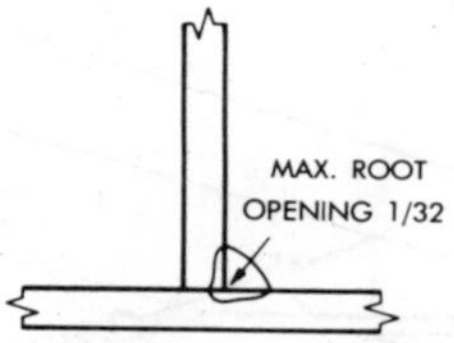

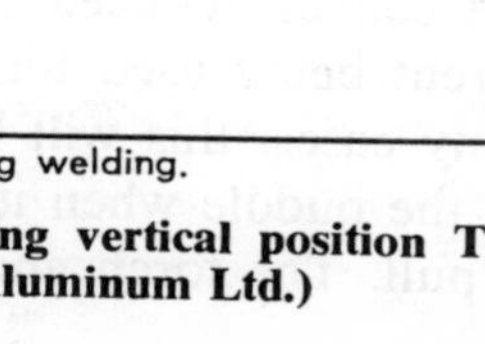

[1]Current should not vary more than 3% during welding.

Fig. 19-41. Manufacturer's specification chart for TIG welding vertical position Tee joints with the alloys shown. (Courtesy of Alcan Aluminum Ltd.)

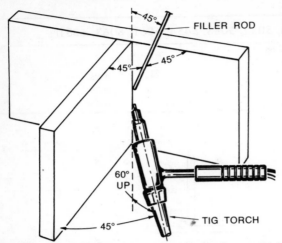

Fig. 19-43. Positioning the TIG torch for vertical up butt joint practice.

other. At that time, the torch is pointed toward the thicker section to ensure equal melting of both pieces.

rod should be held at angles somewhat similar to the angles used for flat welding. A new welder, however, may use the rod and torch angles shown in Fig. 19-44 for beginning practice. This will allow him to easily see the weld puddle. Manufacturer's specification charts for welding lap joints in the overhead position will give complete information for successful overhead TIG welding, such as the chart in Fig. 19-45.

Procedure—With the machine set according to the information outlined for lap or Tee joint overhead welding, the welder can start an arc on the right side of the tacked aluminum coupons. As will all Tee and lap joints, the torch is positioned at a 45° angle to either coupon. See Fig. 19-44. By keeping the size of the puddle slightly smaller than with other practice TIG welds, a new welder

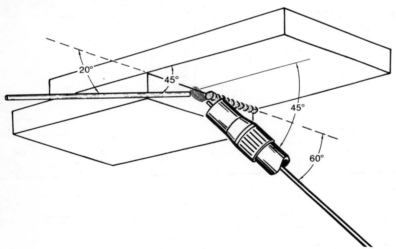

Fig. 19-44. Torch and rod angles for overhead position TIG welding lap joints.

Overhead Position Tee and Lap Joints

Overhead position TIG welding of aluminum is not recommended if the parts can possibly be positioned for flat welding. When this is not possible, these pointers can help make the job easier.

First, the current can be reduced 5% to 10% from the current being used for vertical welding. In many cases, this will help a new welder control the puddle when it is affected by gravity's pull. The torch and filler

will find that the sides of the joints create a cohesive (sticking) effect on the puddle, helping to hold it in place.

The filler rod can then be moved into the molten puddle as with other welding positions. However, with practice, a new welder will be able to move the filler rod from side to side in the molten puddle, bridging the joint gap for better fusion and penetration. During overhead welding, the welder will find that pulling the arc a little farther back each time will let the filler metal solidify

Procedures for Welding Alloys: 1100, 3003, 5052, 5083, 5454, 6061, 6063, 6351

Metal Thickness (in.)	Electrode Dia. (in.)	Filler Rod Dia. (in.)	Tack Spacing (in.)	Alternating Current[1] (amp)	Arc Travel Speed (ipm)	Argon Flow (cfh)	Shielding Gas Nozzle Dia. (in.)	Joint Details	AWS Weld Symbol
1/16	3/32	3/32	3	80	18	10	3/8		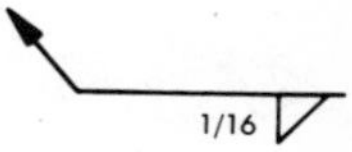 1/16
5/64	1/8	3/32	3	120	12	16	3/8		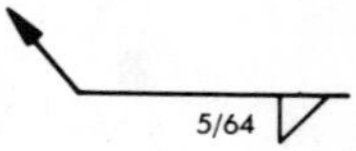 5/64
3/32	1/8	1/8	3	130	11	16	3/8	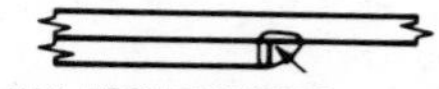 MAX. ROOT OPENING 0	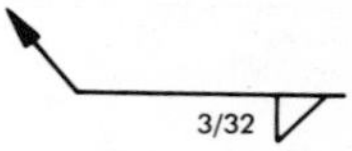 3/32
1/8	3/16	1/8	4	195	10	16	3/8		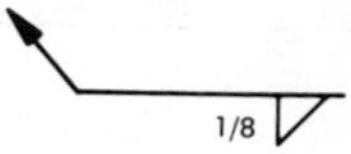 1/8
3/16	3/16	5/32	6	260	10	16	1/2		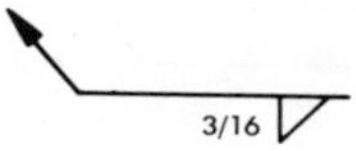 3/16
1/4	1/4	3/16	6	320	8	20	1/2	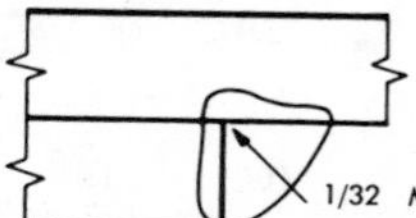 1/32 MAX. ROOT OPENING	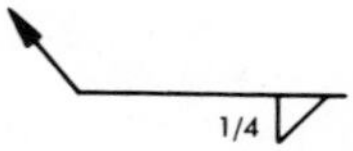 1/4
5/16	1/4	3/16	6	350	7	20	1/2	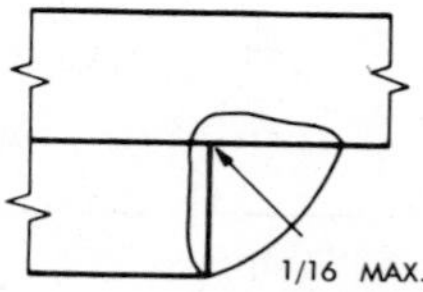	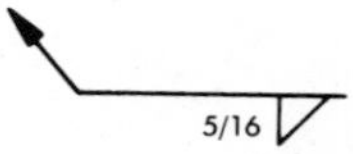 5/16
3/8	1/4	3/16	6	350	6	20	1/2	1/16 MAX. ROOT OPENING	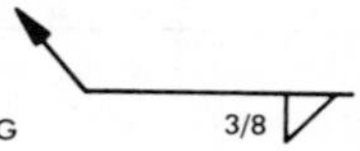 3/8

[1]Current should not vary more than 3% during welding.

Fig. 19-45. Manufacturer's specification chart for welding lap joints in the overhead position. (Courtesy of Alcan Aluminum Ltd.)

Procedures for Welding Alloys: 1100, 3003, 5052, 5083, 5454, 6061, 6063, 6351

Metal Thickness (in.)	Electrode Dia. (in.)	Filler Rod Dia. (in.)	Tack Spacing (in.)	Alternating Current[1] (amp)	Arc Travel Speed (ipm)	Argon Flow (cfh)	Shielding Gas Nozzle Dia. (in.)	Joint Details	AWS Weld Symbol
1/32	3/32	(none)	3	90	30	16	3/8		
3/64	3/32	(none)	3	100	30	16	3/8		
1/16	3/32	(none)	3	120	28	16	3/8		
5/64	1/8	(none)	3	140	24	16	3/8		
3/32	1/8	(none)	3	150	24	16	3/8		
1/8	3/16	1/8	4	180	18	20	3/8		
3/16	3/16	1/8	4	230	16.5	20	1/2		
1/4	1/4	1/8	4	265	12	20	1/2		

[1]Current should not vary more than 3% during welding.

Fig. 19-46. Specifications for TIG welding edge joints on the alloys listed. (Courtesy of Alcan Aluminum Ltd.)

Procedures for Welding Alloys: 1100, 3003, 5052, 5083, 5454, 6061, 6063, 6351

Metal Thickness (in.)	Electrode Dia. (in.)	Filler Rod Dia. (in.)	Tack Spacing (in.)	Alternating Current[1] (amp)	Arc Travel (ipm)	Argon Flow (cfh)	Shielding Gas Nozzle Dia. (in.)	Joint Details	AWS Weld Symbol
1/32	1/32	(none)	2	60	24	16	3/8		
3/64	3/32	(none)	3	110	25	16	3/8	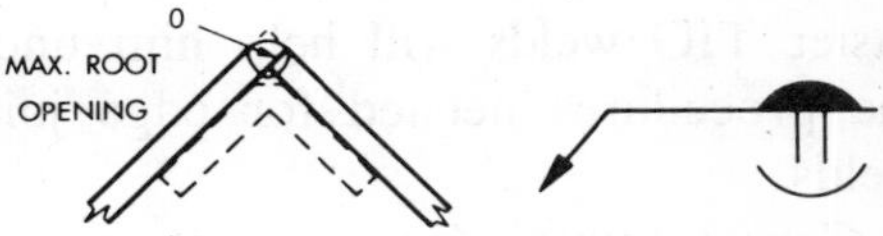	
1/16	1/8	(none)	3	130	30	16	1/2		
5/64	1/8	(none)	3	160	25	16	1/2		
3/32	1/8	(none)	3	170	24	16	1/2	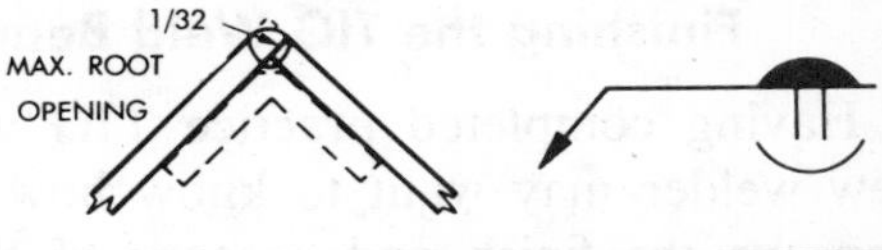	
1/8	1/8	1/8	5	180	12	16	1/2		
3/16	3/16	5/32	4	260	12	16	1/2	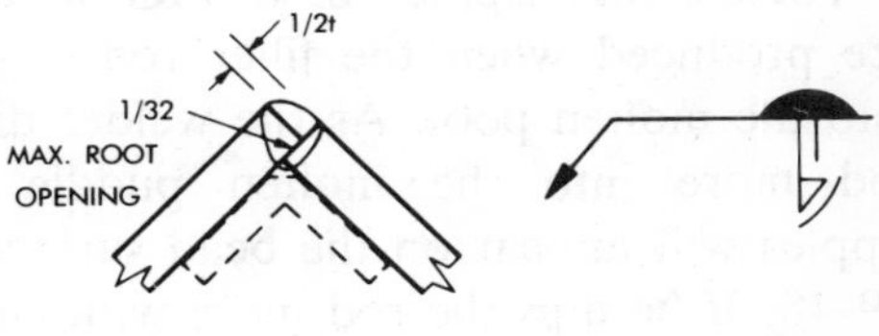	
1/4	1/4	5/32	4	280	8.5	20	1/2		

[1]Current should not vary more than 3% during welding.

Fig. 19-47. Specifications for TIG welding corner joints on the alloys listed. (Courtesy of Alcan Aluminum Ltd.)

faster. That way, it will be less affected by the pull of gravity.

Edge and Corner Joints

Edge and corner joints can be TIG welded in all positions. These joints may be formed with aluminum from $\frac{1}{32}''$ to $\frac{1}{4}''$ thick.

Edge Joints—Edge joints do not usually need a filler rod, except on very thick aluminum stock. Fig. 19-46 shows a manufacturer's specification chart for TIG welding edge joints with the aluminum alloys shown. Following specifications such as these will assure a new welder of successful TIG edge welds. As usual, successful experience with easier TIG welds will help him understand the procedures needed for edge joint TIG welds.

Corner Joints—Corner joints may also not require a filler rod, except on thicker aluminum stock. Fig. 19-47 shows a manufacturer's specification chart for TIG welded corner joints with a temporary backer. Following these instructions will help a new welder easily TIG weld corner joints.

Finishing the TIG Weld Bead

Having completed practice TIG welds, a new welder may want to know how he can improve the finish and contour of the TIG weld bead. This may be done on his future welds by carefully using a filler rod dipping process as follows.

The normal ripples in a TIG weld bead are produced when the filler rod is dipped into the molten pool. As the welder dips the rod more into the molten puddle, more ripples will appear on the bead surface, Fig. 19-48. If he dips the rod more without using enough heat, poor penetration may result. Therefore, the rod should be moved into the molten puddle only as it moves forward along the weld line.

When welding aluminum with the TIG method, there may be a tendency to develop a crater-like hole at the end of the weld. In

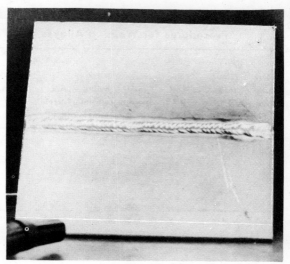

Fig. 19-48. More ripples appear on the bead surface when the rod is dipped more into the molten pool. (Courtesy of Aircom Metal Products, Inc.)

order to eliminate the crater-like hole at the end of the weld, a new welder should add filler rod and slowly ease up on the heat until the machine shuts off.

Some welders prefer to fill a crater hole by first restarting and remelting the puddle. Then, they pump the heat control on and off until the puddle has cooled enough to the point where crater-like holes can no longer form.

Table 19-1. Recommended Welding Current Sources for TIG Welding.

Metal or Alloy	Current Source & Results	
	Alternating Current—High Frequency	Direct Current—Straight Polarity
Aluminum	Excellent	(Not Recommended)
Aluminum Bronze	Good	Excellent
Beryllium Copper	Excellent	(Not Recommended)
Brass	Excellent	Excellent
Copper	Good	Excellent
Inconel	Fair	Excellent
Magnesium	Excellent	(Not Recommended)
Monel	Fair	Excellent
Nickel	Fair	Excellent
Silicon Bronze	Good	Excellent
Stainless Steel	Excellent	Excellent*
Steel	Excellent	Excellent*

*DC straight polarity is recommended on metal over 14 gauge.

Courtesy of Canadian Liquid Air

20

TIG Welding Common Metals

The most common metal welded with the TIG process is aluminum. Earlier, the complete procedure and recommendations for TIG welding aluminum were outlined. A new welder should practice TIG welding aluminum successfully before trying to TIG weld other metals.

Besides aluminum, many other common metals can be successfully welded with TIG equipment. The main feature of TIG welding, the idea of welding in an inert atmosphere, can help a welder make excellent welds on almost all metals. Also, most metals other than aluminum and magnesium do not need the oxide cleaning action of the reverse polarity side or negative half of AC current. These other metals, then, may be welded on straight-polarity DC for a more stable arc and deeper penetration.

TIG Welding Magnesium

A new welder with certain aluminum TIG welding skills in all positions should have little or no difficulty TIG welding magnesium. Aluminum and magnesium have many similar properties that are used when TIG welding. Magnesium and aluminum are alike because they both need the oxide cleaning action of the reverse polarity side or negative half of AC current. TIG welding magnesium can be done by both manual and automatic TIG welding machines.

Until recently, 10 gauge or thicker magnesium was welded by the gas metal-arc process. At the present time, however, manual TIG welding is used for magnesium thicknesses from 20 gauge to ⅜″. The most common magnesium metals welded today are known as ASM-M1 and ASTM-AZ31. The filler metal for ASTM-M1 magnesium should be M1. The filler metal for ASTM-AZ31 should be AZ61X.

Welding Machine

An AC welding machine, such as in Fig. 20-1, is typical for welding magnesium. Good arc stability is maintained with superimposed high-frequency AC current throughout the current range. Also, AC current has the needed oxide cleaning action for good magnesium welds with little distortion. Weld speeds can be increased when using AC current without the fear of oxide contamination. This increased speed, in turn, reduces the weld cost. Table 20-1 gives the recommended AC-HF currents for TIG welding magnesium.

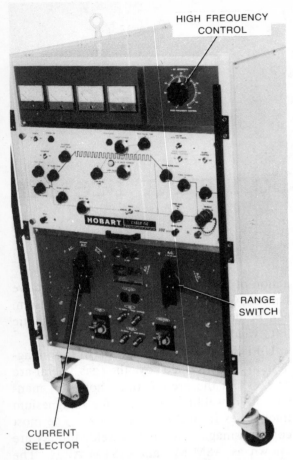

HIGH FREQUENCY
CONTROL

RANGE
SWITCH

CURRENT
SELECTOR

Fig. 20-1. A typical TIG welding machine with AC current and high frequency control available. (Courtesy of Hobart Brothers Company.)

Reverse-polarity DC will also produce good-quality magnesium welds, due to its good oxide cleaning action. However, the low current-carrying capacity of tungsten electrodes can create electrode overheating problems when using DC current. Where weld distortion could be a problem, reverse-polarity DC should be used instead of AC.

Shielding Gas

Generally, helium gas is preferred over argon for shielding magnesium because it better stabilizes the arc. However, argon is still preferred where undercutting could be a problem. The hotter arc produced by using helium shielding gas may create undercutting

and make it harder to weld thin magnesium sheets. On thicker magnesium, the helium allows for greater arc penetration.

When welding thin magnesium, a new welder may find that using pure argon or an argon/helium combination will allow him to better control the possibility of burn-through. When an argon/helium combination is used, the mixture is usually made up of 25% argon and 75% helium. Table 20-1 shows the recommended shielding gas flow for either pure argon or the argon/helium mixture.

Electrode

For TIG welding magnesium, very little difference will be noticed in the arc stability whether the welder is using pure tungsten, zirconium tungsten, or 2% thoriated-tungsten electrodes. Most welders who work on TIG welded magnesium parts prefer to use a zirconium tungsten electrode. Generally, the zirconium tungsten electrode is able to carry higher current than the pure tungsten electrode, making it more popular for those jobs.

Table 20-1 gives the recommended electrode diameters for the nozzle being used. The electrode tip should be balled as for aluminum. See Fig. 20-2.

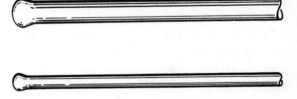

Fig. 20-2. Electrodes for welding magnesium and aluminum have a ball shape on the end.

Positions and Joints

Magnesium can be welded in the flat, vertical, horizontal, and overhead positions, using the current recommended in Table 20-1. When welding is to be done in any position other than flat, the amperage should be reduced 10%-20% from the value listed for good puddle control.

Table 20-1. Recommendations for TIG Welding Magnesium.

Magnesium Thickness (in.)		Type of Weld[2]	Tungsten Electrode Diameter[3]	Filler Rod Diameter[4]	Nozzle Size—Inside Diameter (in.)	Shielding Gas Flow cfh[5]	Welding Current[1] (amps) AC-HF	Number of Passes	Travel Speed (ipm)
20 gauge	.038	Square groove	$\frac{1}{16}$	$\frac{3}{32}$	$\frac{1}{4}$	15	25- 40	1	20
20 gauge	.038	Fillet	$\frac{1}{16}$	$\frac{3}{32}$	$\frac{1}{4}$	15	30- 45	1	20
16 gauge	.063	Square groove	$\frac{1}{16}$	$\frac{3}{32}$	$\frac{1}{4}$	15	45- 60	1	20
16 gauge	.063	Fillet	$\frac{1}{16}$	$\frac{3}{32}$	$\frac{1}{4}$	15	45- 60	1	20
14 gauge	.078	Square groove	$\frac{1}{16}$	$\frac{3}{32}$	$\frac{1}{4}$	15	60- 75	1	17
14 gauge	.078	Fillet	$\frac{1}{16}$	$\frac{3}{32}$	$\frac{1}{4}$	15	60- 75	1	17
12 gauge	.109	Square groove	$\frac{3}{32}$	$\frac{1}{8}$	$\frac{5}{16}$	15	80-100	1	17
12 gauge	.109	Fillet	$\frac{3}{32}$	$\frac{1}{8}$	$\frac{5}{16}$	15	80-100	1	17
11 gauge	.125	Square groove	$\frac{3}{32}$	$\frac{1}{8}$	$\frac{5}{16}$	25	95-115	1	17
11 gauge	.125	Fillet	$\frac{3}{32}$	$\frac{1}{8}$	$\frac{5}{16}$	25	95-115	1	17
$\frac{3}{16}$	.187	Vee groove	$\frac{1}{8}$	$\frac{1}{8}$	$\frac{3}{8}$	25	95-115	2	
$\frac{1}{4}$	.250	Vee groove	$\frac{1}{8}$	$\frac{3}{16}$	$\frac{1}{2}$	25	110-130	2	24
$\frac{3}{8}$	.375	Vee groove	$\frac{1}{8}$	$\frac{3}{16}$	$\frac{1}{2}$	30	135-165	2	20

[1] Increase amperage when backup is used.

[2] Data is for flat position TIG welding. Reduce amperage 10% to 20% when welding in the horizontal, vertical or overhead positions.

[3] For tungsten electrodes: 1st choice—Zirconated EWZr; 2nd choice—Pure Tungsten EWP.

[4] Filler metal most commonly used for popular magnesium alloys is AZ-A61.

[5] Shielding gas used is normally Argon; however, a mixture of 75% Helium + 25% Argon is popular. For heavy thickness 100% Helium is used. Gas flow rates for Helium should be about twice those used for Argon.

The square-groove butt joint can be used on 15-, 18-, or 20-gauge magnesium in the above positions. Tee and lap joints can be made on magnesium from 11 to 20 gauge without a Vee-groove preparation.

Thicker magnesium stock ($\frac{3}{16}$", $\frac{1}{4}$", $\frac{3}{8}$") should have the edges prepared with a Vee before welding either butt, Tee, or lap joints.

Special Procedures

Generally speaking, the techniques used to weld aluminum will work equally well with magnesium. This is especially true for cleaning, since magnesium requires as complete a cleaning as does aluminum before starting the arc. See Fig. 20-3. Usually, the welder should clean both the parent metal and the

Fig. 20-3. Like aluminum, magnesium should be cleaned back $\frac{3}{4}$" from the edge to be welded. The weld procedures are similar for the two metals.

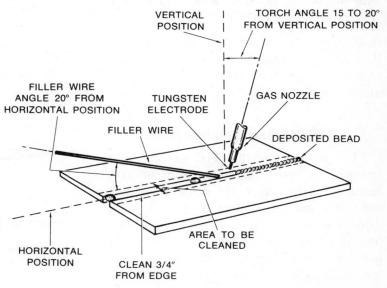

filler rod to insure that any oxides that could cause weld contamination are removed.

When failure of a magnesium TIG weld occurs, it usually happens in the heat-affected zone caused by the grain growth from the welding heat. See Fig. 20-4. When welding heavy magnesium sections, the welder should use 100% helium shielding gas for increased arc penetration. The helium gas

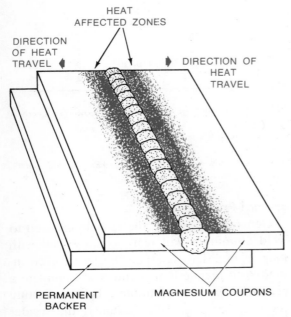

HEAT AFFECTED ZONES

DIRECTION OF HEAT TRAVEL

DIRECTION OF HEAT TRAVEL

PERMANENT BACKER

MAGNESIUM COUPONS

Fig. 20-4. Most of the failure in a magnesium TIG weld will be due to cracks forming in the heat-affected zone.

flow should be adjusted for twice the cfh flow of argon shielding gas. Other welding techniques are very similar to those for aluminum, as in Fig. 20-3.

TIG Welding Stainless Steel

Most welders find that TIG welding stainless steel is easier than aluminum or magnesium, partly due to the color changes in the metal and the welding time needed. Because it dissipates (spreads out) the welding heat very fast, aluminum requires more welding heat and a faster travel speed than stainless steel. Using lower, more controllable

welding heat with stainless steel will usually assure a new welder of a successful TIG weld. The starting end of a stainless steel weld requires very little heat and, although aluminum does *not* give visible color changes with heat, stainless steel *does* help the welder by changing color as it gets hotter. Too much welding heat is being used on stainless steel when puddle ripples are not noticeable and the weld area appears dark purple in color. With experience, the new welder will learn to adjust his heat and welding speed while watching these color changes.

Welding Machine

Straight polarity power is usually preferred for stainless steel TIG welding. Although reverse polarity is used for aluminum and magnesium due to its oxide cleaning ability, TIG welding stainless steel and carbon steel is better done with straight polarity. The superimposed high frequency is often used to start the DC arc.

Shielding Gas

Argon gas is usually used for shielding when TIG welding stainless steel. A mixture of 25% argon and 75% helium may be used to help penetration when welding heavier stainless steel sections. Table 20-2 lists the recommended shielding gas and cfh flow for TIG welding stainless steel.

Electrode

A 2% thoriated tungsten electrode is preferred for welding stainless steel. Electrode diameters of 0.040″ to $^3/_{16}$″ are available to weld stainless steel thicknesses of 24 gauge

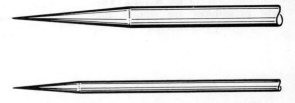

Fig. 20-5. Electrodes for TIG welding carbon steel and stainless steel should be sharpened to a point on the end.

Table 20-2. Recommendations for TIG Welding Stainless Steel.

Stainless Steel Thickness (in.)	Type of Weld[2]	Tungsten Electrode Diameter[3]	Filler Rod Diameter	Nozzle Size—Inside Diameter (in.)	Shielding Gas Flow cfh[4]	Welding Current[1] (amps) DC–SP	Number of Passes	Travel Speed (ipm)[5]
24 gauge	Square groove	.040	1/16	1/4	10	20- 50	1	26
18 gauge	Square groove	1/16	1/16	1/4	10	50- 80	1	22
1/16	Square groove	1/16	1/16	1/4	12	65-105	1	12
1/16	Fillet	1/16	1/16	1/4	12	75-125	1	10
3/32	Square groove	1/16	3/32	1/4	12	85-125	1	12
3/32	Fillet	1/16	3/32	1/4	12	95-135	1	10
1/8	Square groove	1/16	3/32	5/16	12	100-135	1	12
1/8	Fillet	1/16	3/32	5/16	12	115-145	1	10
3/16	Square groove	3/32	1/8	5/16	15	150-225	1	10
3/16	Fillet	1/8	1/8	3/8	15	175-250	1	8
1/4	Vee groove	1/8	3/16	3/8	18	225-300	2	10
1/4	Fillet	1/8	3/16	3/8	18	225-300	2	10
3/8	Vee groove	3/16	3/16	1/2	25	220-350	2-3	10
3/8	Fillet	3/16	3/16	1/2	25	250-350	3	10
1/2	Vee groove	3/16	1/4	1/2	25	250-350	3	10
1/2	Fillet	3/16	1/4	1/2	25	250-350	3	10

[1] Increase amperage when backup is used.

[2] Data is for flat position TIG welding. Reduce amperage 10% to 20% when welding in the horizontal, vertical, or overhead positions.

[3] For tungsten electrodes: 1st choice—2% Thoriated EWTh2; 2nd choice—1% Thoriated EWTh1.

[4] Shielding gas used is normally Argon. A mixture of 75% Helium + 25% Argon should be used for heavier thicknesses.

[5] Travel speed per pass for multi-pass welds.

Table 20-3. Recommended Stainless Steel Filler Metals for TIG and MIG Welding.

AISI TYPE	Typical Rod Composition[1]						APPLICATIONS
	C	Mn	Si	Cr	Ni	Others	
308 L[3]	0.025	1.8	0.40	20.6	9.7	—	For welding types 304, 308, 321 and 347.
308 L Hi Sil[2]	0.025	1.8	0.85	20.6	9.7	—	For welding types 301, 304; particularly piping.
309 L	0.025	1.8	0.40	24.0	13.5	—	For welding 309 and straight-chromium grades when heat-treatment is not possible.
310	0.12	1.8	0.45	26.0	21.0	—	For welding 310, 304 clad and stainless overlay.
316 L	0.025	1.80	0.35	19.5	13.0	2.3 Mo	For welding types 316 L[3]
316 L Hi Sil	0.025	1.80	0.85	19.5	13.0	2.3 Mo	For welding type 316; particularly piping.
317 L	0.025	1.8	0.40	19.0	12.5	3.3 Mo	For welding type 317.
347	0.06	1.3	0.40	19.5	9.5	0.90 Cb	For welding types 321 and 347 where maximum corrosion-resistance is required.

[1] Remainder is iron.

[2] Hi Sil means high silicon content—approximately double.

[3] L means low carbon, 0.03% maximum.

to ½". The electrode tip should be sharpened to a point (Fig. 20-5) before tightening it in the collet.

Filler Metal

Stainless steel filler metal should be the recommended type for the metal being welded. Table 20-3 shows the recommended filler metal for various types of stainless steel. In many cases, a small filler rod will do a good welding job because it needs less heat for melting and fusion. A large rod may soak up too much arc heat and make welding jobs more difficult for a new welder.

Positions and Joints

Stainless steel can be easily welded in all positions, although reducing the amperage 10%-20% is recommended for welding in the vertical, horizontal, and overhead positions for good puddle control. After setting the amperage for the flat position, a new welder should practice other weld positions by slowly reducing the current from 10% to 20%. If good flat-position welding is done at 100 amperes, overhead-position welding should be practiced somewhere between 80 and 90 amperes with adjustments, until a good weld can be made.

Stainless steel less than ¹⁄₁₆" thick is usually welded with a square-butt joint. Fillet and lap joints may be used with stainless steel from ¹⁄₁₆" to ½" thick. A Vee groove is usually not needed on stainless steel less than ¼" thick.

When TIG welding thinner stainless steel, a low heat should always be used even though the time required will increase the welding cost. Stainless steel TIG welds made with low heat are not only stronger than welds made with high heat, but distortion is more easily controlled when low heat and good clamping devices are used.

Special Procedures

A new welder practicing on stainless steel for the first time should tack weld *and* clamp the stainless steel coupons. Since stainless steel will distort more than carbon or other low-alloy steels, good clamping will be needed to offset some of the distortion during welding.

Fillet welding stainless steel is usually done while the stainless steel is clamped to copper back-up plates. The copper backup plates help eliminate fall-through, a problem with stainless steel welding. Fall-through may be an especially difficult problem for a new welder to control. Preheating the back of the copper plates with an oxy-acetylene torch to about 600° F assures the welder of good penetration with equal joint fusion.

When TIG welding stainless steel, as with other metals, the welder must move along while dipping the rod evenly for a smooth, even bead contour. During fillet welding, the puddle only moves up as the torch is moved, unlike the easy upward flow with carbon steel fillet welding. With carbon steel fillet welding, a pushing effect moves the puddle along. Other than this variation, the new welder will find the technique for stainless steel welding to be similar to the techniques for other metals.

By following the stainless steel specifications shown in Table 20-2 and the TIG welding techniques practiced earlier, a new welder should have little trouble successfully TIG welding stainless steel.

TIG Welding Carbon and Low-Alloy Steel

Carbon steels are easier to TIG weld than are stainless steels. By using the techniques for TIG welding other metals and the specifications in Table 20-4, a new welder will be prepared to TIG weld carbon and low-alloy steels in all positions.

Welding Machine

An AC-DC welding machine used for TIG welding aluminum can also be used for TIG welding carbon and low-alloy steel. If the DC setting is used, the machine should be

Table 20-4. Recommendations for TIG Welding Carbon and Low-Alloy Steels.

Carbon or Low-Alloy Steel Thickness (in.)		Type of Weld (All Positions)	Tungsten Electrode Diameter [2]	Filler Rod Diameter [3]	Nozzle Size-Inside Diameter (in.)	Shielding Gas Flow cfh [4]	Welding Current (amps) DC-SP [1]	Number of Passes	Travel Speed (ipm)
20 gauge	.032	Square groove & Fillet	1/16	1/16	1/4	10	75-100	1	13
18 gauge	.040	Square groove & Fillet	1/16	1/16	1/4	10	90-120	1	15
16 gauge	.063	Square groove & Fillet	1/16	1/16	1/4	10	95-135	1	15
3/32	.094	Square groove & Fillet	3/32	3/32	5/16	10	135-175	1	14
1/8	.125	Square groove & Fillet	3/32	1/8	5/16	12	145-205	1	11

[1] Increase amperage when backup is used.

[2] For tungsten electrodes: 1st choice—2% Thoriated EWTh2; 2nd choice—1% Thoriated EWTh1.

[3] Select a filler metal to match the properties and composition of the base metal. Use a deoxidized filler metal rod.

[4] Shielding gas used is normally Argon. A mixture of 75% Helium+25% Argon is sometimes used for heavier thicknesses.

set on straight polarity, as shown in Fig. 20-6. A new welder should use superimposed high frequency for starting the DC arc, although an arc can be started with the high frequency off. High-frequency starting allows the arc to be started without touching the tungsten electrode to the work, thus saving the point of the electrode from coming in contact with the metal being welded.

An electrode for carbon and low-alloy steel TIG welding requires a sharp point, as in Fig. 20-5. If the sharp electrode was allowed to contact the work, it could become contaminated because the molten metal would cling to the electrode's point. A contaminated electrode causes the arc to become unstable or uneven, which makes it difficult to control.

Shielding Gas

Argon is a good shielding gas for TIG welding carbon and low-alloy steels, especially for lighter pieces of the metals. Where heavy sections are being TIG welded, the welder should use a 75% helium and 25% argon mixture. The mixture will allow a hotter arc and better penetration. The recommended shielding gas flow rates are shown in Table 20-4.

Electrodes

EWTh2 (2% thoriated tungsten) electrodes are recommended for carbon or low-alloy steel TIG welds, although some welders use 1% thoriated tungsten electrodes.

The ends of the tungsten electrodes should be sharpened before welding, as shown in Fig. 20-5. If the electrode comes in contact with the molten metal, it should be removed from the torch to avoid contamination and resharpened before further welding. When using straight polarity DC, a sharp point on the electrode will withstand the arc heat for a long time. The sharp point allows the welder to control the arc and the weld bead size more easily than he could with a balled point.

Filler Metal

For practicing TIG welds on carbon or low-alloy steel, use a 1/16″ mild steel filler rod. A high-strength, low-alloy filler rod can be used to give strong, sound TIG welds with carbon and low-alloy steels. All filler rods should be degreased and cleaned since an oil-coated rod can cause weld porosity (pinholes).

The copper-coated, mild-steel rods normally used for oxy-acetylene welding should *not* be used for TIG practice since they have a tendency to contaminate the tungsten electrode by spattering in the weld.

Practice Procedure

Carbon and low-alloy steel from 20 gauge to 1/8″ thick can be welded in all positions when using a butt joint with no preparation. Fillet and lap joints may also be welded in those thicknesses with no preparation other than positioning and cleaning. Unlike the manual shielded-arc welding process, the TIG process requires that the steel have no oil on the surface. During welding, the oil could cause porosity in the weld area. The surface of oily steel can be *etched* (chemically cleaned) with caustic soda.

Set-up—Adjust the equipment for welding 3/32″ carbon or low-alloy steel practice cou-

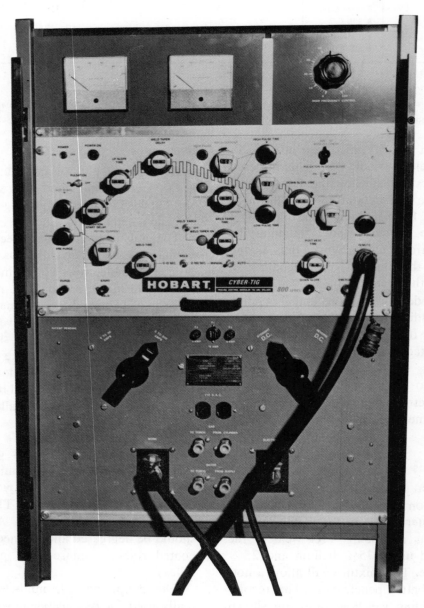

Fig. 20-6. Typical TIG welding machine. The machine is set on straight polarity DC with the high-frequency control turned up. (Courtesy of Hobart Brothers Company.)

pons according to the recommendations in Table 20-4. For this thickness of metal, first try the current setting in the middle of the 135 to 175-ampere range, using argon shielding gas with a flow of 10 cfh. A sharpened 2% thoriated tungsten electrode, $\frac{3}{32}$" diameter, should be used with a $\frac{5}{16}$" inside diameter nozzle.

Butt Welds—With the equipment set up properly, tack the coupons and position them in a fixture, if possible, for flat position TIG welding practice. The flat position square-butt joint should be practiced first. Both the arc and welding are started from the right side of the plate and followed through to the left. The torch and filler rod angle should be the same as used for flat position aluminum welding. See Fig. 20-3.

The *vertical* butt joint is also welded the same as aluminum, with welding being started at the bottom of the plate and going to the top. See Fig. 20-7.

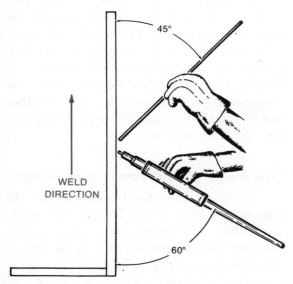

Fig. 20-7. The TIG torch and filler rod should be held at the angles shown for vertical position butt welding carbon or low-alloy steel.

The *horizontal* and *overhead* welding positions are usually easier to TIG weld when low-carbon steel is being welded instead of aluminum. The molten steel puddle can be

kept more plastic than the molten aluminum puddle, and the filler rod can be easily used to move the molten metal into position. The torch and rod angles for these positions should be the same as the angles used to weld on other metals. See Figs. 20-8 and 20-9.

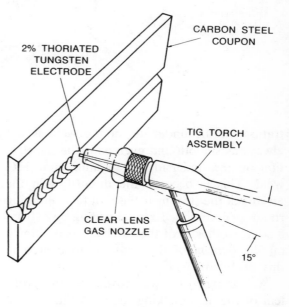

Fig. 20-8. Practicing TIG welding carbon steel coupons on a prepared, horizontal position butt joint.

Tee and Lap Joint Fillet Welds—Carbon and low-alloy steel fillet, Tee, and lap welds can be made in all positions by carefully using the TIG welding techniques discussed for aluminum joints. The specifications shown in Table 20-4 should be followed for the best carbon and low-alloy steel TIG welds. The torch must stay at a 45° angle to each plate during welding unless it is necessary to slant the torch toward a heavier plate to overcome fillet undercut discussed earlier. Slanting the torch will also insure that the heavier piece is equally melted. The torch can also be tilted slightly in the direction of travel to improve the total inert gas shielding and make it easier to see the molten pool.

Special Procedures—When TIG welding carbon or low-alloy steel, the root pene-

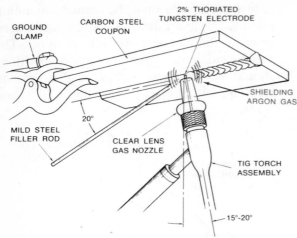

GROUND CLAMP

CARBON STEEL COUPON

2% THORIATED TUNGSTEN ELECTRODE

SHIELDING ARGON GAS

MILD STEEL FILLER ROD

20°

CLEAR LENS GAS NOZZLE

TIG TORCH ASSEMBLY

15°-20°

Fig. 20-9. Practicing TIG welding carbon steel coupons on an unprepared, overhead position butt joint.

tration on the underside of the weld is not always as smooth and even as is the penetration on the underside of an aluminum TIG weld. This is because the impurities in the air attack the steel near the root face, creating rough penetration. Extra shielding gas (from a source other than the torch) can be blown on the *bottom* of the coupons to overcome this problem.

A less expensive way to handle the problem of the air attacking the underside of the root joint is to clamp a grooved copper plate under the steel being welded. The copper plate must be larger than the metal being welded, and the groove in the copper must be directly under the weld line. See Fig. 20-10.

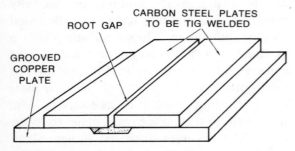

ROOT GAP

CARBON STEEL PLATES TO BE TIG WELDED

GROOVED COPPER PLATE

Fig. 20-10. A grooved copper plate can be used to protect the underside of the root gap for TIG welding.

The copper back–up plates used to protect the root gap must be at least $\frac{3}{16}''$ thick. Preheat the bottom of the copper to about 600°

F after any clamping and tack welding is done. If the copper is not preheated, it will act as a chill bar and soak up the heat that is needed for welding the joint.

The copper may be preheated by turning the clamped copper and steel over to preheat the copper, then turn the work *back* over to weld the coupons.

The rod-dipping technique used for TIG welding aluminum will also work well for carbon or alloy steel TIG welds. Although good heat control is necessary for quality TIG carbon or alloy-steel welds, it is not as important for steel as it is for aluminum. The copper back–up bars may be used for either butt joint welds or, with different preparation, Tee and lap joint fillet welds.

TIG Welding Copper and Copper Alloys

Copper and deoxidized copper alloys are sometimes joined by the TIG welding process. Table 20-5 shows the manufacturer's recommended conditions for welding copper and deoxidized copper alloys.

Welding Machine

When joining copper and deoxidized copper alloys, straight polarity DC should be used. Straight polarity DC current works well for metals such as copper that do not need the oxide cleaning action of reverse polarity DC or AC current.

Table 20-5. Recommendations for TIG Welding Copper and Copper Alloys.

Copper Thickness (in.)[5]		Type of Weld [2]	Tungsten Electrode Diameter [3]	Filler Rod Diameter (in.)	Nozzle Size-Inside Diameter (in.)	Shielding Gas Flow cfh [4]	Welding Current (amps) DC-SP [1]	Number of Passes	Travel Speed (ipm)
¹⁄₁₆	.063	Square groove	¹⁄₁₆	¹⁄₁₆	¹⁄₄	18	100-150	1	12
¹⁄₁₆	.063	Fillet	¹⁄₁₆	¹⁄₁₆	¹⁄₄	18	125-155	1	10
¹⁄₈	.125	Square groove	³⁄₃₂	³⁄₃₂	⁵⁄₁₆	18	170-230	1	10
¹⁄₈	.125	Fillet	³⁄₃₂	³⁄₃₂	⁵⁄₁₆	18	195-245	1	8
³⁄₁₆	.187	Square groove	¹⁄₈	¹⁄₈	³⁄₈	36	185-230	1	10
³⁄₁₆	.187	Fillet	¹⁄₈	¹⁄₈	³⁄₈	36	200-255	1	8
¹⁄₄	.250	Vee groove	¹⁄₈	¹⁄₈	³⁄₈	36	220-275	2	7
³⁄₈	.375	Fillet	¹⁄₈	¹⁄₈	³⁄₈	36	245-285	1	7
³⁄₈	.375	Vee groove	³⁄₁₆	³⁄₁₆	¹⁄₂	45	275-325	2	7
³⁄₈	.375	Fillet	³⁄₁₆	³⁄₁₆	¹⁄₂	36	290-350	2	8
¹⁄₂	.500	Vee groove	¹⁄₄	¹⁄₄	⁵⁄₈	45	370-500	2	6

[1] Increase amperage when backup is used.

[2] Data is for flat position TIG welding. Reduce amperage 10% to 20% when welding in the horizontal, vertical, or overhead position.

[3] For tungsten electrodes: 1st choice—1% Thoriated EWTh1; 2nd choice—2% Thoriated EWTh2.

[4] Shielding gas used is normally Helium. A mixture of 75% Helium + 25% Argon is also very popular on copper and some copper alloys. Argon is usually used to shield bronze.

[5] Preheat ³⁄₁₆" Copper to 200° F; ¹⁄₄" Copper to 300° F; ³⁄₈" Copper to 500° F; ¹⁄₂" Copper to 900° F.

The straight polarity DC arc has deep penetration and a narrow bead. The narrow bead reduces contraction stresses and distortion. By doing this, it will help prevent the hot-short cracking problems that are common when welding copper and its alloys.

Shielding Gas

Copper and copper alloys are usually shielded with pure helium gas, flowing at the rates recommended in Table 20-5. Pure copper and some alloys are better welded with a mixture of 25% argon and 75% helium, depending on the thickness of the metal and the welding heat needed. Bronze metal is best welded in a pure argon atmosphere.

Electrode

The best electrode to use for TIG welding copper and copper alloys is the EWTh1, 1% thoriated tungsten electrode. It is able to withstand increased arc heat, especially if pure helium shielding gas will be used.

Positions, Joints, and Procedures

Copper and copper alloys can be welded in all positions with no joint preparation on

metal up to ³⁄₁₆" thick. A groove preparation is recommended on copper over ³⁄₁₆" thick for equal weld penetration and good strength.

Copper over ¹⁄₈" thick should be preheated before welding. The recommended preheat temperatures for different thicknesses are shown in Table 20-5.

The procedures for TIG welding aluminum will work equally well for copper and copper alloys. Generally speaking, the mechanics of filler rod and torch angles remain the same for copper and aluminum; the detail changes listed in Table 20-5 should help a new welder to easily and successfully TIG weld copper and copper alloys.

TIG Welding Titanium

Since it is not a common metal, a new welder would probably not practice TIG welding on titanium. Table 20-6 shows the recommended TIG welding conditions for welding titanium.

Welding Machine

The welding machine should be set on straight polarity DC for TIG welding titan-

Table 20-6. Recommendations for TIG Welding Titanium.

Titanium Thickness (in.)	Type of Weld	Tungsten Electrode Diameter [1]	Filler Rod Diameter [2]	Nozzle Size-Inside Diameter (in.)	Shielding Gas Flow cfh [3]	Welding Current (amps) DC-SP [4]	Number of Passes	Travel Speed (ipm)
24 gauge	Square groove & Fillet	1/16	None	3/8	18	20- 35	1	6
16 gauge	Square groove & Fillet	1/16	None	5/8	18	85-140	1	6
3/32	Square groove & Fillet	3/32	1/16	5/8	25	170-215	1	8
1/8	Square groove & Fillet	3/32	1/16	5/8	25	190-235	1	8
3/16	Square groove & Fillet	3/32	1/8	5/8	25	220-280	2	8
1/4	Vee groove & Fillet	1/8	1/8	5/8	30	275-320	2	8
3/8	Vee groove & Fillet	1/8	1/8	3/4	35	300-350	2	6
1/2	Vee groove & Fillet	1/8	5/32	3/4	40	325-425	3	6

[1] For tungsten electrodes: 1st choice—2% Thoriated EWTh2; 2nd choice—1% Thoriated EWTh1.

[2] Use filler metal one or two grades lower in strength than the base metal.

[3] Shielding gas used is normally Argon. For higher heat input or thicker titanium, use an Argon-Helium mixture. (See text for more information about shielding gas.)

[4] Without back-up or chill bar, decrease current 20%.

ium. Titanium does not need an oxide cleaning action as do magnesium and aluminum, so there is no need to consider welding titanium with reverse polarity DC, or AC, current.

Titanium is often welded with backup or chill bars, so the recommended TIG welding amperages for titanium should be decreased 20% if titanium is to be welded without backing material.

Shielding Gas

A major difference to consider when TIG welding titanium is its need for additional shielding gas. The shielding gas must always be supplied from the TIG torch, as usual, *and* there must be additional shielding gas blowing on the part of the metal being welded from *behind* the weld, also. This makes TIG welding titanium much more complicated and expensive, compared with other metals.

Argon shielding gas is usually used alone when TIG welding thinner sheets of titanium. An argon/helium mixture may be used for higher heat input on thick pieces of titanium, if needed. Regardless of the gas used, a good deal more gas will be required to shield a titanium TIG weld.

Electrode

A thoriated tungsten electrode should be used for TIG welding titanium. The EWTh2, 2% thoriated tungsten electrode, is recommended if at all possible. If the EWTh2 electrode is not available, or is not preferred for some reason, the EWTh1 electrode may be substituted. The electrode diameter should be chosen based on the recommendations in Table 20-6.

Positions and Joints

Titanium may be TIG welded in any position where enough shielding gas is available on both sides of the weld metal. Titanium as thin as 24 gauge or as thick as 1/2" may be successfully TIG welded.

On titanium metal over 1/4" thick, a Vee-groove preparation is recommended. Back-up

Table 20-7. Recommendations for TIG Welding Silicon Bronze.

Silicon Bronze Thickness (in.)	Type of Weld	Tungsten Electrode Diameter [1]	Filler Rod Diameter[2]	Nozzle Size-Inside Diameter (in.)	Shielding Gas Flow cfh [3]	Welding Current (amps) DC-SP[4]	Number of Passes	Travel Speed (ipm)
1/16	Square groove & Fillet	1/16	1/16	1/4	15	85-125	1	12
1/8	Square groove & Fillet	1/16	3/32	1/4	15	115-150	1	12
3/16	Square groove & Fillet	3/32	3/32	5/16	20	145-195	1	9
1/4	Square groove & Fillet	3/32	1/8	5/16	25	160-225	2	
3/8	Vee groove & Fillet	1/8	1/8	3/8	25	225-290	3	
1/2	Vee groove & Fillet	1/8	1/8	3/8	25	245-295		
3/4	Vee groove & Fillet	1/8	3/16	3/8	25	295-355		
1	Vee groove & Fillet	5/32	1/4	7/16	25	295-360		

[1] For tungsten electrodes: 1st choice—2% Thoriated EWTh2; 2nd choice—1% Thoriated EWTh1.
[2] Use type RCuSIA filler rod.
[3] Shielding gas used is normally Argon. For thicker metal or for higher welding heat, an Argon-Helium mixture should be used.
[4] Increase amperage when back-up is used.

or chill bars may or may not be used, as discussed earlier. However, if they are *not* used, the current setting should be reduced about 20% because there will be no weld heat lost to the chill bar. The current recommendations assume that a chill bar will be used.

TIG Welding Silicon Bronze

Silicon bronze may also be TIG welded and has many characteristics of copper that are important for TIG welding. Table 20-7 shows the recommended information for TIG welding silicon bronze.

Welding Machine

The welding machine should be set on straight polarity DC for TIG welding silicon bronze. Silicon bronze, like titanium and carbon steels, does not need the oxide cleaning action of reverse polarity DC or AC current. Using straight polarity for silicon bronze allows narrow weld beads to be formed. This in turn, will lower the distortion and contraction stresses, helping to eliminate any

problems from hot-short cracking. Hot-short cracking can happen when a metal such as silicon bronze becomes brittle at red heat.

Shielding Gas

The recommendations for shielding gas to be used with silicon bronze welding are similar to other metals. Pure argon is generally used, except on very thick pieces. On thicker sections of silicon bronze, an argon/helium mixture may be used for better penetration, since the helium allows more arc heat to pentrate the metal. The recommended shielding gas flow rates are listed in Table 20-7.

Electrode

As with TIG welding titanium, a thoriated tungsten electrode should be used. The preferred electrode is the EWTh2, 2% thoriated tungsten. However, the 1% thoriated tungsten electrode, EWTh1, may be used as an alternative. Table 20-7 lists the recommended electrode diameters to be used when TIG welding silicon bronze.

Positions and Joints

Silicon bronze may be TIG welded in all four common welding positions, in thicknesses of $\frac{1}{16}''$ to $1''$. Butt, Tee, and lap fillet joints can be successfully welded on silicon bronze. This material is less complicated to weld than titanium because it doesn't require shielding gas on both sides of the weld. Vee-groove preparations should be used if silicon bronze metals $\frac{3}{8}''$ thick, or more, will be welded.

Special Procedures

Silicon bronzes have a tendency to hot-short (become brittle at red heat), so they should be welded with a fairly high travel speed. The welder should use as small a weld puddle as possible, keeping the inner base metal of the silicon bronze relatively cool. In no case should the inner base metal temperature go over 200° F.

As usual, welding equipment manufacturers have spent a great deal of time and money developing recommended settings and procedures for each type of metal to be TIG welded. Table 20-7 lists the recommended procedures for TIG welding silicon bronze. Following these recommendations, a new welder should be able to successfully TIG weld silicon bronze.

Manual TIG Welding Review

Many common metals besides aluminum can be successfully TIG welded after much experience and practice is done on aluminum. The following review pointers are common to most metals and are important for good TIG welding.

A. When TIG welding, dipping the rod will produce bead ripples. The number of bead ripples will be about the same as the number of times the filler rod was dipped. Dipping the rod should be practiced until a good-looking bead has been welded with good penetration.

B. If the arc will not start, first check to see if the work is grounded. If the work is grounded, the spark gap in the high frequency unit may be too wide, or the flow of shielding gas may be too small. In either case, it will be difficult or impossible to start the arc.

C. If the open-circuit voltage on the welding machine is too low, the arc direction will be unstable. An unstable arc may also be caused by the high-frequency unit not operating correctly, or by using an electrode that is not suited for the current being used.

D. Electrode contamination (other metal on the electrode) may be caused by having the electrode extended too far beyond the end of the cup. This and the wrong torch angle may cause the electrode to come into contact with the work. The contact would cause electrode contamination.

E. If the welder is welding with too low a current and/or an unstable arc, carefully dipping the filler rod will not improve the appearance of the weld bead. Or, if the filler rod and torch angle are not correct for the welding position being used, the bead surface will also appear rough.

F. Slow travel speed and/or low current settings will create wide bead profiles. Wide bead profiles can also be caused by using too long an arc or the wrong torch angle.

G. If too much heat is used, *undercut* may be a problem. This can also be caused by using the wrong arc length with the wrong torch angle.

H. *Porosity* may be caused by welding on dirty metal or cooling water leaks in the torch. Also, an air leak in the gas supply or poor quality shielding gas can cause porosity in the weld.

I. Poor penetration can be caused by using a travel speed that is too fast or by adding the filler metal before the molten pool is deep enough. As with manual-shielded arc welding, a long arc may also create poor penetration.

Table 20-8. General Conditions for TIG Welding Common Metals.

MATERIAL	WELDING PROCEDURE	WELDING CURRENT	SHIELDING GAS	TUNGSTEN ELECTRODE
ALUMINUM	Sheets, plates, castings	AC-HF medium penetration	Argon or argon and helium—argon and helium for deeper penetration and faster travel.	Pure or zirconium. Zirconium—X-ray quality welds
	Thick material only	DC-SP deep penetration	Argon or argon and helium—argon and helium preferred.	Thoriated [1]
	Thin material only	DC-RP shallow penetration	Argon	Zirconium or Thoriated [2]
CARBON STEEL	Sheets and plates	DC-SP deep penetration	Argon or argon and helium—argon and helium for extra deep penetration on heavy plate.	Thoriated [1]
	Thin sheets only	AC-HF medium penetration	Argon	Pure or zirconium. Zirconium—longer lasting
COPPER [3]	Sheets and plates	DC-SP deep penetration	Argon or argon and helium—argon and helium preferred for heavy material.	Thoriated [1]
	Very thin material only	AC-HF medium penetration	Argon	Pure or zirconium. Zirconium—longer lasting
COPPER ALLOYS	Material thicker than 0.050	DC-SP deep penetration	Argon or argon and helium—argon ond helium preferred for heavy material except beryllium copper.	Thoriated [1]
	Material thinner than 0.050 Beryllium copper all thicknesses	AC-HF medium penetration	Argon	Pure or zirconium. Zirconium—longer lasting
MAGNESIUM	Sheets, plates, castings	AC-HF medium penetration	Argon	Pure or zirconium. Zirconium—X-ray quality welds
	Thin sheets only	DC-RP shallow penetration	Argon	Zirconium or Thoriated [2]
NICKEL MONEL INCONEL	All Thicknesses	DC-SP deep penetration	Argon	Thoriated [1]
STAINLESS STEEL	Sheets, plates, castings	DC-SP deep penetration	Argon or argon and helium—argon and helium for extra deep penetration on thick material.	Thoriated [2]
	Thin sheets only	AC-HF medium penetration	Argon	Pure or zirconium. Zirconium—longer lasting
TITANIUM	All thicknesses	DC-SP deep penetration	Argon	Thoriated [1]

Courtesy of Hobart Brothers Company.

[1] Grind end to point or near point.
[2] Use with balled end. Slowly increase welding current until ball forms.
[3] Use brazing flux on ¼" or thicker.

J. Weld burn-through is usually caused by a combination of poor fit-up, large root gaps, and a welding current that is too high. Also, an arc that is too short and/or a travel speed that is too slow could cause burn-through.

K. The welder should use as small a shielding gas nozzle as possible, especially where it is hard to see the molten puddle. To produce a good TIG weld bead, the welder must always be able to see the arc and weld pool.

TIG welding any metal has the advantage of the weld and molten metal being shielded from the atmosphere by a "cloud" of inert shielding gas. Since the gas is invisible, the welder can still see the arc and molten pool. Welding conditions vary slightly for TIG welding different metals even though the basic idea is the same. For this reason, a welder must refer to TIG welding tables for each specific metal after he has referred to general specifications, such as Table 20-8.

Fig. 20-11 shows a modern TIG welding production setup. Here, telephone bellboxes are being TIG welded in a small production fixture and the TIG torch is held on a hydraulic piston.

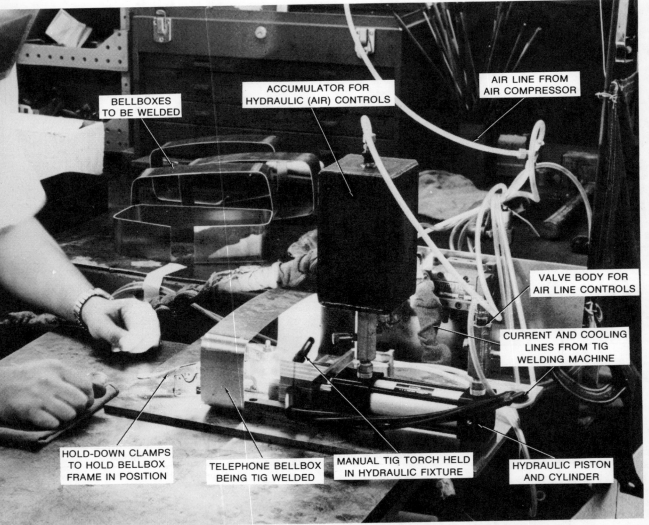

Fig. 20-11. A small production TIG welding setup. (Courtesy of Aircom Metal Products, Inc.)

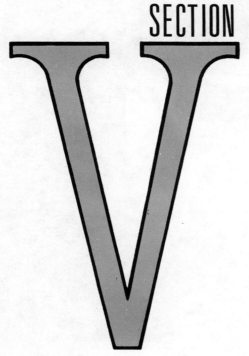

other welding processes

The MIG welding process being used to weld iron plate in a Tee joint.
(Courtesy of Aircom Metal Products, Inc.)

21

Flux-Cored Arc Welding

Flux-cored arc welding is one of the newer shielded arc welding methods. Unlike TIG welding, it uses a consumable electrode and is usually shielded by carbon dioxide gas. Used mostly on production work, the flux-cored arc welding process gets its name from the fact that the electrode wires have a flux *core,* that is, the inside of the electrode wire is packed with flux. The flux-cored electrode wire is a continuous wire wound on spools near the welding machine and is automatically fed into the arc and molten pool to produce the welding process shown in Fig. 21-1.

How It Works

Flux-cored arc welding uses both a shielding gas *and* a flux to produce very clean, sound welds with deep penetration. A complete flux-cored arc welding setup is going to need these three things. See Fig. 21-2.

1. An arc welding machine that is able to produce high currents.

2. A supply of shielding gas to the welding torch.

3. A supply (reel) of flux-cored welding wire, delivered to the torch constantly and automatically.

Wire Electrode and Arc—With flux-cored arc welding, the arc welding equipment pulls the flux core electrode wire off of a spool that is stored on part of the welding equip-

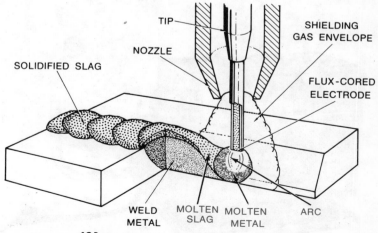

Fig. 21-1. The flux-core arc welding process. (Courtesy of Hobart Brothers Company.)

SOLIDIFIED SLAG

TIP

NOZZLE

SHIELDING GAS ENVELOPE

FLUX-CORED ELECTRODE

WELD METAL

MOLTEN SLAG

MOLTEN METAL

ARC

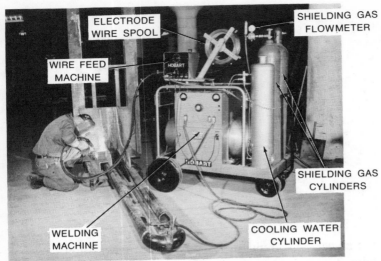

ELECTRODE WIRE SPOOL

SHIELDING GAS FLOWMETER

WIRE FEED MACHINE

SHIELDING GAS CYLINDERS

WELDING MACHINE

COOLING WATER CYLINDER

Fig. 21-2. A water-cooled flux-core arc welding machine. (Courtesy of Hobart Brothers Co.)

ment. See Fig. 21-3. Then, automatic drive wheels *push* the flux-cored electrode wire down the electrode cable to the torch. When the wire gets to the end of the torch, it is melted and becomes part of the weld joint if an arc is working.

The flux-cored electrode wire receives the arc current supplied by the welding machine at the welding gun. The electrode feed rate out of the torch and the welding current are both adjusted by the welder on either the welding machine or the wire feed equipment.

Shielding Gas—Also traveling to the torch inside the welding cable is the *shielding gas.* When the shielding gas gets to the torch tip,

it is directed around the wire electrode. This way, the flux-cored wire electrode is completely surrounded and protected by shielding gas as soon as the electrode leaves the torch tip.

The shielding gas used is usually pure, welding-grade carbon dioxide. In some cases, a mixture of carbon dioxide and argon may be used. In either case, the cfh flow of shielding gas is adjusted at the flowmeter as it was for TIG welding. As usual, manufacturer's specification charts will list the recommended shielding gas and cfh flow along with the other important adjustments for a given flux-cored, arc welding setup.

Fig. 21-3. The flux-core electrode wire is mounted on spools on the flux-core arc welding machine. (Courtesy of Hobart Brothers Company.)

Basic Adjustments—For flux-cored arc welding, then, the welder will have to make three basic adjustments on his equipment as he sets up to weld:

1. He will have to adjust the welding machine *arc current*.
2. He will have to adjust the speed of the *electrode wire feed* to the torch handle.
3. He will have to adjust the rate of *shielding gas flow* to the torch handle.

Flux-cored arc welding torches may be water-cooled. If the torch is water-cooled, the welder will need a permanent or portable water supply for the cooling water. He may

of the shielding gas and the deoxidizing and cleaning action of the flux core. The flux-core elements also act as an arc stabilizer, resulting in smooth welds with a good bead contour. Fillet welds with equal legs can be easily made in the horizontal position, free of undercut and completely fused. See Fig. 21-5.

Range of Thickness—Most steels, including the high-carbon steel shown in Fig. 21-6, can be welded in a wide range of thicknesses from ⅛″ to over 2″. The welding speed may be almost twice as fast as the manual-shielded arc process because of the steady electrode wire feed and the small,

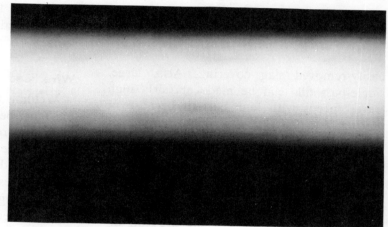

Fig. 21-4. An X-ray of strong weld metal welded with the flux-core arc welding process. (Courtesy of Hobart Brothers Company.)

regulate the cooling water supply on the welding machine if the machine is built with an adjustment for the cooling water. Flux-cored arc welding torches may also be called *guns*.

Advantages of Flux-Core Arc Welding

The flux-cored arc welding process has many advantages over manual shielded arc, submerged arc, and gas metal-arc (TIG and MIG) welding processes.

Strong Welds—Welding with the flux-core arc process assures the operator of strong weld deposits that are able to pass difficult root and face-bend tests. The deposited weld metal is low in hydrogen and it will pass free-bend and ultrasonic tests. The high-quality weld shown in Fig. 21-4 was produced by the flux-core arc welding process. Most of its strength is due to the protection

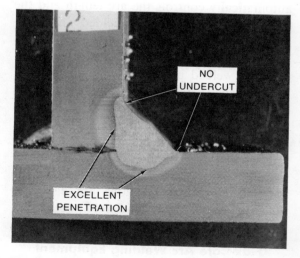

Fig. 21-5. A fillet weld made with the flux-core arc welding process. (Courtesy of Hobart Brothers Company.)

Fig. 21-6. High carbon steel, photographed through a microscope. (Courtesy of Bethlehem Steel.)

easily-removed slag covering. Also, large, single-pass fills can be made at fairly high travel speeds with high current because of the high melting rate in the arc. Horizontal fillets up to 3/8" and flat fillets up to 5/8" can be made in one pass; this will increase production while lowering cost.

Puddle Control—The welder has complete control over the weld puddle in the flux-core arc welding process, since the arc can always be seen. Because the welding machine automatically controls the arc length and the arc is similar to manual-shielded arc welding, most welders will need only a little extra practice in order to run a satisfactory flux-core arc welded bead.

Reduced Distortion—Another advantage of the process is that distortion is reduced in the weldment because of the higher travel speed. Although it is possible to use a high travel speed with the flux-core process in all welding positions, most welders find that the flat and horizontal positions are easier to weld, due to the highly fluid weld pool.

Flux-Core Arc Welding Equipment

As can be seen in Fig. 21-2, many pieces of equipment are needed for the flux-core

arc welding process. The complete setup uses many separate units that are connected to make up the complete flux-core arc welding station. The unit may be fixed or portable.

Power Source

The flux-core arc welding process must have a constant-voltage (CV) power source, either a rectifier machine or a direct current generator machine. If a direct current generator machine is used, it should be set on reverse polarity. Constant-voltage welding machines used for flux-core arc welding range in size from 300 to 900 amperes with voltage outputs of from 25 to 50 volts. All power sources used for this type of welding are rated from 80% to 100% duty cycle.

Wire Feed System

Welding machines for flux-core arc welding must be equipped with a 110-volt alternating current supply to operate the wire feeder. The wire feeder automatically pulls the electrode wire from a spool or coil and then pushes it through the cable assembly and welding gun into the arc. A steady wire feed is needed with a constant-voltage system, and it must be adjustable for different welding currents. See Fig. 21-3.

Drive Rolls—Flat, knurled-Vee, or gear-type drive rolls may be installed in the wire feed machine, depending upon the size of the wire. See Fig. 21-7. Gear-type drive rolls are

KNURLED VEE

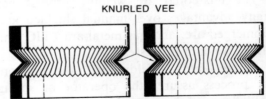

Fig. 21-7. Typical knurled-vee drive rolls for a wire-feed mechanism.

used for large-diameter wires and knurled-Vee rolls are used for medium-size wire. Small-diameter wires may be moved with flat drive rolls.

Wire feeders are usually designed to be operated away from the welding machine. This makes the feeder portable, allowing the welder to cover a large area if needed.

Special Application—There are some applications of flux-core arc welding that will need a microwire, MIG-welding root pass before starting to weld the joint with the flux-core process. A special wire feeder that can feed *two* different electrode wires has been developed for this job. This unit has two wire feed motors, two gun and cable assemblies, and two electrode wire reels. By flipping a switch and picking up the correct welding gun, the welder can use either welding process. See Fig. 21-8.

3. Open the shielding gas and water solenoid valves.

By controlling these factors in order, the circuit will provide the necessary control over *all* the flux-core arc welding equipment.

The control is energized by a single starting switch. It may be the trigger on a semi-automatic welding gun or the start button on an automatic welding machine. In a constant-voltage matched system, the wire feeder connects to the power source with cables.

In semiautomatic flux-core arc welding, the wire feeder gives the welder finger-tip control over the wire feed, gas flow, and welding current, right at the welding gun. A built in circuit-braking switch will stop the wire

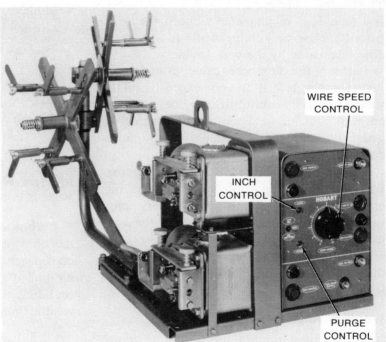

Fig. 21-8. A double-wire feed machine. (Courtesy of Hobart Brothers Company.)

WIRE SPEED CONTROL

INCH CONTROL

PURGE CONTROL

Wire Feed Control Circuit

The wire feed system has a control circuit to start and stop the wire feed drive motor, operate the welding power contactor, and energize the shielding gas and water solenoid valves. With a time delay action, this control circuit will do three jobs in order:

1. Turn on the wire feed drive motor.
2. Turn on the arc welding power.

feed as soon as the welder releases the gun trigger switch.

Purge and Inch Controls—Two buttons on the wire feeder are labeled *purge* and *inch*. They allow the welder to purge the line and control the wire feed. See Fig. 21-8. When the *purge* button is pushed, it will remove air from the shielding gas line by pushing shielding gas through to the gun nozzle without feeding any electrode wire. The *inch* button allows the electrode wire to be fed through the gun or torch before welding. This helps the welder test the drive tension on the wire feed rolls.

Welding Cable Assemblies

The cable assembly carries the electrode wire, welding current, shielding gas, and cooling water from the power source and wire feeder to the welding gun. The parts of the cable assembly are protected in either a molded jacket or a cable sheath.

Flexible Liners—Flexible liners, made of metal or plastic, may run the full length of the cable from the wire feeder to the gun. The flexible liner must be used for electrodes less than $\frac{1}{16}''$ in diameter to keep the wire from buckling or kinking while it is pushed down the cable by the wire feeder.

Because the flux-core cable assembly is very complicated, it is very expensive. The welder has to be careful that he does not step on, kink, cut, or weld too close to the cable assembly. Any of those could easily cause costly damage.

As the welding current is increased, the size of the welding cable must also be increased, especially if the cable is very long before it gets to the wire feeder and then on to the welder's gun. Table 21-1 gives the recommended cable size in relation to the distance between the power source and the wire feeder.

Welding Torches (Guns)

The welding torches used with flux-core arc welding are usually called *guns*. This is because the electrode wire is pushed out of the welding gun, similar to the way that a bullet would be pushed out the barrel of a normal gun.

Gun Cooling—Both water- and air-cooled welding guns are used with the flux-core arc welding process. The type of gun to use depends on how much welding current is needed for the job. Water-cooled guns are usually

Table 21-1.   Recommended Sizes for the Welding Cable from the Welding Machine to the Wire Feeder.

Welding Current (Amperes)	Distance from the Welding Machine to the Wire Feeder (Feet)			
	50	100	150	200
100	1	1	0	0
150	1	1	00	0000*
200	1	0	0000	0000
300	0000	0000		
400	0000	0000		
500	0000(2R)	0000(2R)		

* This may also be written as 4/0.

needed for welding with currents over 500 amperes. Fig. 21-9 shows the inside of a water-cooled gun used for flux-core arc welding.

COOLING WATER IN
COOLING WATER OUT
SHIELDING GAS

POWER CABLE

Fig. 21-9. Water-cooled gun used for flux-core arc welding. (Courtesy of Hobart Brothers Company.)

Carbon dioxide gas is cold before it enters the arc, and so it will help to cool an air-cooled gun. Even so, the *main* cooling for an air-cooled gun comes directly from the surrounding air. Fig. 21-10 shows an air-cooled welding gun used for welding with current up

Fig. 21-10. Air-cooled gun used for flux-core arc welding with no more than 300 amps. (Courtesy of Bernard Welding Accessories.)

to 300 amperes. Fig. 21-11 shows a large, water-cooled welding gun with an outside shielding-gas line.

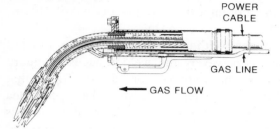

Fig. 21-11. A large, water-cooled welding gun. (Courtesy of Hobart Brothers Company.)

Gun Assembly—The gun assembly must deliver the electrode wire, shielding gas, and welding current from the power source and the welding cable. Hollow tubes of copper alloy, threaded on one end, are screwed into the hollow gun nozzle. They are commonly known as welding *tips*. These tips usually have the electrode size they will take (their inside dimension, bore size) stamped on the tip. Of course, only the correct diameter tip should be installed for the electrode wire diameter being used. Tip cleaners are available to clean the tips for free, even wire feeding.

Straight- and curved-nozzle guns are available, making it easy to weld in hard-to-reach places. The tips on these nozzles may be 4″ to 12″ long and are available for different models of welding guns.

Shielding Gas

The shielding gas has to protect the arc and weld puddle from atmospheric contamination. Carbon dioxide is a cheap, nonflammable (will not burn) gas that works well as a shielding gas for flux-core arc welding.

However, when carbon dioxide is brought to a very high temperature near the welding arc, it becomes active and breaks down into carbon monoxide and oxygen. When this happens, the free oxygen will combine with the other elements in the weld pool, causing them to oxidize.

Because of this, the electrode's flux core must contain deoxidizing elements, such as silicon, aluminum, and/or manganese. These deoxidizing elements have a great attraction for the free oxygen and they will readily combine with it. This helps prevent the oxygen from combining with the carbon or iron in the weld metal, which would give low-quality welds.

Regulating Shielding Gas—For flux-core arc welding, the welder must connect two or more standard carbon dioxide gas cylinders to a manifold for enough shielding gas to maintain the required shielding gas flow rate of 30 to 45 cfh. A calibrated carbon dioxide *flowmeter*, with a needle valve, is required to control the carbon dioxide shielding gas. The *regulator* reduces the gas cylinder pressure to a usable pressure at the flowmeter thus regulating the gas to a constant flow at any desired rate.

Flux-Core Electrode Wire

The flux-core electrode wire is made like a long, mild steel tube that surrounds the core of flux and/or alloy compounds. See Fig. 21-12. The fluxing compounds are chosen for good weld results on whatever type of

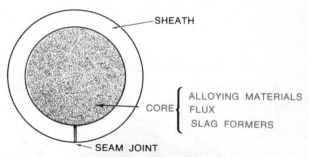

Fig. 21-12. Cross-section of the flux-core electrode.

metal is to be welded with that particular electrode wire. For this reason, different wires may have both different fluxes and different metals making up the wire. The compounds inside the electrode wire do about the same thing as does the coating on a manual shielded arc welding electrode. They act to protect the weld from oxidizing, stabilize the arc, and form inert gas for the best possible shielding.

Electrode wires are available for welding mild steel, low-alloy steel, chrome-moly steel, and hardsurfacing. They are wound around coils or spools, as in Fig. 21-13. As usual, manufacturers publish complete welding charts for each type of flux-core welding. Table 21-2 gives the specifications for the flux-core electrode wire used to weld mild steel. Other tables are available for flux-core arc welding other metals.

smoke and fumes as does manual shielded arc (stick electrode) welding. As mentioned earlier, the smoke and fumes may be irritating, although they are not necessarily harmful.

When welding on zinc- or cadmium-plated steels, however, the welder needs to guard against the poisonous or harmful fumes that these metals give off. It would be a good idea to wear a respirator while welding metals with these coatings. A respirator is a breathing unit to filter out the poisonous gas and

Table 21-2. Recommendations for Flux-Core Arc Welding Mild Steel.

Identification	Welding Conditions			Test Requirements (as Welded)					
				All Weld Metal					
Welding Wire AWS Classification	Current Polarity	Gas Shield	X-Ray Test	Tensile Strength (psi)	Yield Strength (psi)	Elongation %	Transverse Tension (psi)	Guided Bend Test	Impact Test Charpy V
E-70T-1	DC Reverse	Carbon Dioxide	Required	70,000	60,000	22	––	––	20 @ 20° F.
E70T-2	DC Reverse	Carbon Dioxide	Not Required	––	––	––	70,000	Required	Not Required
E70T-3	DC Reverse	None	Not Required	––	––	––	70,000	Required	Not Required
E70T-4	DC Reverse	None	Required	70,000	60,000	20	––	––	Not Required
E70T-5	DC Reverse	Carbon Dioxide	Required	70,000	60,000	22	––	––	20 @ −20° F.
		None	Required	70,000	60,000	20	––	––	

Flux-Cored Arc Welding Safety

Flux-cored arc welding, using a shielding gas, gives off about the same amount of

Fig. 21-13. A spool of electrode wire.

fumes from any welding process. If a respirator is not available, the welding area must be well ventilated with exhaust fans.

Carbon Monoxide—The poison gas *carbon monoxide* is formed when the carbon dioxide shielding gas breaks down at the extreme arc heat, as explained earlier. Fig. 21-14 shows the typical carbon monoxide concentration at certain distances from the carbon-dioxide-gas-shielded welding arc.

The welder will not usually be affected by the carbon monoxide if he operates the welding gun while keeping his face at least 4″ from the smoke cone. For long periods of constant welding, the welder should remain at least 7″ from the smoke cone. When using this and other welding processes, local ex-

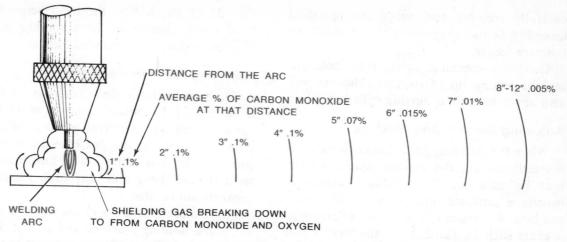

DISTANCE FROM THE ARC

AVERAGE % OF CARBON MONOXIDE
AT THAT DISTANCE

8"-12" .005%

7" .01%

6" .015%

5" .07%

4" .1%

3" .1%

2" .1%

1" .1%

WELDING
ARC

SHIELDING GAS BREAKING DOWN
TO FROM CARBON MONOXIDE AND OXYGEN

Fig. 21-14. Carbon monoxide concentration at certain distances from the carbon dioxide shielded arc. (Courtesy of Hobart Brothers Company.)

haust fans or other ventilating systems should be used.

When using flux-cored arc welding with carbon dioxide shielding gas in areas with poor ventilation, the welder should use a self-contained respirator.

Other Safety Pointers—As with any welding process, the welder should be prepared to face bright light, heat, and hot sparks. These precautions are outlined in the chapter on Welding Safety. Generally speaking, the welder in Fig. 21-2 is well equipped for welding with the flux-cored arc welding process. For the safest possible welding, he should not wear trousers with cuffs, boots with a welt bead around the top of the toe, or shirts that do not have flap-covered pockets. Even with a welding process as carefully controlled as is flux-cored arc welding, hot sparks and slag may still get caught in pant cuffs, shirt pockets, or eventually ruin the top of a good pair of boots.

Setting Up for Flux-Cored Arc Welding

In order to produce high-quality welds, the flux-cored arc welding process, like any other welding process, must be correctly applied and carefully controlled. Although joints designed for other types of welding may be

welded with this process, it is a good idea to modify these joints, if possible, to get the best results. The following information should help a new welder correctly prepare for flux-cored arc welding.

Locating Specifications—As with any other welding process, many adjustments have to be made before the welder will be able to make a successful weld with the flux-cored arc welding process. A chart such as Table 21-3 will give a welder all the information he needs to properly set up the equipment.

Shielding Gas Cylinders

The first step in setting up the equipment is to connect the shielding gas cylinders to the wire feed unit. From the wire feed unit, the shielding gas will be distributed down the welding cable. Here, common safety rules apply when handling the gas cylinders.

In brief, the cylinders must always be kept upright and moved around carefully. After the protective valve caps have been unscrewed, the valves must be "cracked" before the regulators are attached, to clean out the seat. No oil must be used on the regulators and flowmeters, as was outlined with oxygen cylinders. Because two or more carbon dioxide cylinders must be used for enough shielding gas flow, care must be taken to carefully connect and purge the manifold assembly before connecting the manifold to the wire feeder.

Generally speaking, safety rules concerning the setting up of oxygen cylinders will also apply to carbon dioxide cylinders.

Adjusting the Welding Machine

When the shielding gas cylinders have been correctly set up, the welding machine itself may be adjusted. The welding machine is usually a separate unit from the wire feed machine, as seen in Fig. 21-2. By referring to a chart such as Table 21-3, the welder can then set the welding machine voltage for his

Fig. 21-15 shows how the bead height and bead width change as the amperage is increased in the recommended ranges.

Adjusting the Wire-Feed Machine

The wire-feed machine, as shown in Figs. 21-2 and 21-3, actually controls more than the wire feed. The wire-feed machine actually does many jobs in the flux-core arc welding process. The jobs that the wire feed machine must do are listed and may each require a separate adjustment.

A. It must take the *welding current* from the welding machine and pass it down the cable assembly to the gun.

Table 21-3.   Recommendations for Flux-Core Arc Welding Mild Steel with Backup Strips in the Flat Position.

Metal Thickness (Inches)	Type of Joint	Number of Passes	Root Gap Opening (Inches)	Electrode Diameter (Inches)	Welding Power (DC-RP)		Travel Speed (Each Pass, ipm.)
					Volts	Amps	
1/8	Square	1	1/32	3/32	24-26	325	56
3/16	Square	1	1/16	3/32	24-26	350	48
1/4	60° Vee	1	0	3/32	25-27	375	41
3/8	60° Vee	1	0	1/8	25-27	500	24
1/2	60° Vee	1	0	1/8	27-30	550	14
5/8	60° Vee	1	0	1/8	27-30	550	9
3/4	60° Vee	3	0	1/8	27-30	550	18
7/8	60° Vee	4	0	1/8	27-30	550	11
1	60° Vee	6	0	1/8	27-30	550	11

Courtesy of Hobart Brothers Company.

particular welding job, if the machine is equipped with a voltage control. He should also adjust the machine for the recommended current and polarity. Usually, flux-core arc welding requires reverse-polarity DC.

Both bead height and bead width are affected by the welding amperage being used.

B. It may or may not provide an adjustment for the *arc voltage*. This will usually be adjusted at the welding machine and is called the arc voltage adjustment.

C. It must pull the electrode wire from the welding wire spool and push the wire

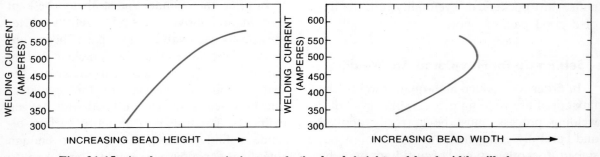

Fig. 21-15. As the amperage is increased, the bead height and bead width will change.

down the cable assembly. This adjustment is called the *wire feed speed* and must be adjusted by the welder for each different welding job.

D. It must control the flow of *shielding gas* through the welding cable. While the flowmeter controls how much shielding gas flows, the wire-feed machine controls whether or not the gas flows on to the welding gun.

E. It must control the flow of *cooling water* through the cable assembly to the welding gun and back to the drain or cooling tank. This may also be an "on-off" control switch.

When all of the connections have been made, tightened, and checked, the welder is ready to turn on the welding machine to open circuit voltage and he should adjust the wire-feed machine carefully, as follows.

Shielding Gas—With the gun trigger depressed and the shielding gas valve open, the welder can first adjust the shielding gas flow. The wire feed machine is usually equipped with a shielding gas "on-off" solenoid valve.

With the valve on, if equipped, the flowing shielding gas is adjusted to 30-35 cfh flow at the flowmeter. The gas connections must be tight and the gun trigger open for the adjustment. Smaller diameter electrode wires will need only 30 cfh, while larger diameter electrode wires will need the heavier 35 cfh.

Wire Feed Speed—Many times, specification tables will not give the wire feed in inches per minute (ipm). Often, wire feed must be adjusted by the welder after a few trial-and-error welds. During the practice welds, the wire feed is too fast if the electrode is being pushed down into the work, welding itself to the work piece. On the other hand, the wire feed is too slow if the electrode disappears up into the shielding gas nozzle, breaking the arc. The wire feed will depend a great deal on the welder's own technique, the welding job being done, the amperage being used, and the electrode wire diameter.

The *approximate* wire feed speed may be determined by comparing the welding current and electrode wire diameter with graphs that manufacturers have calculated for determining the wire feed speed. Fig. 21-16 shows a graph for finding the approximate wire feed speed in inches per minute (ipm) for the welding currents shown. Since thicker electrode wires will be harder to melt, it can be seen that they will need slower speeds and higher amperages than thinner electrode wires.

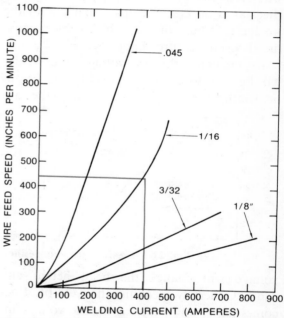

Fig. 21-16. The welding current and electrode wire diameter help the welder decide what wire-feed speed to use. (Courtesy of Hobart Brothers Company.)

As an example, a chart might call for 400 amperes to be used with a $\frac{1}{16}$″ electrode welding wire. To find out what *approximate* wire speed to use, the welder would draw a line *up* from 400 amperes to the $\frac{1}{16}$″ electrode curve in Fig. 21-16. Then, he would go *left* from the point where the 400-ampere line crossed the $\frac{1}{16}$″ welding wire curve. The line going left crosses the wire feed speed line at about 435 inches per minute. See Fig. 21-16. The 435 ipm would be an average, approximate, setting, as mentioned earlier.

When a welder was set up for flux-core arc welding practice, he would only *start* with a wire-feed speed of 435 ipm and then adjust the feed speed either up or down, depending on the results he had with the 435 setting. In some cases, he might find the 435 setting to be "just right," and wouldn't need to change it.

Adjusting the Arc Voltage—After adjusting the wire-feed speed, the welder must adjust the voltage *rheostat* on the front of the wire-feed machine for the right voltage. A rheostat is an electrical part that allows the welder to adjust the arc voltage by turning the rheostat's control knob. See Fig. 21-3.

By changing the arc voltage, the welder will be able to change the weld bead height or width. Fig. 21-17 shows how the arc

welder finds that the arc is uneven, the wire-feed speed should be adjusted to a new current setting. Also, readjusting the voltage control to a different setting may help the welder smooth out the arc.

Adjusting the Travel Speed

The travel speed is one adjustment that is not on either the welding machine or the wire-feed machine. The travel speed refers to *how fast* the welder moves the welding gun along the weld path. The travel speed, then, is one adjustment that cannot be made until the welder is actually welding.

The travel speed is usually given on most specification charts, such as Table 21-3. The travel speed is listed in inches per minute, or ipm. Usually, it will be hard for the welder

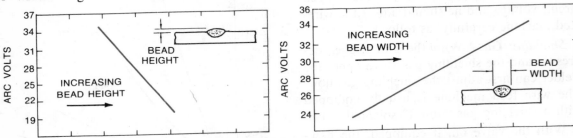

Fig. 21-17. As the arc voltage increases, the bead becomes flatter and wider.

voltage will change the bead height and width. Generally speaking, the weld bead becomes *flatter* and *wider* as the arc voltage increases. See Fig. 21-17.

Although the arc voltage setting depends on the welding conditions, it is usually from 23 to 32 volts, as seen in Table 21-3. A quick

to time himself and measure the inches while he is welding, so he should try a few "dry runs" by moving along the weld path and seeing how long it takes at certain speeds without the arc.

Travel speed will definitely affect the weld bead and height, as seen in Fig. 21-18. Gen-

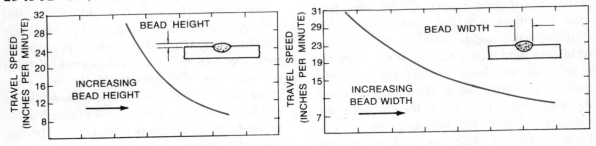

Fig. 21-18. As the travel speed increases, the bead size gets smaller.

practice weld can help show that the wire feed speed and voltage adjustments are giving the right current for a smooth arc. If the

erally speaking, the bead gets *flatter* and *narrower* as the travel speed is increased. Overall, the bead simply gets smaller as the welder increases his speed above the recommended travel speed. See Fig. 21-18.

Welding Gun Angle

When flux-core arc welding, the gun nozzle must be held at one of three positions (angles). Which nozzle position to use depends on the type of joint being welded. During simple forehand welding, the nozzle will usually be held at a 90° angle to the work.

Trailing and Leading Angles—Angles known as *trailing* and *leading* will be useful for hard-to-reach places. The exact angle of either the trailing (pulling) angle or leading (pushing) angle will also depend on the type of welding being done. See Fig. 21-19.

height decreases and the width increases. This will keep happening as the welder moves the gun from the trailing angle range into the leading angle range. When the leading angle is increased too far, the bead will start to become narrow again.

Stickout

Before striking the arc, a new welder needs to know about electrode *stickout*. See Fig. 21-20. Stickout is the *distance* between the *tip* of the gun contact tube and the end of the electrode, before the electrode is melted. Stickout will affect the amount of electrode wire that is preheated (the end of the electrode wire to the contact tube).

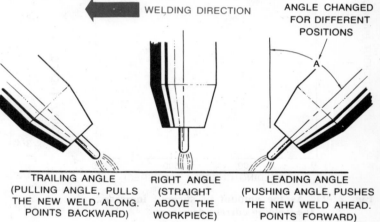

Fig. 21-19. Leading and trailing angles for flux-core arc welding.

The best penetration is obtained when a trailing angle of 10° is used. When the nozzle angle is changed from a 10° trailing angle, penetration decreases. From a trailing angle of 10° to a leading angle of 30°, the relationship between penetration and nozzle angle is almost a straight line. Because of this, the welder can get the best control of penetration in this range. It is not a good idea to use trailing angles greater than 25° during flux-core arc welding.

Changing the nozzle angle can also be used to change the bead height and bead width, because the nozzle angle will change the bead contour (shape). A trailing nozzle angle tends to produce a high, narrow bead. As the trailing angle is reduced, the bead

Before striking an arc, the welder should trigger the welding gun to extend (push) the electrode wire out of the contact tube about 1″. If the wire accidentally pushes out more

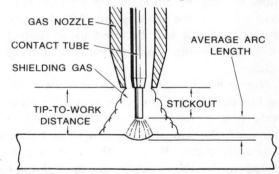

Fig. 21-20. The electrode stickout is the amount of electrode actually exposed from (sticking out) the end of the contact tube.

than 1″, wire cutters can be used to cut the electrode wire off at the desired length.

The amount of stickout and the wire feed speed both affect the weld penetration because they change the welding current slightly. See Figs. 21-16 and 21-21. Fig. 21-21 shows how the stickout affects welding currents in the 375-525-amp range. Increasing the stickout can reduce the welding current by almost 100 amperes. This, of course, will reduce the penetration.

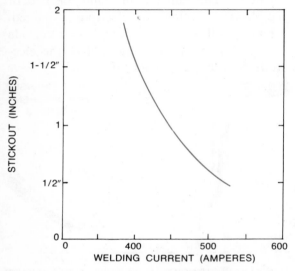

Fig. 21-21. As the amount of stickout increases, the welding current decreases.

With semiautomatic flux-core arc welding, stickout can be adjusted *during* welding. This adjustment makes it easy for the welder to allow for any variation in the joint without having to stop the welding arc. Also because of this, stickout can be used as a control for changing the weld penetration during welding.

How it Works—Stickout changes the welding current by changing how much the electrode wire is *preheated*. As the stickout is *increased*, the wire is preheated more before it actually melts. Because the wire is preheated more, the welding machine doesn't need to put out as much welding current to melt the wire at the feed rate being used. Due to the self-regulating nature of the constant

voltage welding machine, the *overall* result is a decrease in weld current. Since the weld current decreases, the weld penetration also decreases.

On the other hand, as the stickout *decreases*, the wire preheating also decreases. When this happens, the welding machine is forced to put out *more* current to melt the wire at the feed rate being used, because the wire is not preheated as much. The increase in welding current then causes an increase in penetration. Decreased stickout, therefore, will also increase the weld deposit rate. See Fig. 21-22.

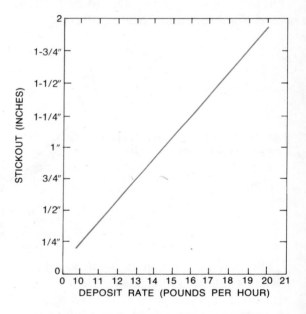

Fig. 21-22. As the stickout is increased, more weld metal is deposited because the electrode wire has received more preheat.

Using Flux-Core Arc Welding

Flat Position—Flux-core arc welding with carbon dioxide shielding gas can be used on joints with an included angle as small as 30° before a root opening is needed. For the best penetration with this joint, however, the welder will have to use a high current and fast travel speed.

A ³⁄₃₂″ diameter electrode wire with about 500 amperes current will penetrate well into

Table 21-4.   Recommendations for Flat Position Groove Welds Using Flux-Core Arc Welding on Mild Steel Plate.

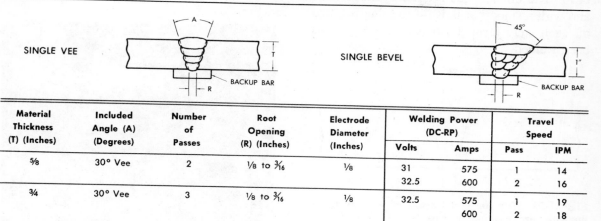

Material Thickness (T) (Inches)	Included Angle (A) (Degrees)	Number of Passes	Root Opening (R) (Inches)	Electrode Diameter (Inches)	Welding Power (DC-RP)		Travel Speed	
					Volts	Amps	Pass	IPM
5/8	30° Vee	2	1/8 to 3/16	1/8	31	575	1	14
					32.5	600	2	16
3/4	30° Vee	3	1/8 to 3/16	1/8	32.5	575	1	19
						600	2	18
						600	3	15
1	30° Vee	4	1/8 to 3/16	1/8	31.5	575	1	21
					32	575	2	11
					32	575	3	15
					32	575	4	12
5/8	40° Vee	2	0 to 1/16	1/8	32	575	1	16
					32	600	2	13
3/4	40° Vee	3	0 to 1/16	1/8	32	575	1	23
					32	575	2	18
					32	600	3	15
1	40° Vee	4	0 to 1/16	1/8	31	575	1	15
					31	575	2	13
					31	575	3	15
					32	600	4	12
1	45° Single Bevel	7	1/8	1/8	32	600	1	17
					32	600	2	24
					32	600	3	18
					32	600	4	15
					32	600	5	16
					32	600	6	21
					28	400	7	4-5

Courtesy of Hobart Brothers Company.

the "insulating" effect of the molten weld pool on thick pieces. Thicker weld beads can be deposited, or narrower included angles used, if a root gap is used in the weld joint. See Table 21-4.

Flux-core arc welds will usually bridge the gap in a joint making it possible to weld many joints without using a back–up strip. The recommended preparation for this type of weld is shown in Fig. 21-23. Standard flux-core arc welds will bridge gaps with slightly less than full penetration. However, when complete penetration is needed, it is hard to control the difference betwen full penetration and burn-through. When complete penetration is needed without back gouging, then, back–up bars are recommended.

Horizontal Position—When welding *horizontal* joints, weld the joint with several bead layers. A single bevel, prepared as shown in Fig. 21-24, will help to support the beads and counteract (work against) the pull of gravity. The first bead is welded at the lowest point in the root gap. Then, the second bead is welded next to the bottom of the first bead. The other beads can be applied easier and with a smoother finish if the slag is taken off each weld layer as soon as the layer has been welded.

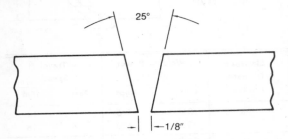

Fig. 21-23. With the metal prepared as shown, the flux-core arc welding process would normally not need a backup bar.

Vertical Position—Vertical butt joints should be welded with a $\frac{1}{16}$" or 0.045" diameter electrode wire. The weld should be made from the bottom of the joint to the top, as for manual shielded arc welding.

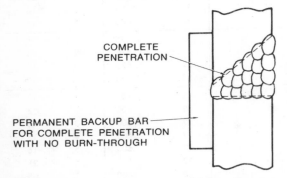

Fig. 21-24. Using a permanent backer for complete penetration.

Overhead Position—When using the flux-core arc welding process in the *overhead* position, use specially designed electrodes. These overhead-position electrode wires contain little flux and are made around a core that contains mostly metal powders and alloys. Back-up strips should be used on all overhead joints with the welding technique being similar to the manual shielded arc process described earlier.

In all flux-core arc welding positions, the gun nozzle must be held at a right angle to the

plate, with the welder moving the gun in the form of a triangular weave pattern, as shown in Fig. 21-25.

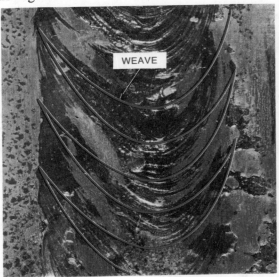

Fig. 21-25. A flux-core arc weld bead with the gun nozzle straight above the weld surface. (Courtesy of Hobart Brothers Company.)

Flux-Core Arc Welding Prepared Joints

Single-Vee—When flux-core arc welding must be done from one side only, the *single-Vee* joint is the best joint to use. If complete penetration is needed, use a backup strip with the single-Vee preparation. This would eliminate the need for back gouging, grinding, or chipping before welding from the other side.

Double-Vee—The *double-Vee* preparation is more economical than the single-Vee for welding plates over 1" thick. Less filler metal is needed in this type of joint and this saves time while giving complete joint fusion. Distortion can be controlled on a double-Vee welded joint if first one side and then the other is welded in turn, as was discussed earlier.

Fillet Welds—These welds, when applied with the flux-core arc welding process, blend smoothly into the base plate. For this reason, many welders prefer this process for welding lap, corner and Tee joints. Penetration with this method is greater than with the manual shielded arc welding process. Because of this, the welder can reduce the fillet size and still have good joint strength. See Figs. 21-26 and 21-27.

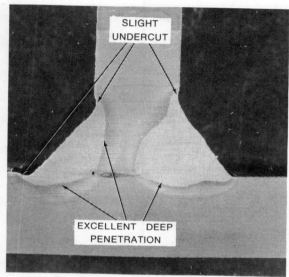

Fig. 21-26. A fillet weld made with the flux-core arc welding process. (Courtesy of Hobart Brothers Company.)

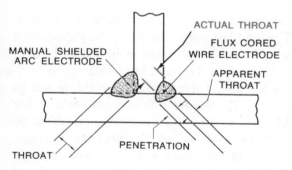

Fig. 21-27. Comparing the fillet welds of the flux core arc welding process and the manual shielded arc welding process. (Courtesy of Hobart Brothers Company.)

Butt Joint Technique

Welding with flux-cored wire electrodes allows for narrower Vee-groove butt joint preparations than when welding with the manual shielded arc process. See Fig. 21-28. Penetration is also deeper with the flux-cored wire electrode process, due to the high welding current density and the penetrating ability

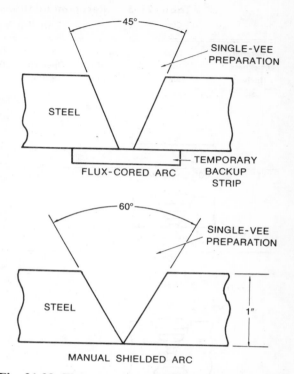

Fig. 21-28. The groove for a single-bevel butt joint preparation does not have to be as wide for a flux-core arc weld as it does for a manual shielded electrode arc weld.

of carbon dioxide gases. The high-current density comes from the small wire diameter and the relatively high welding current. Also, the small wire electrode can penetrate a groove that is too narrow for the manual shielded arc electrode.

When welding the square-groove butt joint, the welder should use a 90° nozzle angle with a 5° to 10° leading angle in the direction of travel, as in Fig. 21-19. A shielding gas flow of 35 cfh and ¾" to 1" electrode stickout should work well for this joint.

Information about the number of passes, electrode diameter, and arc current needed for flux-core arc welding mild steel in the flat position is given in Table 21-3. This information holds true when using either a square-butt joint or a 60° double-bevel joint with a back-up strip. The equipment settings for vertical-position bevel joints are shown in Table 21-5.

Table 21-5.   Recommendations for Flux-Core Arc Welding Mild Steel in the Vertical Position.

Material Thickness (Inches)	Type of Joint	Number of Passes	Root Opening (Inches)	Electrode Diameter (Inches)	Welding Power (DC-RP)		Travel Speed	
					Volts	Amps	Weld Pass	IPM
3/8	60° Single Vee	3	0	0.045	22	180	1	13 down
								7.7 up
								5 up
1	60° Single Vee	5	3/32	0.045	22	180	1	13 down
							2	1.4 up
							3	2.3 up
							4	1.6 up
							5*	11 down
2	60° Double Vee	9	0-1/16	0.045	22	190-200	1	11 down
							2	3 up
							3	3.5 up
							4	2.1 up
							5	2.7 up
							6	2 up
							7	1.8 up
							8	1.4 up
							9	1.3 up

* Joint is back gouged before depositing back weld.   Courtesy of Hobart Brothers Company.

Fillet Weld Technique

For Tee-joint fillet welds, the correct nozzle angle will be 45° from the lower plate, one electrode diameter closer to the bottom plate than to the upright plate. See Fig. 21-29. The recommended shielding gas flow is 35 cfh, with 1″ to 1½″ electrode stickout for this weld joint. See Fig. 21-30.

The welder should follow the information suggested by the manufacturer for preparing fillet weld joints in various positions. Corner, Tee, and lap joints welded in the *flat* position would use the information in Table 21-6.

On corner, Tee, and lap joints in the *vertical* position, Table 21-7, fillet welding will be easier if the welder practices the welds using 21 volts, up to 180 amperes, 30 cfh shielding gas flow, and 3/8″ to ½″ electrode stickout.

On lap-joint fillet welds, the nozzle angle should be 60° from the lower plate and one electrode diameter closer to the bottom plate, similar to the Tee joint example in Fig. 21-29.

Information about welding Tee and lap fillet joints in the horizontal position is listed in Table 21-8.

Flux-Core Arc Welding Problems[1]

Flux-core arc welding, like any other welding process, must be properly applied and carefully controlled to produce high-quality

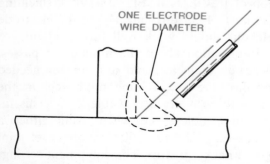

Fig. 21-29. Correct wire position for Tee-joint fillet welds.

ONE ELECTRODE WIRE DIAMETER

[1]Courtesy of Hobart Brothers Company.

Table 21-6.   Recommendations for Flux-Core Arc Welding Mild Steel Fillet Welds in the Flat Position.

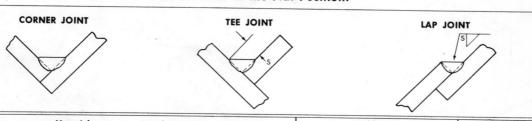

CORNER JOINT   TEE JOINT   LAP JOINT

Weld Size (S) (Inches)	Material Thickness (T) (Inches)	Number of Passes	Electrode Diameter (Inches)	Welding Power (DC-RP)		Travel Speed (IPM Per Pass)
				Volts	Amps	
⅛	⅛	1	³⁄₃₂	24-26	300	53
³⁄₁₆	³⁄₁₆	1	³⁄₃₂	24-26	350	41
³⁄₁₆	³⁄₁₆	1	⅛	24-26	450	40
¼	¼	1	³⁄₃₂	24-26	400	24
¼	¼	1	⅛	25-27	500	25
⁵⁄₁₆	⁵⁄₁₆	1	³⁄₃₂	28-30	550	22
⁵⁄₁₆	⁵⁄₁₆	1	⅛	26-28	460	20
⅜	⅜	1	³⁄₃₂	29-31	575	20
⅜	⅜	1	⅛	29-31	575	20
½	½	1	³⁄₃₂	30-32	625	11
½	½	1	⅛	30-32	625	11
⅝	⅝	3	³⁄₃₂	29-31	475	12
⅝	⅝	3	⅛	27-29	450	14
¾	¾	3	³⁄₃₂	29-31	500	13
¾	¾	3	⅛	28-30	500	12

Courtesy of Hobart Brothers Company.

Table 21-7.   Recommendations for Flux-Core Arc Welding Mild Steel Fillet Welds in the Vertical Position.

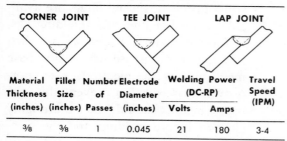

CORNER JOINT   TEE JOINT   LAP JOINT

Material Thickness (inches)	Fillet Size (inches)	Number of Passes	Electrode Diameter (inches)	Welding Power (DC-RP)		Travel Speed (IPM)
				Volts	Amps	
⅜	⅜	1	0.045	21	180	3-4

Courtesy of Hobart Brothers Company.

weld metal. Problems similar to the problems with manual shielded arc electrode welds may also happen with the flux-core process. With training and experience, most welders quickly learn good techniques that help prevent flux-core weld defects. The following information is useful for helping find the causes of weld defects and setting up the recommended correction for common problems with the flux-core process.

Surface and Subsurface Porosity

Figs. 21-31 and 21-32 show examples of surface and subsurface porosity in flux-core arc welding. Surface porosity will appear as holes on *top* of the finished weld, while subsurface porosity will appear as holes *deep in* the weld if the weld is cut apart or X-rayed.

The probable cause of surface and subsurface porosity is contamination from the atmosphere. Moisture, dirt, and oil may cause similar problems if they are on the weld metal. These problems may be caused and corrected as follows:

A. The shielding gas flow may be set too low and, therefore, not pushing away all of the air in the arc area. This can be corrected by increasing the shielding gas flow.

B. The shielding gas envelope may be getting blown away by a wind or draft. It is a good idea to use a shield, or the

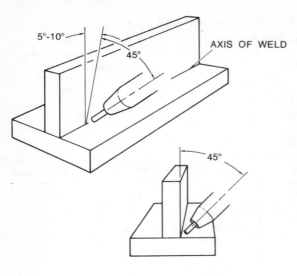

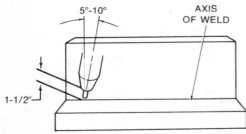

Fig. 21-30. The stickout length and gun nozzle angles for flux-core arc welding.

welder's body, to protect the arc area when the wind is strong.

C. There may be a clogged or defective shielding gas system. This could result from weld spatter clogging the nozzle, a broken gas line, defective fittings in the gas system, bad gas valves,

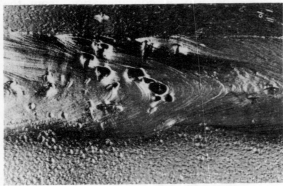

Fig. 21-31. Surface porosity (holes) in a flux-core arc weld. (Courtesy of Hobart Brothers Company.)

Fig. 21-32. Subsurface porosity (holes below the surface) in a flux-core arc weld. (Courtesy of Hobart Brothers Company.)

or a frozen regulator. It is easy to check these things and make the repairs needed.

D. The shielding gas nozzle may be too far from the work. This would allow the shielding gas to spread out before reaching the work. The welder should hold the nozzle closer to the work to correct this shielding problem.

E. A possible cause of porosity is moisture in the shielding gas. Dry, *welding grade* carbon dioxide gas should always be used. If gas contamination is still a problem, special filters are available to further filter the shielding gas.

F. If foreign materials such as oil, dirt, rust, paint, or grease are on the base metal and/or the electrode wire, porosity may result. To prevent this, the base metal should be thoroughly cleaned before welding and the electrode wire should be protected from dirt once it is removed from the carton.

G. Porosity may result if flux-core arc welding is done over a tack weld made with manual shielded arc electrodes. This practice should be avoided for good welds.

Crater Defects and Crater Porosity

A major cause of weld crater defects is removing the welding gun (and shielding gas)

Table 21-8.   Recommendations for Flux-Core Arc Welding Mild Steel Fillet Welds in the Horizontal Position.

TEE JOINT		LAP JOINT		MULTIPLE-PASS TEE JOINT		
Weld Size (S) (inches)	Material Thickness (Inches)	Number of Passes	Electrode Diameter (Inches)	Welding Power DC-RP		Travel Speed (IPM Per Pass)
				Volts	Amps	
1/8	1/8	1	3/32	24-26	350	60
3/16	3/16	1	3/32	24-26	400	41
3/16	3/16	1	1/8	24-26	425	32
1/4	1/4	1	3/32	24-26	400	24
1/4	1/4	1	1/8	25-27	450	25
5/16	5/16	1	3/32	25-27	440	20
5/16	5/16	1	1/8	26-28	460	20
3/8	3/8	1	3/32	26-28	475	15
3/8	3/8	1	1/8	28-30	500	14
1/2	1/2	3	3/32	24-26	400	20
1/2	1/2	3	1/8	25-27	450	18
5/8	5/8	3	3/32	26-28	450	14
5/8	5/8	3	1/8	27-29	450	14
3/4	3/4	6	3/32	28-30	470	20
3/4	3/4	6	1/8	28-30	470	20

Courtesy of Hobart Brothers Company.

before the weld crater has become solid. This allows air to attack the molten crater. See Fig. 21-33. To eliminate crater porosity, the welder should hold the gun in its normal position at the end of the weld until the wire feed and gas flow stop.

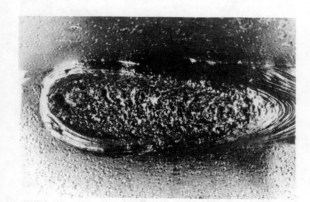

Fig. 21-33. Crater porosity in a flux-core arc weld. (Courtesy of Hobart Brothers Company.)

Burn-Through and Excessive Penetration

Burn-through and too much penetration may result from the heat input in the weld being too high. See Fig. 21-34. This can be corrected by either reducing the wire feed

Fig. 21-34. Burn-through on a flux-core arc weld. (Courtesy of Hobart Brothers Company.)

speed for a lower amperage or increasing the travel speed. Burn-through may also be caused by poor joint design or poor joint preparation. If the root opening is too wide or if the root face is too small, there is usually too much burn-through. This can be corrected by increasing the stickout and using a weaving motion with the welding gun.

When complete penetration is needed, it may be cheaper to use backup bars when welding with mild-steel, flux-core electrodes.

Lack of Penetration

When a weld is cut apart, lack of penetration can often be seen as in Fig. 21-35. Here,

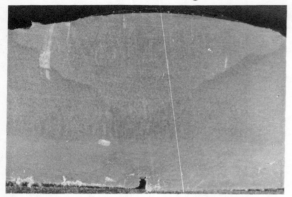

Fig. 21-35. By cutting a weld apart, it may be checked for good penetration. Here, the flux-core arc weld has not penetrated both pieces. (Courtesy of Hobart Brothers Company.)

the weld has not penetrated the bottom of the root joint. A less destructive way to test penetration is to use an X-ray. In Fig. 21-36, the weld has been X-rayed and a clear, unmelted metal edge can be seen. By using an X-ray, the weld doesn't have to be cut open to inspect it. Lack of penetration is often caused

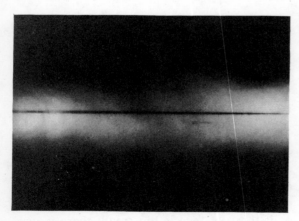

Fig. 21-36. By X-raying the weld, it doesn't have to be cut to be tested. Here, the flux-core arc weld did not penetrate both edges, as seen by the solid edge lines in the weld. (Courtesy of Hobart Brothers Company.)

by too low a heat input in the weld. The welder can correct this by increasing the wire feed speed to get higher amperage. He should also recheck the wire stickout. A normal stickout of 1″ to 1½″ should be used.

Poor welding technique may also cause a lack of penetration. To correct this, the welder should be sure to keep the arc on the *leading* edge of the puddle. Penetration will be deeper when the arc is kept closer to the leading edge of the puddle.

Slag Inclusions and Lack of Fusion

Fig. 21-37 shows a flux-core arc weld with slag inclusions and lack of fusion. This defect is usually caused by poorly cleaning

Fig. 21-37. A flux-core arc weld cut apart to show the slag inclusions and poor penetration. (Courtesy of Hobart Brothers Company.)

the previous weld bead and too low a weld-heat input. The welder can correct this by carefully cleaning all slag from one weld bead before making the next bead pass. The

Fig. 21-38. Flux-core arc weld undercutting, top view. (Courtesy of Hobart Brothers Company.)

heat input must then be high enough to completely fuse both edges of the weld joint.

Weld Undercutting

Weld *undercutting* is not seen as easily in the top view, Fig. 21-38, as it is in the cutaway side view, Fig. 21-39. Usually, a welder can only see the top of a weld, so he should be able to recognize undercutting from the top view.

Fig. 21-39. Flux-core arc weld undercutting, cutaway side view. (Courtesy of Hobart Brothers Company.)

Weld undercutting is often caused by one or more of these factors:

A. Poor welding gun technique. On all passes except the root pass, the welder should move the welding gun from side to side in the joint with a weaving motion, pausing at the edge of the groove before returning to the opposite side. This will evenly distribute the weld metal to each side of the joint.

B. Using too high a travel speed for the voltage and amperage being used. The travel speed should match the voltage and amperage settings. Using a faster travel speed than is recommended for the voltage and amperage settings being used will usually cause undercutting in the weld.

C. Using too high a voltage setting. If the voltage is too high, reducing the voltage should control the weld undercutting.

Other Defects

Correctly applied, flux-core arc welding produces high-quality weld metal. However, the welder is still largely responsible for using the process correctly to make high-quality welds.

Other defects sometimes found in flux-core arc welds may include warpage and rough surfaces, similar to the same types of defects sometimes found in other welding processes. Avoid these defects by using the same normal precautions that were recommended for manual shielded arc welding.

Unequal Leg Fillets—These are caused by poorly positioning the electrode wire when making horizontal fillets. If the wire is too far away from the vertical leg, the weld metal will not "wash up" to the vertical leg as it should. The horizontal leg of the fillet will then be longer than the vertical leg. To correct this defect, the electrode wire should be positioned one wire diameter down toward the horizontal plate, as was shown in Fig. 21-29.

Cracks—Cracking in flux-core arc welds is usually caused by very heavy restraints (clamps) and cooling too fast, the same conditions that cause cracking in manual shielded arc welds. Tests have shown, however, that flux-core arc welds usually resist cracking better than welds that have been made with the manual metal inert gas (MIG) welding process.

Cracking may be caused in flux-core arc welds by depositing small beads with high currents and high travel speeds on very thick, tightly clamped parts without pre-heating. Cracking can be eliminated by using preheating and/or postheating (as when welding with manual metal arc), by using a larger bead for the first pass, and by removing any restraint (clamps, etc.) on the metal, if possible. A convex bead, deposited by lowering the voltage, reduces the danger of cracking.

22

Gas Metal Arc Welding (GMAW)

Similar to TIG welding, another popular, modern welding process is known as *Metal Inert Gas* (MIG) or *Gas Metal Arc Welding* (GMAW). The process is generally known as MIG welding and is a form of arc welding. During MIG welding, the weld is made by an arc between the metal being welded and a continuous, consumable electrode. With TIG welding, of course, the electrode was not consumed. This is the major difference between TIG and Gas Metal Arc Welding.

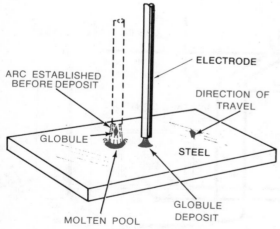

Fig. 22-1. The carbon dioxide GMAW process deposits globules similar to the manual shielded arc welding process shown above.

A mixture of inert gas (carbon dioxide, helium, argon, or combinations of the three) is used to shield the weld process from atmospheric contamination due to elements such as oxygen and nitrogen in the atmosphere. The inert shielding gas principle is the same as was used for TIG or flux-core arc welding.

There are four major classifications of Gas Metal Arc Welding, depending on the shielding gas used and/or how the metal is transferred.

1. *Carbon Dioxide GMAW*—This type of GMAW produces high-speed welds while using carbon dioxide shielding gas. This method is used to weld carbon and alloy steels, and it deposits the filler metal similar to the manual shielded arc method of globular transfer. See Fig. 22-1.

2. *Microwire GMAW* — This GMAW process will weld most thinner-gauge steels than was previously possible with arc welding. The welds deposited by the microwire GMAW process are low in hydrogen content. Microwire is a short-circuiting, metal-arc transfer process for welding in all positions. Small-diameter filler wire, low welding currents, and a shielding gas mixture of 25% carbon dioxide and 75% argon will easily

weld low-alloy, high-strength steels and stainless steels.

3. *MIG-GMAW* — The MIG-GMAW process uses argon as the shielding gas for welding aluminum, aluminum alloys, stainless steel, nickel, nickel alloys, and copper alloys. Many times, when aluminum and aluminum alloys are being welded, argon and helium can be mixed 50/50 or 75/25 for the MIG shielding gas. See Fig. 22-2.

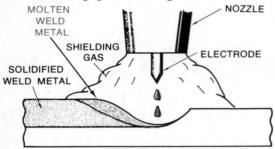

Fig. 22-2. The MIG welding process melts the electrode into the weld pool in an atmosphere of inert shielding gas. (Courtesy of Hobart Brothers Company.)

4. *Spray-Arc GMAW* — The spray-arc transfer is mainly used for welding steels with flat and horizontal fillets. For spray-arc welding, it may be necessary to use about 5% oxygen with the argon shielding gas. This will help give high-speed welds with only little cleanup.

Almost all GMAW welding is MIG welding. In fact, the terms GMAW and MIG are often used to mean the same process. In this chapter, most of the material will refer to MIG welding.

MIG welding is mainly used as a semi-automatic process. However, it can be changed to fully automatic welding on certain jobs. MIG welding gives high-quality welds on most metals and alloys now being used in industry. MIG welding can be done at fairly high travel speeds in all positions. Because of the inert gas shield, MIG welds are usually free of the slag that is often trapped in welds made by the manual shielded arc welding process.

MIG Welding Equipment

The equipment used for MIG welding is almost the same as the equipment described for flux-cored arc welding with carbon di-

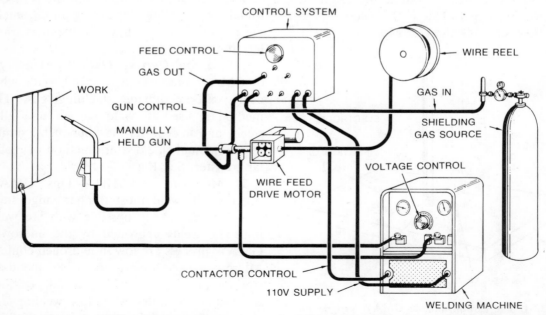

Fig. 22-3. The basic MIG welding equipment set up for MIG welding. (Courtesy of Hobart Brothers Company.)

oxide shielding gas. The major MIG equipment includes a main power source (welding machine), a wire feed machine, controls, shielding gas, cooling water, and many connections. When the machine is used for semiautomatic welding, the welder will use a welding gun attached to the cable assembly. A welding gun is used for fully automatic MIG welding. The supplies of shielding gas and cooling water are attached to the wire feed machine through feeder lines. Fig. 22-3 shows the main parts of a complete MIG welding set-up. The equipment may be mounted on a portable cart, such as the one shown in Fig. 22-4.

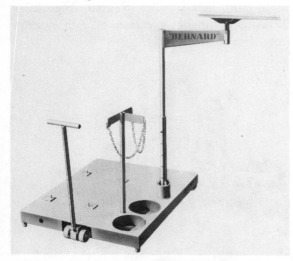

Fig. 22-4. A portable cart used to mount and easily move the complete MIG welding setup. (Courtesy of Bernard Welding Accessories.)

Power Source (Welding Machine)

The power source used for MIG welding should be a constant voltage welding machine, where the output voltage stays about the same regardless of the current being used. (Because constant voltage welding machines do not have a current control, they would be unsatisfactory for manual shielded arc welding.)

The welding *current* output depends on the wire-feed speed, while the *voltage* output is adjusted by a rheostat (variable control) on the welding machine. Reverse polarity DC

(electrode positive) is normally used for semiautomatic MIG welding. The special MIG welding machine in Fig. 22-5 has its own built-in wire feeder.

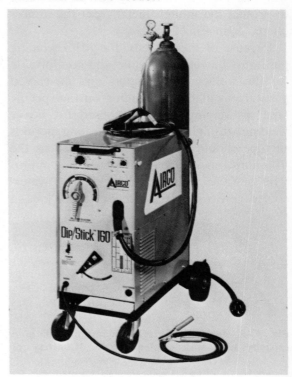

Fig. 22-5. Special MIG welding machine with many built-in features. (Courtesy of Airco Welding Products Division.)

Wire Feed System

Since the constant voltage system depends on the wire burn-off rate for current output, it is necessary to have the wire-feed system matched to the constant-voltage power supply. The amount of welding current supplied to the arc depends on the rate of the wire feed. Therefore, at a given wire-feed speed, the welding machine will be able to supply the proper amount of current for a smooth, steady arc and well-shaped weld beads.

Welding Gun and Cable

As discussed with flux-core arc welding, the cable assembly must carry the electrode wire, welding current, shielding gas, and cooling water to the gun. In turn, the gun must be held close to the work during welding, so that the arc and shielding gas can be controlled around the nozzle and the wire. This will provide the necessary inert gas shield to protect the molten metal from the atmosphere, as in Fig. 22-2.

Because of these needs, the MIG welding gun must be easy to handle and comfortable for the welder. Both water- and air-cooled guns are available, depending on the specific type of welding job. Guns used for heavy welding at high currents over long periods of time should be water-cooled, such as the 700-amp gun in Fig. 22-6. When using microwire or other low-current welding, a water-cooled gun is not required. For those jobs, a

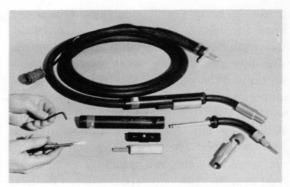

Fig. 22-8. MIG welding gun before and after assembly. Shown assembled with welding cable attached. (Courtesy of Airco Welding Products Division.)

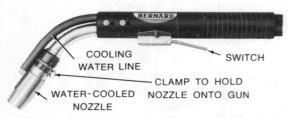

COOLING
WATER LINE   SWITCH

CLAMP TO HOLD
WATER-COOLED   NOZZLE ONTO GUN
NOZZLE

Fig. 22-6. Heavy-duty MIG welding gun. 700 amp capacity with water-cooled nozzle. (Courtesy of Bernard Welding Accessories.)

smaller, air-cooled gun may be used, such as the 300-amp model in Fig. 22-7. Most welding guns are able to be easily taken apart for

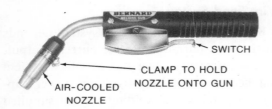

SWITCH

CLAMP TO HOLD
NOZZLE ONTO GUN

AIR-COOLED
NOZZLE

Fig. 22-7. Light-duty MIG welding gun. 300 amp capacity with air-cooled nozzle. (Courtesy of Bernard Welding Accessories.)

repair. Fig. 22-8 shows a MIG welding gun before and after assembly, using only the two tools shown.

Nozzle—The welding gun nozzle is very important. It must be chosen to provide the correct shielding gas flow for the gas and electrode being used, as well as the weld po-

sition and metal being welded. Fig. 22-9 shows a complete selection of gun nozzles from one manufacturer.

SLOTS ALLOW CLAMP   COOLING WATER
TO TIGHTEN NOZZLE   CONNECTIONS FOR A
ON WELDING GUN   WATER-COOLED NOZZLE

Fig. 22-9. MIG welding gun nozzles for different welding jobs. (Courtesy of Bernard Welding Accessories.)

Shielding Gas

For the highest-quality weld metal, the welder should use only *welding grade* shielding gases. These are specially processed to provide a high purity level and low moisture

Fig. 22-10. Argon cylinder with argon shielding gas regulator and flowmeter connected. (Courtesy of Airco Welding Products Division.)

content. The glas flow rates given in specification tables depend on the shielding gas used, the metal being welded, the weld position, the welding speed, and any drafts in the weld area.

Helium, carbon dioxide, and argon are the shielding gases usually used with MIG welding. A mixture may sometimes be used, especially argon and helium. Regardless of the gas used, it must be controlled with a flowmeter regulator, as in Fig. 22-10.

Electrode Wire

Unlike the flux-core electrode wire, the MIG welding wire is solid and uncovered. Carbon steel wire, however, is covered with a thin protective coat of copper to help control oxidation. The MIG welding wire, like flux-core wire, is automatically pushed down into the weld area where it melts into the joint. See Fig. 22-11.

Setting Up GMAW (MIG) Equipment

Setting up the equipment for GMAW (MIG) welding and setting up for flux-core arc welding are very similar. The major difference, of course, between straight MIG welding and flux-core arc welding is in the electrode wire. The wire is solid in the MIG process, whereas the wire center is filled with flux in the flux-core process.

Safety—Common welding safety rules apply when setting up for GMAW (MIG) welding. Regulators, cylinders, and hoses for shielding gases should be handled carefully, as outlined for oxygen equipment.

The wire-feed machine and welding machine must be carefully connected and fully understood before they are adjusted or used. The welding cable must be protected from

Fig. 22-11. The MIG welding electrode wire is automatically pushed down into the molten weld pool.

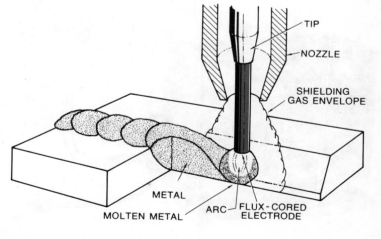

TIP

NOZZLE

SHIELDING GAS ENVELOPE

METAL

MOLTEN METAL   ARC   FLUX-CORED ELECTRODE

welding heat, sparks, and being stepped on or tripped over by the welder. A new welder must understand how to shut the equipment down *before* he begins to practice GMAW. He should have little, if any, trouble setting up and adjusting GMAW (MIG) equipment if he follows the manufacturer's specifications and carefully sets up the equipment in order.

Positioning and Connecting the Equipment

A first step in setting up is to position all the GMAW equipment where it will be safe and can be safely used. Using a cart, Fig. 22-4, is an easy way to locate and set up all the equipment.

Cooling water and shielding gas supplies must be located near the wire-feed machine for easy connections. The welding machine must be near the power outlet unless it is a generating machine. Then, the welding machine should be located as close as possible to the wire-feed machine, to keep the welding power cables as short as possible.

When all the equipment is carefully positioned, it may be connected basically, as shown in Fig. 22-3. Then, adjust the equipment before practicing GMAW (MIG) welding. In Fig. 22-12, the welder has carefully positioned and set up the MIG welding equipment.

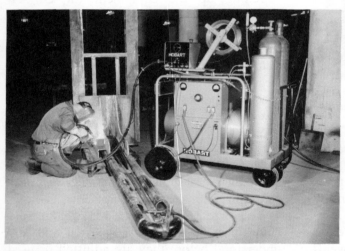

Fig. 22-12. Completely set up MIG welding equipment being used to weld a heavy metal part. (Courtesy of Hobart Brothers Company.)

Fig. 22-13. For a beginning setting, the wire-feed speed should be set on about 5. (Courtesy of Hobart Brothers Company.)

Table 22-1. Recommendations for Microwire GMAW in the Flat Position with Steel Backup Bars.

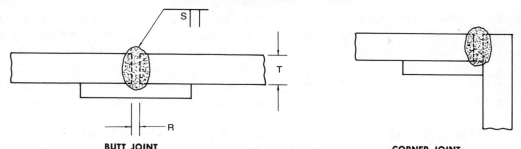

BUTT JOINT   CORNER JOINT

Weld Size (Inches)	Material Thickness (T) Gauge	Material Thickness (T) Inches	Number of Passes	Root Opening (R) (Inches)	Electrode Diameter* (Inches)	Welding Conditions (DC-RP) Arc Volts	Welding Conditions (DC-RP) Amperes	Gas Flow (cfh)	Travel Speed (ipm)
	24	.025	1	0	.030	17-19	90-120	15-20	52-57
	22	.031	1	0	.030	17-19	100-120	20-25	58-62
	20	.037	1	0	.030	17-19	110-130	20-25	58-62
	18	.050	1	0	.035	18-20	160-180	20-25	58-62
1/16	16	.062	1	1/32	.035	19-21	190-210	20-25	68-72
5/64	14	.078	1	1/32	.035	21-23	195-215	20-25	68-72
1/8	1/8	.125	1	1/32	.045	21-23	300-320	20-25	68-72

*Electrode stickout should be 1/4" to 3/8"

Turning on the Electrical Equipment

When the GMAW (MIG) equipment has been carefully set up and connected, the first step for adjusting the equipment is to turn on the electrical machines. The welding machine should first be set for open circuit voltage and then turned on.

Then, the wire-feed machine is turned on. The wire-feed speed control on the wire-feed machine should be adjusted to zero. By doing this, the wire will not be fed through the gun nozzle while making other machine adjustments.

Adjusting the Shielding Gas

The next step for adjusting the GMAW (MIG) equipment is to open the shielding-gas cylinder valve on top of the shielding-gas cylinder. The flowmeter valve can then be opened while the gun trigger is pulled toward the gun handle. The flowmeter valve should be opened very slowly, similar to opening the outlet valve on an oxygen cylin-

der. After adjusting the shielding gas flow rate according to the instructions, the trigger on the gun should be released. The instructions in Table 22-1, for example, give the recommended settings for microwire GMAW in the flat position with steel backup bars.

Adjusting the Wire Feed, Voltage, and Current

After adjusting the shielding gas flow, the welder should adjust the wire-feed speed control. The wire-feed speed helps determine the welding current with GMAW, as it does with flux-core arc welding. A beginning setting would be to turn the control knob to the middle setting of #5. See Fig. 22-13.

The voltage setting, the next adjustment to be made, depends on the GMAW process being used. The voltage will vary from one GMAW process to another. For example, the spray-arc and carbon dioxide, globular transfer voltage settings used for 1/4" thick, low-alloy steel fillet welding are shown in Table 22-2.

Adjusting the Stickout

The stickout (tip-to-work distance) will also vary for each GMAW process being used, as does the voltage. Adjusting the stickout for each GMAW process must be discussed separately. This may be seen on the next page in Fig. 22-14.

A welder who is new to GMAW welding must understand the gun nozzle surface-to-shielding gas tip relationship to be able to make good GMAW welds. The shielding gas tip may be used *flush* with the nozzle for maintaining normal stickout. See Fig. 22-15.

The *extended* tip shown in Fig. 22-16 is not used very often and then only for weld-

Table 22-2. Recommendations for Carbon Dioxide, Globular Transfer and Spray-Arc GMAW on ¼" Low-Alloy Steel.

Method	Electrode Wire Dia. (inches)	Amperage	Voltage	Travel Speed (ipm)
Spray-Arc	1⁄16"	350	25	32
Carbon Dioxide, ⎫ Globular Transfer⎭	5⁄64"	500	32	40

Table 22-3. Recommendations for Microwire GMAW in All Positions.

Material Thickness[1] (Inches)		Electrode Diameter[2] (Inches)	Welding Current DC-RP Amps	Arc Voltage	Wire Feed (ipm)	Shielding Gas Flow[2] (cfh)	Travel Speed (ipm)
Fraction	Decimal						
24 gauge	.025	.030	30- 50	15-17	85 100	15-20	12-20
22 gauge	.031	.030	40- 60	15-17	90 130	15-20	15-22
20 gauge	.037	.035	55- 85	15-18	70 120	15-20	35-40
18 gauge	.050	.035	70-100	16-19	100 160	15-20	35-50
1⁄16	.063	.035	80-110	17-20	120 180	20-25	30-35
5⁄64	.078	.035	100-130	18-20	160 220	20-25	25-30
1⁄8	.125	.035	120-160	19-22	210 290	20-25	15-25
1⁄8	.125	.045	180-200	20-24	210 240	20-25	27-45
3⁄16	.187	.035	140-160	19-22	240 290	20-25	10-19
3⁄16	.187	.045	180-205	20-24	210 245	20-25	18-28
1⁄4	.250	.035	140-160	19-22	240 290	20-25	11-16
1⁄4	.250	.045	180-225	20-24	210 290	20-25	20-24

Courtesy of Hobart Brothers Company.

Data is for flat position. Reduce current 10% to 15% for vertical and overhead positions. All data is for single-pass microwire welds.

[1] For groove and fillet welds. Material thickness also indicates fillet weld size. Use a root opening for square butt joints that is ½ the thickness of the metal being welded.

[2] Shielding gas used should be welding grade carbon dioxide. C25 shielding gas, 25% carbon dioxide and 75% argon, may also be used.

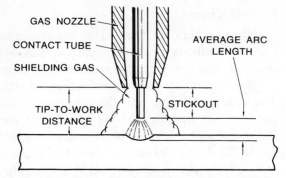

Fig. 22-14. The electrode stickout will vary depending on the type of GMAW being used.

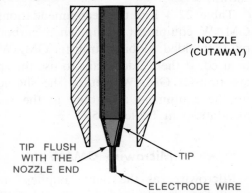

Fig. 22-15. The tip may be flush with the nozzle, using normal electrode stickout.

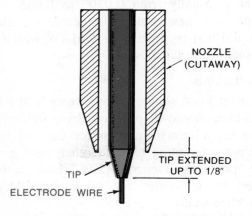

Fig. 22-16. The tip may be extended beyond the nozzle for welding with very low amperage.

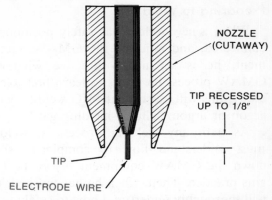

Fig. 22-17. The tip is usually recessed slightly with normal stickout for most welding jobs.

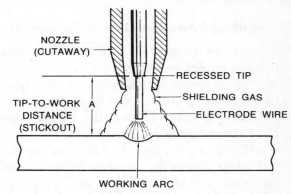

Fig. 22-18. The normally-recessed gun tip with a working arc.

ing with very low amperage. Most welders use the *recessed* tip position for normal GMAW conditions, as in Fig. 22-17. Here, the wire electrode sticks out past the end of the tip to deposit the weld bead. See Fig. 22-18.

Table 22-3 shows several applications of the microwire GMAW process, with amperage settings from 180 to 225 amperes for welding ¼″-thick metal. The new welder should practice GMAW by using a setting in the middle of this range.

Since the ammeter on the welding machine only registers *during actual welding,* a new welder usually has to have someone check the ammeter while he is laying down a practice bead. By increasing the wire-feed speed, the current (amperage) will be increased, and, of course, by slowing down the wire-feed speed, the current will be reduced. This is discussed under flux-core arc welding. Then, after setting the right amperage for good penetration, the arc can be smoothed out by readjusting the voltage rheostat (control). See Fig. 22-13.

Preparing to Weld

When a new welder has safely positioned, connected, and adjusted the GMAW equipment, he is set up to practice whatever GMAW process he wants to learn first. Generally, this will be straight MIG welding with argon or argon/helium shielding gas.

As with any welding process, a welder must understand how to completely shut down the GMAW equipment *before* he begins practice. Then, when he has finished, he will thoroughly understand how to safely turn off the equipment ahead of time.

Shutting Down the GMAW Equipment

Upon the completion of a weld, releasing the gun trigger will stop the wire from feeding. Then, simply withdrawing the wire from the work being welded will break the arc.

Once the arc is broken, the welder must close the shielding gas outlet valve on top of the shielding gas cylinder. Then, pulling the gun trigger (with the wire-feed speed control on zero) will bleed the gas line. This may also be done by using the purge button on the wire-feed machine. To prevent the flowmeter tube from breaking when the shielding gas is turned on again, the valve on the flowmeter must be closed, but only finger tight.

When the shielding gas is shut down correctly, the wire feed machine rollers and the welding machine current should be turned off. The on-off switch on the wire feed is first moved to *off*, shutting down the wire feed. Then, the stop button on the welding machine is pushed to turn the welding machine off. This completes the shutting down.

Gas Metal Arc Welding

A welder should practice GMAW by using the process to weld carbon steel. Normally shielded MIG welding and the other three GMAW metal transfer processes (microwire, carbon dioxide-shielded, and spray-arc) may all be used for welding carbon steel.

Process Charts

Before practicing the three less-common GMAW methods (microwire, carbon dioxide-shielded, and spray-arc), a welder must refer to the manufacturer's specification charts. The charts will give him the information he needs to set up the GMAW equipment for the metal and thickness he is going to weld. Table 22-3, for example, lists the recommendations for microwire GMAW equipment and usage for the types of welds shown.

Table 22-4 lists the recommendations for GMAW equipment set-up when carbon dioxide-shielded, globular-transfer GMAW will be used. If the welder plans to use the spray-arc transfer GMAW method, he should set up the equipment according to the recommendations in Table 22-5.

Microwire GMAW

Microwire, as the name implies, means small-diameter electrode wire. In this GMAW process, therefore, the electrode wire diameter is usually from 0.030″ to 0.045″. As with any new welding process, a welder should first try microwire GMAW with butt joints in the flat position.

Butt Joints

Butt joint, square-groove welds in the flat position should be practiced by a welder using the microwire process for the first time before attempting to weld other weld joints or positions. Generally, ⅛″-thick carbon steel stock is used for microwire GMAW practice welds.

Setting Up Equipment—The welder can use either 0.035″- or 0.045″-diameter electrode wire for welding ⅛″-thick carbon steel. When using 0.035″ wire, a reverse-polarity DC current from 120 to 160 amperes should be used. A voltage setting from 19 to 22 volts should produce a smooth arc action when microwire welding is done with those amperages.

Table 22-4. Recommendations for the Carbon Dioxide, Globular Transfer GMAW Process When Used on Carbon and Low-Alloy Steel.

Material Thickness (Inches)		Type of Weld[1]	Electrode Diameter (Inches)	Welding Current DC-RP (Amps)	Arc Voltage	Wire Feed (ipm)	Travel Speed (ipm)	Shielding Gas Flow[2] (cfh)
Fracttion	Decimal							
18	.050	Fillet	.045	280	26	350	190	25
		Square groove	.045	270	25	340	180	25
16	.063	Fillet	.045	325	26	360	150	35
		Square groove	.045	300	28	350	140	35
14	.078	Fillet	.045	325	27	360	130	35
		Square groove	.045	325	29	360	110	35
		Square groove	.045	330	29	350	105	35
11	.125	Fillet	1⁄16	380	28	210	85	35
		Square groove	.045	350	29	380	100	35
3⁄16	.188	Fillet	1⁄16	425	31	260	75	35
		Square groove	1⁄16	425	30	320	75	35
		Square groove	1⁄16	375	31	260	70	35
1⁄4	.250	Fillet	5⁄64	500	32	185	40	35
		Square groove	1⁄16	475	32	340	55	35
3⁄8	.375	Fillet	3⁄32	550	34	200	25	35
		Square groove	3⁄32	575	34	160	40	35
1⁄2	.500	Fillet	3⁄32	625	36	160	23	35
		Square groove	3⁄32	625	35	200	33	35

Courtesy of Hobart Brothers Company.

[1]On square groove welds a backing is required. The backing may be either copper or steel.
[2]Carbon dioxide, welding grade.

The shielding gas used for microwire welding practice should either be welding-grade carbon dioxide or a mixture of 75% argon and 25% carbon dioxide. The shielding gas flow should be from 20 to 25 cfh.

The wire-feed speed should be adjusted for 210 to 290 ipm. When using this wire-feed speed and amperage, a bead travel speed of 15 to 25 ipm should work well for practice. Depending on results, the welder may have to adjust the voltage, wire-feed speed, and travel speed after a couple of practice welds.

When the 0.045″-diameter electrode wire is being used, the welder will need to change the amperage setting according to the recommendations in Table 22-3. With this electrode, the gun nozzle should be held about 1⁄4″ to 3⁄8″ from the work, using a recessed tip. Various stickouts should be tried, ranging from 1⁄4″ to 5⁄8″, depending on the results. On 1⁄8″-thick mild steel, a 1⁄4″ stickout will create deep penetration and may burn

through. Generally speaking, a longer stickout will give less weld penetration.

Coupon Positioning—The welder should tack weld two pieces of 1⁄8″-thick carbon steel stock. For practice, each piece should be at least 4″ long and 2″ wide. A 1⁄16″ root opening is needed, as in Fig. 22-19.

Flat-Position Procedure—The gun nozzle should be held at a 90° angle to either cou-

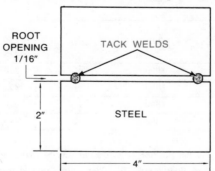

Fig. 22-19. The practice coupons should be tacked and positioned as shown for microwire GMAW practice.

pon, as in Fig. 22-20. After lowering the helmet for eye protection, the gun trigger can be squeezed. When the gun trigger is squeezed, it immediately turns on the controls and starts the arc. Then, the welder should travel from right to left for a flat-position butt joint. The gun is held with a 90° nozzle angle to either coupon and about a 5° angle in the direction of travel. In this position, the GMAW gun is "pulling" the weld, as dis-

Table 22-5. Recommendations for the Spray-Arc Transfer GMAW Process When Used on Carbon and Low-Alloy Steel.*

Material Thickness[1] (Inches)	Type of Weld[2]	Electrode Diameter (Inches)	Welding Current DC-RP (Amps)	Arc Voltage	Wire Feed (ipm)	Shielding Gas Flow[3] (cfh)	Travel Speed (ipm)	Number of Passes
1/8	Fillet	1/16	300	24	165	40-50	35	1
3/16	Fillet	1/16	350	25	230	40-50	32	1
1/4	Butt	1/16	325	24	210	40-50	30	2
			375	25	260			
1/4	Butt	3/32	400	26	100	40-50	35	2
			450	29	120			
1/4	Fillet	1/16	350	25	230	40-50	32	1
1/4	Fillet	3/32	400	26	100	40-50	32	1
3/8	Butt	1/16	325	24	210	40-50	24	2
			375	25	260			
3/8	Butt	3/32	400	26	100	40-50	28	2
			450	29	120			
3/8	Fillet	1/16	350	25	230	40-50	20	2
3/8	Fillet	3/32	425	27	110	40-50	20	1
1/2	Butt	1/16	325	24	210	40-50	24	3
			375	26	260			
			375	26	250			
1/2	Butt	3/32	400	26	100	40-50	30	3
			450	29	120			
			425	27	110			
1/2	Fillet	1/16	350	25	230	40-50	24	3
1/2	Fillet	3/32	425	27	105	40-50	26	3
					110			
3/4	Butt	1/16	325	24	210	40-50	24	4
			375	26	260			
			350	25	230			
3/4	Butt	3/32	400	26	100	40-50	24	4
			450	29	120			
			425	27	110			
3/4	Fillet	1/16	350	25	230	40-50	24	5
3/4	Fillet	3/32	425	27	110	40-50	26	4
1	Fillet	1/16	350	25	230	40-50	24	7
1	Fillet	3/32	425	27	110	40-50	26	6

Courtesy of Hobart Brothers Company.

*Data is for the flat position only. Do not use this data for positions other than flat.

[1] For groove and fillet welds. The fillet weld size and material thickness should be the same.

[2] Use a Vee-groove preparation for material 1/4" to 1/2" thick. Use a double-Vee groove preparation for material 1/2" thick and over.

[3] Argon shielding gas should be used with 1% to 5% oxygen.

cussed earlier. It is said to have a 5° pulling angle, as does the welding gun and position shown in Fig. 22-20.

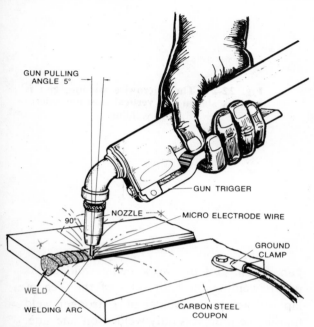

GUN PULLING ANGLE 5°

GUN TRIGGER

NOZZLE

MICRO ELECTRODE WIRE

90°

GROUND CLAMP

WELD

WELDING ARC

CARBON STEEL COUPON

Fig. 22-20. The microwire welding gun is held at a 90° angle to either plate and "pulls" the weld at about a 5° angle.

A welder who is just beginning to microwire weld should use a root opening on square butt joints with steel over $\frac{3}{16}''$ thick. After some practice with microwire welding, however, he will need little, if any, root opening for metal less than $\frac{1}{4}''$ thick.

Vertical and Overhead Procedure—For vertical and overhead position butt welds, it will be necessary to reduce the recommended flat position welding current 10% to 15% for good microwire weld beads.

During vertical position microwire welding, the nozzle should be held near the 90° angle to each plate being welded, with a slight pushing angle as in Fig. 22-21. The recommended welding procedure is from the bottom to the top. However, the welder should practice vertical position welding on light-gauge metal from the bottom up *and* from the top down while using different stickout lengths from $\frac{1}{4}''$ to $\frac{5}{8}''$. During practice,

then, the welder will realize the effect of stickout on penetration. As discussed earlier, less stickout gives greater penetration. With carefully adjusted wire-feed speed and stickout length, the experienced welder will be able to weld light-gauge metal in the vertical position from the top down, as well as from the bottom up.

When welding plates $\frac{1}{8}''$ or thicker, the plates should be spaced with a root gap for vertical position welding. With a stickout of about $\frac{3}{8}''$, a slight weaving motion on the first pass will help keep the molten metal from running down the root gap. By using the information for various thicknesses of metal shown in Table 22-3 and some practice, most welders will have little trouble depositing sound microwire weld metal with good penetration in the vertical position.

Horizontal Position Procedure—Horizontal position microwire butt welding will cause little trouble after the welder has learned to control the problems caused by gravity's pull. When welding microwire butt joints in the horizontal position, the nozzle is held about 5° down from being straight in front of the plates being welded. Welding is usually done from the left to right, with the nozzle being held at the pulling angle as shown in Fig. 22-22.

Lap and Tee Joints

When using the microwire GMAW process to weld lap and Tee-fillet joints, manufacturer's specifications, similar to those shown in Table 22-3, should be followed. The fillet-weld size usually depends on the thickness of the material being welded.

Preparation—When using microwire GMAW in all positions, the welder should be sure that the base metal is thoroughly clean before welding on any filler metal. The stickout, gun angle, gas flow, voltage, and amperage must be set according to the manufacturer's specifications for the thickness of metal being welded. Welding should be done where there are no drafts. This will help the shield-

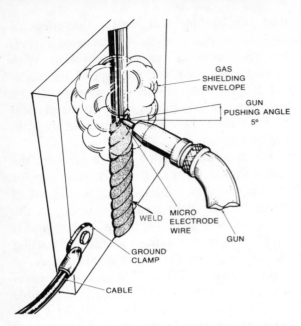

GAS
SHIELDING
ENVELOPE

GUN
PUSHING ANGLE
5°

WELD

MICRO
ELECTRODE
WIRE

GUN

GROUND
CLAMP

CABLE

Fig. 22-21. The microwire welding gun is held as shown for vertical position microwire welding.

ing gas do a good job of keeping the atmosphere's nitrogen and oxygen out of the weld.

Procedure—When welding lap and Tee-fillet joints, the welder should hold the microwire gun nozzle similar to the method recommended for manual shielded arc welding fillet joints. The gun should be held at a 45° angle to either plate and 10° from the direction of travel (a 10° pushing angle). When the angles have been adjusted and the welding helmet lowered, the gun trigger is squeezed to turn the controls on and start the arc.

After the arc is started, the welder should keep a distance of ¼″ to ⅜″ between the gun nozzle and the work being welded. As usual, welding on both sides of the Tee or lap will assure the welder of a stronger weld than if the joint was only welded on one side.

Spray-Arc GMAW

With the spray-arc GMAW process, metal is added to the weld in the form of a fine spray of molten metal droplets. This is different from most other processes because the spray-arc has the molten metal in very small droplets, not the single globules or large drops that other processes send across the

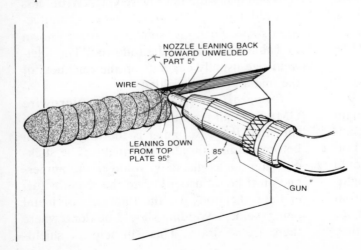

NOZZLE LEANING BACK
TOWARD UNWELDED
PART 5°

WIRE

LEANING DOWN
FROM TOP
PLATE 95°

85°

GUN

Fig. 22-22. When microwire welding in the horizontal position, the gun and nozzle should be held as shown.

arc. Spray-arc GMAW is used mostly for fast, low heat input welding on thinner sheet-metal joints.

Setting Up

The GMAW equipment should be set up for spray-arc GMAW according to the specifications in Table 22-5. As can be seen in Table 22-6, the differences between the spray-

has to move along much faster. Otherwise, the torch angles and techniques for micro-wire and spray-arc GMAW are very similar.

Carbon Dioxide-Shielded GMAW

This process, as the name implies, uses welding grade carbon dioxide shielding gas

Table 22-6. Principal Differences Between Microwire and Spray-Arc GMAW Processes.

Process	Metal Thickness (Inches)	Electrode Diameter (Inches)	Amperage	Voltage	Wire Feed (ipm)	Shielding Gas Flow (cfh)	Travel Speed (ipm)
Microwire	$\frac{3}{16}$	.035	120-160	19-22	210-290	20-25	15-25
Spray-Arc	$\frac{3}{16}$	.035	350	25	230	40-50	32

arc and microwire GMAW processes are generally the amperage and travel speed. These make the spray-arc a hotter, faster welding process.

For spray-arc welding, metal from 1/4" to 1/2" thick should be prepared with a single-Vee preparation. Metal 1/2" thick, or over, should be prepared with a double-Vee preparation. Argon shielding gas with 1% to 5% oxygen should be used to shield spray-arc GMAW.

Procedure

When using spray-arc GMAW, the welder must remember that the current and travel speed are much higher than with microwire. For this reason, the entire welding process

to protect the molten metal from the atmosphere. Butt and fillet welds can be easily made with this process on materials from 18 gauge to 1/2" thick, or more.

Setting Up

Setting up for carbon dioxide-shielded GMAW is similar to setting up for the other GMAW process. Before welding, of course, a specification chart should be referred to for the correct GMAW settings and adjustments for this particular process. Table 22-4 gives the correct settings for common GMAW with the carbon dioxide process. Table 22-7 compares the welding specifications and settings for carbon dioxide-shielded and spray-arc GMAW processes.

Table 22-7. Principal Differences Between Spray-Arc and Carbon Dioxide-Shielded GMAW Processes.

Process	Metal Thickness (inches)	Electrode Diameter (inches)	Amperage	Voltage	Wire Feed (ipm)	Shielding Gas Flow (cfh)	Travel Speed (ipm)
Spray-Arc	1/2 (Fillet)	$\frac{3}{32}$	425	27	105-110	40-50	26
Carbon Dioxide-Shielded	1/2 (Fillet)	$\frac{3}{32}$	625	36	160	35	23

Procedure

The welding procedures for carbon dioxide-shielded GMAW are basically the same as for the spray-arc process. For practicing carbon dioxide-shielded GMAW, however, a new welder should practice with different stickout and nozzle lengths.

The nozzle angle for most position welding will be about the same as the angle used for manual shielded arc welding, except that the pulling or pushing angle should never be slanted down more than 10° from straight up, especially in the flat position. This is similar to Fig. 22-20. For a square-groove butt welds, a copper or steel backing strip will make it easier to weld with carbon dioxide shielding gas. For this reason, backing strips should be used with this process, when possible. Before practicing groove welds, it is a good idea to be able to weld good fillet and lap welds. As with other processes, the welder must refer to the weld specifications for the thickness of metal that he will be welding for good carbon dioxide-shielded GMAW. See Table 22-4.

Fillet Welding with Carbon Dioxide GMAW

Adjustments—For fillet welds with carbon dioxide-shielded GMAW, the rheostat should be set at about 23 volts for practice welds. Then, it can be adjusted for a smooth arc after some weld samples have been made. The wire-feed speed control can then be adjusted, followed by adjusting the welding machine current for 260 to 275 amperes. The shielding gas flow should be adjusted to about 23 cfh.

With these settings made, the welder is prepared to practice Tee- and lap-joint fillet welds on ¼″ metal stock with the carbon dioxide shielded-GMAW process.

Procedure—The nozzle of the gun should be held from ⅜″ to ½″ from the work. Then, the gun trigger can be pulled to start and establish the arc. The plates are then tacked for either lap- or Tee-joint practice.

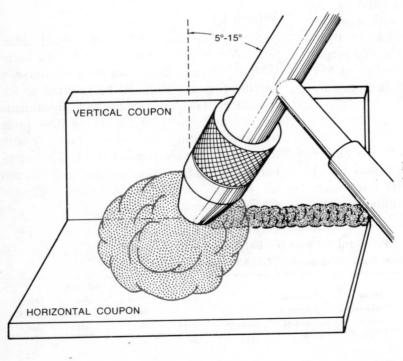

5°-15°

VERTICAL COUPON

HORIZONTAL COUPON

Fig. 22-23 Welding with the micro-wire GMAW gun nozzle held at the correct angles.

Welding can then be done in a single pass from left to right. While welding, the gun nozzle should stay at 45° to either coupon, and the gun should be held about 10° to 15° in the direction of travel. See Fig. 22-23.

While trying out different travel speeds, it may be noticed that reduced travel speed makes a larger weld bead. See Fig. 22-24.

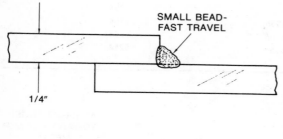

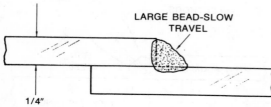

Fig. 22-24. As the welding travel speed increases, the bead size decreases.

Since the thickness of the metal being welded is related to the bead size, the welder may have to adjust his travel speed to obtain the proper bead size. This may be especially necessary on lap-joint fillet welds.

MIG Welding Aluminum

Once a welder is able to do a good job of welding mild- and low-alloy steels by the different GMAW processes, he will be able to comfortably join aluminum, magnesium, and stainless steel with the standard MIG welding procedure.

Manual MIG welding is commonly used to join ⅛″ or thicker aluminum. Aluminum thinner than ⅛″ is more difficult to weld because of problems controlling penetration and distortion in the weld metal. Manual

MIG welding is usually replaced by automatic if the work can possibly be positioned in either the flat or horizontal position. Manual *TIG* (Tungsten Inert Gas) welding is easier to use than MIG welding when making aluminum welds that require quick direction or position changes. This is usually noticed when different kinds of pipes are being joined. Today, however, MIG welding is preferred for aluminum welding in general and particularly for production welding, since flux is not required and little or no finishing is necessary for good-looking weldments.

Filler Metal

When MIG welding aluminum, the welder should refer to specifications such as Table 22-8 for correct filler metal electrode wire. Using the wrong filler metal alloy could result in a weld that is both weak and/or brittle. In all cases, the manual MIG aluminum electrode wire must be very clean and must have the composition and mechanical properties of the aluminum being welded. Spools of aluminum electrode wire should be kept covered and stored at an even temperature in a dry place.

Preparation

For good MIG aluminum welds, fit-up and cleanliness are very important. The edges being joined must be clean and parallel all along the weld line.

Because of this, it is important to clean the joint by using either mechanical or chemical methods. Abrasive cloth and fine steel wool, for example, may be used as *mechanical* cleaners.

A *chemical* method that is good for cleaning aluminum parts before welding is done by dipping the parts in a 5% sodium hydroxide solution that has been heated to 150° F. After the parts have been dipped for about one minute, they may be removed and rinsed in cold water. Then, they are dipped in a 50% nitric acid solution at room temperature. After one minute in the nitric acid bath,

Table 22-8. Recommended Aluminum Filler Alloy Electrode Wires for MIG Welding Common Aluminum Alloys.

Aluminum Alloy	Major Filler Alloy (Recommended Filler Alloy)	Alternate Filler Alloy
1100	1100	4043
3003	1100	4043
4043	1100	4043
5052	5356	5554 or 5254
5254	5254	5554
5454	5554	5254 or 5356
5554	5554	5254
5356	5356	5254
5083	5356	5254
6351	4043 or 5356	5254
6061	4043 or 5356	5254
6063	4043 or 5356	5254
X7004	5356	

the parts should be rinsed in boiling water and dried.

Positioning

When MIG welding aluminum, it is usually necessary to use clamps or other simple fixtures to hold the parts in place for tacking and welding. Sometimes, using both a fixture *and* tack welding are necessary to hold the parts in the correct position.

When using temporary flat aluminum backing strips, the welder should be sure that the strips don't become fused with the joint being welded. Backing strips should be removed if a reverse pass is to be made after gouging. *Permanent* backing strips should be made from the same composition as the parent metal and filler alloys, to insure complete fusion with the weld joint.

Torch Angle

When MIG welding aluminum, the gun is held at an angle like the angle used for different types of steel. The angle for butt joints is about 90° to either plate, Fig. 22-20.

The gun angle used for Tee and lap joints should be about 45° toward either coupon. However, when the pieces being welded are not the same thickness, the gun should be tilted slightly toward the heavier member. This will insure good, equal melting of each part. See Fig. 22-25.

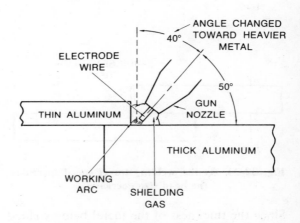

Fig. 22-25. When heavy aluminum stock forms one side of a lap weld, the torch must be angled to point more toward the thicker stock.

Gun Technique

Because the MIG gun adds its own filler metal, the welder can have *both* hands on the gun. Using both hands on the gun while welding will help the welder easily run a steady bead. (With TIG welding, of course, the welder needs one hand for the filler rod.)

In order to keep the welding gun fairly cool, the arc should be completely outside the shielding gas nozzle at all times. To see the arc without disturbing the inert gas shield, the welder should hold the gas nozzle about ½″ from the work.

The MIG welding arc is powerful and puts a great deal of heat into the work during most steel welding. For this reason, the welder has to guard against a *cold start*.

To prevent a cold start, the welder should start the arc on a piece of metal near where the joint is to be welded. The arc should also be allowed to stop on another piece of metal at the other end of the weld bead. The two pieces of metal used to start and stop the arc are known as *run-on* and *run-off* blocks. See Fig. 22-26.

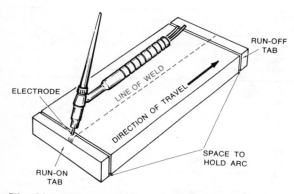

Fig. 22-26. Run-on and run-off blocks may be used to avoid a cold start when MIG welding aluminum.

Aluminum MIG Welding Procedure

Due to the number of aluminum alloys being used today, it is always necessary to refer to the manufacturer's specifications before MIG welding aluminum. See Fig. 22-27. Differences in filler alloy composition will have a definite effect on the burn-off rate. This, in turn, helps decide what procedure should be used. Good specifications will answer these questions for the welder.

A short, low-voltage MIG arc is used to penetrate the aluminum and produce small fillet welds. The short arc can also be used to deposit root passes before applying the top weld passes. For applying a *cover* pass, the arc is lengthened to give less penetration.

Butt Joints—Using the technique discussed for MIG welding alloy steel and following the manufacturer's specifications, a welder

should have little trouble successfully MIG welding aluminum butt joints.

When a butt joint is to be welded from both sides, however, the welder should gouge the reverse side to provide a smooth, even groove for the reverse pass. Removing metal from the back of the root pass will remove incomplete fusion defects *and* oxide inclusions. See Fig. 22-28.

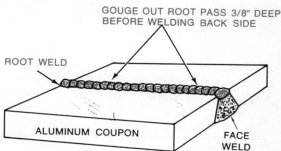

Fig. 22-28. Back-gouging before rewelding will improve the strength of the aluminum MIG weld if it is done correctly.

Vertical-position butt joints should be welded from the bottom up. Like flat and horizontal butt joints, vertical-position butt joints can be welded with or without permanent or temporary backing strips. Manual MIG welding butt joints in the overhead position is not recommended unless backing is used. If it is not possible to put backing on the opposite side of the aluminum in the overhead position, the work should be welded in the flat position.

Tee and Lap Joints—As with butt joints, Tee and lap joints can be MIG welded by using the technique used to weld the same joints made with mild steel.

After the settings have been made according to the manufacturer's specifications, the aluminum plates can be tacked as shown in Fig. 22-29. With the nozzle held at a 45° angle to either plate, the arc is started and the joint welded from left to right. Holding the gun nozzle at about 15° in the direction of travel should also help the welder produce good MIG welds on aluminum plate. See Fig. 22-30.

Butt Joints—Flat With Temporary Backing

Procedures for welding with **4043** filler alloy
Parent alloys: 6351, 6061, 6063, 1100, 3003

Metal Thickness (in.)	Weld Pass Number[1]	Electrode Wire Dia. (in.)	Direct Current[2] (amp)	Arc Voltage[2,3] (volts)	Arc Travel Speed (ipm)	Argon Flow (cfh)	Shielding Gas Nozzle Dia. (in.)	Details and Order of Welding	AWS Weld Symbol
1/8	1 B2R	0.047	155 155	22 23	36 48	30	1/2	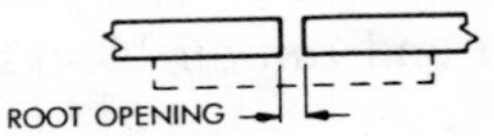 ROOT OPENING 1/16 MAX. 3/32	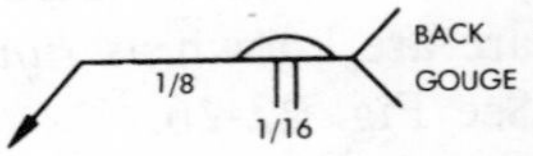 1/8 / 1/16 BACK GOUGE
3/16	1 B2R	0.062	200 200	22 23	24 48	30	1/2	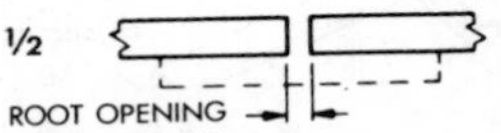 ROOT OPENING 3/32 MAX. 1/8	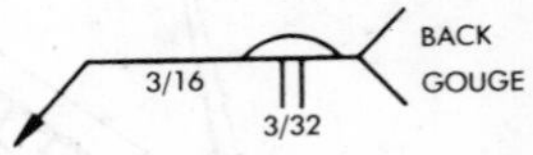 3/16 / 3/32 BACK GOUGE
1/4	1 2 B3R	0.062	230 230 230	23 25 25	48 36 36	35	5/8		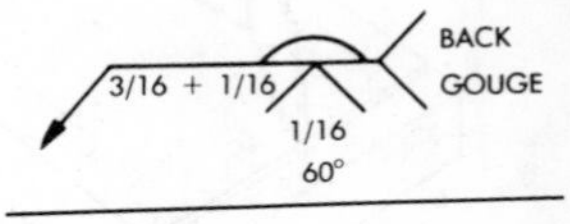 3/16 + 1/16 / 1/16 / 60° BACK GOUGE
3/16	1 2 B3R	0.062	250 250 250	23 25 25	36 32 41	40	5/8	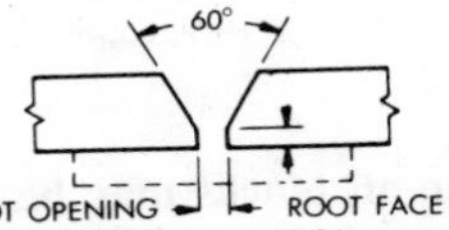 60° ROOT OPENING 1/16 MAX. 3/32 ROOT FACE 1/16	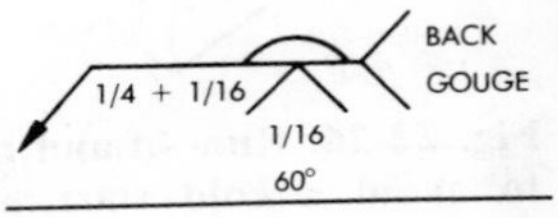 1/4 + 1/16 / 1/16 / 60° BACK GOUGE
3/8	1 2 B3R	0.062	270 270 270	23 25 25	24 16 24	40	5/8		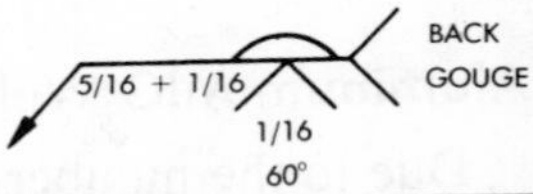 5/16 + 1/16 / 1/16 / 60° BACK GOUGE
1/2	1 2 3 B4R	0.062	270 270 270 270	23 24 24 24	23 23 20 30	40	5/8	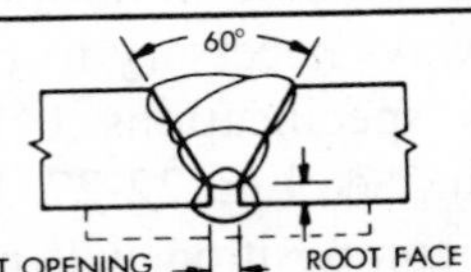 60° ROOT OPENING 1/16 MAX. 3/32 ROOT FACE 1/16	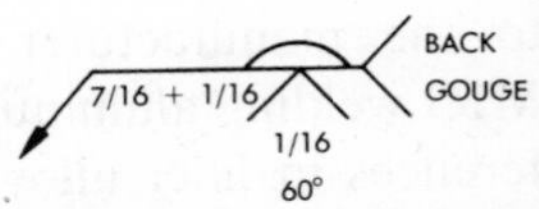 7/16 + 1/16 / 1/16 / 60° BACK GOUGE

[1]B indicates back gouge before making pass.
R indicates weld pass made on the reverse of
the side on which the first pass was made.

[2]Current and arc voltage should not vary more than 3% during welding.
[3]Measured between the contact tube and the work.

Fig. 22-27. Typical MIG welding specifications. (Courtesy of Alcan Aluminum Ltd.)

After tacking the coupons, the welder should start the arc with the gun nozzle about 3" *ahead* of the weld starting point. There, the nozzle should be held about ¾" from the joint, as in Fig. 22-31. Then, the trigger is pulled as the gun is moved toward the starting point. The welder should quickly move the gun toward the starting point while he gradually lowers the gun nozzle, so that the gun nozzle will be only ⅜" from the joint when the welder actually gets to the joint starting point. See Fig. 22-31.

With the working arc at the starting point, the welder must wait until a ⅜" bead face is built up before moving along the weld joint. After the bead face has been estab-

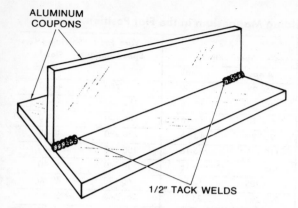

Fig. 22-29. A Tee joint, tack welded before MIG welding.

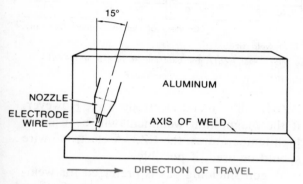

Fig. 22-30. When MIG welding aluminum Tee joints, the gun nozzle should be pointed about 15° toward the direction of travel.

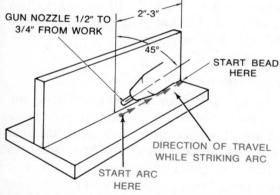

Fig. 22-31. Starting the arc and bead with a MIG gun on an aluminum Tee joint.

lished, the nozzle should be moved smoothly in the weld direction at the right travel speed for an even bead width.

When the weld is completed, the gun is backed up about ⅜″. There, a short pause will cause the crater made by the arc to fill

up. When the crater is filled, the gun trigger is released and the gun is left *in* the welding position until the arc has stopped. This will prevent excessive welding wire stickout.

With Tee joints, the welder may need to use alternate beads to help control overheating and warping the weldment. See Fig. 22-32.

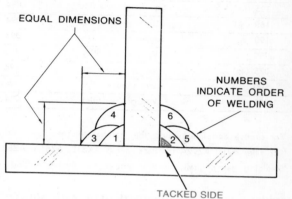

Fig. 22-32. MIG welded aluminum Tee joint with alternate beads to control overheating and distortion.

MIG Welding Magnesium

MIG welding magnesium is fairly similar to MIG welding aluminum. By using the manufacturer's recommended specifications, Table 22-9, and the same welding procedures as for aluminum, most welders will have little, if any, trouble MIG welding magnesium.

Magnesium from 24 gauge to 1″ thick can be easily welded with the MIG process. However, the welder should use a Vee-groove preparation for MIG welding magnesium ¼″ thick or over.

Methods Used — Magnesium over ³⁄₁₆″ thick is welded by the GMAW/MIG spray-arc process. Metal transfer by the spray-arc process must be done with at least 200 amps and over 20 volts. For MIG welding aluminum with amperage and voltage below these figures, metal transfer must be done with the microwire process.

The microwire process deposits metal only when the wire shorts to the base metal, so a

Table 22-9. Recommendations for MIG Welding Magnesium in the Flat Position.

Material Thickness[1] (Inches)	Type of Weld	Wire Diameter[2] (Inches)	Arc Specifications		Wire Feed Speed (ipm)	Shielding Gas Flow[3] (cfh)
			Amps	Volts		
.025	Square groove fillet	.040	26- 27	13-16	180	40-60
.040	Square groove fillet	.040	35- 50	13-16	250-340	40-60
.063 (1/16)	Square groove fillet	.063	60- 75	13-16	140-170	40-60
.090 (3/32)	Square groove fillet	.063	95-125	13-16	210-280	40-60
.125 (1/8)	Square groove fillet	.094	110-135	13-16	100-130	40-60
.160 (5/32)	Square groove fillet	.094	135-140	13-16	130-140	40-60
.190 (3/16)	Square groove fillet	.094	175-205	13-16	160-190	40-60
.250 (1/4)	Vee groove fillet	.063	240-290	24-30	550-660	50-80
.375 (3/8)	Vee groove fillet	.094	320-350	24-30	350-385	50-80
.500 (1/2)	Vee groove fillet	.094	350-420	24-30	385-415	50-80
1.000 (1)	Vee groove fillet	.094	350-420	24-30	385-415	50-80

Courtesy of Hobart Brothers Company.

[1]For groove and fillet welds, the material thickness also indicates fillet weld size.
[2]Use filler metal AZ92A for base metals AX31B, AZ61A, AZ80A. Use filler metal AZ61A for base metals AZ31B and AZ61A.
[3]Shielding gas normally used in argon. For thicker sections, an argon/helium mixture is recommended.

colder arc is actually used than with spray-arc. As with aluminum welding, the short arc transfer allows magnesium from 0.025 to 3/32″ thick to be easily welded, because the molten puddle quickly becomes solid. The filler metal and base metal weld by the micro-wire process gives an excellent metallurgical bond, as shown in Fig. 22-33.

Weld Conditions—Magnesium base metals AX31B, AZ61A, and AZ80A should be welded with an AZ92A filler metal electrode wire. AZ31B and AZ61A base metals use AZ61A filler metal electrode wire. However, if the magnesium weldment will be used at high temperatures, the EZ33A electrode wire should be used, if possible.

Argon shielding gas is preferred for welding magnesium up to 1/8″ thick. For pieces over 1/8″ thick, an argon/helium shielding gas mixture will allow better penetration and better shielding. Table 22-10 shows the effect of different shielding gases on MIG-welded AZ31B alloy magnesium.

Magnesium Butt Welds

Using the microwire process, magnesium butt joints can be welded with complete penetration. If welded from both sides, the root of the first weld is usually back gouged by a round-nose chisel before applying the rear weld bead. Back-gouging, as with aluminum welding, assures the welder a weld deposit free of porosity. Even if complete penetration is not required, the same cleaning procedures apply. See Fig. 22-34.

Generally speaking, all the MIG welding procedures for aluminum butt welds also apply to magnesium butt welds, except as noted above.

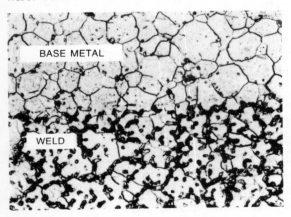

Fig. 22-33. Good base metal and filler metal bond on MIG welded magnesium. (Courtesy of The Dow Chemical Company.)

Table 22-10. Effect of Different Shielding Gases on Weld Appearance and Operating Limits for Welding 0.063-Inch AZ31B Magnesium Alloy.

Shield Gas %			Arc		Typical Weld Cross Section
Argon	Helium	Chlorine	Volts	Amps	
100	0	—	14.9-16.4	52-59	
90	10	—	13.4-15.3	56-65	
80	20	—	13.5-15.3	56-67	
75	25	—	14.3-15.3	55-60	
70	30	—	13.7-15.4	53-65	
25	75	—	14.6-16.7	47-56	
0	100	—	14.4-16.2	46-54	
99	—	1	14.3-16.8	50-57	*
97	—	3	14.5-16.3	51-61	*

*Film formation on weld retards flow or spread of the weld bead.   Courtesy of the Dow Chemical Company.

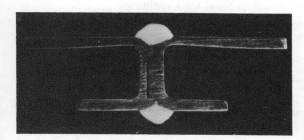

Fig. 22-34. MIG welded butt welds on two extruded magnesium pieces. Back gouging is not possible, so the metal must be very clean when welded. (Courtesy of The Dow Chemical Company.)

Magnesium Fillet Welds

The carbon dioxide shielded, globular metal transfer GMAW/MIG welding process is used for good root penetration and concave bead shape on magnesium fillet welds. See Fig. 22-35. Due to the lower current

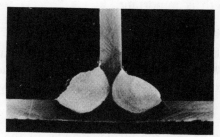

Fig. 22-35. Carbon dioxide shielded, globular transfer GMAW/MIG weld on 1/8″ magnesium stock. (Courtesy of The Dow Chemical Company).

used in microwire GMAW/MIG welding, a convex bead with less penetration results when that process is used to weld magnesium, as shown in Fig. 22-36. Spray-arc may be used on magnesium over 0.160″ thick to reduce the weld spatter commonly found with the globular transfer process and to give better penetration and bead contour.

Fig. 22-36. Microwire MIG weld made on 1/8″ magnesium stock. (Courtesy of The Dow Chemical Company.)

Other than choosing the process to be used, magnesium fillet welding procedure is similar to aluminum. Magnesium lap joints, Fig. 22-37, may be GMAW/MIG welded by

Fig. 22-37. Magnesium lap joint, MIG welded. (Courtesy of The Dow Chemical Company.)

either the microwire or globular transfer methods. For complete root penetration, the carbon dioxide shielded, globular transfer process will be the most satisfactory. Even with the globular transfer process, welding on both sides is recommended where total strength is important.

Magnesium Corner Welds

Corner welds can be welded by the microwire MIG process. Weld penetration may be controlled by using back-up bars as shown in Fig. 22-38. However, the welder will have fewer buildup problems if the joints are designed so that the inside surfaces of the base

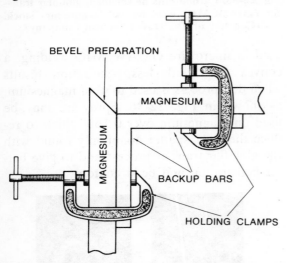

Fig. 22-38. With one magnesium coupon beveled on one end and both coupons positioned as shown, a strong MIG corner weld can be made on magnesium stock.

members meet to form the inside corner, as shown in Fig. 22-39.

MIG Welding Stainless Steel

Microwire GMAW/MIG welding is usually used to fabricate stainless steel units. This is due to two factors: the increased welding speed of microwire GMAW, and the appearance and quality of the welds produced. The techniques and procedures for GMAW/MIG welding stainless steel are the same, generally, as the techniques recommended for the other metals that may be MIG welded.

As usual, the welder must carefully follow the manufacturer's recommendations for the set-up differences between stainless steel and other metals. The stainless steel GMAW/MIG specifications shown in Table 22-11 may be used with the MIG welding techniques described earlier for other metals. Together, these will help any welder develop skill in all-position MIG welding stainless steel.

Shielding Gas

The shielding gas used for stainless steel GMAW/MIG welding has a definite effect on the microstructure of the MIG-welded stainless steel. Because of this unusual condition, the welder must know about the type of shielding gas needed for each welding position and other weld conditions.

Table 22-11. Recommendations for Stainless Steel MIG Welding Fillet and Groove Welds in the Flat Position.

Material Thickness[1] (inches)		Electrode Diameter (inches)	Welding Current DC-RP (amps)	Arc Voltage	Wire Feed (ipm)	Shielding Gas Flow[2] (cfh)	Travel Speed (ipm)	No. of Passes
Fraction	Decimal							
1/16	.063	.035	60-100	15-18	90-190	12-15	15-30	1
3/32	.093	.035	125-150	17-21	230-280	12-15	20-30	1
1/8	.125	.035	130-160	18-24	250-290	12-15	20-25	1
5/32	.156	.045	190-250	22-26	200-290	15-18	25-30	1
1/4	.250	.045	225-300	22-30	260-370	25-30	25-30	1

[1]Material thickness also indicates fillet weld size.

[2]Recommended shielding gas depends on weld conditions. See text.

Courtesy of Hobart Brothers Company.

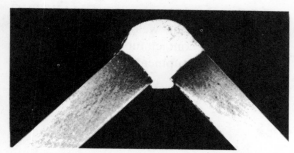

Fig. 22-39. The inside corners of these magnesium coupons were touching before they were MIG welded, reducing build-up problems. (Courtesy of The Dow Chemical Company.)

Argon and 1% to 5% oxygen shielding gas can only be used for high-frequency, flat position welding. Welding with this gas mixture creates a more convex bead than is usually required, especially for fillet welds.

An argon/carbon dioxide mixture (75% argon and 25% carbon dioxide) may be used for all position welding. Welds made with this gas mixture, however, are slightly convex and heavily oxidized. Also, the argon/carbon dioxide shielding gas mixture may try to pick up carbon while welding. If carbon is picked up, especially in multipass welding, the carbon could reduce the corrosion-resistance of the weld metal. For this reason, the argon/carbon dioxide shielding gas mixture should only be used if corrosion-resistance is not too important.

Overall, for clean welds in all positions (especially Tee and lap joints), the shielding gas recommended is the following mixture: 90% helium, 7½% argon, 2½% carbon dioxide.

Troubleshooting MIG Welding

Like other welding processes, GMAW/MIG welding may present some problems if the equipment is not set up and used correctly. Many welding problems are common to both TIG and MIG welding. MIG welding may have additional problems because the electrode is moving down the cable to be consumed at the arc.

Reviewing a few troubleshooting problems will help any welder better understand MIG welding.

GMAW/MIG Reminders

A. If the arc is not properly started, the filler wire may fuse to the work. This will cause the electrode wire to break or bend at the end of the contact tube, as in Fig. 22-40. When this

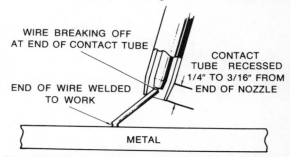

Fig. 22-40. If the MIG arc is not started correctly, the electrode wire may weld itself to the work and break off inside the contact tube.

happens, the electrode may "freeze" to the tube by welding itself to the contact tube's wall. After turning off the welding machine, the welder should be sure that the electrode wire in the contact tube is free before trying to again start the arc and wire feed.

B. The contact tube (tip) should be recessed in (pushed down into) the gun nozzle about ¼″ to ³⁄₁₆″. See Fig. 22-17.

C. All metals to be MIG welded should be cleaned before welding. This will help eliminate contaminants and dirt that could enter the molten pool and cause weld problems.

D. The specifications for voltage, amperage, and shielding gas flow settings must be carefully followed. Too little a shielding gas flow may cause the weld bead to pick up contamination from the air. These contaminants may include oxygen, nitrogen, or solid particles.

E. Too high a voltage setting and/or too long an arc length may cause the wire to burn up into the contact tube. This is especially true when the feed is turned off to stop the weld.

F. For the best possible GMAW/MIG welding, the welder should keep the nozzle, contact tube, and other parts of the equipment clean and properly adjusted.

23

Pipe Welding

Modern pipes are manufactured from both ferrous and nonferrous metals, as well as from different types of plastics. Pipes made from ferrous metals (wrought iron, low and high alloy steels, stainless steels) are still in great demand. Although *wrought iron* pipe is used when corrosion will be a problem, *plastic* pipes appearing on the market are being used, where possible, in place of wrought iron. Plastic pipe is lighter than wrought iron, easier to handle, and, in most cases, meets the requirements for strength and corrosion-resistance. Plastic pipes can either be joined by adhesive bonding or a special plastic welding process.

Pipes containing *chromium* and *molybdenum* are still used in power plants where they may have to hold up under temperatures as high as 700° F. *Stainless steel* pipe is produced in austenitic grades for use with chemicals and low or high temperature fluids. It may be welded in a manner similar to steel pipe. However, the electrodes, welding specifications, and techniques mentioned earlier for stainless steel welding should also be used for stainless steel pipe welding.

Aluminum and *copper* alloy pipe welding techniques are also similar to the procedures used for welding steel pipe. Because of these similarities, the steel welding techniques can be easily adapted to fit the requirements for welding aluminum and copper pipe.

Qualifications for Pipe Welding

Until recently, pipe welding qualifications have varied in different parts of North America. To stabilize pipe welding qualifications and pipe system testing, standards have been set up by the American Petroleum Institute (API).

To be approved, pipe welders must pass certain tests. Today, the tests normally depend on the pipe system codes being used by pipe system owners, engineers, and local and/or federal authorities.

Welders of subcritical mild steel pipe, for example, are often expected to pass rigid tests before attempting to weld even subcritical systems. However, because the lower working pressure of subcritical pipe systems requires less tensile strength than high pressure pipe lines, a lower certification standard may be made for subcritical pipe welders.

Advantages of Welded Pipe

Material Saving — Welded pipe fittings, Fig. 23-1, are now available for air conditioning, heating, refrigeration, and other plumbing jobs at a cost that is often less than threaded fittings. Therefore, using welded

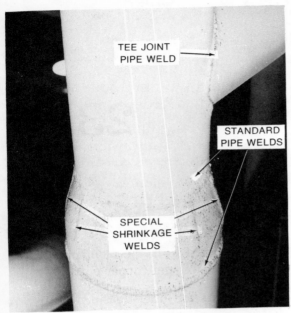

Fig. 23-1. Typical welded pipe fittings.

pipe on these systems would save up to 50% of the material cost alone. Since threaded pipe is no longer required for these jobs, thinner-walled pipe can be used without the loss of strength that comes from thread cutting. See Figs. 23-2 and 23-3.

In Fig. 23-2, the welded pipe fitting has the same outside and inside diameter as the pipe section. If properly welded, the joint will be as strong as the rest of the pipe. In Fig.

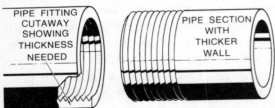

Fig. 23-2. Threaded pipe fittings must have thick walls because the threads will be cut in the pipe wall.

23-2, where threaded fittings are used, the wall must be thick enough to allow the material to be cut out to form the threads needed for attaching the fitting.

Labor Saving—Installing a welded pipe system is quicker and easier than installing a system using all threaded pipe. Pipes can be cut to length for welded systems without having to allow for any pipe overlap in which to cut threads. See Fig. 23-2. Installing the welded system is also easier because the pipe being used with a welded system is lighter. Table 23-1 illustrates the approximate time needed to complete a joint using each of the two systems.

Strength—A properly welded pipe system is strong, leak proof, and will usually outlast a similar system that has been put together with threaded fittings. When properly welded, the fittings become an integral part of the welded pipe system. For this reason, vibration and other stresses do not affect the connection as easily as they do threaded systems. See Fig. 23-4.

Fittings properly welded to pipes seldom leak. Threaded fittings, however, especially those using gaskets, may require extra time for eliminating some leaks when the joints are tested under high pressure.

Easier Flow—Turbulence, erosion, and clogging are greatly reduced in the welded system, especially where the inside diameter of the fitting and pipe fit up to eliminate any pocket. These pockets cannot be eliminated in a threaded system because of variations made by the threaded joint. See Fig. 23-5.

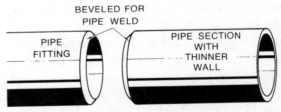

Fig. 23-3. Welded pipe fittings may have thinner walls because there will be no threads cut in the pipe wall.

Insulation and Appearance — Insulating welded systems is simple and easy, since the outside diameters of the pipes and fittings are identical. Because the outside surface is smooth, either "wrap-around" or tubular insulation can be easily and quickly installed.

Table 23-1. Time Comparison Chart for Welded and Threaded Pipe Fittings.

NOMINAL Size of fitting	6″	5″	4″	3″	2½″	2″	1½″	1¼″	1″
MINUTES FOR THREADING (Includes: Cutting and Threading the Pipe and Tightening the Fitting.)	51	39	28	20	15	10	8	6	5
MINUTES FOR WELDING (Includes: Cleaning Ends of the Pipe, Welding, and Cleaning the Weld.)	26	21	16	12	10	8	7	6	5

Courtesy of Lincoln Electric Co.

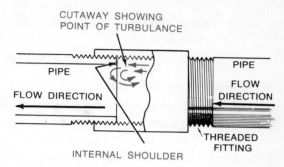

Fig. 23-5. The normal internal shoulders in a threaded fitting may create turbulence and cause clogging.

Nesting the pipes against walls and corners will give a better job appearance and save space when a welded system is used. Welded fittings are more compact, and the space needed for welding is less than the space needed for using a wrench on threaded fittings. For these reasons, installing pipe in compact spaces is both possible and easy with welded pipe and fittings.

Methods of Welding Pipe

Both oxygen-gas welding and several forms of arc welding may be used to weld subcritical pipe systems made of pipe from 2″ to 6″ in diameter. Pipe welding with the oxygen-gas method is described in another chapter.

Manual shielded arc welding was used for pipeline welding as early as 1928. Since that time, its use has increased, and today almost all transmission pipeline is arc welded, using one of several methods. As pipe transmission lines come to be used for moving many more commodities (goods) besides oil and gas, a welder will have to know more techniques used for welding transmission pipelines.

Various forms of arc welding are gradually taking over the subcritical pipe joining process, too, so welders should be prepared to weld these systems with acceptable standard procedures. This chapter will outline the welding technique used to arc weld both critical and noncritical pipe systems.

Types of Pipe Systems

The pipe industry can be divided into four different groups for welding purposes. Each pipe system has its own welding requirements and specifications. The four groups include:

1. Subcritical Pipe Welding
2. Critical Pipe Welding (Pressure and Power Systems)

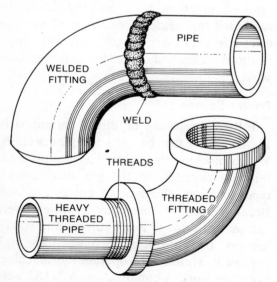

Fig. 23-4. Similar pipe joints for comparison, welded and threaded.

3. Cross-Country Transmission Pipeline Welding
4. Water Transmission Pipe Welding

Subcritical Pipe Welding

With subcritical pipe welding, most of the pipe and fittings are fairly small in diameter and have standard wall thicknesses. Low-pressure heating systems, air conditioning, different service installations, and simple water supply systems are all classified as subcritical pipe systems. Welding subcritical pipe systems is classified as low-pressure pipe welding because there is little or no human danger if a defective weld is made. For this reason, special welding qualifications are not often required for subcritical pipe welding in many places.

Critical Pipe Welding (Pressure and Power Systems)

Welding pressure and power systems are classified as critical because of the high pressure that the system is usually required to withstand. Unlike subcritical pipe welding, welders working on pressure and power pipe systems must have special qualifications meeting any pipe code requirements for the area where the welding is being done. Critical pipe welding is required for pipe systems in refineries, generating plants, chemical processing plants, petroleum plants, and other places where weld defects could be both costly and dangerous to human life.

Cross-Country Transmission Pipeline Welding

Cross-country transmission pipelines are usually used to transport natural gas or petroleum products. In the near future, however, it is expected that other commodities (goods) will be easily and cheaply moved by using cross-country pipelines.

The pipe used for cross-country pipelines varies from about 10″ diameter branch lines to over 36″ diameter main lines. Weld qualification standards for transmission pipeline welding in the United States are set up by the American Petroleum Institute (API).

Water Transmission Pipe Welding

Water transmission pipe welding can be done by any welder, without special welding qualifications. If the water transmission pipe is under government jurisdiction (area of control), however, the welder must be certified to weld on that particular job.

Water transmission pipelines come in sizes as small as 1″ in diameter for branch lines. Main lines, on the other hand, come in sizes up to 8′ in diameter and may be cement-coated. Water transmission lines for crop irrigation are often made of galvanized iron to withstand corrosion. Unlike cement-coated pipes, corrosion-resistant galvanized pipes are fairly easy to install and weld.

Manual Arc Welding Subcritical Pipe

If a new welder has practiced the various welds and positions outlined earlier, he should have little difficulty manually arc welding *subcritical* pipe. In the beginning, however, the welder should test each weld to find any weaknesses in his pipe welding techniques. A type of practical test, outlined later, will easily tell the welder if he has used the right electrode angle, arc length, and weld specifications mentioned earlier for a strong, sound weld.

Power Source

Either alternating or direct current welding machines can be used for pipe welding. Which machine the welder will use on a certain job depends on the job specifications, type, location, and type of power available. Because rectifier machines provide careful arc control through exact current and voltage settings, they are often used when the right current is available to operate the machines. For shop work, a combination AC/DC welding machine is often chosen because of its light weight and portability. See Fig. 23-6.

Fig. 23-6. Typical AC/DC arc welding machine for general shop use on pipe welding and other jobs. (Courtesy of Lincoln Electric Co.)

Fittings

Manufacturers make different sizes of valves and fittings that are prepared for *butt* or *socket* welding. See Fig. 23-7. The fittings

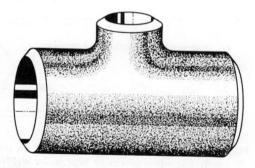

Fig. 23-7. Welded pipe fittings, already prepared with a bevel end for welding.

prepared for *butt* welding are smoothly beveled at the right angle for the type and size of pipe being used. The pipe, in turn, can be cut and beveled by automatic tool- or flame-cutting machines. Manual (hand) cutting will

usually require some grinding for a notch-free joint.

Socket welding fittings, Fig. 23-8, are usually used for pipe up to 2″ in diameter. These fittings can be quickly and easily soldered to copper and copper alloy pipe.

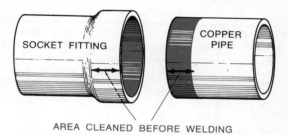

SOCKET FITTING   COPPER PIPE

AREA CLEANED BEFORE WELDING

Fig. 23-8. Socket fittings are usually used on copper pipe up to 2″ in diameter.

Electrodes

Subcritical pipe can be successfully welded by using different electrodes that are compatible (work well) with the power source available.

AWS E6010 and E7010 electrodes are excellent for use on subcritical pipe systems where DC current is being used. These electrodes should be used with reverse polarity for horizontal position and vertical-up pipe joint welding. The penetrating characteristic of reverse polarity is very desirable for these welding jobs.

When welding vertical-down stringer beads on subcritical pipe, straight polarity should be used, especially where large gaps in the joint need to be filled. Straight polarity will give less penetration and help eliminate burn-through. The Shield-Arc 85, for example, is a typical electrode for such a job. It is manufactured by the Lincoln Electric Company and is available in $\frac{3}{32}$″ diameter for welding pipe 3″ in diameter or less, as well as for fine, spray-arc transfer welding.

AWS E6010 and E6011 electrodes will also produce excellent subcritical pipe welds if the manufacturer's specifications for the electrodes are carefully followed. E6011 electrodes will weld subcritical pipe systems on either AC or DC current. Table 23-2 shows the welding specifications for welding pipe with certain wall thicknesses from 1″ to 10″ in diameter.

Table 23-2. Typical Pipe Welding Specifications.

Pipe Dia. (in.)	Wall Thick- ness (in.)	Electrode Dia. (in.)	Current (amps.)	Welding Time (min. per joint)	Pounds of Electrode Used per Joint
1	0.133	3/32	65	5	0.15
2	0.154	3/32	70	8	0.25
2½	0.203	3/32	70	10	0.3
3	0.216	3/32	70	12	0.4
4	0.237	1/8	80	16	0.6
5	0.258	1/8	85	21	0.8
6	0.280	1/8	85	26	1.0
8	0.322	5/32	170	32	1.3
10	0.365	5/32	170	40	1.8

Subcritical Pipe Welding Procedure

By fabricating (welding) sections of a pipe system in a shop with good pipe welding fixtures, labor can be saved and the cost of the total system will be lowered. Unlike threaded pipe, welded pipe does not have any definite order in which it has to be installed into the total system. Once it is properly installed and welded, additional labor will not be needed to stop the leaks that may happen when using threaded pipe.

Aligning Pipe—Welding time can be reduced by making an aligning fixture, Fig. 23-9, especially where several pipes of the same size or many straight fittings are to be welded. The speed and ease of welding will more than make up for the time spent making the aligning fixture.

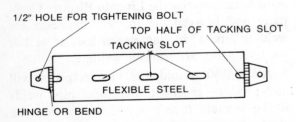

1/2″ HOLE FOR TIGHTENING BOLT
TOP HALF OF TACKING SLOT
TACKING SLOT
FLEXIBLE STEEL
HINGE OR BEND

Fig. 23-9. Flexible steel aligning fixture, used for holding pipes in alignment while tack welding.

The metal used for the aligning fixture must be flexible enough to allow it to tighten around the pipe for accurate setups. Different sizes of aligning fixtures can be made that would be handy when many joints of different size pipe are to be welded.

When the aligning fixture is placed around the pipes to be welded, tack welding can be done by striking an arc on the pipes through the holes in the aligning fixture. See Fig. 23-10.

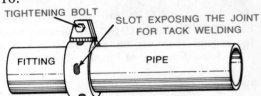

TIGHTENING BOLT
SLOT EXPOSING THE JOINT FOR TACK WELDING
FITTING   PIPE

Fig. 23-10. Using the flexible steel aligning fixture for tack welding the pipes to be welded.

Basic Procedure—The fitting is first butted into the prepared pipe that is held in a fixture. Many times, small diameter pipe can be aligned, as was a broken shaft, by using a piece of angle iron for alignment and support. See Fig. 23-11. A small gap or space is usually left between the butted portions, as in Fig. 23-11. The gap is usually as wide as the electrode wire is thick.

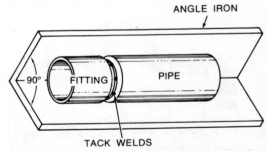

ANGLE IRON
90°   FITTING   PIPE
TACK WELDS

Fig. 23-11. A piece of angle iron may be used to easily align two pieces of pipe to be welded.

The joint should first be tack welded on top (the flat position). Then the unit is turned and checked for correct spacing before it is tacked in the second place, opposite the first. Tacks are then placed on opposite sides of the joint, halfway between the first and second tacks. See Fig. 23-12.

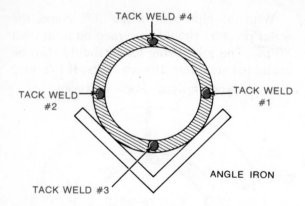

Fig. 23-12. Cross-section view of a tack-welded pipe joint.

Welding Technique—Welding can be done in either the vertical position (Fig. 23-11) or the horizontal position (Fig. 23-13). When welding pipe in the vertical position, many welders like to weld the joint from the bottom up, using at least three passes on large

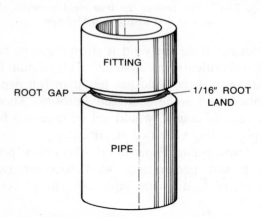

Fig. 23-13. Pipe prepared to be welded in the horizontal position.

pipe. Smaller pipe can also be welded the same way, but two passes are usually enough for the smaller subcritical pipe. It is also possible to weld subcritical pipe from the top to the bottom, for sizes less than 2″ in diameter.

Horizontal position pipe welding will usually need more passes than if the same size pipe was being welded in the vertical position. If pipe is welded in the horizontal position, it will need at least three passes for pipe up to

2″ in diameter. Four passes are usually used on 3″ diameter pipe and as many as 5 passes (Fig. 23-14) may be needed on pipe up to 6″ in diameter.

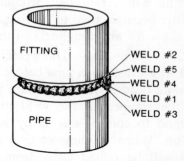

Fig. 23-14. Bead sequence (order of welding) for thick, horizontal position pipe.

Using the correct electrode inclination (angle) with a close arc will help control many problems that could arise. The electrode angle recommended for each position on a common pipe weld is shown in Fig. 23-15. As mentioned earlier, the welder should clean the starting and finishing points

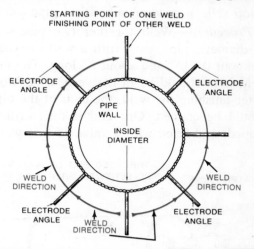

Fig. 23-15. As the side of a pipe is welded, the electrode angle must change as the welder goes up the pipe's side.

of the first half of the pipe weld bead before starting the weld bead on the opposite side. See Fig. 23-15. For the remaining weld beads, the welder should strike the arc on either side of the point where the first bead was started.

This will ensure that the starting point of one bead is not the starting point of another.

Roughness in the land or bevel preparations that are left over from manual flame cutting will create few problems once the welder becomes used to moving the electrode to allow for any roughness in the pipe joint.

Controlling Slag—The welder should chip and wire brush each pass before applying additional beads. This will ensure that slag will not be trapped in the weld area. Another method to control slag is by welding vertical-up. When welding vertical-up, flux and slag will be pulled away from the cleaned bead area by gravity.

Subcritical Pipe Weld Test

The following simple test will help determine the strength and other characteristics of practice subcritical pipe welds. Using the test, a welder will be able to recognize several common pipe weld faults. Practice, followed by more testing, will allow the welder to develop skill in welding subcritical pipe joints.

Procedure—Weld together two pieces of 1″ diameter pipe, each with a wall thickness of about 0.133″ and about 8″ long. The first weld joint should be in the vertical position. After finishing the weld, the ends of the pipe should be capped. One end is then drilled, tapped, and fitted with a valve so that water

With the pipe bent in the "U" shape, the water pressure should be turned on to at least 80 psi. The joint being tested should then be examined very carefully for leaks. If the weld

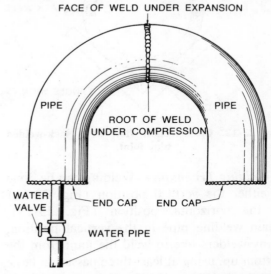

Fig. 23-17. For testing the true weld strength, the pipe is first bent into a U-shape.

does not leak, the weld is strong enough for all subcritical pipe systems. This would be true even if the weld has less than full penetration and some scattered porosity. Those defects, of course, would not be accepted for pipe welding classified as critical.

Porosity and Penetration—To check porosity and penetration, weld together two pieces of 4″ diameter pipe, 2½″ to 3″ long,

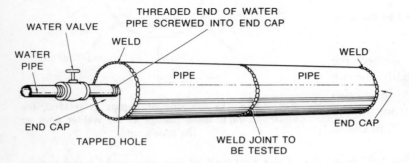

Fig. 23-16. Setup for testing a weld joint. The water pipe is screwed into one end cap.

can be supplied at about 80 to 90 psi pressure. See Fig. 23-16. Then, the pipe is bent by a manual or power bender into a "U" shape, as shown in Fig. 23-17.

in the vertical position. After welding, the pipe should be allowed to cool in a draft-free area. This will ensure that there will be no air-quenching or hardening.

The pipe is then cut as shown in Fig. 23-18, using a hand or power hacksaw. Each newly-cut weld face is filed smooth and then

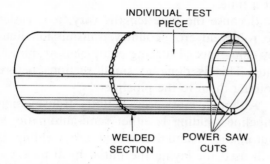

Fig. 23-18. **When testing for good penetration and possible porosity, the welded pipe must be cut as shown.**

etched with 6% nital. Through a chemical reaction, the nital will indicate any defects near the root, face, and/or interior of the weld.

Then, after visual inspection, the four coupons can be placed in a vise and fully bent over, as in Fig. 23-19. Two of the four coupons should have root bend tests, while the

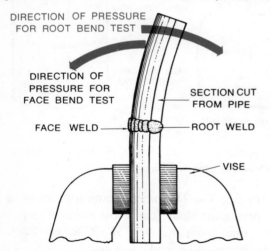

Fig. 23-19. **Testing the weld coupons made in Fig. 23-18.**

other two coupons have face bend tests. If the coupons or welds break, the welder will be able to see the problem that caused the failure by looking at the break. By continued welding and testing practice, he should be

able to eliminate or control any further defects.

The beginning welder can then practice welding pipe in the horizontal position, using the number of beads recommended for the position. Testing can then be done as recommended for other subcritical pipe welds.

Before welding any pipe for commercial purposes, the welder should check with local authorities to see if special welding codes are in effect and/or qualifications are required for the specific welded-pipe system being planned.

Manual Arc Welding Critical Pipe Systems

Welders who will have to weld on *critical* pipe systems (pressure pipe lines or other pressure systems) must take special welding tests to qualify as a certified pressure-system welder. The welder must pass individual tests specified by the code that applies to the system being built. The specifications and methods of administering the tests will vary for different codes. Information is available from the National Certified Pipe Welding Bureau or the American Petroleum Institute, concerning qualifications needed for welders who plan to weld cross-country or in-plant critical pipe systems.

Pipelines today are inspected by sensitive radiographic (x-ray like) equipment that has created higher standards for critical pipe welds. These standards have been established under either the ASA or the ASME welding codes. Reasons for rejecting critical pipe welds, together with suggestions on how to correct the problems, are presented later in the chapter.

Both pressure pipe systems and cross-country transmission pipelines are classified as critical pipe welding jobs. Special tests and qualifications are required to do either type of welding. Although listed separately, the materials and techniques used for cross-country lines and pressure pipe systems are similar enough to be discussed together in this section. Where differences apply, the different divisions will be discussed separately.

Installing Critical Pipelines

Pressure pipeline welding may also be referred to as power pipe welding. Pressure pipe welding is usually done in one of three positions, as shown in Fig. 23-20:

1. Pipe free in the flat position. If the pipe is not fastened (fixed) in the flat position, the pipe may be rolled as it is being welded. This gives the welder an advantage of welding the entire joint from one position and is called *roll welding*. See Fig. 23-20A.

2. Pipe fixed in the flat position. If the pipe is fixed in the flat position, it cannot be rolled to be easily welded. This is called *fixed horizontal* or *fixed-flat position* welding. See Fig. 23-20B.

and welding skills, one of which is known as the *stove pipe method*. Here, each length of pipe is joined to the section laid before it, one at a time.

Because the pipe lengths vary, it is easier to roll weld cross-country transmission lines by using special turning equipment at the job site. This is done to small sections before joining them to the pipeline.

Although most cross-country transmission pipeline welding is either semi-automatic or automatic, manual shielded arc welding is still used for laying the initial bead pass.

Types of Pressure Pipe

Before welding on critical pipeline systems, the welder should know something about the chemical analysis of the pipe, so that the correct welding practices can be used to weld that particular pipe.

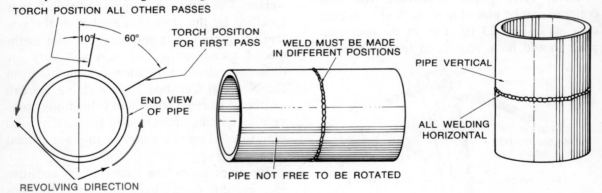

A. *Pipe free in the flat position.*   B. *Pipe fixed in the flat position.*   C. *Pipe in the vertical position.*

Fig. 23-20. The three common positions for pipe welding.

3. Pipe in the vertical position. When the pipe is standing up on end, the pipe itself is in the *vertical position*. Then, the weld is made in the horizontal position. See Fig. 23-20C.

Pipe systems installed in steam generating plants, chemical processing plants, and refineries are usually welded in either the vertical position or the fixed-flat position.

Cross-country transmission pipeline welding techniques use many special procedures

API 5L–X—The most common pipes used for petroleum and gas cross-country pipelines come from the API 5L–X series. Each letter in the pipe code follows:

1. API—This stands for American Petroleum Institute.
2. 5L—This is the standard specification for line pipe.
3. X—This letter stands for the pipe grade. In this case, it stands for high-test pipeline pipe.

4. Two final numbers—The complete pipeline grade specification has the *X* followed by two final numbers. These two numbers are the first two digits of the *minimum yield strength* of the pipeline metal, as shown in Fig. 23-21.

The API chemical anaylsis requirements for –42, –46, and –52 series 5L-X cross-country transmission pipe are shown in Fig. 23-22.

The chemical analysis (make up) of X46 and X52 alloys can be as high as 0.35% carbon and 1.40% manganese. However, that

As shown in Fig. 23-22, the pipe's sulfur content is kept below 0.05%. As mentioned earlier, a slight increase in the pipe's sulfur content can cause the welded pipe to crack or become porous. Most pipe manufacturers try to keep the pipe's sulfur content quite low, usually about 0.035%.

Power Source

Either an AC or DC welding machine can be used for welding critical pipe with the manual shielded arc process. As with subcritical pipe welding, welders may prefer one

Fig. 23-21. Physical strength requirements for API 5LX series cross-country transmission pipe. (Courtesy of Union Carbide Canada Ltd.)

Grade	Yield Strength Min., psi	Tensile Strength Min., psi	0.500 to 0.312 incl.	% Elongation in 2 in. (min.) Tabulated Wall Thickness, In.			
				0.281	0.250	0.219	0.188
X42	42,000	60,000	25.00	23.75	22.50	21.25	20.00
X46	46,000	63,000	23.00	21.50	20.00	18.50	17.00
X52	52,000	66,000	22.00	20.00	18.00	16.00	14.00

particular chemical composition might create welding problems, especially in cold weather. For cold weather welding, pipe with a lower carbon and manganese content is usually

type of machine over another, possibly depending on the available power. Rectifier AC machines can be used only if the critical high pressure system is being installed near an

Fig. 23-22. Chemical requirements for API 5LX series cross-country transmission pipe. (Courtesy of Union Carbide Canada Ltd.)

Process of Manufacture	Grade	Carbon Max. %	Mang. Max. %	Phos. Max. %	Sulfur Max. %
SEAMLESS					
Electric furnace, open-hearth, basic-oxygen or killed deoxidized basic-bessemer:					
Non-Expanded	X42	0.29	1.25	0.04	0.05
Non-Expanded	X46, X52	0.32	1.35	0.04	0.05
Cold-Expanded	X42, X46, X52	0.29	1.25	0.04	0.05
Killed, deoxidized, acid-bessemer, or killed deoxidized basic-bessemer:					
Non-Expanded	X42	0.24	1.25	0.10	0.05
Non-Expanded	X46, X52	0.27	1.35	0.10	0.05
Cold-Expanded	X42, X46, X52	0.24	1.25	0.10	0.05
WELDED					
Electric furnace, open-hearth, basic-oxygen or killed deoxidized basic-bessemer:					
Non-Expanded	X42	0.28	1.25	0.04	0.05
Non-Expanded	X46, X52	0.31	1.35	0.04	0.05
Cold-Expanded	X42, X46, X52	0.28	1.25	0.04	0.05
Killed deoxidized acid-bessemer, or killed deoxidized basic-bessemer:					
Non-Expanded	X42	0.23	1.25	0.10	0.05
Non-Expanded	X46, X52	0.26	1.35	0.10	0.05
Cold-Expanded	X42, X46	0.23	1.25	0.10	0.05
Cold-Expanded	X52	0.24	1.25	0.10	0.05

used. High-strength pipe, such as X56 or X60, usually requires preheating before welding. This will be especially true for good high-strength pipe welds when welding outdoors in cold weather.

available electrical source. Gasoline- or diesel-powered welding machines are usually used for welding cross-country transmission pipelines, although some contractors use propane-powered welding machines.

Critical Pipe Fittings

Valves and fittings, similar in appearance to those used for subcritical pipe systems, are available for small, medium, and large critical pipe systems. The bevel preparations on all critical pipe and pipe fittings are carefully prepared to be free of notches and defects. This will assure that all welding, whether manual or automatic, can be done at an even rate while producing a quality weld.

Electrodes

Critical pipe can also be successfully welded by using different electrodes that are compatible with the current being used, either AC or DC.

AWS E6010 and E7010 are the most popular electrodes used for welding critical pipe, either for pressure systems or cross-country transmission lines. These electrodes should be used with reverse polarity DC current for welding most critical pipe joints. For very critical welds where maximum strength is needed, the E8010 electrode may also be used.

It is not unusual to weld the stringer bead of a pipe joint with an E6010 electrode and then use a low-hydrogen electrode for the filler beads. This may be done when a special fitting with a reduced gap is used or when using a backup ring is undesirable. The E6010 electrode is able to penetrate well into the root of the joint, even if there is a small root gap. If the E6010 electrode is needed for that reason, using the electrode to weld vertical-down will allow a higher welding speed.

If a low-hydrogen electrode is used for *all* weld passes, a backup ring should always be used and welding should be done in either the horizontal or vertical position. The welder should always use a short arc with this electrode for a strong, porosity-free weld.

After finding the specifications for the pipe to be welded, the welder should refer to the electrode manufacturer's specification chart to find the right electrode for the job. The chart will list the physical properties for each electrode having a recommended AWS number for pipe welding. See Table 23-3.

Critical Pipe Welding Introduction

Cross-country transmission-pipeline welding and in-plant pressure pipe welding will vary slightly. Selecting the actual procedure to be used on the project usually depends on

Table 23-3. Physical Properties of Common AWS Electrodes Used for Pipe Welding.

	Physical Properties			
	As Welded		Stress Relieved	
AWS Electrode Classification	Tensile Strength (psi)	Yield Strength (psi)	Tensile Strength (psi)	Yield Strength (psi)
E6010	62,000-72,000	50,000-60,000	60,000-70,000	46,000-56,000
7010A1 & 7010G	70,000-78,000	57,000-63,000	70,000-75,000	57,000-62,000
E7018	72,000-80,000	60,000-70,000	70,000-78,000	57,000-67,000
E9018-G	100,000-105,000	88,000-93,000	90,000-95,000	77-000-82,000
E11018-G	119,000-128,000	111,000-117,000	110,000-119,000	97,000-103,000

Courtesy of Lincoln Electric Co.

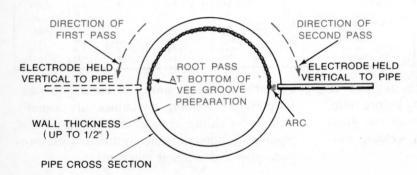

Fig. 23-23. **Vertical-down welding technique, as used for cross-country transmission pipeline welding.**

the welder's skill. For this reason, each type of critical pipe welding has been covered separately.

Cross-country transmission-pipeline welding on pipe with up to ½" wall thickness is usually done in the vertical-down position, as in Fig. 23-23. Critical pressure systems with in-plant pipe welding, on the other hand, are usually welded in the vertical-up position, as in Fig. 23-24. If cross-country transmission pipe has over ½" wall thickness, vertical-up welding is also used.

any notches, cuts, and/or foreign matter (dirt, scale, grease). Foreign matter on the face of pipe or fitting bevels may cause porosity (holes) in the weld and/or incomplete fusion.

When welding pipe in any position, the pipe should be clean and carefully aligned, as shown in Fig. 23-25. Usually, when welding cross-country pipe, each section to be welded is lined up and kept in the correct position for welding by using *inside* alignment fixtures, as in Fig. 23-26. External

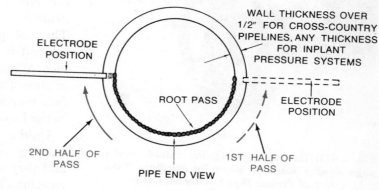

Fig. 23-24. Vertical-up welding technique. Used for in-plant pressure pipe welding and transmission pipeline welding with wall thickness over ½".

Cross-Country Pipeline Welding

When welding cross-country pipe, it is important to have the bevel preparation free of

aligning fixtures, Fig. 23-27, may be used on smaller diameter pipe. If the pipe section needs to be bent for a certain position, a pipe bending machine may be used for careful, accurate bends. See Fig. 23-28.

23-26. Internal pipe aligning fixture, used to align cross-country pipelines before welding.

Procedure—The first weld pass is usually made by two welders, working on opposite sides of the pipe and dragging the electrode. The electrode is pulled down around the root gap with the electrode coating moving

Fig. 23-25. Pipe joint preparation with a 30° bevel angle. This is a typical preparation for cross-country pipeline welding.

A. Open

B. Closed

Fig. 23-27. External pipe-aligning fixture, used to align smaller diameter pipes for welding. (Courtesy of Proline Pipe Equipment Ltd.)

Fig. 23-28. For bending cross-country transmission pipe, a large pipe bender, as shown, may be used. (Courtesy of Proline Pipe Equipment Ltd.)

against the bevel as the electrode melts down. With the right gap between the pieces to be welded and the right amperage setting, good welders will be able to penetrate the root joint and completely fuse the base metal. Since the first bead (the *stringer* bead) is the most important, burn-through or undercut must be controlled for a smooth inside surface. With some practice, most welders are able to weld the root pass with good penetration and fusion, *without* burning away the side walls of the groove. If the bevel walls of the groove are accidentally burned away, the burns are sometimes known as *wagon tracks.*

After applying the stringer bead, the slag must be removed so that it will not later cause slag inclusions. Bead cleaning can be done by using chipping hammers, wire-brushing and/or grinding.

The second pass, known as the *hot* pass, is then welded in place, using higher amperage. This will penetrate the stringer bead and burn

out any wagon tracks that may have accidentally been burned in during the first pass. The hot pass will allow any slag that is trapped in the first pass to float to the surface. When applying the hot pass, the welder should slow the travel rate to reduce surface pinholes, slag inclusions, and roughness in the beads.

Unlike the stringer and hot passes, the remaining passes are known as *filler* passes, and they are applied with slight side-to-side weaving. The slight weaving insures that the groove is being well filled and that the walls are being well fused to the first two beads. See Fig. 23-29. Improper bead cleaning,

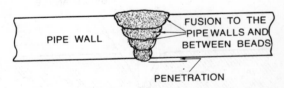

Fig. 23-29. Filler passes on a cross-country pipe weld help to build up the joint and melt in the root pass.

travel speed, and electrode movement while welding the filler passes, however, may cause porosity in the weld.

The final pass, known as the *cover pass,* eter of the pipe $\frac{1}{32}''$ to $\frac{1}{16}''$, while overlapping the original groove by about $\frac{1}{16}''$ on both sides. To control undercut while welding the cover pass, the welder should hesitate (stop) briefly at the end of each weave. See Fig. 23-30.

If any pinholes are seen in the finish pass, they may be caused by too high an amperage or too much width in the finish pass. Excessive build-up and too much width in the finish pass are not needed for cross-country pipeline welding.

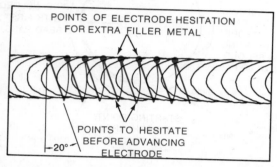

Fig. 23-30. To control the appearance and undercut on the final pass, the welder should hesitate at each end of the final bead weaves.

Vertical-Up Welding Procedure—For vertical-up pipe welding, the prepared ends on the pipe or fittings must be cleaned as they were for vertical-down welding. Poor fusion and porosity may result when vertical-up welding is done on metal surfaces covered with foreign matter.

Critical vertical-up pipe welding, as with other pipe welding, may be done with a backup ring, as shown in Fig. 23-31. The

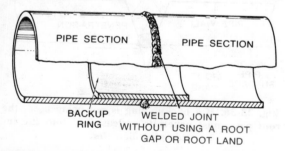

Fig. 23-31. The backup ring fits inside the welded pipe joint to help prevent fall-through and keep the pipes in alignment.

backup ring helps to control burn-through and helps align the pipe pieces for accurate welding. Some welders and contractors prefer to always use backup rings when welding critical high-pressure pipe, especially when

the pipe is made from alloy metals. See Fig. 23-31.

When vertical-up welding without a backup ring (Fig. 23-32), the pipe pieces should

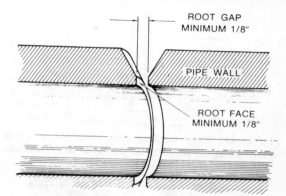

Fig. 23-32. If the pipe joint is welded without a backup ring, it should have at least a ⅛″ root gap and a ⅛″ root face.

be separated at least ⅛″ around the entire pipe, with a root face of at least ⅛″. The included angle, when the pipes are butted together, should be about 60°. Then, when the joint is welded with a stringer bead from an iron-power E6010 electrode, it should have good penetration and fusion.

Alloy cross-country pipe is often welded vertical-up, using a low-hydrogen electrode for the first pass. Welders using this technique for critical cross-country pipe welding must also use a backup ring to eliminate the root gap and root face shown in Fig. 23-32. When welding alloy pipe with a low-hydrogen electrode, even *with* a backup ring, the welder will need a root gap around the joint equal to the diameter of the electrode and its coating. See Fig. 23-33.

In all cases, cross-country transmission lines must be cleaned and inspected after they have been welded. Fig. 23-34 shows a typical machine used to clean and wrap completed cross-country pipeline welds.

In-Plant Pressure and Power Pipe Welding

Generally speaking, in-plant pipe welding techniques can also be used for cross-coun-

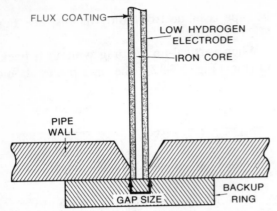

Fig. 23-33. When welding alloy pipe, the gap size should be equal to the total width of the electrode.

Fig. 23-34. When cross-country transmission lines have been welded, a cleaning and wrapping machine, as shown, may be used to finish the joint. (Courtesy of Proline Pipe Equipment Ltd.)

try pipeline welding, if it is required by the contractor or job supervisor.

Procedure—Pipe Fixed in the Flat Position—Before welding pipes fixed in the flat position, the welder will have to align the pipes by using the aligning fixture discussed earlier for subcritical pipe systems. After the pipes have been cleaned, aligned, and gapped, they may be tack welded in position with at least four tack welds. See Fig. 23-12. The root pass can then be welded by starting at the bottom of the pipe and welding toward the top. After welding to the top, the welder should thoroughly chip and clean the weld starting point.

Then, the arc can be restruck and the second half of the pipe welded, as in Fig. 23-35. The end of the first pass, of course, must be cleaned before the last half of the second

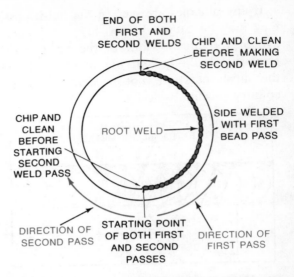

Fig. 23-35. Typical procedure for in-plant pipe welding when the pipe is fixed in the horizontal position.

bead is melted in at the top. Otherwise, poor fusion, porosity, and slag inclusions may result.

Where a backup ring is *not* being used, the root pass must weld together both sides of the bevel near the Vee bottom *and* penetrate at least $\frac{1}{16}''$ below the inside pipe thickness. See Fig. 23-36.

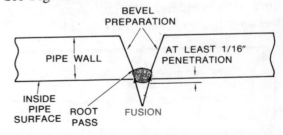

Fig. 23-36. If a backup ring is not being used, the root pass must penetrate at least $\frac{1}{16}''$ into the inside pipe diameter.

When a backup ring, Fig. 23-33, *is* being used, the root pass must penetrate and fuse the bottom sides of the preparation, *and* penetrate and fuse the back-up ring. When this is done correctly, both sections and the backup ring will be combined into a single unit.

To eliminate any possible slag inclusions, the welder should thoroughly clean the root pass by using a chipping hammer and/or wire brush. After cleaning the root pass, the second pass should be started near the bottom, but not at the same place where the first pass was started. The amperage must be adjusted so that the arc will burn well into the beveled preparation and root pass *without* the filler metal being pulled down by gravity while welding is being done in the vertical or overhead positions. As with subcritical pipe welding, the electrode must be held at the right angle for pipe welding. Each filler bead is then welded on top of the previously cleaned bead.

The final pass should rise above the outside surface of the pipe about ¹⁄₁₆″ while covering the edge of the groove up to ⅛″.

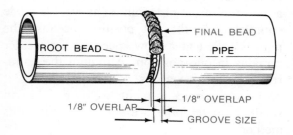

Fig. 23-37. The final bead should overlap the groove width by ⅛″ on either side.

See Fig. 23-37. Undercutting can be controlled by allowing the electrode to hesitate (stop) for a moment at the end of each weave, as for other forms of pipe welding.

Procedure-Pipe in the Vertical Position— Many in-plant pressure-pipe systems require that welds be made on pipe in the vertical position. When the pipe is in the vertical position, the weld will be entirely horizontal, as shown in Fig. 23-38. The dimensions used for these horizontal position welds are the same as used for vertical-up pipe welds on fixed, horizontal position pipe. The root weld penetration should have ¹⁄₁₆″ reinforcement around the inside diameter of the pipe, the same as when welding pipe in the horizontal position. The root bead surface should be smooth and even, with no pockets or inden-

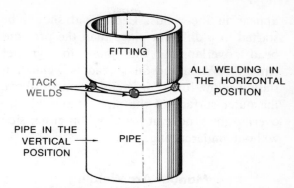

Fig. 23-38. When pipe is in the vertical position, the weld is in the horizontal position.

tations that could cause turbulence in the liquid moving through the pipe.

The welder may be required to join lengths of pipe in the vertical position with or without a backup ring. If a backup ring is being used, the bevel preparation will not require a root face, as in Fig. 23-32.

Thick pipe—The root pass and up to four filler passes are usually welded on the vertical pipe as bead passes. A ³⁄₃₂″ electrode and a current setting of about 90 amperes may be used for practice. For very thick pipe, the next 6 beads are welded with ⅛″ electrodes and an extra 20 to 25 amps. Then, if needed, another 10 or 12 beads can be welded in to fill the groove. These extra beads are usually welded with a ⁵⁄₃₂″ electrode and a current setting of 160 amps.

Each bead in the sequence, Fig. 23-29, must be cleaned before applying the next bead to ensure that slag inclusions will not

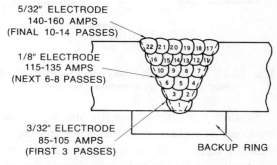

Fig. 23-39. When welding thick pipe in the vertical position, the electrode size and machine amperage used should be changed for different passes.

appear in the welds. Each bead should be started in a different place than the previous bead, overlapping the joint for greater strength. When the final bead is welded, it should allow for a $\frac{1}{16}''$ reinforcement above the outer surface of the pipe. It should also overlap the groove at least $\frac{1}{8}''$ on either side without undercutting.

Manual Arc Welding
Water-Transmission Pipelines

The qualification tests for welding water-transmission pipelines are not usually as difficult as are the qualification tests for welding critical high-pressure lines. Butt welds, similar to those used on critical or subcritical pipelines, are also frequently used to join water-transmission pipelines.

Lap Collars—Lap joints are normally the best joints to use for joining pipe with a wall thickness of at least $\frac{3}{8}''$. *Lap collars* have an inside diameter equal to the outside diameter of the pipe, allowing the pipes to be butted together and welded into a single unit with fillet welds. See Fig. 23-40.

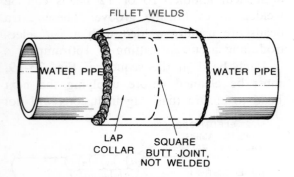

FILLET WELDS

WATER PIPE

WATER PIPE

LAP COLLAR

SQUARE BUTT JOINT, NOT WELDED

Fig. 23-40. A lap collar may be used for easily joining water transmission pipelines.

Smaller diameter water-transmission pipelines can be welded with the same techniques used for critical and subcritical pipe systems. Generally speaking, the prepared fittings, electrodes, and positions for other pipe welding will work for water pipelines.

Alternate Methods of Pipe Welding

Many welding methods can be used to weld pipe besides manual shielded arc and oxygen-gas. Almost any special arc welding method may be used to weld pipe, with excellent results in most cases. This section will describe how arc welding processes, other than manual shielded arc, may also be used to weld various types of pipe.

Roll Welding

A welder who is interested in pipe welding should be able to weld pipe in a process known as *roll welding*.

Before roll welding, the pipes are first tack welded, as outlined earlier. The pipe is then placed on rollers, and the welder strikes an arc on top of the pipe joint. The weld is made completely in the flat position because the pipe is rolled at a slow speed as the welder remains in one position. Although the hand-roll welding process is not as popular as it once was, some in-plant welding jobs on small diameter pipe are still done with this method.

Automatic Double-Ending

When two short lengths of pipe are welded together into one longer length, the process is called *double-ending*. This is frequently used on cross-country pipeline welding, since shorter pipe lengths must be transported to the job site where they will be welded into the line. The double-ending process is also used, if possible, for welding in-plant pipe systems.

Many pipeline companies and contractors have developed fully automatic double-ending equipment by using the submerged-arc process. When using these fully automatic methods for double-ending, the pipes to be welded are automatically rotated by rolling machines, similar to simple hand-roll welding. This allows the welding machine to stand still at the top of the pipe joint while the weld is being made, as in Fig. 23-41.

Fig. 23-41. The automatic double-ending pipe welding process.

Various automatic double-ending methods have been used successfully on different sizes of pipe and for different job applications. There are three basic examples.

1. In the first double-ending method, the root face is ground to ¼", and the pipe sections are then butted together. A backup ring is always used with this method to protect against burn-through. The butted pipes are then rotated, and two weld passes are made on the outside of the joint.

    When the final cover pass has been welded, an *internal* (inside) backup bead is applied by an automatic welding head mounted on a *gantry boom* (swivel framework) riding inside the pipe. This process cannot be used on pipe less than 22" in diameter due to the room needed for the boom and the welder, to line up welding head.

2. In the second double-ending method, the root face is first ground to ⅜". The pipes are then butted together, held in alignment by a line-up clamp, and tack welded on the outside. The tack welds are made manually, and, if necessary, sealing is done when the joint has a poor fit-up. After tacking, the clamps are removed.

An internal bead is then welded by the automatic welding head, mounted on the boom extending inside the pipe. After the internal bead has been welded, the outside bead is welded in a *single pass,* similar to the method shown in Fig. 23-41.

3. The third method of double-ending uses a flexible fiberglass sleeve that is held firmly under the joint by a copper chill ring and an internal alignment clamp. The fiberglass sleeve regulates the internal shape and size of the joint, assuring good penetration without burn-through. Also, it prevents the copper ring from being burned and eliminates the need for an internal bead.

Flux-Core Pipe Welding

Although the flux-core arc welding process may also be used to weld transmission pipeline systems, it is usually used to fabricate pressure-pipe systems on in-plant jobs. In many cases, flux-core pipe welding is preferred over manual shielded arc, submerged arc, and microwire welding methods.

Flux-core arc welding is usually done with a rolling pipe fixture, although it can be done on fixed-position pipe welded in the field.

Flux-core pipe welding will usually meet the welding requirements of most welding codes, even those using an x-ray inspection. The welder should know the inspection code to be used on a piping system before beginning to weld the system.

There are several advantages of flux-core arc welding over the other processes used to fabricate pipe.

A. The welder is always able to see where he is welding the beads, since the arc can always be seen.

B. The flux-core arc welding method is more continuous than methods using coated electrodes. Coated electrodes, of course, would require more stops and starts to change electrodes as they were consumed.

C. There is less slag and flux to remove from the weld beads, so the welding speed is increased.

Flux-core Roll Welding—For large diameter pipe, it is usually necessary to use a special fixture to roll the pipe so that all the welding can be done in the flat position. During roll welding (using either manual or automatic welding machines), it will be necessary to weave the bead inside the width of the weld groove. The weaving procedure during manual shielded arc welding will depend on the welder's skill. A mechanical *oscillator* (a type of automatic weaver) is used to hold and weave the torch head at the top of the pipe during the automatic flux-core pipe welding process. For field welding cross-country transmission pipelines, automatic devices known as *double ending rigs* are used.

Flux-Core Pipe Welding Setup and Equipment—Usually, the contractor will require that backup rings be used for welding pressure pipes with automatic flux-core, flat position roll welding. The backup ring becomes a permanent part of the joint after the joint is welded. See Fig. 23-42. Copper backup rings are sometimes preferred and they may also meet the code specifications of steel rings. Fig. 23-42 shows flux-core pipe welding

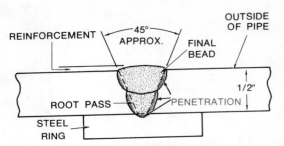

Fig. 23-42. Flux-core pipe welding with a steel backup ring on ½″ wall thickness pipe.

passes on ½″ wall thickness pipe, using a steel backup ring.

The core wire used for both beads with ½″ wall thickness pipe should be ⅛″ diameter, if available. Both passes can be made with reverse polarity DC current, set at about 400 amperes and 23 volts. The carbon dioxide shielding should be set at 35 cfh for welding indoors. Due to air currents while welding in the field, it may be necessary to increase the shielding gas flow to 50-70 cfh for good outdoor shielding.

Procedure—For joining pipe with flux-core arc welding, the root gap in the joint should be ³⁄₁₆″. The included angle in the beveled pipe joint should be about 45°. Then, the pipes are held in place with an aligning fixture and tack welded in four places, as done with other pipe welding processes.

For wall thicknesses of pipe up to ½″, the joint can usually be welded in two passes. The first pass is welded on the top of the pipe as the pipe revolves. When welding heavier-walled pipe, it may be necessary to use more than two passes. If more than two passes are being made, the wire size, gas flow, and welding current are the same as for ½″ wall thickness pipe.

Smooth Inside Procedure — Where a smooth, inside surface is needed on the pipeline, the reinforcement backup ring must be replaced. Here, a special technique is used to eliminate the turbulence caused by the backup ring while giving good strength in the weld joint.

For this technique, the microwire GMAW process is used for the first and second weld passes. After the microwire GMAW passes are welded in the joint, the other weld passes can be made with the flux-core process.

For the smooth inside surface, the pipes to be joined should be first butted together with an included angle from 60° to 75°. The root gap between the butted ends of the pipe should be from $\frac{3}{32}''$ to $\frac{1}{8}''$. For the first and second (microwire) passes, the welder should set the gas flow for about 20 cfh, the welding amperage at 170, and the voltage at about 22. The filler wire size for the microwire passes should be 0.035''.

The flux-core wire used for the other weld passes should be at least $\frac{3}{32}''$ diameter. For this wire size, 350 amps, 28 volts, and 35 cfh shielding gas flow should be used.

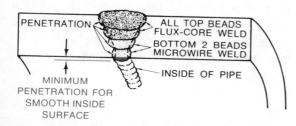

Fig. 23-43. For a smooth inside pipe surface, the welds must be made as shown with two different welding processes.

Fig. 23-43 shows the process needed for each weld bead in order to give good pipe strength without backup rings, for a smooth inside surface. As the pipe is turned, the torch position for the microwire passes should be about 50° below the top position of the pipe. For the flux-core passes, the gun should be held about 20° below the top of the pipe. See Fig. 23-44. Some welders prefer to make the first pass with the TIG method.

For pipe with wall thicknesses over $\frac{1}{2}''$, the welder may have to apply more passes. If more passes *are necessary,* the welding conditions for the fourth pass can be repeated. For welding outdoors, it may be necessary to increase the shielding gas flow for good, complete atmospheric shielding.

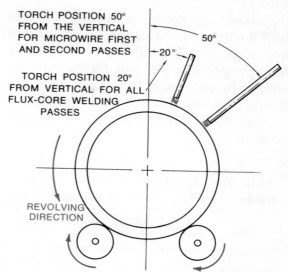

Fig. 23-44. When using two welding processes for a smooth inside pipe surface, the torch angle must be changed for each process.

Fixed Position Flux-Core Welding—For welding pipe systems in the fixed position using the flux-core method, the welder will have to use small diameter wires with the same properties as the base metal of the pipe being welded. When welding fixed-position pipes, it is necessary to align and hold them with fixtures so that they can be tack welded without disturbing the root gap necessary for good penetration and fusion.

Solid welding wires were developed that could match the critical properties of low chrome-moly pressure pipe. These, however, created welding problems, and welding equipment companies had to develop small diameter flux-core wires to be used for critical classification high-pressure pipe systems. When using these special flux-core wires on chrome-moly alloy steel pipe welds, the vertical-up technique should be used. The root pass, through, is often applied vertical-down.

The specifications for welding pipe that is fixed in the *horizontal* position are shown in Table 23-4. For welding pipe that is fixed in the *vertical* position, the welder should follow the specifications outlined in Table 23-5.

Before welding any type of pipe, the welder should check the position of the torch and the current and flowmeter settings. This will help insure that the weld will have good root and bevel surface penetration. Also, before attempting to weld pipe with the flux-core arc welding process, the welder should review the information about flux-core arc welding.

Then, successful flux-core pipe welds may be made.

Table 23-4. Specifications for Flux-Core Arc Welding Pipe Fixed in the Horizontal Position.

Bead Number	Electrode Wire Diameter (in.)	Welding Current (amps.) DC-RP	Welding Voltage (volts)	Shielding Gas Flow[1] (cfh)	Weld Travel Direction
1 (root pass)	0.045	100	20	25	Vertical-down
All Others	0.045	120	20	25[1]	Vertical-up

[1]Shielding gas flow may be increased up to 100% for welding outdoors.

Table 23-5. Specifications for Flux-Core Arc Welding Pipe Fixed in the Vertical Position.

Bead Number	Electrode Wire Diameter (in.)	Welding Current (amps.) DC-RP	Welding Voltage (volts)	Shielding Gas Flow[1] (cfh)
1	0.045	145	21	25
2	0.045	200	21	25
3	0.045	200	21	25
4	0.045	200	21	25
5	0.045	200	21	25
6	0.045	170	20	25

[1]Shielding gas flow may be increased up to 100% for welding outdoors.

Microwire Pipe Welding

Microwire pipe welding can be up to twice as fast as manual shielded arc welding because there is no slag on a finished weld made with microwire welding. Microwire welding is often used for making good test weld joints on carbon steel pipe. Adopting this small-diameter wire welding for joining both carbon steel and low-chromium molybdenum pipe has been made possible by using the right electrode filler wire and shielding gas.

The shielding gases most commonly used with microwire pipe welding are either argon and carbon dioxide mixtures or pure welding-grade carbon dioxide. The argon and carbon dioxide mixture is usually made up of 75% argon and 25% carbon dioxide. This mixture should not be used or stored at temperatures below freezing because the gases tend to separate at freezing temperatures.

Power Source—The welding machine used for microwire pipe welding can be either a rectifier or generator machine to supply direct current. Either machine should be equipped with a line voltage compensator to allow a constant arc voltage, even when the line voltage changes slightly. Regulated hot start (RHS) power supplies allow a welder to select a higher voltage for a given time period. Using a hot start at the beginning of a pass will help the welder apply a smooth bead over the entire length of the work. Power supplies with a single-phase input are better when roll welding in the flat welding position.

Electrode Wire—For welding low-carbon steel pipe with the microwire process, a 0.035″ diameter electrode wire works well in most cases, especially when puddle control is important. Filler wires with diameters up to $3/64$″, although available, are not recommended for all-position microwire pipe welding. The puddle made with the $3/64$″ diameter wire is usually large and fluid and is difficult to control when making critical pipe welds.

For microwire welding mild carbon steel and low-alloy, high-strength steel pipes, one of three basic types of electrode filler wires may be used, as shown in Table 23-6. For microwire welds with x-ray quality, type 18 (low-alloy, high-strength) electrode wire is often used.

Electrode wire composition (makeup) will vary with each welding equipment manufacturer. Table 23-7, for instance, shows the chemical composition of the microwire electrodes from one manufacturer. Special wire

Table 23-6. Characteristics of Electrode Filler Wires Used for Microwire Pipe Welding Mild-Carbon and Low-Alloy High-Strength Steel Pipes.

Chemical Analysis (%)	Electrode Wire Type					
	18		25		28	
	Wire	Deposit	Wire	Deposit	Wire	Deposit
Carbon	0.12	0.10	0.16	0.12	0.11	0.10
Manganese	1.90	1.40	1.20	0.70	1.65	1.20
Silicon	0.80	0.55	0.50	0.20	1.12	0.60
Phosphorus	0.02	0.02	0.02	0.02	0.021	0.02
Sulfur	0.02	0.02	0.19	0.02	0.024	0.025
Other Elements	Molybdenum 0.5 Nickel 0.1	Molybdenum 0.50	(none)	(none)	(none)	(none)
Wire Usage	High-Quality Microwire Pipe Welding (X-ray quality)		Standard Micro wire Pipe Welding (general purpose)		High-Efficiency Microwire Pipe Welding	
Mechanical Properties						
Tensile Strength (psi)	100,000		75,000		88,000	
Yield Strength (psi)	85,000		63,000		68,000	
Elongation in 2″ (%)	22		26		28	

Courtesy of Hobart Brothers Company.

electrodes are available to match the composition of popular types of chrome-moly pipe, as shown by the specifications in Table 23-8. These wires are usually used for the first and

Table 23-7. Typical Filler Wire Composition for Common Microwire Electrodes

Chemical Composition (%)	Wire Type		
	Linde 65	Linde 82	Linde 83
Carbon	0.04	0.09	0.12
Manganese	1.20	1.00	1.90
Silicon	0.50	0.45	0.73
Sulfur	0.02	0.024	0.035
Phosphorus	0.017	0.017	0.025
Molybdenum	(none)	(none)	0.55
Other	Aluminum 0.10 Zirconium 0.07 Titanium 0.10	(none)	(none)

Courtesy of Union Carbide Canada Ltd.

second passes on pressure piping to eliminate the backup ring.

Preparation—One of the most important things for a welder to learn before welding pipe is the edge-beveling process. Beveling is usually done by a rotating *cutting head* that is set up to cut the pipe from the outside at the correct bevel angle for the job being done. Other forms of cutting equipment remain stationary while the *pipe* is rotated for the bevel cut.

Besides machine cutting, the welder can also use an abrasive cutting wheel to prepare the edge bevel (Fig. 23-45), or he may use oxy-acetylene cutting equipment. Regardless of the cutting equipment being used, the welder must try to eliminate any notches in the bevels. Since a critical quality code weld can be made in many different joint designs, the angles on the pipe bevels are not too critical. Although included angles from 55° to

80° have been used successfully, most welders prefer to use an included pipe bevel of about 37½° when welding to meet the ASME Boiler Code. See Fig. 23-46.

Table 23-8. Composition of Electrode Filler Wires Used for Microwire Pipe Welding Chrome-Moly Pipe.

Element %	Wire Type and Application					
	5% Chromium ½% Molybdenum Pressure Pipe		1¼% Chromium 1½% Molybdenum Pressure Pipe		2¼% Chromium 1% Molybdenum Pressure Pipe	
	Wire Analysis	Deposit Analysis	Wire Analysis	Deposit Analysis	Wire Analysis	Deposit Analysis
Carbon	0.06	0.06	0.06	0.08	0.05	0.08
Manganese	0.49	0.40	0.51	0.40	0.49	0.40
Silicon	0.31	0.25	0.40	0.25	0.29	0.15
Phosphorus	0.02	0.02	0.02	0.02	0.022	0.020
Sulfur	0.019	0.020	0.022	0.02	0.020	0.020
Other Elements	Chromium 5.5 Molybdenum 0.50	Chromium 5.0 Molybdenum 0.50	Chromium 1.4 Molybdenum 0.50	Chromium 1.20 Molybdenum 0.50	Chromium 2.3 Molybdenum 1.0	Chromium 2.10 Molybdenum 1.0

Courtesy of Hobart Brothers Company.

Fig. 23-45. One method of preparing any metal for welding is grinding. Here, a welder is preparing a large pipe flange for welding. (Courtesy of The Norton Company)

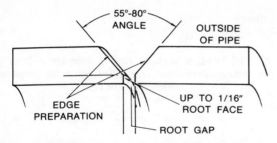

Fig. 23-46. The bevel preparation for good pipe welding should have the dimension shown.

After beveling the pipe joint edges, the welder should clean each side of the pipe bevel at least one inch back from the bevel joint, all around the pipe. Paint, rust, scale, or oil could penetrate the weld if they were not removed and could cause the weld to be weak and porous.

Positioning—For microwire pipe welding, the root face should not be greater than $\frac{1}{16}''$, as in Fig. 23-46. If the root face is larger than $\frac{1}{16}''$, root bead penetration will probably be too poor for good critical code welding. This would be especially true if the root gap was kept at the minimum size of about $\frac{1}{16}''$. If the welder has to weld a pipe joint without a root face (Fig. 23-47), he should reduce the root gap to help prevent weld fall-through.

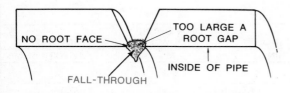

Fig. 23-47. If too large a root gap is used with no root face, weld fall-through may be a problem.

After beveling and cleaning, the pipe sections to be welded can then be butted together, leaving the root opening needed by the type of shielding gas being used. A $\frac{3}{32}''$ root opening, for example, would be used with a 65° to 70° included bevel angle for carbon dioxide shielding gas. If the same bevel joint was to be used with an argon and carbon dioxide shielding gas mixture, the root opening should be opened an additional $\frac{1}{32}''$-$\frac{1}{16}''$.

The dimensions and specifications for these joints are set up for critical code pipe welding by the ASME (American Society of Mechanical Engineers). Sections of the ASME Boiler Code outline weld specifications for fabricated welds such as power boilers, low-pressure and heating boilers, and unfired pressure vessels. An example of ASME weld specifications is shown in Fig. 23-48.

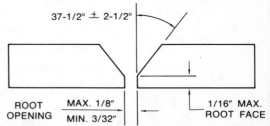

Fig. 23-48. A typical pipe joint specification for ASME pipe welding. (Courtesy of Union Carbide Canada Ltd.)

For API (American Petroleum Institute) code pipeline welding, the welder is required to prepare the pipe joint as shown in Fig. 23-49. Root openings for API pipeline welding must allow the recommended burn-through in the root, using a higher welding current than for other code work.

Backup rings are not usually used for microwire pipe welding. However, if the welder is required to use a backup ring, a bevel preparation should be used without a

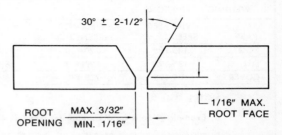

Fig. 23-49. A typical pipe joint specification for API pipeline welding. (Courtesy of Union Carbide Canada Ltd.)

root face. Without a root face, the backup ring will be well fused to the joint by the root bead. See Fig. 23-50.

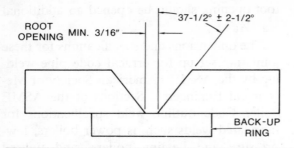

ROOT
OPENING MIN. 3/16″

37-1/2° ± 2-1/2°

BACK-UP
RING

Fig. 23-50. If a back-up ring is used for a joint to be microwire welded, the joint must be prepared without a root face.

Horizontal Fixed Microwire Pipe Welding Procedure.

Weld Conditions:

    A. CODE — ASME VIII & IX

    B. MATERIAL — LOW CARBON STEEL

    C. PIPE SIZE — $\frac{3}{16}$″ TO ¾″ WALL, ALL DIAMETERS

    D. FILLER METAL — LINDE 65, 0.035″ DIA.

    E. SHIELDING GAS — C-25, 20 - 30 CFH

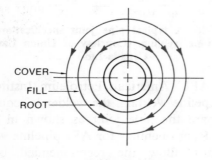

COVER
FILL
ROOT

Welding Technique

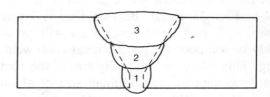

Welding Order

Weld Specifications

SVI-300 POWER SUPPLY SETTINGS

SLOPE	VOLTAGE	INDUCTANCE	OPEN CIRCUIT VOLTAGE	HOT START
FLAT 6-7½	LOW 5½	LOW - MIN.	25	+2V FOR 1 SEC.

WELDING CONDITIONS

PASS	WIRE FEED SPEED I.P.M.	WELDING CURRENT DCRP	WELDING VOLTAGE	WELDING SPEED IPM	DIRECTION	TOTAL ARC TIME
ROOT	185-195	120-130	18½-19½	12½	DOWNHILL	8 MIN. FOR
FILL	185-195	120-130	18½-19½	7½	DOWNHILL	6″ DIA. SCH. 40
COVER	185-195	120-130	18½-19½	6½	DOWNHILL	.280″ WALL

NOTES:

    a. Example given is for 6 in. diameter schedule 40 pipe (.280″ wall thickness).

    b. Fill passes may be added if required for heavier walls.

    c. Recommended maximum pass width is ½″.

    d. The fill pass, No. 2, may be stopped at 5 o'clock and 7 o'clock to reduce the reinforcement in the cover pass in the overhead position.

Courtesy of Union Carbide Canada Ltd.

Fig. 23-51. Sample weld specification chart.

Aligning and Tacking—When the pipes have been positioned with the correct root opening, they must be aligned with one of several aligning fixtures. After the pipes are aligned, tack welds should be welded in at least four places, equally spaced around the outside of the pipe and in the bottom of the bevel preparation. The root space should be rechecked before each new tack weld is made, to be sure that the weld pull has not changed the correct spacing.

When a welder is working on code-welded pipe systems, he should grind the beginning and end of each tack. This will create a taper on each end of the tack, removing any defects where the tack was started and stopped. Since the ends of the tack are tapered, an even root bead with good fusion can be welded over the tack welds as the root pass is welded around the entire pipe.

Basic Techniques—The ASME Boiler Code welding requirements can be met by a capable welder using the microwire welding method in the flat welding position. Using this technique, a welder will have to keep the arc on the leading edge of the puddle to help control *cold lapping* (laminating due to incomplete fusion). This is especially a problem with vertical welds.

Different welding equipment manufacturers have, of course, published complete weld technique specifications for making good microwire pipe welds. These are often available at welding equipment supply houses or from equipment manufacturers. The tech-

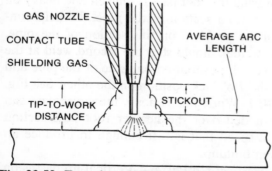

Fig. 23-52. For microwire pipe welding, the stickout should be from ¼″ to ⅜″.

nique specifications will give all the information needed for a given weld position and job. While it would be difficult to publish all these sheets in one book, a typical specification sheet is shown in Fig. 23-51.

Microwire Pipe Welding in the Fixed Horizontal Position—For microwire pipe welding is this position, the welder should use a ¼″ to ⅜″ stickout. See Fig. 23-52. The root pass is started in the middle of the top tack weld and then moved down either side of the pipe to the bottom. See Fig. 23-53. The gun position for remelting the tack to begin the root pass should be about 50° from the

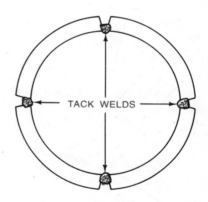

Fig. 23-53. Both microwire root passes are started at the top tack weld.

perpendicular (straight up), as shown in Fig. 23-54. If the welder has trouble starting the arc when practicing this weld, the stickout can be increased to as much as ⅝″. This will usually make the arc easier to start.

The gun can be weaved back and forth during the root pass, especially if the root gap is fairly wide. Even so, the welder should keep the root opening small enough to allow complete penetration without having to weave the microwire gun.

When the tack and first ½″ of pipe are fused, the welder can reduce the gun angle as shown in Fig. 23-55. As the welder reaches

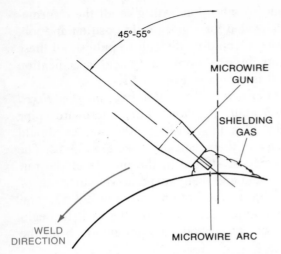

Fig. 23-54. Correct torch position for the beginning root passes when microwire pipe welding.

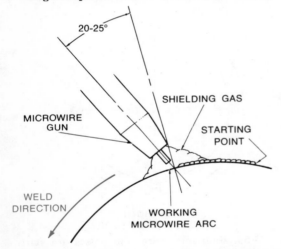

Fig. 23-55. When the tack has been remelted and the root pass begun, the microwire gun angle can be reduced for the rest of the root pass.

the bottom of the pipe (the overhead welding section), the backhand gun angle is usually increased, as in Fig. 23-56.

The root pass should not be stopped once microwire pipe welding has begun, unless it is possible to stop on a tack. If the root pass is stopped at a place *other* than on a tack, crater-cracking may take place in the weld.

When the root pass has been welded on the fixed horizontal pipe, the filler and cover passes can be welded with either the vertical-up or vertical-down welding techniques.

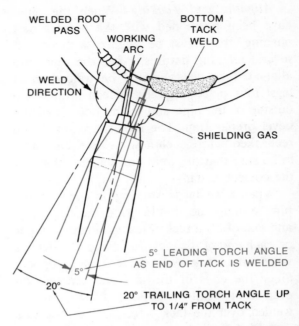

Fig. 23-56. As the bottom tack weld is melted in, the gun must be angled more toward the tack weld.

Vertical-Down Passes—After cleaning the root pass, the second pass is started about ½" from where the root pass was started. Weld passes should not be started exactly where other welds have been started if strong, x-ray quality welds are being made. The gun is usually held at a 20° angle, and, when the arc is working, the torch should be weaved from side to side, covering the previous bead. Cold lapping can be controlled if the weld speed is kept steady and the arc always kept ahead of the puddle.

Many times, welders find that the filler pass covering the root pass is uneven and heavy on the bottom section of the pipe. To help control the problem on this section of the pipe, the welder should stop the second weld at the 5 o'clock position on one side of the pipe and at the 7 o'clock position on the other. See Fig. 23-57. When this is done, the cover pass can be added over the heavier root pass section at a speed that is fast enough to produce an even buildup.

Vertical-Up Passes—Vertical-up passes on pipe are only about half as fast as are verti-

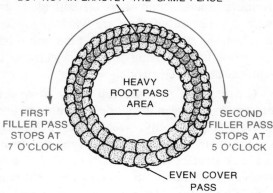

BEGIN BOTH FILLER PASSES NEAR THE TOP, BUT NOT IN EXACTLY THE SAME PLACE

HEAVY ROOT PASS AREA

FIRST FILLER PASS STOPS AT 7 O'CLOCK

SECOND FILLER PASS STOPS AT 5 O'CLOCK

EVEN COVER PASS

Fig. 23-57. To control build-up on the bottom of the pipe weld, the welder should weld that area separately.

cal-down passes. The wire feed speed should be reduced (which in turn reduces the amperage) according to the specifications given for the pipe being welded. Vertical-up microwire welding is started at the middle of the bottom tack and continued up one side of the pipe to the top, where the weld is stopped. The other side of the pipe can then be welded the same way, with the starting and stopping points of both welds completely fused together.

Then, once the root pass has been made, the other passes can be welded with the weave motion shown in Fig. 23-58. The filler pass should be wide enough to cover the root pass and it should be started at the bottom of the pipe, as was the root pass. If the arc is always

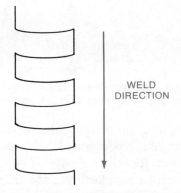

WELD DIRECTION

Fig. 23-58. For good microwire filler passes in the vertical-up position, the weave pattern shown above should be used.

kept working on the bevel pipe joint, cold lapping can be more easily controlled.

Microwire Welding in the Vertical Position—When welding pipe in the vertical position, the entire weld is horizontal, as was shown in Fig. 23-20C. The root pass should be welded with a gun trailing angle of about 20°, and the gun should be pointed up toward the top piece of the pipe about 10°. See Fig. 23-59. By using the gun in this position,

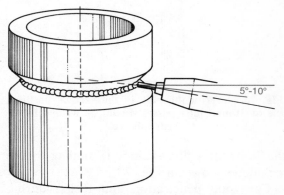

5°-10°

Fig. 23-59. When welding vertical pipe, the gun should have a trailing angle of 20° and should be pointing toward the top pipe at about a 10° angle.

the welder will have the extra arc control needed to prevent burn-through.

Although the gun should normally not be weaved when welding the root pass, the weave shown in Fig. 23-60 may be used if

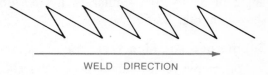

WELD DIRECTION

Fig. 23-60. For an uneven or wide root gap, the root pass for a microwire vertical pipe weld can be weaved as shown.

the root gap is uneven or too wide. When welding the root pass near the tack welds, the gun angle should be changed from the 20° trailing angle to a 5° leading angle for good fusion at the tack weld.

After the root pass has been welded, the welder can apply the first filler pass to the lower section of the bevel, as shown in Fig.

23-61. Both the first and second filler passes must fuse well into the other pass *and* the joint bevel. The gun weave pattern used for

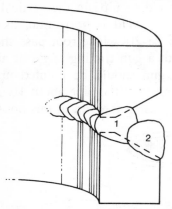

Fig. 23-61. With the pipe in the vertical position, the horizontal filler passes should be welded in the order shown.

the horizontal filler passes is similar to the technique shown in Fig. 23-58.

For good penetration and fusion during the third pass (the top filler pass), the welder must leave enough space between the bottom fill pass and the bevel edge on the top pipe. See Fig. 23-61. This will allow the third bead to be welded well into the top bevel and into the other two beads. When welding pipe in the vertical position, the gun should be held at a 15° trailing angle and upward angle of 5° to 10°, similar to the positions shown in Fig. 23-59.

Submerged Arc Pipe Welding

For many years, submerged arc welding has been used to weld the longitudinal (top) joint of pipe manufactured from formed steel plate. See Fig. 23-62. By using submerged arc welding, it is possible to make the weld

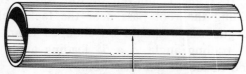

JOINT TO BE WELDED

Fig. 23-62. Pipe formed from steel plate will have a joint to be welded. The weld seam is often made with submerged arc welding.

joint economically because it has a faster welding speed than manual shielded arc welding. The submerged arc welding process for welding the longitudinal pipe joint has progressed from using a single electrode to using a double electrode.

Currently, a new method using *three* AC electrodes has been developed, capable of producing high-quality welds at extra-high speeds. The three-wire, submerged-arc process was developed to help control the weld problems of undercutting and poor weld contour that were common with the two-wire process, especially when the weld speed was increased.

The three-wire process, Fig. 23-63, allows faster welding speeds, so that the manufacturer will be able to weld the pipe seam more economically. The two-wire process speed can be increased up to 75% when using the three-wire process. Steel plates from ¼″ to ⅝″ thick can be welded with the three-wire process by varying the welding speed for the penetration required.

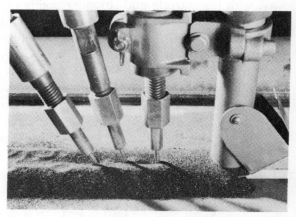

Fig. 23-63. The three-wire, submerged-arc welding process being used to weld pipe seams.

Three-wire submerged arc welds are usually completely penetrated, with good depth and fusion for clear x-rays. The three-wire process uses a standard flux for arc shielding. However, the flux is ground to a finer consistency to allow good, smooth, submerged arc welding on pipeline seam welds. This will provide good fusion, penetration, and bead contour (shape).

A track for the guide wheel is used to weld thin plates and the guide wheel, in turn, centers the three electrode heads in the pipe joint groove. On thicker plates, the bevel still acts as the electrode guide track and keeps the molten metal at a certain reinforcement height and width.

To increase the welding speed, of course, it is necessary to increase the heat in the weld zone. With either the one- or two-wire submerged arc process, increasing the heat for increasing the welding speed will usually cause a poor bead contour with excessive undercutting. By spacing the electrodes closer together, as is done in the three-wire process, each electrode's arc heat is effective in a common area on that part of the pipe being joined, reducing the heat loss.

AC welding current is used with the three-wire process to help control arc blow, a common problem when the one-wire submerged arc process is used with DC. Although the three-wire process was first developed to weld line pipe at speeds up to 125 ipm, it can also be used for applying multipass welds on heavy plate joints.

Pipe Welding Reminders

Pipe welding is a very important welding process for industry and construction. Because of its importance, and for people's safety, pipe welding is carefully controlled and regulated by pipe welding codes. The codes insure that only qualified welders do certain pipe welding jobs, and that certain materials, procedures, and tests are used in the welding process.

For these reasons, pipe manufacturers and various associations publish detailed literature and weld specifications for virtually all pipe welding jobs. By carefully following these specifications, qualified welders can successfully produce pipe welds to pass the test that applies to their welding job.

The welds on the pipe system below had to be made by certified welders. Fig. 23-64 shows a flange welded to a pipe so that the pipe may be bolted to a valve. In Fig. 23-65, a welder has just left the repair on a heating and cooling system pipe.

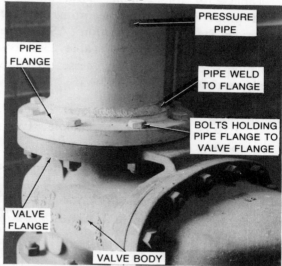

Fig. 23-64. Pipe welded to a flange connection.

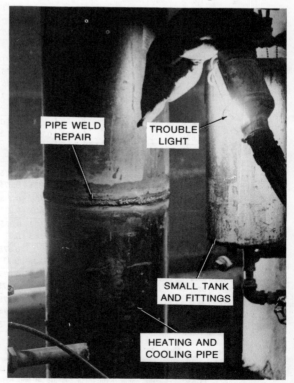

Fig. 23-65. A new weld for a pipe repair.

Pipe Specification Charts—Before welding any pipe, the welder should refer to a pipe specification chart to find the wall thickness and other physical properties of the pipe to be welded.

Table 23-9, for example, shows the standard properties of pipe according to the American Standards Association (ASA) pipe codes B36.10 and B36.19. Table 23-9 lists the pipe sizes and wall thicknesses that are established standards as follows:

A. The traditional weights are: Standard, Extra Strong, and Double Extra Strong.

B. The wall thickness specifications for carbon-steel and alloy-steel pipe *other than* stainless steels. These specifications are known as ASA standards B36.10.

C. The wall thickness specifications for corrosion-resistant materials. These are the specifications listed in ASA B36.19 code and the ASTM specification A409.

Before attempting any pipe welding job, pipe specifications (such as Table 23-9) and weld specifications (such as Fig. 23-51) should be carefully reviewed for the job being done.

Table 23-9. Common American Standards Association (ASA) Pipe Specifications.

NOMINAL PIPE SIZE	OUT-SIDE DIAM.	NOMINAL WALL THICKNESS FOR														
		SCHED. 5S*	SCHED. 10S*	SCHED. 10	SCHED. 20	SCHED. 30	STAND-ARD†	SCHED. 40	SCHED. 60	EXTRA STRONG§	SCHED. 80	SCHED. 100	SCHED. 120	SCHED. 140	SCHED. 160	XX STRONG
1/8	0.405	.....	0.049	.....	.....	.....	0.068	0.068	.....	0.095	0.095	.....	.....	.....	.....	.....
1/4	0.540	.....	0.065	.....	.....	.....	0.088	0.088	.....	0.119	0.119	.....	.....	.....	.....	.....
3/8	0.675	.....	0.065	.....	.....	.....	0.091	0.091	.....	0.126	0.126	.....	.....	.....	.....	.....
1/2	0.840	0.065	0.083	.....	.....	.....	0.109	0.109	.....	0.147	0.147	.....	.....	.....	0.188	0.294
3/4	1.050	0.065	0.083	.....	.....	.....	0.113	0.113	.....	0.154	0.154	.....	.....	.....	0.219	0.308
1	1.315	0.065	0.109	.....	.....	.....	0.133	0.133	.....	0.179	0.179	.....	.....	.....	0.250	0.358
1 1/4	1.660	0.065	0.109	.....	.....	.....	0.140	0.140	.....	0.191	0.191	.....	.....	.....	0.250	0.382
1 1/2	1.900	0.065	0.109	.....	.....	.....	0.145	0.145	.....	0.200	0.200	.....	.....	.....	0.281	0.400
2	2.375	0.065	0.109	.....	.....	.....	0.154	0.154	.....	0.218	0.218	.....	.....	.....	0.344	0.436
2 1/2	2.875	0.083	0.120	.....	.....	.....	0.203	0.203	.....	0.276	0.276	.....	.....	.....	0.375	0.552
3	3.5	0.083	0.120	.....	.....	.....	0.216	0.216	.....	0.300	0.300	.....	.....	.....	0.438	0.600
3 1/2	4.0	0.083	0.120	.....	.....	.....	0.226	0.226	.....	0.318	0.318	.....	.....	.....		
4	4.5	0.083	0.120	.....	.....	.....	0.237	0.237	.....	0.337	0.337	.....	0.438	.....	0.531	0.674
5	5.563	0.109	0.134	.....	.....	.....	0.258	0.258	.....	0.375	0.375	.....	0.500	.....	0.625	0.750
6	6.625	0.109	0.134	.....	.....	.....	0.280	0.280	.....	0.432	0.432	.....	0.562	.....	0.719	0.864
8	8.625	0.109	0.148	.....	0.250	0.277	0.322	0.322	0.406	0.500	0.500	0.594	0.719	0.812	0.906	0.875
10	10.75	0.134	0.165	.....	0.250	0.307	0.365	0.365	0.500	0.500	0.594	0.719	0.844	1.000	1.125	1.000
12	12.75	0.156	0.180	.....	0.250	0.330	0.375	0.406	0.562	0.500	0.688	0.844	1.000	1.125	1.312	1.000
14 O.D.	14.0	0.156	0.188	0.250	0.312	0.375	0.375	0.438	0.594	0.500	0.750	0.938	1.094	1.250	1.406	.....
16 O.D.	16.0	0.165	0.188	0.250	0.312	0.375	0.375	0.500	0.656	0.500	0.844	1.031	1.219	1.438	1.594	.....
18 O.D.	18.0	0.165	0.188	0.250	0.312	0.438	0.375	0.562	0.750	0.500	0.938	1.156	1.375	1.562	1.781	
20 O.D.	20.0	0.188	0.218	0.250	0.375	0.500	0.375	0.594	0.812	0.500	1.031	1.281	1.500	1.750	1.969	
22 O.D.	22.0	0.188	0.218	0.250	0.375	0.500	0.375	.....	0.875	0.500	1.125	1.375	1.625	1.875	2.125	
24 O.D.	24.0	0.218	0.250	0.250	0.375	0.562	0.375	0.688	0.969	0.500	1.218	1.531	1.812	2.062	2.344	
26 O.D.	26.0	.....	.....	0.312	0.500	.....	0.375	.....		0.500						
28 O.D.	28.0			0.312	0.500	0.625	0.375	.....		0.500						
30 O.D.	30.0	0.250	0.312	0.312	0.500	0.625	0.375	.....		0.500						
32 O.D.	32.0	.....	.....	0.312	0.500	0.625	0.375	0.688		0.500						
34 O.D.	34.0	.....	.....	0.312	0.500	0.625	0.375	0.688		0.500						
36 O.D.	36.0	.....	.....	0.312	0.500	0.625	0.375	0.750		0.500						
42 O.D.	42.0	.....	.....	.....	.....	.....	0.375	.....		0.500						

All dimensions are given in inches.

The decimal thicknesses listed for respective pipe sizes represent nominal or average wall dimensions. The actual thicknesses may be as much as 12 1/2 % under the nominal thickness because of mill tolerance.

*Schedules 5S and 10S are available in corrosion-resistant materials and Schedule 10S is also available in carbon steel.

†Thicknesses shown in italics are also available in stainless steel under the designation Schedule 40S.

§Thicknesses shown in italics are also available in stainless steel under the designation Schedule 80S.

24

Hardsurfacing

Today, it is important to know when it is economical to weld hardsurfacing material on metal parts, either before or after the wearing surfaces of the parts are worn out. Also, the welder may have to determine what conditions created the wear and which available material may be used to help keep the parts from wearing so fast in the future. By hardsurfacing, the welder can reduce the cost of replacement parts and keep the old parts working well for a longer period of time.

To decide about hardsurfacing a part, the welder must know what kind of part is to be hardsurfaced, what the hardsurfacing will have to do, and what type of wear the part will see. Knowing these factors will help make choosing a hardsurfacing material somewhat easier. Regardless of the type of wear problem, there is usually a hardsurfacing material available to help control wear, increasing the part's life. Selecting the right hardsurfacing material and the correct method of depositing the material is also easier when the welder knows the working conditions for the part being hardsurfaced.

Most welders find that hardsurfacing is not too difficult if they already know how to weld for routine maintenance and general production welding. However, hardsurfacing materials react differently to heating and cool-

ing during welding than do common carbon steels. In almost all hardsurfacing jobs, the material cost is the smallest part of the total cost.

Types of Wear

Before attempting hardsurfacing (hardfacing, as it is sometimes called), a welder should know something about the different kinds of wear, or how a surface is broken down. The most common kind of wear is *abrasion,* although *corrosion* and *impact* will also reduce the life of the part. Each type of wear may require different hardsurfacing material than will slow cooling.

Fig. 24-1. A grinding wheel removes metal by an abrasive action.

563

Abrasion

Parts being moved through different kinds of soil, especially sand and gravel, will be affected by the grinding action of the material. When metal parts roll, rub, or slide against each other, especially in dry, dirt-filled air, the parts will tend to wear away due to this *abrasive* action. Fig. 24-1 shows a worker using an abrasive grinding wheel while working on a truck frame. Abrasion may take place under either low or high pressure.

Corrosion

This is basically a simple chemical reaction, usually where iron is slowly oxidized by the air's oxygen. *Corrosion* results when different chemicals act to eat away parts of certain metals, such as iron and steel. Nonferrous metals (lead, copper, stainless steel, galvanized iron) are not as likely to rust as are ferrous metals (metals containing iron). Rusting, scaling, and oxidation cause metals such as iron and steel to deteriorate (be eaten away) very rapidly. The deterioration is in-

creased even more when the materials are heated, such as the rusty materials heated by the sun in Fig. 24-2. Special materials may be used on the parent metal to help the metal stand up against these conditions.

Impact

Repeated pounding, either *with* a metal part or *on* the metal part, is called *impact*. Impact, whether light or heavy, can deform

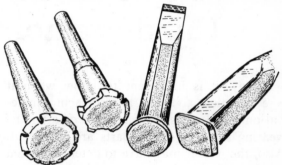

Fig. 24-3. Constant impact has caused these tool heads to be "mushroomed" over on the top.

the surface of the part and create cracking, chipping, or, if continued, breaking. The tools shown in Fig. 24-3 have been pounded over

Fig. 24-2. The combination of air, water, and heat will cause most ferrous metals to quickly rust unless they are protected.

on top by the constant impact of a hammer. Many welders are often faced with the possibility of a part wearing away due to a *combination* of these common wear factors, acting together to wear down a working piece of metal. To control or repair the damage, the welder must understand which of the factors caused the wear to begin with. Then he will be able to apply the best possible hardsurfacing material to help control any further wear.

Hardsurfacing Metallurgy

When alloyed steels are made up of as high as 50% of the alloying material, the

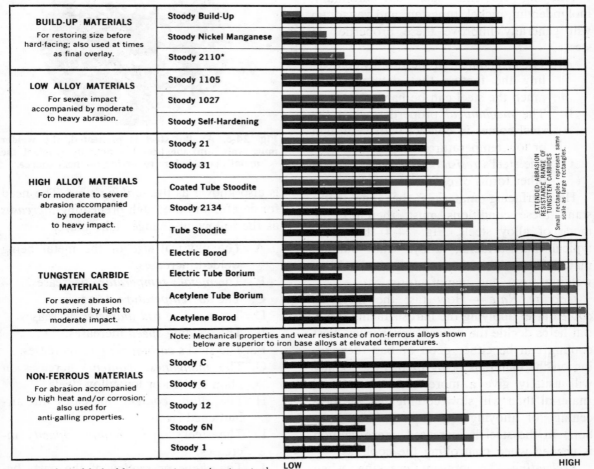

Fig. 24-4. A manufacturer's chart of available hardsurfacing materials. (Courtesy of Stoody Company.)

properties of the steel are changed. The alloying material may improve hardenability and/or resistance to chemical attack. Alloying material may also raise the oxidation temperature. Since high-carbon steel decreases in ductility as the carbon content is increased, the impact it can absorb will also decrease as the carbon content is increased. The more carbon in the steel, the more brittle it becomes, and the more likely it is to crack. For this reason, high-carbon steels cannot generally be hardsurfaced. When very strong properties are required for hardsurfacing, only alloyed steels can produce the properties.

Alloying elements such as chromium, molybdenum, nickel, tungsten, and vanadium will generally improve these properties of steel:

A. The *mechanical* properties, by controlling the factors that determin how much the steel can be hardened.

B. The *mechanical* properties when the steel comes in contact with very high or low temperatures.

C. The steel's *resistance* to oxidation and other forms of chemical attack.

Hardsurfacing materials that can withstand these conditions must, therefore, be made of alloy steels. Fig. 24-4 shows the typical hardsurfacing alloys and buildup materials available for manual shielded arc welding from one manufacturer.

Cooling Rate — How fast the metal is *cooled* after applying the hardsurfacing material helps decide the deposit's properties. Fast cooling can be either good or bad, depending on the materials being used. Fast cooling will usually give a more abrasion-resistant material than will slow cooling. As the percentage of alloying elements (such as carbon) in the steel is increased, more caution is needed when controlling the cooling.

Preheating—To prevent cracking when hardsurfacing, the welder may need to *preheat* the base metal. The exact preheat temperature depends on both the carbon and alloy content of the base metal. When hardsurfacing is to be done on base metal with a high-carbon content, the welder should use a higher preheat temperature to guard against cracking. Generally, high alloy metal will require a greater preheat temperature than will a similar metal with less alloy content.

When actually preheating, the metal must be brought to the required temperature and held there until the heat has reached the metal's core. See Fig. 24-5. This holding time

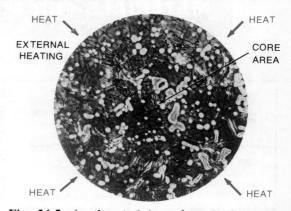

HEAT   HEAT

EXTERNAL HEATING   CORE AREA

HEAT   HEAT

Fig. 24-5. As the steel is preheated, the welder must wait until the heat has actually reached the metal's core before removing the heat source.

is also known as the *soaking time*. The need for careful and thorough preheating *increases* as the following changes take place.

A. The physical *size* of the metal being welded increases.

B. The *metal's temperature* decreases.

C. The *air temperature* decreases.

D. The *welding rod diameter* decreases.

E. The *welding speed* increases.

F. The steel's *carbon content* increases.

G. The *manganese content* increases in plain carbon or low-alloy steels.

H. The *alloy content* increases in air-hardened steels.

I. The steel's *air-hardening capacity* increases.

J. The *shape* or *section* of the parts becomes more complicated.

To be able to use the right preheat temperature, the welder must find out the base me-

Table 24-1. Recommended Preheat Temperatures for Hardsurfacing the Metals Listed.

Metal Group	Metal Type	Carbon Content (%)	Preheat Temperature (°F)
Plain Carbon Steels	Plain Carbon Steel	Below 0.20	Up to 200
	Plain Carbon Steel	0.20-0.30	200-300
	Plain Carbon Steel	0.30-0.45	300-500
	Plain Carbon Steel	0.45-0.80	500-800
Carbon Moly Steels	Carbon Moly Steel	0.10-0.20	300-500
	Carbon Moly Steel	0.20-0.30	400-600
	Carbon Moly Steel	0.30-0.35	500-800
Manganese Steels	Silicon Structural Steel	0.35	300-500
	Medium Manganese Steel	0.20-0.25	300-500
	SAE T 1330 Steel	0.30	400-600
	SAE T 1340 Steel	0.40	500-800
	SAE T 1350 Steel	0.50	600-900
	12% Manganese Steel	1.25	Usually Not Required*
High Tensile Steels (Also See Steels Below)	Manganese Moly Steel	0.20	300-500
	Manten Steel	0.30 Max.	400-600
	Armco High Tensile Steel	0.12 Max.	Up to 200
	Mayari R Steel	0.12 Max.	Up to 300
	Nax High Tensile Steel	0.15-0.25	Up to 300
	Cromansil Steel	0.14 Max.	300-400
	Corten Steel	0.12 Max.	200-400
	Yoloy Steel	0.05-0.35	200-600
	Jalten Steel	0.35 Max.	400-600
	Double Strength No. 1 Steel	0.12 Max.	300-600
	Double Strength No. 1A Steel	0.30 Max.	400-700
	Otiscoloy Steel	0.12 Max.	200-400
	A. W. Dyn-El Steel	0.11-0.14	Up to 300
	Cr-Cu-Ni Steel	0.12 Max.	200-400
	Cr-Mn Steel	0.40	400-600
	Hi-Steel	0.12 Max.	200-500
Nickel Steels	SAE 2015 Steel	0.10-0.20	Up to 300
	SAE 2115 Steel	0.10-0.20	200-300
	2½% Nickel Steel	0.10-0.20	200-400
	SAE 2315 Steel	0.15	200-500
	SAE 2320 Steel	0.20	200-500
	SAE 2330 Steel	0.30	300-600
	SAE 2340 Steel	0.40	400-700
Medium Nickel Chromium Steels	SAE 3115 Steel	0.15	200-400
	SAE 3125 Steel	0.25	300-500
	SAE 3130 Steel	0.30	400-700
	SAE 3140 Steel	0.40	500-800
	SAE 3150 Steel	0.50	600-900
	SAE 3215 Steel	0.15	300-500
	SAE 3230 Steel	0.30	500-700
	SAE 3240 Steel	0.40	700-1000
	SAE 3250 Steel	0.50	900-1100
	SAE 3315 Steel	0.15	500-700
	SAE 3325 Steel	0.25	900-1100
	SAE 3435 Steel	0.35	900-1100
	SAE 3450 Steel	0.50	900-1100
Moly Bearing Chromium and Chromium Nickel Steels	SAE 4140 Steel	0.40	600-800
	SAE 4340 Steel	0.40	700-900
	SAE 4615 Steel	0.15	400-600
	SAE 4630 Steel	0.30	500-700
	SAE 4640 Steel	0.40	600-800
	SAE 4820 Steel	0.20	600-800
Low Chrome Moly Steels	2% Cr.-½% Mo. Steel	Up to 0.15	400-600
	2% Cr.-½% Mo. Steel	0.15-0.25	500-800
	2% Cr.-1% Mo. Steel	Up to 0.15	500-700
	2% Cr.-1% Mo. Steel	0.15-0.25	600-800
Medium Chrome Moly Steels	5% Cr.-½% Mo. Steel	Up to 0.15	500-800
	5% Cr.-½% Mo. Steel	0.15-0.25	600-900
	8% Cr.-1% Mo. Steel	0.15 Max.	600-900
Plain High Chromium Steels	12-14% Cr. Type 410	0.10	300-500
	16-18% Cr. Type 430	0.10	300-500
	23-30% Cr. Type 446	0.10	300-500
High Chrome Nickel Stainless Steels	18% Cr. 8% Ni. Type 304	0.07	Usually do not require preheat but it may be desirable to remove chill
	25-12 Type 309	0.07	
	25-20 Type 310	0.10	
	18-8 Cb. Type 347	0.07	
	18-8 Mo. Type 316	0.07	
	18-8 Mo. Type 317	0.07	

Courtesy of The Tempil Corporation

*When welding outdoors in extremely cold weather, 11% to 13% manganese steel parts should be warmed to 100°-200° F. Under normal conditions, 11% to 13% manganese steel should not be preheated and welding temperatures over 500° F should be avoided for prolonged periods.

tal composition. Carbon steels, for example, are magnetic except when they are heated to their critical temperature. Austenitic manganese steel is not magnetic unless is has been work-hardened. Scarifier teeth (ripper teeth used on road construction equipment for digging up asphalt surfaces and other materials) are often made of austenitic manganese steel. These teeth may be magnetic because of work-hardening. For this reason, a welder who has to hardsurface a dragline bucket or other kinds of teeth should test the teeth with a magnet to find out where work-hardening has occurred. By using the magnet, the welder will be able to tell the difference between carbon and austenitic manganese steel. Once the metal composition has been decided, the welder can use a preheating chart (Table 24-1) to find the recommended preheating temperature.

Types of Hardsurfacing Structures

As with other forms of metallurgy, certain grain structures in the metal will give the metal the properties needed for each hardsurfacing job. Understanding each of the grain structures and their physical properties will help a welder be able to do a better job of figuring out any hardsurfacing application.

Martensitic Structure—This is a hard grain structure with little ductility but fairly good abrasion-resistance. *Martensite,* Fig. 24-6, forms when there is a high percentage of carbon and alloy in the material and the

Fig. 24-6. **Martensite 4340 steel, magnified 500 times. (Courtesy of United States Steel.)**

cooling rate above the critical temperature is fast enough. The critical temperature is about 1600° F.

If controlled cooling from above the critical temperature takes place in a furnace, the metal may become ductile and machinable. Then, if it is rapidly reheated above the critical temperature and cooled, it will become a hard, structured material. Therefore, the cooling rate will change the hardness and abrasion-resistance of the deposits.

Austenitic Structure—*Austenitic* deposits have a soft, tough material structure. Although this material, Fig. 24-7, will work-harden at the surface, it remains tough and

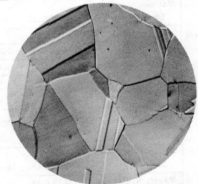

Fig. 24-7. **Austenite structure in an alloy steel that was rapidly cooled from the temperature range where austenite is stable. Magnified 500 times. (Courtesy of United States Steel.)**

soft below the surface. It is not magnetic and has good abrasion-resistance. A good example of high-carbon austenite is manganese steel.

Unlike manganese steel, low-carbon austenitic stainless steel is not as abrasion-resistant and will not easily become work-hardened. It is especially important to remember that the hardness in austenite deposits does not necessarily indicate the material's abrasion-resistance.

Semiaustenitic Structure — This structure consists of both martensitic and austenitic materials. Semiaustenitic structures have good abrasion-resistance. The abrasion-resistance, important to the metal's hardness, is affected very little by the cooling rate.

Since the austenite is not very stable, the material's surface can be transformed into martensite with very little cold working. Since the martensite has a high carbon content, the material will have its abrasion-resistance increased. Here, remember that hardness does not always indicate the abrasion-resistance of semiaustenitic material.

High-carbon, Chromium Carbide Structure—Materials with a high carbon, chromium carbide structure are high in carbon and produce a carbide deposit that has high abrasion-resistance and fair impact-resistance. If the carbon is alloyed with the chromium, the resulting alloy will be a hard matrix of either austenitic or martensitic structures. These structures will be in the form of very hard chromium carbide that does not respond very well to heat treatment. See Fig. 24-8.

Fig. 24-8. Martensite and retained austenite. Multiplied 500 times.

Preparing For Hardsurfacing

Hardsurfacing may usually be done with almost any arc welding method. Because worn parts may be located away from the shop and/or oddly shaped, manual shielded arc welding with special hardsurfacing electrodes is the method most often used to hardsurface parts. When hardsurfacing is done during production, it may be done with one of the automatic arc welding methods.

Common arc welding safety rules also apply to hardsurfacing. Other arc welding procedures and techniques also apply to hardsurfacing, so the subjects outlined here will be those that *are* different from manual shielded arc welding.

Hardsurfacing Power Source

For hardsurfacing, the welder can use either an AC or DC welding machine. When using AC, however, the welder should use only electrodes designated for that particular current and job. In most cases, the amperage must be kept low to help control the dilution (mixing) that takes place between the molten, softer, parent metal and the hardsurfacing material. To be sure that there is maximum hardness where the part will contact the abrasion and/or impact, it is usually a good idea to apply *two* layers of hardsurfacing material. Deep penetration, resulting from using reverse polarity DC current, should only be attempted when reverse polarity DC is specified for the electrode being used. Weld specifications, showing recommended amperages, polarities, and other data, are usually listed on hardsurfacing electrode containers.

Hardsurfacing Electrodes

Hardsurfacing alloys and buildup materials are usually applied, whether in the shop or in the field, by one of three basic arc welding processes.

The three arc welding processes that are usually used for welding hardsurfacing material are:

1. Manual Shielded Arc. Used with either a standard electric arc or a shielded inert arc.
2. Semiautomatic Methods. Used with submerged arc, shielded inert arc, or electric open arc.
3. Automatic Methods. Used with either an open or submerged arc.

After finding out the right procedure for a given hardsurfacing job, the welder should refer to a manufacturer's electrode specification chart to double-check the application

Table 24-2. Electrode Specifications for High-Alloy Hardsurfacing Materials.

Electrode Information	Mechanical Properties (All hardness readings are based on Rockwell C scale, designated as R_c)	Deposit Properties	Recommended Uses
STOODY 21 Coated for AC or DC. Bare for DC. Diameters: Coated: 1/8", 5/32", 3/16", 1/4" Bare: 3/16", 1/4"	Hardness: 2 Passes—Manganese Steel ....46-50 Rc 2 Passes—Medium Carbon Steel ......52-56 Rc 2 Passes—Cast Iron ....55-59 Rc (500° F. Interpass Temperature) Melting Point ....2450° F Specific Gravity 7.75	Bonds well with carbon or alloy steel, including manganese and cast iron. Surface checks relieve stresses and help prevent warpage. Slightly magnetic on carbon and low alloy steels; non-magnetic on manganese. Not machinable or forgeable.	Recommended as an overlay on new or worn parts to resist severe abrasion and impact. Used primarily in construction, rock products (brick and clay), mining, and agriculture.
STOODY 31 Coated for AC or DC. Diameters: 1/8", 5/32", 3/16", 1/4"	Hardness: 2 Passes (weave beads) Medium Carbon Steel ....47-49 Rc 2 Passes (weave beads) Manganese Steel ....45-48 Rc Melting Point ....2400° F Specific Gravity ....7.80 For hot wear applications up to 850° F.	Bonds well with carbon or alloy steels and manganese. Deposits polish to mirror finish. Slightly magnetic on carbon steels; non-magnetic on manganese steels. Not machinable or forgeable. Multi-layer deposits are extremely sound. Can be heat-treated.	An outstanding high alloy electrode with exceptional welding characteristics. Recommended where sound deposits with low coefficient of friction are desired in construction, cement, and steel industries. Deposits provide excellent bearing surfaces on friction type guides, cement mill gudgeons, etc.
Coated Tube Stoodite Coated for AC or DC. Diameters: 1/8", 5/32", 3/16", 1/4", 3/8"	Hardness: 2 Passes (weave beads)—Medium Carbon Steel ....57-61 Rc 2 Passes (weave beads)— Manganese Steel ....47-51 Rc Melting Point ....2350° F Specific Gravity ....7.90 For hot wear applications up to 900° F.	Bonds with carbon or low alloy steels, including manganese and cast iron. Surface checks relieve stresses and reduce warpage. Slightly magnetic on carbon and low alloy steels; nonmagnetic on manganese steels. Not machinable or forgeable.	An outstanding medium-priced high alloy material used chiefly in construction, brick and clay, mining, rock products, and cement industries.
Tube Stoodite Bare for Oxy-Acetylene or TIG Application. Diameters: 5/32", 3/16", 1/4"	As Deposited Hardness ....56-58 Rc Melting Point ....2275° F Specific Gravity 7.45	Deposits polish to a mirror finish under earth abrasion; low coefficient of friction. Not forgeable or machinable; magnetic on carbon and low alloy steels.	Ideal for metal-to-metal wear and earth abrasion. Particularly suited where thin deposits are required. Recommended for all types of farm implements.
STOODY 2134 Coated for AC or DC. Diameters: 5/32", 3/16", 1/4"	Hardness: 2 Passes (weave beads)—1045 Plate As Welded ....56-60 Rc Lime-cooled from 1750° F. ....48-51 Rc Water-quenched from 1750° F. ....63-65 Rc 2 Passes (weave beads)—Manganese Steel as Welded ....45-50 Rc Deposits may work-harden 5 to 6 points. For hot wear application up to 950° F.	Bonds readily to carbon, low alloy and manganese steels. Magnetic on carbon and low alloy steels; nonmagnetic on manganese steel. Deposits are sound and dense. Surface checks relieve stresses and reduce warpage. Not machinable or forgeable.	Developed primarily for earth-working equipment, crushing, and similar tools subject to severe abrasion with moderate to heavy impact; compressive strength is very high.

Table 24-2. Electrode Specifications for High-Alloy Hardsurfacing Materials (continued).

Information Electrode	Mechanical Properties (All hardness readings are based on Rockwell C scale, designated as Rc)	Deposit Properties	Recommended Uses
STOODY 1105 Coated for AC or DC. Diameters: ⁵⁄₃₂", ³⁄₁₆", ¼"	Hardness: (½" weave beads—air cooled) 2 Passes—Mild Steel (Reverse Polarity) ....36-39 Rc 2 Passes—1045 Mild Steel (Reverse Polarity) ....38-42 Rc 2 Passes—Mild Steel (Straight Polarity) ....37-40 Rc 2 Passes—1045 Steel (Straight Polarity) ....42-44 Rc 2 Passes—Mild Steel (AC) ....39-42 Rc 2 Passes—1045 Steel (AC) ....42-45 Rc 4 Passes—1045 Steel (Straight Polarity) ....44-47 Rc	Bonds readily to carbon, low alloy and manganese steels; not recommended for cast iron. Deposit properties are the same as those of Stoody 105. Forgeable and machinable with Grade 883 Carboloy or equivalent. Magnetic on carbon and low alloy steels; nonmagnetic on manganese.	Recommended for hardfacing tractor rollers, idlers, arch wheels, shovel rollers and idlers, sprockets, drive tumblers, churn drills, charging car wheels and similar parts involving high impact, abrasion, and metal-to-metal wear.
STOODY 1027 Coated for AC or DC. Diameters: ⅛", ⁵⁄₃₂", ³⁄₁₆", ¼"	Hardness: 2 Passes (weave beads)—Medium Carbon Steel ....45-49 Rc 2 Passes (weave beads)—Manganese Steel ....15-19 Rc 2 Passes (weave beads)—Cast Iron ....48-50 Rc (500° F. Interpass Temperature.) Melting Point ....2600° F. Specific Gravity 7.8	Deposits are sound and smooth with minimum spatter loss; graphitic coating eliminates slag removal problems. Magnetic on carbon, low-alloy steels, and cast iron; nonmagnetic on manganese. Forgeable at red heat; subject to heat treatment.	For surfacing new or worn parts to resist impact and moderate abrasion. Also for applications. Also for appli-to-metal wear. Best choice for cast iron parts.
STOODY SELF-HARDENING Coated for AC or DC Diameters: ⅛", ⁵⁄₃₂", ³⁄₁₆", ¼"	Hardness: All Weld Metal ....54-58 Rc 2 Passes (weave beads)— Mild Steel ....52-56 Rc Water-quenched from 1700° F. ....56-59 Rc Furnace-cooled from 1700° F. ......19-23 Rc 2 Passes (weave beads)— 1045 Steel ....54-58 Rc Water-quenched from 1700° F. ...56-60 Rc Furnace-cooled from 1700° F. ......19-23 Rc Melting Point ....2525° F. Specific Gravity 7.8	Can be applied to plain or alloy steels; magnetic on carbon or low alloy steels. Can be forged at red heat; not readily machinable.	Slightly higher alloy content gives this material more wear resistance than Stoody 1027, although both materials are generally recommended for similar applications.

Courtesy of Stoody Company

information and electrode specifications. A typical hardsurfacing electrode specification chart is shown in Table 24-2.

Regardless of the process used, the welder should refer to the electrode manufacturer's specification chart for the type of electrode (buildup or hardsurfacing) recommended for the part to be welded. Electrode manufacturers usually list the recommended hard-

surfacing electrode for a given part to be hardsurfaced, as in Fig. 24-9. Each electrode has been developed with special properties for meeting the requirements of a certain hardsurfacing job.

For example, an electrode that is suitable for resisting maximum abrasion would not usually be suitable for resisting impact; likewise, the reverse may be true. For this rea-

PART DESCRIPTION	WELDING PROCESS	BASE METAL	RECOMMENDED BUILD-UP MATERIAL	RECOMMENDED HARD-FACING MATERIAL	WELDING PROCEDURE
TRACK ROLLERS 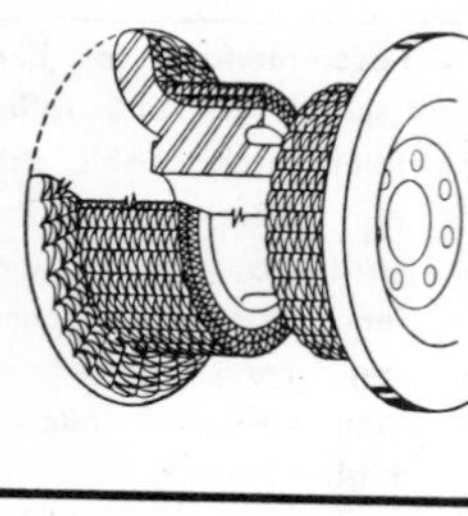	Manual	Carbon Steel	Stoody Build-Up	Stoody 1105	Mount roller on jig for downhand welding. Apply transverse beads on running face and flange.
	Semi-Automatic	Carbon Steel	Stoody Build-Up Wire		
	Automatic	Carbon Steel	Stoody 104	Stoody 105 Stoody 107	Reconditioned most economically by the automatic process. Contact your tractor dealer or a custom shop for details or write Stoody Company for equipment requirements and welding procedure.

Fig. 24-9. Hardsurfacing specifications from one manufacturer, used to repair track rollers. (Courtesy of Stoody Company.)

son, the welder must carefully examine each job's requirements and then match the hardsurfacing electrode to that job.

Most hardsurfacing electrodes manufactured today are known by their trade names, rather than by the American Welding Society specifications. Almost all hardsurfacing electrodes may be classified into one of three general groups: (1) Ferrous-Based Alloys; (2) Tungsten-Cobalt Alloys; (3) Tungsten-Carbide Alloys.

Ferrous-Based Alloys—These electrodes are basically made of iron, with added hardening alloys such as carbon, chromium, and manganese. *Ferrous-based* alloy electrodes have a wide range of hardness and toughness, and, if welded to a properly prepared surface, will withstand fairly severe impact. Welded hardsurfacing layers of manganese alloy will withstand both abrasion *and* severe impact.

Tungsten-Cobalt Alloys—Deposits welded with *tungsten-cobalt* alloy electrodes are very hard and brittle. After deposits have been welded to the parent metal, the deposited material cannot be flame cut, drawn, or forged. The hardness will even be able to stand up against red heat and will hold up under severe abrasion and minor impact.

Tungsten-Carbide Alloys—The combined tungsten and carbon hardsurfacing materials (tungsten-carbide alloys) can be classified into three groups: (1) composite tungsten-carbide rod; (2) cast tungsten-carbide inserts; and (3) cemented tungsten-carbide inserts.

1. Composite Tungsten-Carbide Rod — These rods, shown in Fig. 24-10, are made of a mild steel tube, filled with

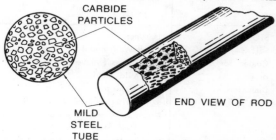

CARBIDE PARTICLES

END VIEW OF ROD

MILD STEEL TUBE

Fig. 24-10. The composite tungsten-carbide rod contains small pieces of tungsten carbide.

different-sized pieces of tungsten carbide. A rod that is classified as "8-10" will be made with particles that *will* pass through a screen with 8 openings per inch, but *will not* pass through a screen with 10 openings per inch. See Fig. 24-11. The size of the pieces in the mild steel tubes (10-20, 20-30, etc.) can be varied to satisfy certain job needs.

The tungsten-carbide pieces do not melt when they are heated. Rather,

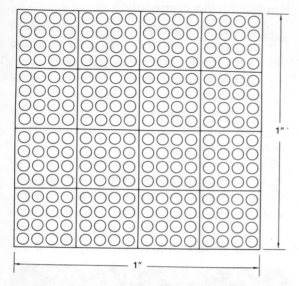

Fig. 24-11. A screen is used to size the tungsten-carbide particles in a composite rod. This enlarged 1″ screen has 16 openings per inch, or 256 per square inch.

they are held together and bonded to the metal being welded when the mild steel tube around the particles melts and then becomes solid. As the mild steel tube is worn away by abrasion on the built-up part, the hardened tungsten-carbide pieces are left on the worn-down surface. These can then stand up under more abrasion. As more of the steel wears away and the tungsten-carbide pieces drop from the wearing surface, more tungsten-carbide pieces in the steel are exposed to take over the job of resisting more abrasion. These deposits will withstand severe abrasion and the steel matrix (tube) will withstand some impact.

2. Cast Tungsten-Carbide Inserts—These carbide inserts are cast (poured) into the size and shape needed for the job. Although they are very hard and brittle, they can withstand only little impact by themselves. When surrounded by tougher, stronger metal, they do a good job of resisting abrasive wear. Fig. 24-12 shows several

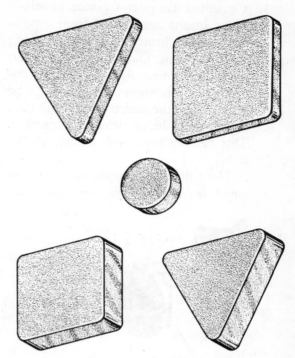

Fig. 24-12. Replaceable tungsten-carbide cutting tips.

examples of cast tungsten-carbide inserts.

Parts that surround cast tungsten-carbide inserts can be rebuilt with tungsten-carbide composite rods (discussed earlier) when the inserts are worn down and cannot serve their purpose. New inserts, surrounded by the composite rod material, will rebuild cast tungsten-carbide inserts to their original strength. Fig. 24-13 shows a cast insert clamped onto the end of a lathe tool holder.

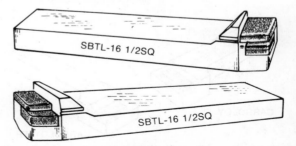

SBTL-16 1/2SQ

SBTL-16 1/2SQ

Fig. 24-13. Adjustable tool holders with tungsten-carbide cutting tips clamped in place.

3. Cemented Tungsten-Carbide Inserts— These inserts are formed from a powder and then heated in an oven, making a hard cutting material. The powder, mixed with other elements such as cobalt and titanium-carbide, can be pressed into the desired shape and then heated until the particles fuse together. This heating process is known as *sintering*.

The finished products, one of which is *carballoy*, may be used to tip

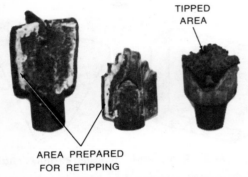

Fig. 24-14. Drill bits used to drill rock, shown at left as prepared for retipping and at right as completely retipped.

the drilling bits in Fig. 24-14 or cemented to the tip of cutting tool holders for lathes or shapers, as in Fig. 24-15. These inserts can be renewed in the same way as tungsten carbide inserts.

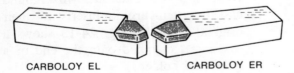

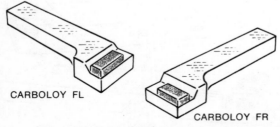

Fig. 24-15. Tungsten-carbide cutting tips, cemented to ordinary cutting tools.

Hardsurfacing Preparation

Preparing for hardsurfacing includes cleaning the work, preparing the surface to be built up, filling depressions (if necessary), and, if needed, preheating. To help control problems such as spalling and porosity, the welder should remove dirt, grease, oil and rust from the surface of the piece to be hardsurfaced or built up.

Rebuilding—Some base metals work-harden, squash, or crack in service. Any damage due to these factors should be repaired before welding hardsurfacing materials onto the metal. One method of preparing the surface is to grind off the thin top layer of worn-

Fig. 24-16. One way of preparing a worn-out surface for hardsurfacing is to grind a thin layer of metal off the top. (Courtesy of The Norton Company.)

out metal, as shown in Fig. 24-16. Another method is to weld a layer of mild- or low-alloy steel in place before welding on the hardsurfacing material.

Some areas of working surfaces may wear away faster than other areas. If a smooth, even surface is needed on the finished part, the depressions should be rebuilt before welding hardsurfacing material to the part.

When parts show excessive wear, the welder should build up the surface to within $\frac{3}{16}''$-$\frac{3}{8}''$ of its original size before welding on any type of hardsurfacing material. The buildup material and the hardsurfacing materials are different for different jobs.

The material used for building up must have the mechanical strength needed for the structure. The buildup material must also be able to withstand being deformed and mushroomed under heavy impact or high compression loads. Using buildup materials without these qualities will later cause the hardsurfacing material to spall (break off) because there is no firm support under the hardsurfacing material.

Positioning—The part to be hardsurfaced should be placed in the flat position, if possible, to allow for "downhand" hardsurfacing. This allows the welder to work at faster speeds for more economical hardsurfacing. If it is necessary to hardsurface a part that is *not* in the flat position, the welder should only use hardsurfacing electrodes specially designed for other positions.

Fixtures—Using work-holding tools and fixtures may often speed up hardsurfacing production. These special tools, such as those in Fig. 24-17, will help the welder be able to rebuild and hardsurface in the flat welding position. This, of course, will allow him to hardsurface easier and faster.

Basic Hardsurfacing Procedure

Hardsurfacing the wearing surfaces on common parts in the heavy construction, mining, agriculture, and/or petroleum industries have many procedures in common. Although the detail differences are discussed later in the chapter, many common procedures can be discussed together.

To begin hardsurfacing cast iron, for example, a welder should realize that cast iron is very crack-sensitive, although it is sometimes necessary to hardsurface the metal when it will be subject to common abrasion.

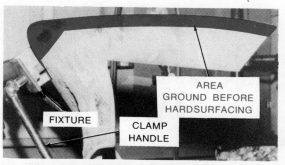

Fig. 24-17. A cultivator shovel ready for hardsurfacing. The special fixture is holding the shovel in position for easy hardsurfacing.

To do a good job of hardsurfacing cast iron, the metal must be preheated to a dull red. The dull red will appear at about 1000° to 1200° F. Slow cooling and peening will help to relieve stresses in the cast iron and eliminate cracking.

Cross-Checking—To successfully hardsurface other metals, the welder should understand the difference between cross-checking

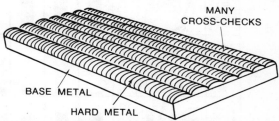

Fig. 24-18. When the hardsurfacing material has been correctly welded to the base metal, the cross-checks will be numerous, tight, and closely-spaced. They will not extend into the base metal. (Courtesy of Stoody Company.)

and cracking due to hardsurfacing. The cross-check pattern shown in Fig. 24-18 is very desirable because it reduces the residual (leftover) stresses. If cross-checking does *not* happen when the metal has been hardsurfaced, the residual stresses, when combined with external stresses of service, may cause cracking or spalling. See Fig. 24-19.

A welder should induce (try to cause) checking if it does not naturally happen during hardsurfacing. Many hardsurfacing welders will induce checking by laying a wet rag on the heated surface, or, during cooling,

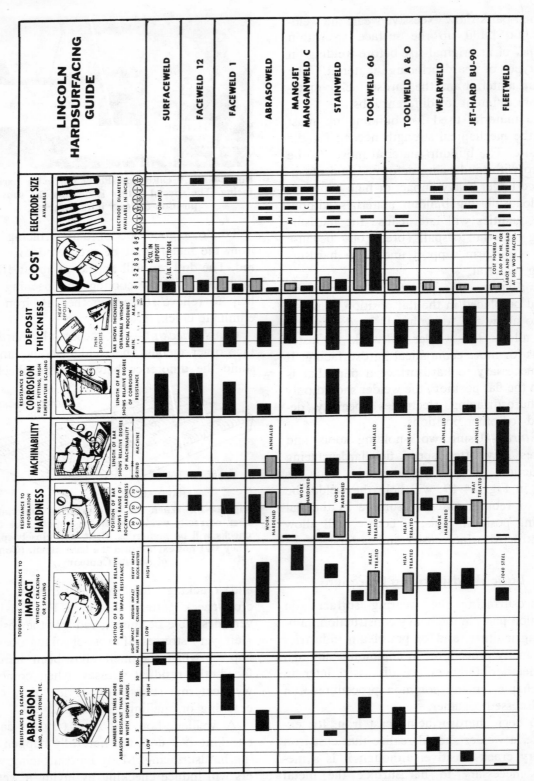

Fig. 24-20. A complete hardsurfacing guide, used to choose production hardsurfacing electrodes from one manufacturer. (Courtesy of Lincoln Electric Co.)

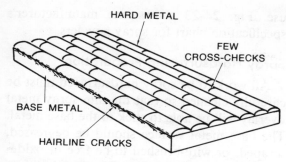

Fig. 24-19. When hardsurfacing has been incorrectly welded to the base metal, there will be little or no cross-checking. Hairline cracks will appear where the deposit is welded, and may cause spalling in service. (Courtesy of Stoody Company.)

striking the hardsurface deposit with a peening hammer.

Sometimes, it is necessary to have a check-free hardsurface on certain hardsurfacing jobs. This can be done by carefully following the exact preheat and postheat recommendations and using a recommended alloy hardsurfacing electrode, available from many hardsurfacing electrode manufacturers.

Hardsurfacing New Parts

In many cases, different industries will hardsurface new parts before they are used for abrasion-impact service. By doing this, the part's life will be extended and the "downtime" required to remove and replace the worn part will be reduced. Fig. 24-20, for example, shows how one manufacturer's electrodes may be used during production to hardsurface parts before they are used.

In Fig. 24-20, the *position* of each bar in the column generally indicates how good the electrode is in that classification. Usually, the farther to the right the bar is positioned, the better the electrode is. The width of the bar indicates how much the welding procedure affects the deposit. The wider the bar, the greater the effect of the welding procedure used.

Hardsurfacing Technique

A common bead technique is usually used to build up and fill a surface before hard-

surfacing. This may be used for low places, patching, or overall buildup.

Hardsurfacing can then be applied to the built-up surface by weaving, instead of laying many hardsurfacing beads. See Fig. 24-21. If the layer of hardsurfacing alloy is too

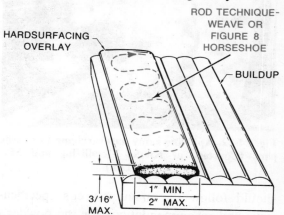

Fig. 24-21. Recommended weave techniques for a hardsurfacing overlay on built up weld beads. (Courtesy of Stoody Company.)

thick, cracking or chipping may happen when the part comes into contact with abrasion or impact. In most cases, two weave layers of hardsurfacing material are enough for a good, wear-resistant surface that is free of any mixture with the base metal.

Spray Hardsurfacing

Spray hardsurfacing combines metal spraying and welding for easily welding wear-resistant overlays on parts subject to abrasion, heat, corrosion, or combinations of the three. It is usually used when *thin* deposits of hardsurfacing material are needed. Spray hardsurfacing may be applied with guns or hand-held torches. See Fig. 24-22.

Spray hardsurfacing is done with fine metal-spray *powders,* giving a smoother and more uniform hardsurface deposit than would be possible with manually applied hardsurfacing electrodes.

For successful spray hardsurfacing, the welder should know about the different types of spray hardsurface powders available and

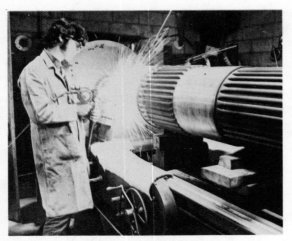

Fig. 24-22. Applying spray hardsurfacing to a large part. (Courtesy of Kierks Metallizing and Machine, Ltd.)

should follow the manufacturer's specifications and recommendations for the powder's use. Fig. 24-23 shows one manufacturer's specification chart for spray powders.

Spray Hardsurfacing Preparation

All work to be spray hardsurfaced must be carefully cleaned so that the thin spray metal overlay will be able to stick to the base metal. The base metal surface should be degreased, scraped, or wire-brushed to be free of oxides and other "chemical dirt." Blasting the piece with an iron grit is recommended to roughen-up the surface and give the liquid metal powder a porous surface on which to cling.

When the metal has been properly cleaned, it is usually mounted in the chuck of a lathe or other machine. This will allow the part to be turned so that the welder will simply have to aim the spray gun to correctly apply the molten hardsurfacing powder.

DESCRIPTION	TYPICAL ANALYSIS AND HARDNESS		GENERAL DESCRIPTION AND RECOMMENDED USES
STOODY 60 **STOODY 60 T. G.**	C — 0.66 Si — 4.25 Cr —14.00 B — 3.15	Fe — 4.50 Ni — Bal. 56-62 Rc	A nickel-base alloy with outstanding resistance to abrasion, corrosion and high heat; impact resistance is good. Low coefficient of friction. Use for pump components, shaft sleeves, thrust collars, guides, bushings. Finish by grinding.
STOODY 63 **STOODY 63 T. G.**	C — 0.03 Si — 2.5 B — 2.0	Fe — 0.30 Ni — Bal. 54-59 Ra	A nickel-base alloy with high ductility and workability. Deposits can be hand ground. Used primarily in the glass industry to protect parts against abrasion, heat and corrosion.
STOODY 64 **STOODY 64 T. G.**	C — 0.14 Si — 3.29 Cr — 9.53 B — 1.88	Fe — 3.54 Ni — Bal. 36-41 Rc	Nickel-base alloy recommended for application to parts requiring precise finish machining. Can also be ground. Deposits provide excellent corrosion and heat resistance; very good impact resistance. Use on pump parts, shafts, valves, dies.
STOODY 65 **STOODY 65 T. G.**	C — 0.38 Si — 3.54 Cr —11.19 B — 2.14	Fe — 3.38 Ni — Bal. 46-51 Rc	Has greater ductility than Stoody 60. Deposits can be machined with carbide tools or ground. Nickel-base alloy for pump parts, dies, extrusion screws, wear rings.
STOODY 90 **STOODY 90 T. G.**	C — 0.05 Si — 1.88 Cr —22.09 B — 2.3	W — 4.5 Fe — 0.40 Cobalt--Bal. 46-51 Rc	A cobalt-base alloy with deposit properties similar to those of Stoody 6 welding electrodes. Outstanding hot hardness properties and excellent resistance to abrasion, corrosion and impact. Especially useful where nickel is unacceptable. Can be machined with carbide tools and ground.
STOODY 85 T. G.	45% cast tungsten carbide (Borium) 55% nickel-base powder (Stoody 60)		This is a composite material developed specifically for use with hand spray torches. Deposits are heterogeneous with undissolved Borium particles imbedded in a high strength, wear-resistant matrix. Ideal for tillage tools, some types of hammers, bits, augers, etc.

Fig. 24-23. A manufacturer's recommendation chart for metal-sprayed hardsurfacing powders. (Courtesy of Stoody Company.)

The welder should follow the manufacturer's recommendations for spray gun settings and the description and recommended use of each hardsurfacing spray powder, so that the part being hardsurfaced will not crack or lift while in service. Fig. 24-24 shows the recommended spray hardsurfacing equipment set-up specifications from one equipment manufacturer.

Metco 2P Gun

Oxygen	15 psi	24 cfh
Acetylene	12 psi	28 cfh
Powder Valve	8 to 10 clicks	
Metering Valve	#2 orifice	
Tip to Work Distance	8 to 9 inches	

Colmonoy D Gun

Oxygen	24 psi	28 cfh
Acetylene	15 psi	45 cfh
Air	30 psi	
Tip to Work Distance	9 to 10 inches	
	Neutral Flame	

Fig. 24-24. Specifications for spray hardsurfacing equipment from one manufacturer. (Courtesy of Stoody Company.)

Spray Hardsurfacing Technique

With the work piece turning in a lathe or other revolving machine, the recommended spray powder is first melted in the gun and then blown out of the gun nozzle. When the spray hardsurfacing is done, the entire hardsurface area is evenly heated, using a reducing flame, to about 800°F. When the surface is *completely* heated, the torch is held two inches from the work piece to completely fuse the base metal and molten hardsurfacing powder.

Basic Hardsurfacing Jobs

It would probably be impossible to cover in one book all the hardsurfacing done to equipment and parts used in different industries. However, reviewing the hardsurfacing techniques used by several basic industries will help give any welder an understanding of the hardsurfacing techniques that are fre-

quently used. While the coverage here is limited to arc welding hardsurfacing materials, hardsurfacing may also be done with an oxy-acetylene torch and special filler rods.

Regardless of a welder's hardsurfacing experience, it is a good idea to contact a welding equipment company when faced with a new hardsurfacing problem. Because of constant research, the equipment company may have developed a new hardsurfacing alloy electrode, rod, or powder for a given job.

Hardsurfacing Heavy Construction Parts

The heavy construction industry must hardsurface many parts. As tools dig into rocks and earth, they should be able to resist abrasion and impact or they will wear out much too soon. Time and money will be lost as the equipment must be shut off and the worn-down tools repaired or replaced. Because of this, several common heavy construction tools may need to be hardsurfaced.

Top Carrier Rolls—Both carbon steel and cast iron top carrier rolls, Fig. 24-25, can be

Fig. 24-25. Top carrier rolls. One of the many parts to be hardsurfaced by the heavy construction industry. (Courtesy of Stoody Company.)

built up and hardsurfaced after the bushings have been removed and the surface cleaned. Cast iron carrier rolls should be preheated to about 1200° F and then placed in a fixture for flat position welding.

Depressions and general buildup can then be done on the cast iron by using a recommended hardsurfacing electrode. Although carbon steel top carrier rolls are positioned as are cast iron rolls, they may be built up without any preheating. Carbon steel rolls can, after positioning, be built up by running narrow beads across the face of the rolls with a recommended electrode.

Drive Sprockets—These parts, shown in Fig. 24-26, are prepared and positioned as

Fig. 24-26. Hardsurfaced drive sprocket. (Courtesy of Stoody Company.)

were the top carrier rolls. Hardsurfacing alloy is then welded downhand, from left to right.

So that the hardsurface overlay doesn't change the sprocket's teeth size, the welder should make a sheet metal *template* (sizing gauge) to check the tooth size after hardsurfacing. By using the template, the welder will be able to mark the areas between the teeth that need to be ground, so that the sprocket teeth will all be the same. See Fig. 24-27.

Ripper Teeth—Carbon steel ripper teeth usually require little, if any, buildup material

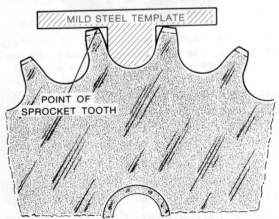

Fig. 24-27. When hardsurfacing a drive sprocket, a mild steel template may be used to make sure that all the sprocket teeth are the same size. (Courtesy of Stoody Company.)

before welding on hardsurfacing material. The top and sides of ripper teeth can be covered with stringer beads by using a recommended electrode for those places on the teeth. For the top and side surfaces of each tooth, a stronger hardsurfacing material is usually recommended in the area about 2″ from the end, as shown in Fig. 24-28. Electrodes made with a slightly softer material

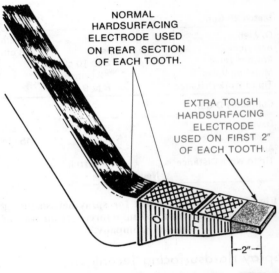

Fig. 24-28. Ripper teeth may need a different hardsurfacing electrode on the very front edge, where the worst abrasion takes place.

can be used on the rest of the shank. (That part of the shank may be in contact with the abrasion only part of the time.)

A *waffle* pattern, Fig. 24-29, can be used on the top and sides of large ripper teeth shanks, using the same hardsurfacing electrodes as were used for protecting a wearing surface.

Dozer Blades—These blades, shown in Fig. 24-29, do not usually need any buildup before the hardsurfacing is welded on. After cleaning the worn dozer blade, the welder should weld hardsurfacing beads across the outer corner and along the edge in the flat welding position. Knowing about the conditions where the bit will be used will help the welder be able to choose the best hardsurfacing material to withstand the wear or impact.

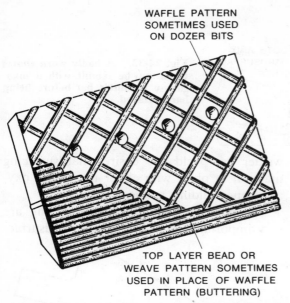

WAFFLE PATTERN
SOMETIMES USED
ON DOZER BITS

TOP LAYER BEAD OR
WEAVE PATTERN SOMETIMES
USED IN PLACE OF WAFFLE
PATTERN (BUTTERING)

Fig. 24-29. Dozer blades may be reinforced with a bead or waffle pattern, or both. (Courtesy of Stoody Company.)

Hardsurfacing beads welded side by side and properly applied will withstand almost any impact. This is known as *buttering*, shown in Fig. 24-29. Where moderate impact and severe abrasion will be a problem, the welder should use an electric boron electrode or a similar electrode, welded on the dozer blade as were the other electrodes.

Shovel Teeth—The teeth on power shovels, like many other heavy construction or earthmoving tools, should be hardsurfaced *before* being used. The type of bead pattern applied to the teeth may have a definite effect on the tool's service life.

Teeth attached to the shovel, normally working in rock, should be hardsurfaced with beads running the *length* of the tooth, as shown in Fig. 24-30. When the shovel is used, large gravel will roll up on the hard bead material slightly above the parent metal surface. This bead pattern can be applied more quickly when buttering and, therefore, will lower the overall hardsurfacing cost.

Fig. 24-31. Hardsurfacing material is welded *across* the shovel tooth if the tooth will be used in clay, dirt, or sand. (Courtesy of Stoody Company.)

Fig. 24-30. Hardsurfacing material is welded *up* the shovel tooth if the tooth will be used in rock. (Courtesy of Stoody Company.)

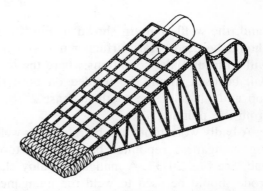

Fig. 24-32. If the shovel tooth will be used in rock *and* dirt, clay, and sand, hardsurfacing material is welded in a waffle pattern.

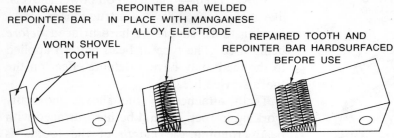

MANGANESE
REPOINTER BAR

WORN SHOVEL
TOOTH

REPOINTER BAR WELDED
IN PLACE WITH MANGANESE
ALLOY ELECTRODE

REPAIRED TOOTH AND
REPOINTER BAR HARDSURFACED
BEFORE USE

Fig. 24-33. A badly worn shovel tooth may be rebuilt with a manganese repointer bar before being hardsurfaced.

Teeth working in clay, dirt, or sand, on the other hand, should have hardsurfacing beads welded *across* the narrow face of each tooth, as shown in Fig. 24-31. When the teeth are in contact with both rock *and* dirt and

welder can weld the desired hardsurfacing material in place with either the arc or oxyacetylene welding methods.

The position and method of welding are very important, regardless of the hardsurfac-

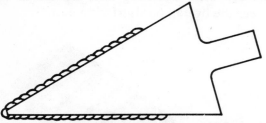

In most cases, hardsurfacing both the top and bottom ultimately results in a dull tooth.

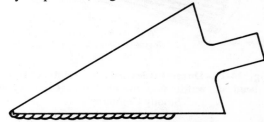

In most cases, hardsurfacing the bottom does not produce good shovel life.

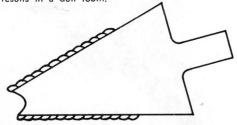

After the hard metal is worn away from the tip, the base metal will cavitate and hardsurfacing will chip off, producing a dull tooth.

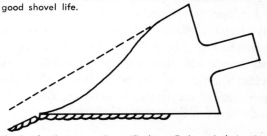

The hardsurfacing overlay will chip off through lack of support as the unprotected top wears away.

Fig. 24-34. Common shovel tooth problems when poor hardsurfacing techniques have been used. (Courtesy of Stoody Company.)

sand, the waffle pattern shown in Fig. 24-32 should be used. Hardsurfacing material that will withstand the conditions where the teeth will be used should be welded on both the top and sides of the tooth at least 2″ back from the point.

A badly worn tooth can be rebuilt by welding manganese repointer bars on the tooth end. See Fig. 24-33. A manganese-alloy electrode should be used to weld the manganese repointer bars in place. After welding the repointer bar in place on the tooth end, the

ing electrode used. Fig. 24-34 shows the problems that may arise if shovel teeth are not hardsurfaced correctly. Fig. 24-35 shows the generally recommended hardsurfacing procedures to use for avoiding the problems developed by the poor techniques used in Fig. 24-34.

Hardsurfacing Mining Industry Parts

Hardsurfacing is especially important in the mining industry. Here, tools must constantly dig into the earth and be exposed to

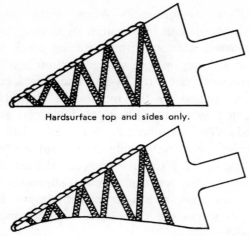

Hardsurface top and sides only.

Self-sharpening action results as unprotected bottom wears. After initial wear, hard metal will retard base metal erosion.

Fig. 24-35. Recommended hardsurfacing procedure for shovel teeth, to eliminate common tooth problems. (Courtesy of Stoody Company.)

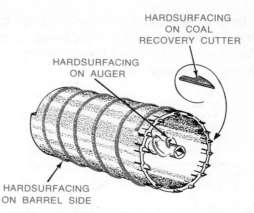

HARDSURFACING ON COAL RECOVERY CUTTER

HARDSURFACING ON AUGER

HARDSURFACING ON BARREL SIDE

Fig. 24-37. The cutters and barrel sides of coal recovery core barrels should be hardsurfaced for long life.

abrasion and impact. Also, the time and expense of removing tools is greater in the mining industry, where the equipment may be located far underground. Several common mining tools will need to be hardsurfaced for long life.

Baffle Plates—Ore chutes, also known as *baffle plates,* are made of carbon steel and should always be hardsurfaced before being used. A typical hardsurfacing electrode should be welded to the baffle plate in the flat welding position. The hardsurfacing beads should be welded in the direction that the material flows. The distance between each hardsurface bead depends on the type of material being transported, but the beads are usually welded from ½″ to 1½″ apart.

Coal Recovery Augers—These tools, Fig. 24-36, are usually used to move coal inside the mine shaft. Because coal recovery augers and coal recovery core barrels are manufactured from carbon steel, they must be hardsurfaced on the spiral edges to protect them from wear. The cutters on the core barrels must also be hardsurfaced to help control cutter wear. See Fig. 24-37.

While the barrel or auger is being revolved, the welder can weld the hardsurfacing material to the *flight* (barrel's outside). After removing the coal recovery cutters from the core barrel, they can be built up with the recommended hardsurfacing electrode.

Hardsurfacing Agricultural Parts

In agriculture, tools are almost always used in contact with the dirt and some clay and sand in the soil. These conditions may not have as much impact as rock, but the constant abrasion will wear down the farmer's tools if they are not hardsurfaced. For this reason, many common agricultural tools should be hardsurfaced for longer wear.

Agricultural tools used to till the soil usually have thin cutting edges that are often re-

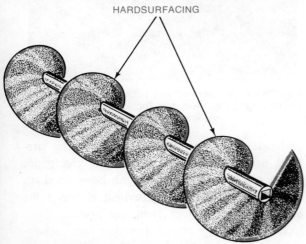

HARDSURFACING

Fig. 24-36. The edges of a coal recovery auger should be hardsurfaced for long life.

built and hardsurfaced with the oxy-acteylene process. Some agricultural tools, however, are repaired and hardsurfaced with electric arc welding, as discussed here.

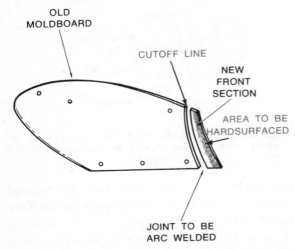

Fig. 24-38. An old moldboard, prepared for repair.

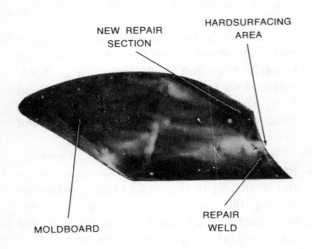

Fig. 24-39. Completely repaired moldboard, ready for service. (Courtesy of Stoody Company.)

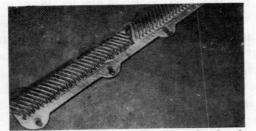

Fig. 24-40. An agricultural rasp bar, hardsurfaced for long service life. (Courtesy of Stoody Company.)

Plow Moldboards—When badly worn, moldboards can be repaired by cutting away the work section with a cutting torch and then welding on a new fitted section made of plow steel. See Fig. 24-38. The new section being joined to the moldboard should be welded in place with a low-hydrogen electrode. The moldboard's point can then be hardsurfaced by welding on several hardsurfacing beads. Using the correct hardsurfacing electrode to resist maximum abrasion will help keep abrasive wear to a minimum. Fig. 24-39 shows the completely repaired moldboard.

Rasp Bars—These tools, shown in Fig. 24-40, are especially subject to wear near the middle of the bar. Electric tube borium, 30-40 particle size, will do a good job of building up rasp bars to increase their service life. The leading corners (faces of the cutting edge) should be kept square. The welder can use a piece of straight iron as a straight edge to make sure that all the rasps are the same height.

Subsoiler Teeth—The teeth of subsoilers, Fig. 24-41, must be hardsurfaced on the top

Fig. 24-41. An agricultural subsoiler, hardsurfaced on the tip, sides, and top for long life. (Courtesy of Stoody Company.)

and sides with stringer beads across the surface, using a milder hardsurfacing electrode. Then, electric tube borium can be welded on as a solid hardsurfacing deposit at the point and about 2″-3″ back from the point.

Hardsurfacing spray powders are sometimes used in place of either the electric arc or the oxy-acetylene methods, especially where a tool's thin edge must carry a cutting edge. Many of the spray powders now available can resist much of the wear faced by agricultural tools.

Hardsurfacing Petroleum Industry Parts

To successfully hardsurface petroleum industry parts, the welder should understand the welding current requirements when using tube borium. When using tube borium, the welder should use the lowest possible amperage, since higher arc heat would create more surface area dilution and dissolve some of the borium particles. However, since electric arc welding is faster than oxy-acetylene welding and dilution is controlled by using stringer beads, many petroleum industry parts can be repaired and hardsurfaced by using borium tube electrodes.

Drill Collars—Worn drill collars should be built up and hardsurfaced to resist abrasion during drilling. Many exploratory seismographic (earth vibration testing) companies manually hardsurface their new drilling tools and equipment before using them.

Since drill collars usually contact abrasives while drilling, a layer of 20-30 particle size electric tube borium is usually welded on the collar's outer surface. This is usually done with a weave motion, as shown in Fig. 24-42.

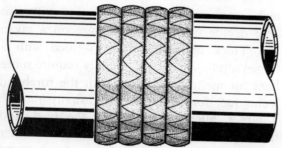

Fig. 24-42. A weave motion is usually used to build up petroleum industry drill collars with hardsurfacing. (Courtesy of Stoody Company.)

Before the borium is welded on the collar, the welder must preheat the collar to 600° F. This temperature must be maintained while the hardsurfacing is welded in place. Then,

after welding, the collar is allowed to slowly cool while wrapped in asbestos or covered with lime. Slow, rotating fixtures will allow the welder to hardsurface the drill collar from the top in a flat welding position.

Kelly Bushings—These special bushings become worn by the sliding action that is part of raising and lowering a tool called a *Kelly* during drilling. See Fig. 24-43. Worn Kelly

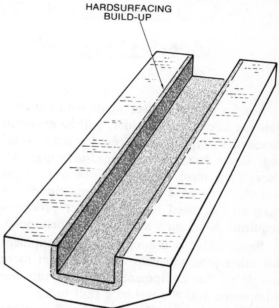

HARDSURFACING BUILD-UP

Fig. 24-43. As a Kelly bushing used in the petroleum industry becomes worn down, it may be built up with hardsurfacing.

bushings should be cleaned before they are built up or hardsurfaced. These bushings can be welded in the flat position without preheating. Since the weld buildup must be ground to fit the square-shaped Kelly, the welder should weld enough hardsurfacing buildup to allow for grinding when the weld is finished. (Kellys also come in shapes other than square.)

Buildup electrode can be welded along the length of the bushing, giving the buttering affect that was shown in Fig. 24-29. The built-up bushings can then be either machined or ground to fit the shaft and allow it to slide freely between the bushing ways (guides).

25

Common Low-Temperature Welding Processes

Besides common steel welds with the oxygen-gas torch, many welders will be expected to use the oxygen-gas process for braze welding, silver brazing, and hardsurfacing. Although the most common use of oxygen-gas welding is to weld steel, the same equipment may also be used for many other processes requiring heat, such as those just mentioned.

Being able to use oxygen-gas equipment for other jobs will help the welder get more use from the equipment, making both the equipment and the welder's skill more valuable. In each case, the heat of a flammable gas burning in oxygen is used to work a given material. Usually, acetylene is used.

Plastic Welding — Another low-temperature welding process is plastic welding. While plastics are often joined by chemicals, they may also be joined by melting their edges together in a plastic weld. This is done with an electric torch that heats an inert gas (or air) being blown across the heating element. The heated gas or air is then used to melt the edges of the plastic pieces being welded, sometimes with a plastic filler rod.

Tip Selection

As for other oxygen-gas welding processes discussed earlier, the welder must always be able to choose the correct tip size for the job and, usually, the filler rod size needed.

The type of material being heated will usually help decide whether a large or small tip should be used. Some materials, of course, melt at a lower temperature than others and would therefore need only to be heated with a small tip. Regardless of the tip size, similar torch flames will have the same temperature, but smaller flames may give off less heat. Materials that conduct heat better than others will, in many cases, require a larger tip to keep the material hot enough to melt.

The thickness of the material, the welding technique, and the filler rod size will also help decide what tip size to use on a given job. Thick material can be welded with a larger torch, since it will usually require more heat for melting. Welding with the forehand technique, for example, will usually require a larger filler rod than will the backhand technique. Backhand welding, on the other hand, usually requires a larger tip because the filler rod deflects some of the welding heat.

The weld position is another factor that may influence the welder's choice of tip size. Most welders prefer to do overhead and vertical welds at a slightly lower temperature than flat position welds. Therefore, welds in those positions could be made with a smaller tip. As usual, weld specification tables will help the welder choose the correct tip and rod for the type and thickness of the material being joined. This is necessary in any oxygen-gas welding process.

Braze Welding

Braze welding is used for groove, fillet, and other weld joints where the filler metal does not depend on *capillary* action. Capillary action is interaction between the contacting surfaces of liquid and a solid. During barze welding, the weld bond depends on the forces at the inside face, between the copper alloy and the base metal. For this reason, it is very important that the metal surfaces are clean. When the filler metal comes into contact with the parent metal, the filler metal tries to open up the grain of the parent metal. When the grain is opened, it allows the filler metal to penetrate the parent metal along the grain boundaries, creating a strong bond in the braze-welded parts.

Fig. 25-1. Grinding a cast iron casting to prepare it for braze welding. (Courtesy of The Norton Company.)

Metals Joined—Copper, brass, steel, and cast iron are the most common metals joined by braze welding. Extra preparation is sometimes needed for cast iron before braze welding. Cast iron Vee joints are sometimes *seared* (quickly burned on the surface) by using an oxidizing flame on the edges of the joint. This helps to remove the carbon and graphite particles near the metal's surface. After searing the joint edges, the welder may find that *tinning* (initially coating the cast iron) will be easier, and the tinning will better adhere (stick) to the cast iron.

Joint Selection

For braze welding, butt joint preparations are usually used. The bevel edges sometimes used for butt joints can be formed by flame cutting, chipping, or grinding, as shown in Fig. 25-1. When a cutting torch has been used for beveling the joint edges, sandblasting or grinding should be used to clean the rough edges before braze welding.

Cleaning

The welder should thoroughly clean the parts for successful braze welding. Grease, paint, oil or other foreign matter can be removed by using solvents, brushes, or grinding. *Flame cleaning,* shown in Fig. 25-2, is

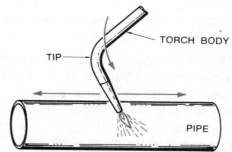

Fig. 25-2. Sweeping the torch flame across a piece to burn off any surface dirt is called *flame-cleaning.*

a method used to "burn off" surface dirt. Many welders prefer to sandblast the parts if sandblasting equipment is available.

Wire-brushing is usually good enough for cleaning if the material is fairly clean. Otherwise, the welder may need to grind the parts as shown in Fig. 25-1.

Filler Rod and Flame

The filler rod used for most braze welding should be a bronze rod that melts at about 1600° F. Manganese bronze or high zinc-brass welding rods can also be successfully used for braze welding.

A highly oxidizing *flux* is recommended. It must be able to remove any carbon or graphite from the prepared surface of the cast iron. The oxidizing agent in braze welding fluxes is usually manganese dioxide. When heated, the manganese dioxide flux will prepare the surface of the parent metal to accept the bronze rod.

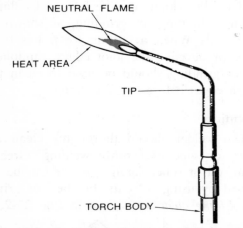

Fig. 25-3. A neutral flame, as used for braze welding.

The flame used for braze welding should be either neutral (Fig. 25-3) or slightly reducing (Fig. 25-4). Using either of these flames will help the welder easily control the welding heat and guard against adding too much oxygen to the braze weld.

Tacking and Positioning

The prepared parts should be tack welded in position for good braze welding, as for other types of welding. The part to be welded should be held by a fixture for flat position welding. When braze welding, welding positions other than flat are very difficult. The

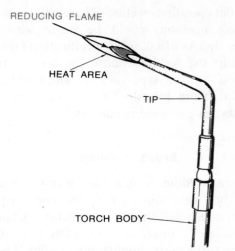

Fig. 25-4. A reducing flame, as used for braze welding.

parts may or may not be preheated. In many cases, however, preheating will increase the welding speed.

Technique and Procedure

By using the backhand welding technique with a tip size twice that recommended for welding the same thickness of steel, the welder will be able to make a successful braze weld.

During braze welding, the welder should never allow the parent metal to become hot enough to melt. Usually, the base metal is heated to about 700° F (a dark red). While heating the base metal, the end of the bronze rod should also be heated. The bronze rod is hot enough if, when dipped in the flux, the flux sticks to the bronze rod.

When the metal being braze welded is dark red, the fluxed rod is allowed to melt in the joint so that a thin coat of the rod melts into the edges of the joint. This process of applying a thin coat of molten filler metal to the joint and gently rubbing it along both sides of the break is known as *tinning* the joint. Only by doing a good job of tinning the joint surfaces can a strong weld be made. If the base metal is heated too much, the filler metal will bubble on the base metal, and tinning will be almost impossible. If the parent metal is not heated enough, it will be too cold and the filler metal will "ball up" and not tin at all.

After tinning, the welder can finish filling the joint with molten filler metal in one or more passes. The filler metal should not be reheated after it has cooled because it may break down when reheated.

Cleaning and Cooling

When the braze welding has been completed, the slag on the surface of the weld should be removed by either wire-brushing or chipping. Usually, the welded parts (especially cast iron) should be packed in ashes or lime, or wrapped in asbestos. This will allow slower cooling, creating more ductility in the welded material.

Brazing

Brazing is a process where joining is made possible by first heating the parts to be joined to temperatures *above* 800° F. Then, a

Table 25-1. Brazing Recommendations for Common Copper Alloys[1]

General Group	Coppers	Yellow Brasses	Leaded Brasses	Tin Brasses	Phosphor Bronzes
Principle Types	Electrolytic Tough Pitch. Phosphorous Deoxidized. Oxygen Free, High Conductivity.	Low Brass, Cartridge Brass, Yellow Brass, Muntz Metal.	Leaded Commercial Bronze, Low Leaded Brass, Medium Leaded Brass, High Leaded Brass, Free Cutting Brass, Free Cutting Muntz Metal, Architectural Bronze.	Admiralty, Naval Brass, Manganese Bronze.	Phosphor Bronze (A, C, D, E).
Nominal Composition	Copper: 99.90% Min. Oxygen: 0.04% Copper: 99.90% Min. Phosphorous: 0.02% Copper: 99.92% Min.	Zinc: 20%-40% Copper: Balance	Zinc: 9.25%-40% Lead: 0.5%-3% Copper: Balance	Zinc: 28%-39% Tin: 0.75%-1% Admiralty-As-0.04% Manganese Bronze: Iron: 1.4% Manganese: 0.1% Copper: Balance	Tin: 1.25%-10% Phosphorus: 0.01%-0.50% Copper: Balance
Principle Uses	Electrical conductors, auto radiators, plumbing, dairy and heat exchanger tubing, busbars and wave guides.	Musical instruments, lamp fixtures, hinges, locks, plumbing accessories, flexible hose, radiator cores, bellows.	Screw machine parts, pump cylinders and liners, plumbing accessories, gears, wheels, pinions, forgings, extrusions.	Condenser and heat exchanger tubes and plates, marine hardware, pump rods, shafts and valve stems.	Chemical hardware, Bourdon tubing, electrical contacts, flexible hose, pole line hardware.
Recommended Brazing Alloys[2]	Sil-Fos, Sil-Fos 5 Easy-Flo, Easy-Flo 45, Easy-Flo 35, Easy-Flow 43, Easy-Flo 30, Braze RT, Braze 202, Braze NT, Braze DE, Braze ETX, Braze 603, Fos-Flo	Easy-Flo, Easy-Flo 45, Easy-Flo 35, Easy-Flo 43, Easy-Flo 30, Braze RT, Braze 202, Braze NT, Braze DE, Braze ETX, Sil-Fos, Sil-Fos 5	Easy-Flo, Easy-Flo 45, Easy-Flo 35, Easy-Flo 43, Easy-Flo 30, Braze 560, Braze 603, Sil-Fos, Sil-Fos 5.	Easy-Flo, Easy-Flo 45, Easy-Flo 35, Easy-Flo 43, Easy-Flo 30, Braze RT, Braze 202, Braze NT, Braze DE, Braze ETX, Sil-Fos, Sil-Fos 5.	Easy-Flo, Easy-Flo 45, Easy-Flo 35, Easy-Flo 43, Easy-Flo 30, Braze RT, Sil-Fos, Sil-Fos 5, Braze 202, Braze NT, Braze DE, Braze ETX.
Recommended Flux	None required with Sil-Fos, Sil-Fos 5, or Fos-Flo. Handy Flux or Handy Flux Type LT with Easy-Flo alloys and Braze alloys.	Handy Flux or Handy Flux Type LT.	Handy Flux.	Handy Flux or Handy Flux Type LT.	Handy Flux or Handy Flux Type LT.

Recommended Atmosphere (Type and Maximum Dew Point)	Lean or Rich Exogas $+20°$ F. Reacted Endogas $+20°$ F. Dissociated Ammonia $+20°$ F.	Purified, Lean Exogas $-40°$ F. Reacted Endogas $-20°$ F. Dissociated Ammonia $+20°$ F.	Purified, Lean Exogas $-40°$ F. Reacted Endogas $-20°$ F. Dissociated Ammonia $+20°$ F.	Purified, Lean Exogas $-40°$ F. Reacted Endogas $-20°$ F. Dissociated Ammonia $+20°$ F.	Lean or Rich Exogas $+20°$ F. Reacted Endogas $+20°$ F. Dissociated Ammonia $+20°$ F.
Notes	To avoid embrittlement, electrolytic tough pitch copper should not be brazed in hydrogen-containing atmospheres. Handy Flux Type LT is beneficial for long furnace brazing cycles. Fos-Flo is suitable for joining copper to copper and copper to copper alloys where critical impact or vibration stresses are not encountered in service.	In furnace brazing, flux may be used with the atmosphere for good "wetting" by the brazing alloy. Easy-Flo alloys are preferred for furnace brazing to avoid having the zinc removed from high zinc brasses.	When furnace brazing, flux may be used with the atmosphere for good "wetting" by the brazing alloy. Keep brazing cycles short to minimize lead pickup in the brazing alloy. Leaded brasses must be stress-relieved before brazing to avoid intergranular cracking. Heat uniformly. The Easy-Flo alloys (Braze 560 or Braze 603) are preferred for furnace brazing to avoid having the zinc removed from high zinc brasses. Brazing of leaded brasses containing more than 5% lead is not recommended.	When furnace brazing, flux may be used with the atmosphere for good "wetting" by the brazing alloy. Easy-Flo alloys are preferred for furnace brazing to avoid having the zinc removed from high zinc brasses.	The dew point and carbon dioxide content of the recommended atmospheres are not critical for phosphor bronzes but flux may be required with the atmosphere for good "wetting" by the brazing alloy.

Courtesy of Handy and Harman of Canada, Ltd.

[1]Note: This table is intended to cover only a few typical applications. Many specific cases exist other than those listed.
[2]Listed in order of preference.

nonferrous filler metal is added to the joint. This filler metal must have a melting point below the melting point of the metal being joined.

During brazing, the welder must fit the pieces being joined so that the filler metal can flow between the parts through a process known as *capillary attraction*. Capillary attraction takes place when the contacting surfaces of a liquid and a solid interact so that the liquid surface is no longer flat. Capillary attraction is the reason why a *carefully* filled glass of water will not spill over as the liquid water reaches the top of the solid glass.

Uses of Brazing—By using low temperature brazing, the welder can produce high-strength joints that are able to resist fatigue and corrosion. Pieces of steel, cast iron, nickel, inconel, monel, copper, brass, bronze, cast bronze, magnesium, and aluminum can be brazed together by using the right fitup, temperature, and brazing filler metal. Combinations of dissimilar metals, such as nickel alloys and other metals that can be silver brazed, can be easily joined with brazing. Copper-alloy base metals can also be easily brazed to dissimilar metals. For all brazing jobs, however, the welder should refer to the manufacturer's specification charts for filler rods and important brazing information. Table 25-1 gives the information for the coppers, brasses, and bronzes shown.

Properties of Brazing—Because it is possible to braze with low temperatures, distortion will be well controlled. Little, if any, finishing will then be needed to produce a good-looking brazing job.

If the brazed job needs to be corrosion-resistant, the welder will have to use a filler metal with corrosion-resistant properties. Color match with low-temperature brazing rods is usually difficult, since the filler metal usually ranges from silver-white to copper-red. For this reason, brazed parts may often be painted or plated for complete finishing.

Braze Metal Temperatures—When brazing metals together, it is important that the liquidus temperature of the filler metal be *lower* than the solidus temperature of the base metal. The *liquidus* temperature is the *lowest* temperature at which the filler metal (or any metal) is completely liquid, not just starting to melt.

The *solidus* temperature, on the other hand, is the *highest* temperature that the base metal (or any metal) can reach and still remain completely solid, just before it starts to melt. Charts such as Table 25-2 are available from metal manufacturers to show both the liquidus and solidus temperatures of different metals.

Filler Metal and Brazing Flux

Filler metal for brazing is available in different shapes: wires, rods, sheets, and washers. The exact shape to use depends on the type of joint being welded. For example, pieces of filler metal can be cut from thin sheets and then placed between the surfaces to be joined, as in Fig. 25-5.

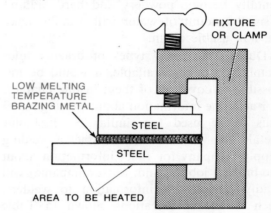

Fig. 25-5. Brazing may be done by placing a piece of low melting temperature alloy between the pieces of metal to be brazed.

Washers and rings, on the other hand, are usually used for joining pipe in pipe joints. Here, the ring or washer is placed around the joint area before the brazing heat is applied. See Fig. 25-6. Powdered silver brazing alloys are also available for special brazing jobs.

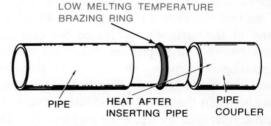

Fig. 25-6. Brazing pipe may be done by using a low melting temperature alloy ring in the pipe joint.

A good *flux* is important for successful brazing. The flux may be a coating on the brazing rod, or it may be in paste form. Either way, the flux specifications are available from welding supply houses. It is im-

Table 25-2. Common Brazing Metal Temperatures.

Metal	Solidus Temperature Melting Point (°F)	Liquidus Temperature Flow Point (°F)
Copper (tough-pitch)	1949	1981
Copper (oxygen-free)	1981	1981
Everdur®	1866	1930
Manganese-Bronze	1590	1630
High-Zinc Bronze	1630	1650
Phosphorus-Copper	1300	1530
Sil-Fos®	1185	1300
Easy-Flo® 35	1125	1295
Easy-Flo® 45	1125	1145
2S Brazing Material	1070	1165
Nickel-Chromium	1850	1950

portant to realize that brazing fluxes for brazing one type of metal should *not* be used for brazing other metals.

All paste fluxes, regardless of type, should be brushed over the entire joint for the best coverage. Mixed fluxes should be carefully sealed after they have been used, or they will become powdery and hard. If they *do* accidentally become powdery and hard, adding a small amount of water will usually make the flux usable again.

Due to the many types of brazing filler metals and fluxes available, it would be impossible to cover all of them in one chapter. For accurate information about brazing materials to be used for similar or dissimilar metals, the welder should contact a welding supply company for exact information about the brazing job at hand. Most companies will gladly furnish this information to welders, such as the specifications shown in Table 25-1.

To review the basic techniques used for low-temperature brazing, the common process of silver brazing can be outlined. Silver brazing procedures are common to other types of brazing.

Preparation

Grease, oil, oxides, and other foreign matter should be removed from the surface, regardless of the type of material to be brazed. Cleaning the joint surfaces may be done with any of the methods mentioned for cleaning steel, with one important exception. Harsh grinding or chipping should not be done on surfaces to be brazed. This may cause deep scratches in the metal and would prevent the close tolerance between the surfaces that the capillary action needs for strong joints. High polish, scale or other foreign matter on the part's surface may also be removed by pickling or using fine emery cloth.

Positioning

After the surfaces to be brazed are cleaned, they can be positioned with either a lap or

butt joint. However, the lap joint is preferred if at all possible. Band saw blades, for example, have been successfully butt welded, but the blades are usually cut on an angle to allow for more surface-holding area. See Fig. 25-7.

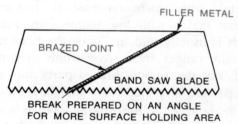

Fig. 25-7. When a brazed joint is made on a diagonal preparation, the joint has more holding area.

Joint surfaces can be prepared on a magnetic table, a large grinder (See Fig. 24-5), a milling machine, shaper, or other automatic cutting machine. In Fig. 25-8, the casting is being prepared by grinding on a large surface grinder. By grinding or machining, the joint surfaces can be kept parallel and at the same distance from each other.

During most low-temperature brazing, the parts to be joined are spaced from 0.003″ to 0.006″ apart. High joint strength with low-temperature brazing depends, of course, on both the preparation and cleaning, and also on the space left for the brazing filler material. Thin-brazed joints, properly prepared, are usually stronger than those that were made with large root gaps.

An experienced welder can usually do brazing in any position. However, it is usually easier to practice brazing in the flat position, with the parts held in a fixture to make it easy to heat underneath the metal plates. See Fig. 25-5. As the metal cools after brazing, the welder should loosen the fixture to relieve any strain on the brazed joint.

Procedure

When the metal has been cleaned and positioned, the joint is ready to be brazed. Usu-

Fig. 25-8. A surface grinder being used to prepare a casting for brazing. (Courtesy of The Norton Company.)

ally, a reducing flame (2½ X or 3X, Fig. 25-9) may be used for silver brazing, with the tip size depending on the thickness and melting point of the filler metal. Using the correct tip size will eliminate overheating, a common problem when just beginning to braze.

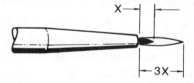

Fig. 25-9. Silver brazing is often done with a 2½ X or 3X reducing flame.

For successful brazing, the joint must first be evenly heated, after the flux has been placed in the joint. A flux reaction will take place to indicate that the flux and joint metal are at the right temperature to accept the filler rod. When heated correctly, the wet flux will boil at 212° F, leaving a white powder that turns puffy at 600° F. From the white powder, the flux turns into a clear, viscous fluid at about 800° F. At a temperature of 1100° F, the flux again becomes watery and dark, indicating that any more

increase in temperature will flow the filler metal. After adding molten filler metal to *this* temperature of flux, the parts are allowed to cool.

While brazing, the torch should always be kept in motion. The flame should be "played" on the base metal to allow the heat in the joint to melt the filler metal. The flame's direct heat should *not* melt the filler metal during brazing. Only the base metal heat should be used to melt the filler metal.

If it is necessary for the welder to feed brazing filler rod into the joint, the joint surfaces must be heated enough so that the brazing filler rod will flow freely when it is touched to the joint. Melting points of most common brazing filler rods vary from about 1125° F. to 1760° F, depending on the alloy.

Cleaning

Depending on the alloy, base metal, and/or flux used, there may be a light slag or discoloration on the brazed joint. This can be easily removed with cleaners or a wire brush. In severe cases, it may be necessary to use a chipping hammer to clean off the slag.

Brazing is a fairly easy-to-use oxygen-gas welding process. Its main advantage is that the base metal does not get very hot during brazing. Because of this, brazing fairly thin sheets of metal may be done with little distortion.

Oxygen-Gas Hardsurfacing

Many parts that are not thick enough to be hardsurfaced by arc welding may be easily hardsurfaced by using the oxygen-gas (oxyacetylene) flame.

The purpose of hardsurfacing, of course, is to protect a part against wear by welding an overlay of hardsurface metal to the area on the part that will be exposed to the most wear. This may be done either before or after the part is put in service. Before outlining oxygengas hardsurfacing, it would be a good idea to review the basic types of wear.

Types of Wear—Although *abrasion* is the most common type of wear, *impact* and *corrosion* also limit the service life of many machine parts. Hardsurfacing filler metal that is able to resist one type of wear may not be able to resist another type of wear. While abrasion, impact, and corrosion can separately wear out machine parts, they may also *combine* with each other to cause even more trouble. Therefore, it is necessary to know both *how* the material should be applied and *which* type of hardsurfacing material will help control the problem.

Choosing a Hardsurfacing Material—Knowing what the part is, why the hardsurfacing is being done, and the type of wear to which the part will be exposed will simplify choosing a hardsurfacing material.

In most cases, two or more factors are wearing down the part at the same time. Because of this, it is necessary to figure out which wear factor is doing the most damage. For this reason, the final selection of a hardsurfacing material is usually a compromise. This is because, generally speaking, hardsurfacing materials that have maximum abrasionresistance have minimum impact-resistance,

and so on. Therefore, if a part to be hardsurfaced will mostly undergo abrasion, it will be necessary to choose a hardsurfacing material with maximum abrasion-resistance *and* enough impact-resistance to withstand whatever pounding (impact) the part will receive in service.

As the carbon content of steel is increased, the steel decreases in ductility and will be able to absorb less impact. Also, as the carbon in the steel is increased, the steel becomes more brittle and more likely to crack. For this reason, high-carbon steels cannot generally be hardsurfaced. When extra-strong properties are required for hardsurfacing, alloy steels will usually be able to produce the properties.

Cooling Rate—After applying the hardsurfacing material, the cooling rate will influence (help decide) the properties of the deposit. Fast cooling can be either good or bad, depending on the hardsurfacing materials used. Fast cooling will usually give a more abrasion-resistant material than will slower cooling. However, more caution is needed when controlling the cooling as the number of alloy elements in the steel increase.

Oxy-Acetylene Hardsurfacing Procedure

The procedures used for most oxy-acetylene hardsurfacing jobs are fairly similar. Different types of hardsurfacing filler metals are manufactured for use with oxy-acetylene welding. The welder should refer to the manufacturer's specifications to learn the hardsurfacing purposes and techniques for each type of hardsurfacing filler rod. Information about fairly common oxy-acetylene hardsurfacing jobs, including their filler metals and flames, can be outlined as an introduction to oxy-acetylene hardsurfacing.

During oxy-acetylene hardsurfacing, the difference between the oxy-acetylene flames is a very important part of the total operation. Fig. 25-10 shows the basic oxy-acetylene flames used for hardsurfacing. As discussed earlier, adding carbon to steel increases the steel's hardness. When steel is heated red hot with a carburizing oxy-acetylene flame, the carbon from the carburizing flame will penetrate the steel and increase the steel's hardness.

Even a slightly oxidizing flame, on the other hand, may reduce the steel's hardness. This is because an oxidizing flame will attack the carbon in the steel (and eventually the steel itself), creating weak metal. For this reason, a welder doing hardsurfacing work may have to use an excess acetylene (carburizing) flame with an acetylene feather two to three times as long as the inner cone. See Fig. 25-10.

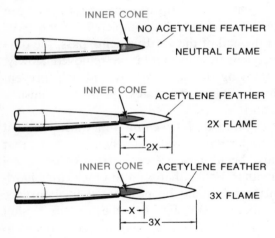

Fig. 25-10. The basic oxy-acetylene flames used for oxy-acetylene hardsurfacing. (Courtesy of Union Carbide Canada Ltd.)

Cleaning—Before any type of hardsurfacing work is done, the part's surface should be cleaned to remove any dirt, grease, oil, or other foreign matter. The cleaning may be done by wire-brushing or grinding. For most oxy-acetylene hardsurfacing repairs, the quality of cleaning will help decide how well the material will later stand up to wear.

Positioning—When the cleaned parts are ready to be hardsurfaced, they may be placed in a fixture to hold them in position. The cultivator shovel in Fig. 25-11, for example, would be placed in a fixture so that it could be turned to complete the hardsurfacing in the flat position. Although oxy-acetylene

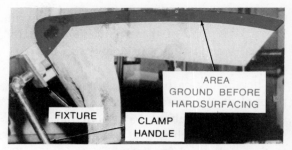

Fig. 25-11. A cultivator shovel positioned in a fixture for easier hardsurfacing.

hardsurfacing can be done in any position, most welders find that, if possible, the parts are most easily worked in the flat position. In the flat position, gravity will help the welder control the molten hardsurfacing material.

Preheating—In many cases, it is a good idea to preheat the parts to be hardsurfaced. The preheat temperature should be from about 300° to 450° F. Heavier sections to be hardsurfaced, however, may require higher preheating temperatures, even up to 900° F. Preheating may be done by using either an oxy-acetylene torch or a furnace. A furnace should be used to preheat larger and heavier parts for good, even preheating throughout the piece.

Technique—Either the *backhand* or *forehand* technique can be used to hardsurface metal with the oxy-acetylene torch. Many experienced welders, however, prefer the backhand technique, especially on thick, heavy parts. A larger tip than would normally be used for welding is often used to apply hardsurfacing. The flame is then adjusted for

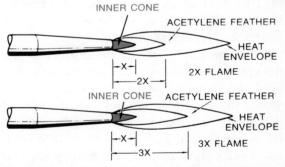

Fig. 25-12. For most oxy-acetylene hardsurfacing, either a 2X or 3X flame is used.

either a 2X or 3X acetylene feather, as in Fig. 25-12. One of these flames will usually be used for most hardsurfacing with the oxy-acetylene torch.

The slightly carburizing flame is normally pointed at the surface of the material, with the inner cone about ⅛" from the metal's surface. The torch is held at an angle similar to the angle used for flat position steel welding. With experience, however, many welders find that the torch angle depends on the thickness of the hardsurfacing layer being applied. When hardsurfacing light, knife-edge

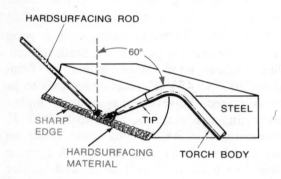

Fig. 25-13. The torch should be held at the 60° angle shown for hardsurfacing thin edges.

pieces, for example, the torch should be held in a position about 60° down from the vertical, as in Fig. 25-13. In this position, the flame pressure will flow the hardsurfacing metal along the edge being hardsurfaced *without* overheating the edge. When heavy sections are being hardsurfaced, the torch may be held at an angle closer to vertical. Usually, this angle would be about 30°, as shown in Fig. 25-14.

With the torch held at the correct angle, the area to be hardsurfaced should be heated until the metal's surface begins to "sweat." Then, the hardsurfaced rod can be pushed down into the flame where it will melt on the sweated area. For a smooth surface area, the welder should "puddle" the deposit and allow the hardsurface rod to melt into the puddle.

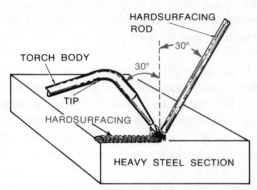

Fig. 25-14. For hardsurfacing thicker pieces, the torch angle can be reduced to as small as 30°.

Once a hardsurfacing pass has been started, it should be completed in one pass. At the end of the hardsurfacing pass, the oxy-acetylene flame should be removed slowly. Removing the flame slowly will help prevent shrink holes and cracking in the finished deposit.

Cooling and Cleaning—In most cases, it is important to very slowly cool parts that have been hardsurfaced. High-carbon steel or other metals that tend to crack after hardsurfacing should be postheated in a furnace to about 1200° F. Then the furnace is shut off and the parts are left in the furnace to cool slowly.

The hardsurfacing slag, if any, can be removed by chipping or wire brushing. Many types of hardsurfacing have slag that is difficult to grind smooth. For these hardsurfacing types, emery stones are available that can be attached to different grinding tools. Then the slag may be removed by using the emery stone.

Using Oxy-Acetylene Hardsurfacing on Common Tools

Many common tools can be rebuilt and hardsurfaced by the oxy-acetylene method. To successfully repair a worn tool or part, it is important that any worn piece be repaired, forged, and/or rebuilt as close to its original shape as possible before hardsurfacing is applied. Tools to be hardsurfaced should be positioned in a fixture or on a fire-brick table, so that the hardsurfacing can be done in the flat position.

Oxy-acetylene hardsurfacing should always be used when a thin, highly wear-resistant surface is needed. The oxy-acetylene process allows a good bond between the parent metal and the hardsurfacing material, with very little of the hard overlay material mixing with the softer parent metal. Also, tools can be sharpened before hardsurfacing with little danger of burning away the thin edge of the parent metal. After hardsurfacing, for example, the soil's abrasion on the soft side of a tool would keep the leading edge in good cutting condition. See Fig. 25-15.

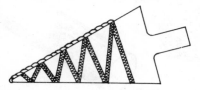

Hardsurface top and sides only.

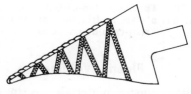

Self-sharpening action results as unprotected bottom wears. After initial wear, hard metal will retard base metal erosion.

Fig. 25-15. When the hardsurfacing material is correctly applied, the natural soil abrasion will help to keep the tool sharp.

Since many welding equipment companies supply good hardsurfacing alloy rods, the welder should select the type of rod he can best use for a given job and then follow the specifications for that rod's use. Fig. 25-16 shows one equipment manufacturer's speci-

fications for the hardsurfacing filler rods shown.

Farm Tools—Most farm tools are either low-alloy steel or straight carbon steel. When a welder needs to hardsurface these low- or medium-carbon steels, the job will be fairly easy since no special preparation (other than cleaning) is required. The carbon content of the steel in a given farm tool can be judged

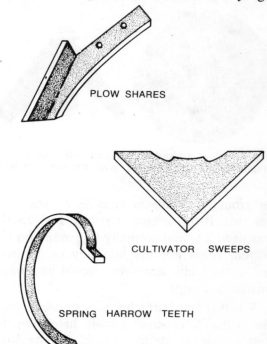

PLOW SHARES

CULTIVATOR SWEEPS

SPRING HARROW TEETH

Fig. 25-17. Typical farm tools to be hardsurfaced.

for hardsurfacing by using a spark test. Fig. 25-17 shows several typical farm tools to be hardsurfaced for longer wear.

High-carbon and medium-alloy materials should be preheated to at least 400° F before applying the hardsurfacing material. After hardsurfacing the part, cooling should be controlled by placing the part in a box of

Alloy Number	Wear Factor	Impact Factor	Rockwell C Hardness		Deposit Properties
Harfac 6038	60	38	60-63	Oxy-acetylene	Smooth deposit of high chrome alloy. Cannot forge or heat treat. For severe abrasion.
Harfac 8732-A	87	32	Beyond Scale	Oxy-acetylene	Extremely hard carbide alloy deposit. Cannot forge or heat treat. For extreme abrasion.

Fig. 25-16. Specifications for one manufacturer's oxy-acetylene hardsurfacing filler rods. (Courtesy, Hardfacing Alloys of Canada, Ltd.)

lime or by covering it with an asbestos blanket. Fig. 25-18 shows a grain-drill disc and cultivator sweep, two farm tools that can be restored by oxy-acetylene hardsurfacing.

New Tools—When hardsurfacing a *new* tool, grind it clean at least 1″ back from the cutting edge to remove any dirt or protective coating. For example, if the grain-drill disc in Fig. 25-18 was a new tool, it would

Fig. 25-18. A grain-drill disc and cultivator sweep, prepared for hardsurfacing with an oxy-acetylene torch.

be ground around the outside on the convex side. If the cultivator sweep, Fig. 25-18, was new, it would normally be ground on the bottom edge. For use in sandy or abrasive soil, a new cultivator sweep would be ground on the top edge.

Whether hardsurfacing new or used tools, the welder's experience will help him be better able to decide where the hardsurfacing should be applied. In all cases, the most important conditions to know for hardsurfacing a new tool are the conditions in which the tool will be used.

Oil Industry Tools—Oil industry tools, especially the drill bits in Fig. 25-19, are often built up with tungsten-carbide composite rods. These hardsurfacing filler rods have a mild steel matrix tube enclosing different-sized particles of tungsten carbide. See Fig. 25-20.

To repair the oil industry drill bits shown in Fig. 25-19, they should first be rebuilt with a tube-chromface or other buildup rod. Then, they may be hardsurfaced with a 30-40 grade tungsten-carbide hardsurfacing rod. The 30-40 grade rod contains fairly

fine tungsten-carbide particles that *will* pass through a 30-grit screen but *not* through a 40 grit screen.

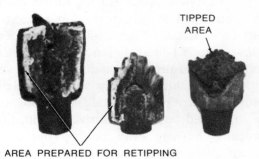

Fig. 25-19. Oil industry drill bits to be hardsurfaced.

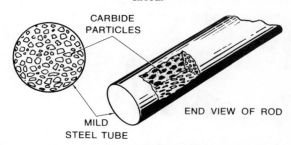

Fig. 25-20. The tungsten-carbide hardsurfacing filler rod has a steel matrix tube around the tungsten-carbide particles.

Plastic Welding

Welding plastic materials, as in welding metal, requires a heat source and, usually, a filler rod. Joint design and preparation for plastic welding are similar to those used for metal *arc* welding. During plastic welding, however, only the outer surfaces of the base metal and filler rod become molten for welding. When the heat is withdrawn from a plastic weld joint, a strong, permanent bond of solidified filler rod plastic and parent plastic should result.

Plastic welding is actually a combination of electric and gas welding processes. To weld plastic, electricity is used to provide the heat but a gas (compressed air or inert gas) is used to *carry* the heat from the plastic welding torch to the plastic surfaces being welded. Fig. 25-21 shows a typical plastic welding unit.

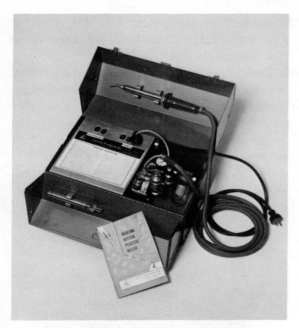

Fig. 25-21. A typical plastic welding unit. (Courtesy of Laramy Products Co.)

Classification of Plastics

Plastics can be divided into two basic groups: thermosetting and thermoplastic. Thermosetting plastics will actually *harden* when exposed to either a chemical reaction or heat, so they cannot be reformed by either catalysts or heat. Therefore, thermosetting plastics are considered to be unweldable.

Thermoplastics, on the other hand, *can* be welded. Thermoplastics soften when heated and solidify when cooled, without any other chemical reaction taking place. Through heat and pressure, thermoplastics can be formed, molded, and welded many times without changing the chemical makeup of the material.

Not only are thermoplastics lighter than steel, they also resist normal corrosion. As a structural material, many thermoplastic materials are being used in place of steel. (The

Table 25-3. Typical Thermoplastic Weld Specifications.

	PVC Type I	PVC Type II	PVC Plasticized	Polyethylene Regular	Polyethylene Linear	Polypropylene	Chlorinated Polyether	FEP Fluorocarbon	Acrylic
Welding Temperature[1] (° F)	500 to 550	475 to 525	500 to 800	500 to 550	550 to 600	550 to 600	600 to 650	550 to 650	600 to 650
Welding Gas	Air	Air	Air	Inert	Inert	Inert	Air	Air	Air
Butt-Weld Strength (%)	75-90	75-90	75-90	80-95	50-80	65-90	65-90	80-95	75-85
Maximum Continuous Service Temperature (° F)	160	145	150	140	210	230	250	250	140
Bending & Forming Temperature	250	250	100	245	270	300	350	550	280
Cementable	yes	yes	yes	no	no	no	no	no	yes
Specific Gravity	1.35	1.35	1.35	.91	.95	.90	1.4	2.15	1.19
Support Combustion	no	no	no	yes	yes	·yes	no	no	yes
Odor Under Flame	HCL	HCL	HCL	Wax	Wax	Wax	Sweet Chlorine	Pungent	Sweet
Color[2]	Grey	Light Grey	Black	Translucent or Black	White or Black	Cream to Amber	Olive Drab	Bluish Translucent	Transparent

[1] Measured ¼″ from welding tip.   Courtesy of Laramy Products Co.

[2] These are most commonly used colors and are subject to change. For complete details refer to the specifications of individual plastic manufacturers.

Table 25-4. Recommended Thermoplastic Butt Weld Conditions.

Sheet Thickness (in.)	Root Gap (in.)	Rod Size (in.)		Number of Following Beads	
		Root Bead	Following Beads	Single-V	Double-V (each side)
1/16	1/64	3/32	(none)		not practiced
3/32	1/64-1/32	1/8 Single Vee	(none)		
		3/32 Double Vee	(none)		
1/8	1/32	3/32 Single Vee	3/32 or 1/8	2 or 1	
		1/8 Double Vee			(none)
5/32	1/32-1/16	3/32 Single Vee	1/8	2	
		1/8 Double Vee			(none)
3/16	1/16	1/8 Single Vee	5/32	2	
		5/32 Double Vee	(none)		
1/4	1/16-3/32	1/8 Single Vee	1/8	5	
		1/8 Double Vee	1/8		2
3/8	3/32	5/32 Single Vee	5/32 or 3/16	9 or 5	
		5/32 Double Vee	5/32		2

Courtesy of Laramy Products Co.

tensile strength limits of most thermoplastic materials is about 9000 psi.)

Because thermoplastics are fairly light weight, they are easier to handle and can be more easily shipped than other materials, and at a lower cost. Also, thermoplastics can be riveted, bolted, threaded, and machined much like steel. With high tensile strength cements, thermoplastics can be easily and quickly fabricated by both gluing and welding. Polyethylene and polyvinyl chloride, two of the many thermoplastics on the market, are the more commonly welded thermoplastics due to their wide melting range.

Plastic Welding Equipment

Plastic welding equipment is unlike common equipment used to weld steel. Plastic welding equipment generally uses air or inert gas to transfer electrically-created heat to the plastic surface.

Plastic Specifications — For successful welding with *any* material, the welder must have complete specifications for the material being welded. Both plastic manufacturers and plastic welding equipment manufacturers list complete weld specifications for most common thermoplastics.

Table 25-3, for example, lists the properties of most common types of thermoplastics. These may be used to help choose a suitable plastic for a given job. Table 25-4 lists the recommended weld specifications for butt welds on common thermoplastic sheet.

Torches — Electrically heated torches, shown in Fig. 25-22, are used to weld sheets,

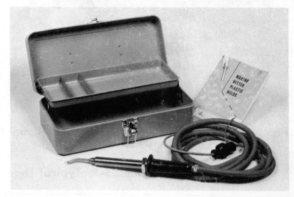

Fig. 25-22. An electrically-heated plastic welding torch. (Courtesy of Laramy Products Co.)

pipes, and other shapes of plastic material. The torch uses a gas (compressed air or inert gas) to pass over a heated source. The electrically heated source raises the temperature of the welding gas to about 450°-800° F.

Torch Tips—Torch tips, such as those shown in Figs. 25-23, 24, and 25, are used to tack plastic welds and complete the welding. Round tips, Figs. 25-23 and 24, are

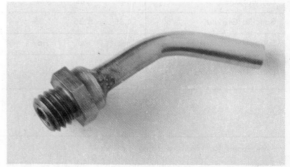

Fig. 25-23. A round, plastic welding torch tip. (Courtesy of Laramy Products Co.)

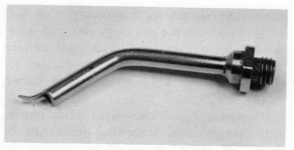

Fig. 25-24. A common type of round, plastic welding torch tip. (Courtesy of Laramy Products Co.)

Fig. 25-26. A heat gun used to warm large sections of plastic and for preheating before welding. (Courtesy of Eddy Products Corp.)

used for general plastic welding and close, confined work.

Speed welding tips, Fig. 25-25, can weld up to four times faster than a standard tip.

Fig. 25-25. A speed welding torch tip used for plastic welding. (Courtesy of Laramy Products Co.)

Speed tips are used for welding pipe over 4″ in diameter and for welding long, straight welds in tank and duct work. *Combination* round and tacking tips are also available for general plastic welding.

Gas and Power Control Unit—This small unit reduces the amount of inert gas needed for plastic welding. The control unit does this by automatically switching from inert gas to compressed air when the torch is not being used for a few moments.

Heat Gun—This is actually an electrically heated blower, as in Fig. 25-26. The heat gun is not actually used for welding plastic. Instead, it is used for softening large areas of plastic for bending, forming, and, if needed, preheating before welding.

Plastic Welding Rods—These rods are available for welding most thermoplastic materials. Plastic welding rods are usually

available in diameters from $\frac{3}{32}$″ to $\frac{3}{16}$″. Close-tolerance plastic welding rods must be used with speed welding tips so that the rod will not stick in the preheated tube. See Fig. 25-27.

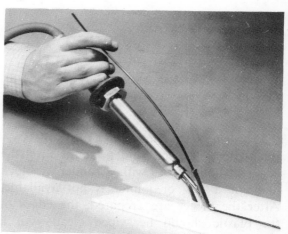

Fig. 25-27. Close-tolerance plastic filler rods must be used with a speed welding tip so that the filler rod doesn't stick to the tip nozzle. (Courtesy of Laramy Products Co.)

Preparing Different Types of Plastic Welds

Either metal or plastic pieces will, basically, use the same type of welds. Some of the more commonly used welds for plastics are shown in Fig. 25-28. Preparing the plastic surfaces for welding is just as important as for welding steel. The plastic must be clean, carefully positioned, and well prepared for a good weld.

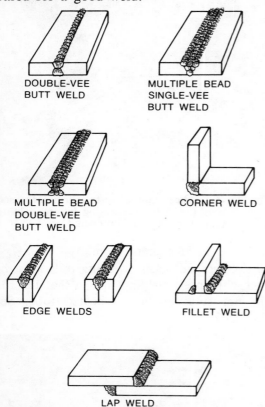

Fig. 25-28. Types of welds commonly used for plastic welding. (Courtesy of Laramy Products Co.)

Preparing Butt, Edge, and Corner Welds— For these welds, the welder should bevel the edges of both pieces with a sander or block plane if a saw or jointer is not available. After the pieces are prepared, the edges of the plastic should be wiped clean with a piece of lint-free cloth. Solvents should not be used to clean the beveled edges since they may soften the edges and cause a poorly finished weld.

The pieces should be prepared and positioned with a $\frac{1}{32}''$ land. When placed together, the two pieces should have a Vee groove angle of about 60°. If the pieces are to be tacked together with a tacking rod,

the welder should leave a root gap of $\frac{1}{64}''$ to $\frac{1}{32}''$ between the pieces. A root gap is not needed when a tacking tip is used to tack weld the pieces. The bevel preparations for different plastic welding joints are shown in Fig. 25-29.

*Preparing Lap Welds—*Lap welds require little preparation, since the pieces to be joined are placed on top of one another. As with other plastic welds, the lap joint should be clean and free of all contaminants. When the plastic pieces have been cleaned and positioned for a lap weld, "C" clamps or tack welds are usually used to hold the pieces together.

*Preparing Fillet Welds—*As with lap welds, little preparation is needed for fillet welds. The plastic pieces to be joined must be clean, free of all dirt, and held securely in position by clamps, tack welds, or special fixtures. When making fillet welds on pieces where one or both edges are beveled, a root gap from $\frac{1}{64}''$ to $\frac{1}{32}''$ should be used if a tacking tip is not available.

General Plastic Welding Torch Reminders

A few general reminders can help a welder successfully operate an electric welding torch for the first time. They are good points to check, also, if problems arise while welding plastics.

A. Be sure that the welding gas supply is *clean*. Moisture and/or oil in the welding gas may prevent a good welding bond and/or cause a short circuit in the torch's heating element.

B. The electricity must always be turned off *before* the welding gas is turned off. A good idea to remember is to always turn the gas *on* first and *off* last.

C. The torch should always be grounded. This will help prevent a short circuit that could cause electric shock and damage to the heating element.

D. The *volume* of welding gas passing over the heating element determines

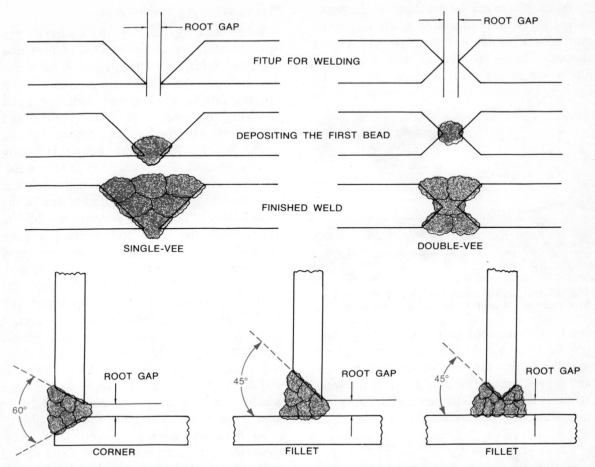

Fig. 25-29. Bevels and preparations used for common plastic welds. (Courtesy of Laramy Products Co.)

the welding temperature. To *increase* the welding gas temperature, *decrease* the welding gas volume. Likewise, to decrease the welding gas temperature, increase the welding gas volume. To determine the correct temperature of the heated welding gas, hold a thermometer ¼" from the end of the welding tip while the gas and electricity are on.

E. The torch barrel or welding tip will be *very hot* when the torch is turned on. For this reason, they will burn the welder if either part is accidentally touched.

F. To give the heating element long life, always use the recommended welding temperature for a given torch model.

G. Thoroughly read and understand the manufacturer's operating instructions before using plastic welding equipment for the first time.

Tack Welding Plastic Joints

As during metal welding, tack welding plastic joints is one way of temporarily holding the pieces in position for final welding. When plastic welding, this may be done by using either a small diameter rod or a tacking tip on the welding torch.

The advantage of using a tacking tip is that it is fast and neat. Using a tacking tip will also help eliminate a possible source of weakness in the completed weld. The weakness could result if the rod tacks were left in place.

When pieces to be welded are large or bulky, short tacks are often made in important places (such as corners) and evenly spaced. Then, when these partially tacked pieces are in the right position, the tacking tip may be drawn around the entire joint to

Fig. 25-30. Laying down the plastic filler rod before welding. (Courtesy of Laramy Products Co.)

create a continuous seal. The resulting tack will be able to hold large pieces of material together so that they can be carefully handled and moved without coming apart. If the welder needs to position the pieces differently, the tack weld may be broken and remade.

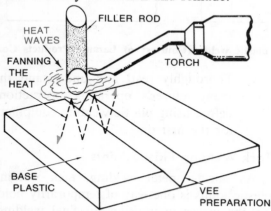

Fig. 25-31. To start the weld, the end of the filler rod and the base plastic should be heated until they become shiny and tacky.

When the pieces are tacked and positioned, the welder is ready to make the final weld. Since the pieces being welded are held in place by the tack welds, no fixtures or clamps are necessary and the welder has both hands free for good welding control.

Plastic Welding Techniques

When welding plastics, the materials are fused together by a *combination* of heat and pressure. With the conventional *hand welding* method, the combination comes from pushing on the welding rod with one hand and using the other hand to heat the rod and base plastic with the hot gas from the welding torch. In Fig. 25-30, the welder is laying down a bead of filler plastic before final welding. Successful welds require that both the pressure and heat be kept constant and in the right balance. Too much *pressure* on the rod may stretch the bead and produce a poor weld. Too much *heat,* on the other hand, may char, melt, or distort the plastic being welded.

Starting the Weld—To start the weld, the welder should hold the torch tip about ¼″ to ¾″ above the plastic. In this position, a starting area is preheated on the base plastic and rod until the base plastic appears shiny and becomes tacky. See Fig. 25-31. The rod should be held at a 90° angle to the base plastic and moved up and down slightly, so that the rod barely touches the base plastic. (On materials such as polyethylene, polypropylene, and fluorocarbons, the rod should be held at a 45° angle for the best results.)

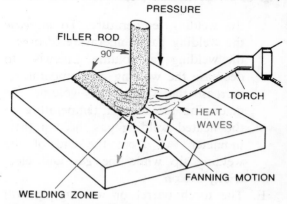

Fig. 25-32. Correct technique for laying the bead after the weld is started.

For the correct heat balance, the welder should fan or weave the torch up and down, as in Fig. 25-31. This will heat both the rod and the base plastic equally. Then, when heated enough, the rod will easily stick to the base plastic and the actual weld can begin.

The filler rod should be pressed into the base plastic with a slight downward pressure. When a molten curve starts where the rod meets the plastic, the rod should be bent naturally and then moved forward, as in Fig. 25-32. If the rod is accidentally overheated, it will become rubbery and it will be very difficult for the welder to apply even pressure. Overheating the base material, on the other hand, will cause it to char or melt, creating a poor plastic weld.

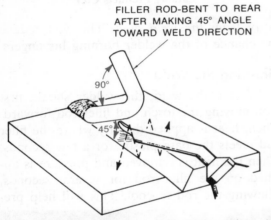

FILLER ROD-BENT TO REAR
AFTER MAKING 45° ANGLE
TOWARD WELD DIRECTION

90°

45°

Fig. 25-33. When using a polyethylene filler rod, the rod must be looped back 90° after leaning 45° in the weld direction.

Laying the Bead—Once the weld has been started, the torch should be fanned from the rod to the plastic about twice every second, as in Fig. 25-32.

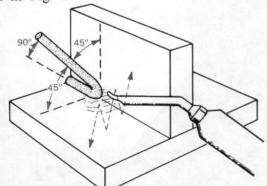

90°   45°

45°

Fig. 25-34. For plastic fillet welds, the filler rod should be fed into the fillet from equal angles to either weld piece.

Because the welding rod is thinner than the base plastic, the rod will heat up quicker. To allow for this difference in heating rate, the arc should be fanned toward the base plastic about 60% of the time when using a 1/8″ filler rod. If a thicker 5/32″ filler rod is being used, the arc should be fanned on the base plastic only about 40% of the time. In either case, the fanning should heat about 1/2″ of the welding rod and about 3/8″ ahead of weld on the base plastic.

Correct Filler Rod Angle—For welding most thermoplastics, the filler rod should be held at a 90° angle to the base material. Although welding can be done faster by leaning the rod backward (away from the weld direction), this will usually stretch the rod and produce checks and cracks in the finished weld after it cools. In order to put enough pressure on a *polyethylene* rod, it must be fed into the weld at a 45° angle *in* the weld direction; however, the upper part of the filler rod should loop *away* from the weld direction. See Fig. 25-33.

For fillet welds, the rod should be held midway between the two welded surfaces, as in Fig. 25-34. As usual, it is important to preheat all the surfaces being welded for the best plastic weld results.

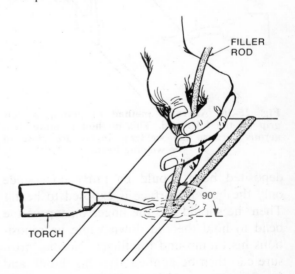

FILLER
ROD

90°

TORCH

Fig. 25-35. Recommended method of getting a new grip on the filler rod.

Feeding the Filler Rod—While welding the plastic joint, the rod will be melted down into the joint, making it necessary for the welder to get a new grip higher on the rod. Unless this is done carefully, the sudden release of pressure may cause the filler rod to lift away from the weld. When this happens, air may be trapped under the weld, causing a poor weld and, sometimes, complete weld failure.

To easily get a new grip on the filler rod, the welder should put his fourth and fifth fingers on opposite sides of the rod while he repositions his thumb and forefinger, Fig. 25-35. A second easier method may also be used.

In the second method, the welder should put either his third or fourth finger on *top* of the deposited bead, as in Fig. 25-36. (The

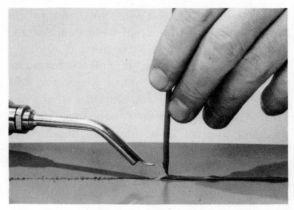

Fig. 25-37. Joining two filler rods. (Courtesy of Laramy Products Co.)

away from the weld area. This will reduce the chance of the welder burning his fingers.

Finishing the Weld

To stop the weld, the welder should first stop moving the torch and filler rod forward. Then, heat is applied directly where the filler rod meets the base plastic for a few seconds. Then the heat is removed and pressure is applied on the filler rod for several seconds, allowing the rod to cool. This will help pre-

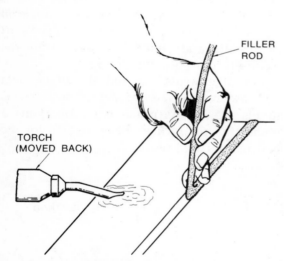

Fig. 25-36. Alternate method of getting a new grip on the filler rod. This method is more dangerous because the welder's fingers are close to the welding heat.

deposited bead should be fairly cool since only the bottom surface is exposed to heat.) Then, he should use the finger on top of the bead to hold the rod down while he repositions his thumb and forefinger. Normal pressure can then be applied after his thumb and forefinger are repositioned. *Important: When using this method, the torch must be turned*

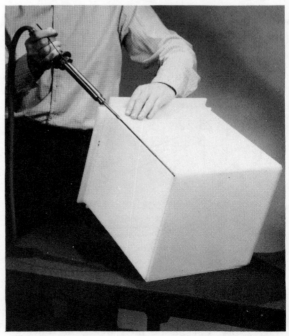

Fig. 25-38. A plastic box being welded with a speed tip. (Courtesy of Laramy Products Co.)

vent the possibility of the bead being pulled up from the weld joint. Lastly, the pressure is released and the filler rod is twisted until it breaks.

If the weld is to be continued, the deposited bead should be broken by *cutting* the filler rod with a sharp knife at a 30° angle, after the rod is allowed to cool under pressure for several seconds.

When joining one filler rod to another in the same weld joint, the new rod should first be cut at a 60° angle. Then, the 60° surface of the new rod is heated and welded on the 30° end of the old rod. See Fig. 25-37. Using this method, the filler rods almost appear to be one piece when joined together. In any case, welds should never be spliced by overlapping them side by side.

Common Plastic Welding Problems

During any welding process, problems may come up as a result of complications in the process itself. Plastic welding is no different and has many possible problems common to other forms of welding: porosity, poor penetration, warping, distortion, cracking, etc.

Generally speaking, four factors will usually decide the quality of a plastic weld. Before any plastic welding is done, these four items must be determined from weld specification charts for a given plastic weld:

1. Welding *temperature*
2. Welding *speed*
3. Filler rod *pressure*
4. Filler rod *angle*

Table 25-5 lists the more common plastic welding problems with their probable causes and corrections. With a good knowledge of plastic welding, a welder will be able to easily weld large plastic pieces. The plastic box being welded with a speed tip in Fig. 25-38 is typical of the large plastic pieces that may be welded.

Table 25-5. Common Plastic Welding Defects and Their Normal Corrections.

POOR PENETRATION

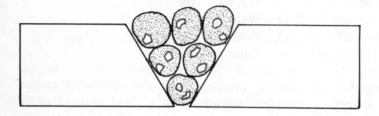

Cause:

1. Porous weld rod
2. Too much heat on rod
3. Welding too fast
4. Rod too large
5. Improper starts or stops
6. Improper bead crossing
7. Rod stretching

Correction:

1. Inspect rod, use good weld rod
2. Use proper fanning motion
3. Check welding speed
4. Weld beads in proper sequence
5. Cut rod at an angle, cool before releasing
6. Stagger starts and overlap splices ½″

POROUS WELD

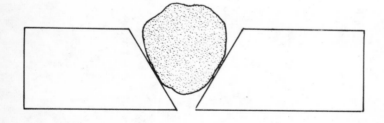

Cause:

1. Faulty preparation
2. Rod too large
3. Welding too fast
4. Not enough root gap

Correction:

1. Use 60° bevel preparation
2. Use small rod for root pass
3. Check for flow lines while welding
4. Use tacking tip or leave ⅟₃₂″ root gap and clamp the pieces

SCORCHING

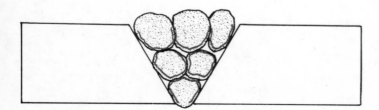

Cause:

1. Temperature too high
2. Welding too slow
3. Uneven heating
4. Material too cold

Correction:

1. Increase air flow
2. Hold constant speed
3. Use correct fanning motion
4. Preheat material in cold weather

DISTORTION

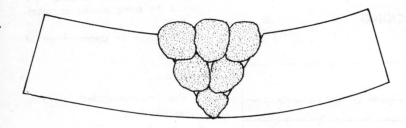

Cause:

1. Overheating at joint
2. Welding too slow
3. Rod too small
4. Improper sequence

Correction:

1. Allow each bead to cool
2. Weld at constant speed—use speed tip
3. Use larger-size or triangular-shaped rod
4. Offset pieces before welding
5. Use double-Vee or backup weld
6. Backup weld with metal

WARPING

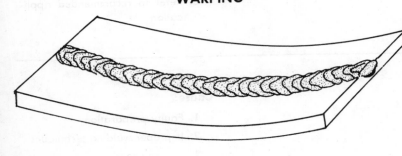

Cause:

1. Material shrinkage
2. Overheating
3. Faulty preparation—too much root gap
4. Faulty clamping of parts

Correction:

1. Preheat material to relieve stress
2. Weld rapidly — use backup weld
3. For multilayer welds — allow time for each bead to cool
4. Clamp parts properly—backup to cool

POOR APPEARANCE

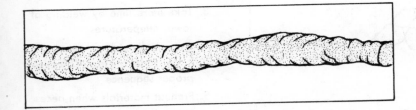

Cause:

1. Uneven pressure
2. Excessive stretching
3. Uneven heating

Correction:

1. Practice starting, stopping, and finger moving on rod
2. Hold rod at proper angle
3. Use slow, uniform fanning motion and heat both rod and material
   (For Speedwelding: use only moderate pressure, constant speed, and keep shoe free of residue.)

STRESS CRACKING

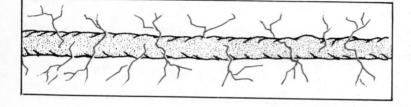

Cause:

1. Improper welding temperature
2. Undue stress on weld
3. Chemical attack
4. Rod and base material are not the same plastic composition
5. Oxidation or degradation of weld

Correction:

1. Use recommended welding temperature
2. Allow for expansion and contraction
3. Stay within known chemical resistance and working temperatures of material
4. Use similar materials and inert gas for welding
5. Refer to recommended application

POOR FUSION

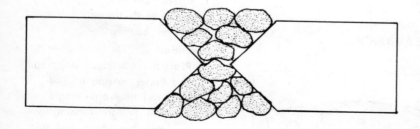

Cause:

1. Faulty preparation
2. Improper welding techniques
3. Wrong speed
4. Wrong filler rod
5. Wrong temperature

Correction:

1. Clean materials before welding
2. Keep pressure and fanning motion constant
3. Take more time by welding at lower temperatures
4. Use small rod at root and large rods at top — practice proper sequence
5. Preheat materials when necessary
6. Clamp parts securely

Courtesy of Laramy Products Co.

Index